T0137976

Nanotechnology in the Life Sciences

Series Editor

Ram Prasad
Department of Botany
Mahatma Gandhi Central University
Motihari, Bihar, India

Nano and biotechnology are two of the 21st century's most promising technologies. Nanotechnology is demarcated as the design, development, and application of materials and devices whose least functional make up is on a nanometer scale (1 to 100 nm). Meanwhile, biotechnology deals with metabolic and other physiological developments of biological subjects including microorganisms. These microbial processes have opened up new opportunities to explore novel applications, for example, the biosynthesis of metal nanomaterials, with the implication that these two technologies (i.e., thus nanobiotechnology) can play a vital role in developing and executing many valuable tools in the study of life. Nanotechnology is very diverse, ranging from extensions of conventional device physics to completely new approaches based upon molecular self-assembly, from developing new materials with dimensions on the nanoscale, to investigating whether we can directly control matters on/in the atomic scale level. This idea entails its application to diverse fields of science such as plant biology, organic chemistry, agriculture, the food industry, and more.

Nanobiotechnology offers a wide range of uses in medicine, agriculture, and the environment. Many diseases that do not have cures today may be cured by nanotechnology in the future. Use of nanotechnology in medical therapeutics needs adequate evaluation of its risk and safety factors. Scientists who are against the use of nanotechnology also agree that advancement in nanotechnology should continue because this field promises great benefits, but testing should be carried out to ensure its safety in people. It is possible that nanomedicine in the future will play a crucial role in the treatment of human and plant diseases, and also in the enhancement of normal human physiology and plant systems, respectively. If everything proceeds as expected, nanobiotechnology will, one day, become an inevitable part of our everyday life and will help save many lives.

More information about this series at http://www.springer.com/series/15921

Nilesh Sharma • Shivendra Sahi

Editors

Nanomaterial Biointeractions at the Cellular, Organismal and System Levels

Editors
Nilesh Sharma
Department of Biology
Western Kentucky University
Bowling Green, KY, USA

Shivendra Sahi
Department of Biological Sciences
University of Sciences
Philadelphia, PA, USA

ISSN 2523-8027   ISSN 2523-8035   (electronic)
Nanotechnology in the Life Sciences
ISBN 978-3-030-65794-9   ISBN 978-3-030-65792-5   (eBook)
https://doi.org/10.1007/978-3-030-65792-5

This Springer imprint is published by the registered company Springer Nature Switzerland AG
The registered company address is: Gewerbestrasse 11, 6330 Cham, Switzerland

Preface

Engineered nanomaterial applications have expanded from catalysis, electronics, and filtration to therapeutics, diagnostics, agriculture, and food because of the unique properties and potentials of different nanoparticles and nanomaterials, particularly belonging to metals. As the roles of nanomaterials are diversifying rapidly in every field cited above, the concern for their safety has also heightened in recent times. While novel applications emerge every year, research focusing on their unintended effects on living organisms and the natural environment that supports biota has also intensified. The available evidence—for and against—is strong, as it has been seen in every case of technological development. The solution thus lies in refining our knowledge and understanding the intricate modes of interactions that may occur between nanomaterials, organisms, and ecosystems. A fuller understanding sets conditions for the judicious application of technological innovation. Developing a holistic and balanced approach based on high-quality research may help to harness the full potential of nanotechnology in the twenty-first century.

Current research shows that nanoparticles can interact with an organism at the cellular, physiological, biochemical, and molecular levels. However, our knowledge of how they affect these changes, selectively or generally, locally or globally, in diverse organisms or ecosystems is very limited and far from satisfactory. Data indicates that the biological function largely depends on the concentration, shape, size, and surface characteristics of nanoparticles, as well as the cellular, genomic, and the epigenetic environments of the organism.

Therefore, this compilation focuses on the body of original work carried out by distinguished investigators using a range of nanomaterials and living organisms in different conditions. In the above backdrop, the book is organized in four parts: Part I containing chapters on interaction studies based on model organisms, Part II containing chapters on interaction studies based on plants (crop species) and agricultural systems, Part III containing chapters on interaction studies based on human cells and microbes, and Part IV containing chapters focusing on general mechanisms of interactions.

This collection includes specific case studies as well as general review articles highlighting aspects of multilayered interactions between nanomaterials and organisms and serves not only research and academic scholars but the concerned industry and policymakers as well. As this book presents the latest overview of the interface between nanomaterials and a variety of organisms, it can be ideal reference material for undergraduate or graduate-level courses in nanotechnology, nanotoxicology, or bionanotechnology.

We both acknowledge the contribution of the International Society of Trace Element Biogeochemistry (ISTEB) for hosting The International Conference on the Biogeochemistry of Trace Elements (ICOBTE) with a dedicated special section for the study of "Nanomaterials: Applications & Impacts" at ETH Zurich (Switzerland) in 2017. This was the occasion when the seed for the book was sown in discussions with the fellow researchers.

We acknowledge the continued support of our host institutions in our academic pursuits.

Bowling Green, KY, USA   Nilesh Sharma
Philadelphia, PA, USA   Shivendra Sahi

Contents

Part IV   General Mechanisms of Interaction

Contributors

Mataz Alcoutlabi Department of Mechanical Engineering, University of Texas, Rio Grande Valley, Edinburg, TX, USA

Suzanne A. Apodaca Environmental Science and Engineering PhD Program, The University of Texas at El Paso, El Paso, TX, USA

University of California Centre for Environmental Implications of Nanotechnology, The University of Texas at El Paso, El Paso, TX, USA

Rajesh Prabhu Balaraman Department of Biological Sciences, Sam Houston State University, Huntsville, TX, USA

Jacob Beer Department of Microbiology & Immunology, Midwestern University, Downers Grove, IL, USA

Sinilal Bhaskaran Fatima Mata National College, Kollam, Kerala, India

T. Bhuvaneswari Department of Biotechnology, Periyar University, Salem, Tamil Nadu, India

Jesus Cantu University of California Centre for Environmental Implications of Nanotechnology, The University of Texas at El Paso, El Paso, USA

Chemistry and Biochemistry Department, The University of Texas at El Paso, El Paso, TX, USA

Pabel Cervantes Bren School of Environmental Science and Management, University of California, Santa Barbara, CA, USA

Pranav Chandra Valley Children's Healthcare, Madera, CA, USA

Madhusudan Choudhary Department of Biological Sciences, Sam Houston State University, Huntsville, TX, USA

Keni Cota-Ruiz Department of Chemistry and Biochemistry, The University of Texas at El Paso, El Paso, TX, USA

UC Center for Environmental Implications of Nanotechnology (UC CEIN), The University of Texas at El Paso, El Paso, TX, USA

Ashley Cox Department of Biomedical Sciences, Marshall University School of Medicine, Huntington, WV, USA

M. d'Amora Nano Carbon Materials, Istituto Italiano di Tecnologia (IIT), Genoa, Italy

Guadalupe de la Rosa Departamento de Ingenierías Química, Electrónica y Biomédica, División de Ciencias e Ingenierías, Universidad de Guanajuato, León, Guanajuato, Mexico

Center for Environmental Implications of Nanotechnology (UCCEIN), University of California, Santa Barbara, CA, USA

R.K. Dearth Department of Biology, University of Texas, Rio Grande Valley, Edinburg, TX, USA

Chaoyi Deng Environmental Science and Engineering PhD Program, The University of Texas at El Paso, El Paso, TX, USA

University of California Centre for Environmental Implications of Nanotechnology, The University of Texas at El Paso, El Paso, TX, USA

Elliott Duncan Environmental Contaminants Group, Future Industries Institute, University of South Australia, Mawson Lakes, SA, Australia

Astha Dwivedi Department of Biochemistry, University of Allahabad, Prayagraj, India

Lauren Flores Department of Biological Sciences, Sam Houston State University, Huntsville, TX, USA

Ma. Concepción García-Castañeda Departamento de Ingenierías Química, Electrónica y Biomédica, División de Ciencias e Ingenierías, Universidad de Guanajuato, León, Guanajuato, Mexico

Jorge L. Gardea-Torresdey Environmental Science and Engineering PhD Program, The University of Texas at El Paso, El Paso, TX, USA

University of California Center for Environmental Implications of Nanotechnology, The University of Texas at El Paso, El Paso, TX, USA

Chemistry and Biochemistry Department, The University of Texas at El Paso, El Paso, TX, USA

N. Geetha Department of Botany, Bharathiar University, Coimbatore, TN, India

S. Giordani School of Chemical Sciences, Dublin City University, Glasnevin, Dublin, Ireland

Jose A. Hernandez-Viezcas Department of Chemistry and Biochemistry, The University of Texas at El Paso, El Paso, TX, USA

UC Center for Environmental Implications of Nanotechnology (UC CEIN), The University of Texas at El Paso, El Paso, TX, USA

Environmental Science and Engineering Ph.D. program, The University of Texas at El Paso, El Paso, TX, USA

Nazanin Nikoo Jamal Environmental Contaminates Group, Future industries Institute, The University of South Australia, Mawson Lakes, SA, Australia

School of Natural and Built Environments, The University of South Australia, Mawson Lakes, SA, Australia

Amit Kumar Department of Botany, Hansraj College, University of Delhi, Delhi, India

Qingqing Li Stockbridge School of Agriculture, University of Massachusetts Amherst, Amherst, MA, USA

Martha L. Lopez-Moreno Department of Chemistry, University of Puerto Rico at Mayaguez, Mayaguez, PR, Puerto Rico

Chuanxin Ma Key Laboratory for City Cluster Environmental Safety and Green Development of the Ministry of Education, Institute of Environmental and Ecological Engineering, Guangdong University of Technology, Guangzhou, China

Hardik Majmudar Department of Microbiology & Immunology, Midwestern University, Downers Grove, IL, USA

Jovinna Mendel Department of Biological Sciences, Sam Houston State University, Huntsville, TXUSA

Rashmi Mishra Department of Biotechnology, Noida Institute of Engineering and Technology, Greater Noida, India

Yogendra Mishra University of Southern Denmark, Mads Clausen Institute, NanoSYD, Sønderborg, Denmark

Gary Owens Environmental Contaminants Group, Future Industries Institute, University of South Australia, Mawson Lakes, SA, Australia

Abhay Kumar Pandey Department of Biochemistry, University of Allahabad, Prayagraj, India

Akhilesh Pandey Department of Neurology, Texas Tech University Health Sciences Centre, Garrison Institute on Aging, Lubbock, TX, USA

Prabhash Kumar Pandey Department of Biochemistry, University of Allahabad, Prayagraj, India

J.G. Parsons Department of Chemistry, University of Texas, Brownsville, TX, USA

Jose R. Peralta-Videa Department of Chemistry and Biochemistry, The University of Texas at El Paso, El Paso, TX, USA

UC Center for Environmental Implications of Nanotechnology (UC CEIN), The University of Texas at El Paso, El Paso, TX, USA

Environmental Science and Engineering Ph.D. program, The University of Texas at El Paso, El Paso, TX, USA

Naleeni Ramawat Amity Institute of Organic Agriculture, Amity University Uttar Pradesh, Noida, India

Swati Rawat Environmental Science and Engineering PhD Program, The University of Texas at El Paso, El Paso, TX, USA

University of California Centre for Environmental Implications of Nanotechnology, The University of Texas at El Paso, El Paso, TX, USA

Shivendra Sahi Department of Biological Sciences, University of Sciences, Philadelphia, PA, USA

Amit Kumar Sharma Department of Biochemistry, University of Allahabad, Prayagraj, India

Ananya Sharma Vanderbilt University School of Medicine, Nashville, TN, USA

Nilesh Sharma Department of Biology, Western Kentucky University, Bowling Green, KY, USA

Deepak Shukla Department of Ophthalmology & Visual Sciences, University of Illinois at Chicago, Chicago, IL, USA

Akanksha Singh Amity Institute of Organic Agriculture, Amity University Uttar Pradesh, Noida, India

Amit Kumar Singh Department of Biochemistry, University of Allahabad, Prayagraj, India

Aniruddha Singh Western Kentucky Heart and Lung, Bowling Green, KY, USA

University of Kentucky College of Medicine, Bowling Green, KY, USA

Vanderbilt University School of Medicine, Nashville, TN, USA

Archna Singh Department of Botany, Hansraj College, University of Delhi, Delhi, India

Indrakant Singh Molecular Biology Research Lab, Department of Zoology, Deshbandhu College, University of Delhi, New Delhi, India

Vaibhav Tiwari Department of Microbiology & Immunology, Midwestern University, Downers Grove, IL, USA

Carolina Valdes Department of Chemistry and Biochemistry, The University of Texas at El Paso, El Paso, TX, USA

Edgar Vázquez-Núñez Departamento de Ingenierías Química, Electrónica y Biomédica, División de Ciencias e Ingenierías, Universidad de Guanajuato, León, Guanajuato, Mexico

P. Venkatachalam Department of Biotechnology, Periyar University, Salem, Tamil Nadu, India

Yi Wang University of California Centre for Environmental Implications of Nanotechnology, The University of Texas at El Paso, El Paso, TX, USA

Chemistry and Biochemistry Department, The University of Texas at El Paso, El Paso, TX, USA

Jason C. White Department of Analytical Chemistry, The Connecticut Agricultural Experiment Station, New Haven, CT, USA

Baoshan Xing Stockbridge School of Agriculture, University of Massachusetts Amherst, Amherst, MA, USA

Ye Yuqing UC Center for Environmental Implications of Nanotechnology (UC CEIN), The University of Texas at El Paso, El Paso, TX, USA

About the Editors

Nilesh Sharma is a senior instructor in the Department of Biology at Western Kentucky University. His research interests include environmental biotechnology, bionanotechnology and nanotoxicology. He has presented his research widely in national and international research meetings and published dozens of peer-reviewed manuscripts.

Shivendra Sahi is a Professor of Bionanotechnology in the Department of Biological Sciences at the University of the Sciences, Philadelphia. He is the author of over a hundred peer-reviewed manuscripts and the recipient of research grants worth millions.

Part I
Nanomaterial Interactions in Model Organisms

Chapter 1
Application of Titanium Dioxide Nanoparticles in Consumer Products Raises Human Health Concerns: Lessons from Murine Models of Toxicity

Ashley Cox, Pranav Chandra, and Nilesh Sharma

Contents

1 Introduction

Titanium dioxide (TiO_2) is a naturally occurring substance that is found in many different minerals, such as anatase, rutile, and brookite. Over the natural sources, nanotitania is a significant part of the gigantic marketplace of engineered nano-materials (ENMs) that have entered the food chain and other consumer products. Figure 1.1 exhibits the use of nanotitania across various sectors of industry. One estimate shows that the production of TiO_2 pigment in the United States alone for 2015 was approximately 1.16 million tons (U.S. Geological Survey 2016).

A. Cox
Department of Biomedical Sciences, Marshall University School of Medicine, Huntington, WV, USA

P. Chandra
Valley Children's Healthcare, Madera, CA, USA

N. Sharma (✉)
Department of Biology, Western Kentucky University, Bowling Green, KY, USA
e-mail: Nilesh.sharma@wku.edu

© Springer Nature Switzerland AG 2021
N. Sharma, S. Sahi (eds.), *Nanomaterial Biointeractions at the Cellular, Organismal and System Levels*, Nanotechnology in the Life Sciences, https://doi.org/10.1007/978-3-030-65792-5_1

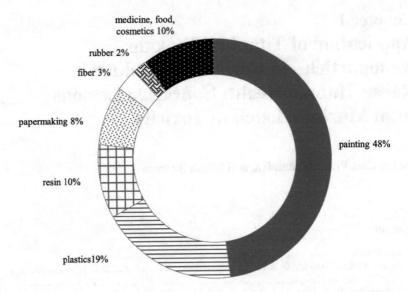

Fig. 1.1: Application of TiO$_2$ NPs (%) in industry (reproduced from Baranowska-Wójcik et al. 2020)

The production and consumption of TiO$_2$ in the United States has consistently stayed above 1100 thousand tons per year (gross weight) since 1997 despite it has been classified as *possibly carcinogenic to humans (Group 2B)* by the International Agency for Research on Cancer (IARC 2010). TiO$_2$ is often used as a whitening agent and ultraviolet (UV) blocker due to its brightness and high refractive index, respectively. For this purpose, titanium nanoparticles (TiO$_2$ NPs) are used in a wide array of personal care products such as sunscreens, cosmetics, toothpastes, shampoos, conditioners, and deodorants (Grand and Tucci 2016). Food-grade TiO$_2$ NPs are used in a large number of edible products such as milk, gums, candies, donuts, pastries, and other enhanced foods (Baranowska-Wójcik et al. 2020). Titanium additives are also used for their antibacterial effects, which extend the shelf life of foodstuffs. Their biomedical applications include pharmaceuticals and medical devices. In the agriculture industry, they are used in the production of fertilizers and pesticides that can significantly affect soil fertility, growth of plants, and crop yield. According to a 2015 report compiled by the Nanotechnology Consumer Product Inventory (CPI), the global market produced 1814 products—before this compilation—based on nanotechnology, including 117 in the food and beverage category (Vance et al. 2015).

The food products with the highest content of TiO$_2$ included candies, sweets, and chewing gums in a sampling study (Weir et al. 2012). Among personal care products, toothpastes and select sunscreens contained 1% to >10% titanium by weight (Fig. 1.2). Looking at this figure, one can see that three sunscreens had the highest concentrations (>10% by weight of the product) of any PCPs followed by toothpastes. The Ti content ranged from below the detection limit (0.0001 µg Ti/mg) to a

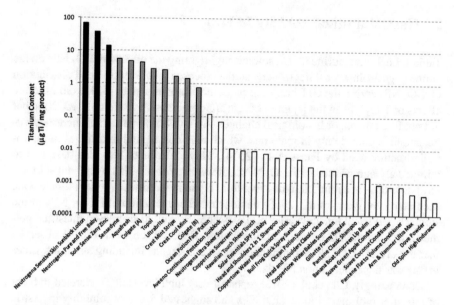

Fig. 1.2. Total titanium concentration for PCPs. Black bars are sunscreens with TiO_2 listed on the label. Gray bars are toothpastes with TiO_2 listed on label. Open bars are for products whose labels did not reference TiO_2 (reproduced from Weir et al. 2012)

high of 0.014 µg Ti/mg for several high-consumption pharmaceuticals. Electron microscopic analysis of food-grade TiO_2 (E171) suggests that approximately 36% of the particles are less than 100 nm in at least one dimension, while the higher bound of sizes exceeds up to 200 nm in some products. It was also estimated that a child could consume even 2–4 times more TiO_2 NPs per 1 kg of body weight a day than an adult person. In Great Britain, children under the age of 10 consume about 2–3 mg TiO_2/kg body weight/day, while adults consume about 1 mg TiO_2/kg body weight/day (Baranowska-Wójcik et al. 2020).

In recent years, the effect of nanoparticles on human health has sparked off debates and scientific scrutiny. In the USA, TiO_2 NPs can be used in food if its content does not exceed 1% of the total weight of the product containing $nTiO_2$. In Europe, it can be used in amounts not exceeding the intended target level (Baranowska-Wójcik et al. 2020). As the permitted limit for edible products is high, a plethora of studies has examined the effect of nanotitania exposures via various routes using animal models. This article presents an update of recent studies on rodent models, particularly focusing on Ti accumulation, transport, and toxicity in the GI tract, liver, kidney, lung, and brain of animals. Many findings from the representative studies are indicative of human scenario and warrant careful examination on the role of nanotitania on human health.

2 Gastrointestinal Toxicity *In Vivo*

Table 1.1 exhibits murine model studies, highlighting the effects of TiO_2 NPs on the animals' gastrointestinal tract. Many studies focused on absorption and distribution of TiO_2 NPs across the GIT barrier in particular. Tassinari et al. (2015) did examine effects of TiO_2 NPs in the jejunum of rats after oral administration of TiO_2 NPs, but did not find any major histological changes. An interesting pattern of increased villus height occurred only in male rats, which may correspond to changes in microvilli structure seen by Faust et al. (2014). This experiment was also designed to mimic daily human intake of TiO_2 NPs, but was only carried out over 5 days. Long-term, low-dose accumulation may therefore be more worthwhile to investigate. One such study by Wang et al. (2016) demonstrated accumulation of TiO_2 NPs in the stomach and small intestine after oral exposure, as well as decreased intestinal permeability and effects on nutrient absorption. Hendrickson et al. (2016) observed that the overall TiO_2 NP accumulation in several organs including small intestines in rats was dependent upon NP size.

Contrastingly, Cho et al. (2013) reported very high levels of Ti excreted in feces of rats after oral ingestion of TiO_2 NPs and suggested that low solubility in gastric fluid may cause the NPs to be excreted as opposed to being absorbed. There was also very little Ti excreted through urine and in tissues, supporting the hypothesis that the TiO_2 NPs were unable to be absorbed. However, Brun et al. (2014) showed TiO_2 NPs were capable of translocating the regular ileum epitheliums and follicle-associated epithelium (FAE) in murine models. This is interesting given that during the *in vitro* portion of this study TiO_2 NPs were found to accumulate within, but never completely cross, the epithelial cell model. This indicates there is some interactions allowing these NPs to cross the barrier *in vivo*. As stated previously, this may be due to an interaction with microvilli, which are associated with absorption in the gut, or contact with the gastric fluid that causes significant aggregation of NPs. This is interesting to note that Brun et al. (2014) had also reported moderate changes in the expression of genes, involved in epithelium structure and maintenance, after analyzing tissues. This indicates a possible connection between TiO_2 NPs and signaling molecules or gene expression. TiO_2 accumulations in mouse gut tissues were low in this study, and they could not be quantified by micro-particle-induced X-ray emission (μPIXE) which has a detection limit of 20–30 ppm. Similar results of zero to low oral bioavailability of TiO_2 NPs have also been reported more recently (Geraets et al. 2014; MacNicoll et al. 2015; Donner et al. 2016).

One factor that may contribute to the disparity of results from various experiments with respect to absorption of TiO_2 NPs in the GIT is a lack of NP characterization. Since NP diameter can vary greatly from manufacturer specifications once they are in solution it is important to observe hydrodynamic diameter and size in NP studies. This is one area where much of the research is not standard, with some experiments characterizing factors such as hydrodynamic diameter and some that do not. Hydrodynamic diameter can also vary greatly depending on the medium that carries NPs. For example, Janer et al. (2014) showed that hydrodynamic diameter of

Table 1.1 Effects of TiO$_2$ NPs on gastrointestinal tract *in vivo*

Model	NP primary size	Hydrodynamic/ agglomerate size	NP structure	Treatment parameters	Route	Results	Reference
BI 57/6 mice	66 nm	Not specified	Anatase	100 mg/kg once a day for 10 days	Oral	Increased CD4+ cells and inflammatory cytokines; hypertrophy and hyperplasia in mucosal epithelium of small intestine	Nogueira et al. (2012)
Sprague Dawley rats	21 nm	37.8 ± 0.4 nm	Anatase/ rutile	0–1041.5 mg/ kg per day for 13 weeks	Oral	Very high Ti levels excreted in feces, indicating little to no absorption in GIT; no significant tissue distribution	Cho et al. (2013)
Mice	12 ± 3 nm	H2O: 132 ± 0.8 nm Gastric fluid: 218.4 ± 2.9 nm Intestinal fluid: >1000 nm	Anatase	12.5 mg/kg one time	Oral	Ti-rich regions in epithelial layer and deep GIT tissues; increased paracellular permeability	Brun et al. (2014)
CRL:WI (WU) Wistar rats	NM100: 42–90 nm NM101: 6 nm NM102: 20 nm NM103: 20 nm NM104: 20 nm	0–250 nm for all types NM102 showed additional mode 250–2500 nm	Anatase Anatase Anatase Rutile Rutile	8.4–14.1 mg/kg one time (IV) Repeated cumulative 42.3–71.9 mg/ kg (IV) 6.8–8.6 mg/kg one time (oral) Repeated cumulative 34.1–59.9 mg/ kg (oral)	IV Oral	Minor absorption by GIT in oral study; slow tissue elimination of Ti	Geraets et al. (2014)

(continued)

Table 1.1 (continued)

Model	NP primary size	Hydrodynamic/ agglomerate size	NP structure	Treatment parameters	Route	Results	Reference
Sprague-Dawley male rats	18 ± 8 nm	202 nm at 1 mg/ml 117 nm at 10 mg/ml	Mainly anatase	100 mg/kg	Oral	No increase in Ti levels in small intestine; considerable accumulation in one Peyer's patch section; freely distributed NPs in cytoplasm	Janer et al. (2014)
Sprague-Dawley rats	Nanoamor:40 nm Nanoamor: 120 nm Sigma: 40–50 nm Sigma: up to 5 μm	Majority of particles 600–700 nm	Anatase Anatase Rutile Rutile	5 mg/kg one time	Oral	No significant absorption by GIT	MacNicoll et al. (2015)
Sprague-Dawley rats	20–60 nm	In H$_2$O: 284 ± 43 nm In DMEM: 220 ± 68 nm	Anatase	0–2 mg/kg once per day/5 days	Oral	No change in histological analysis of jejunum; villus height increased in male rats	Tassinari et al. (2015)
Crl:CD (SD) rats	Uf-1: 31 nm Uf-2: 20 nm Uf-3: 25 nm Pg-1: 191 nm Pg-2: 206 nm Pg-3: 88 nm	43 nm 42 nm 47 nm 153 nm 195 nm 213 nm	Anatase/ rutile Anatase Rutile Anatase Rutile Rutile	0–2000 mg/kg one time	Oral	Lack of genotoxic effects observed in organs due to inability of GIT to absorb NPs	Donner et al. (2016)
Sprague-Dawley rats	NM-101: 5–10 nm NM-25: 20–25 nm	Approx. 100 nm Larger than 100 nm	Anatase Anatase	250 mg/kg once a day for 28 days	IG	Size dependent absorption in small intestine; no mortality or toxicity signs	Hendrickson et al. (2016)
C57BL/6J mice—WTand*Nlrp3*$^{-/-}$	30–50 nm	Not specified	Anatase	0–500 mg/day/ kg for 7 days	Oral	Severe colitis and disruption of mucosal epithelium in WT mice; *Nlrp3*$^{-/-}$ and no pre-existing colitis mice showed no differences	Ruiz et al. (2017)

Model	NP primary size	Hydrodynamic/ agglomerate size	NP structure	Treatment parameters	Route	Results	Reference
BALB/c mice— WT and colitis associated cancer (CAC) models	Mostly less than 100 nm with some peaks above	300 nm	Not specified	5 mg/kg for 5 days/10 weeks	IG	Increased tumors and NF-κB in CAC mice; non-CAC showed morphological changes	Urrutia-Ortega et al. (2016)
Sprague-Dawley rats	75 ± 15 nm	Not specified	Anatase	0–200 mg/kg per day for 30 and 90 days	Oral	NPs found in stomach and small intestine; decrease in intestinal permeability; NPs affected nutrient absorption; genotoxic effects (general)	Wang et al. (2016)

TiO_2 NPs with a primary particle size of 18 ± 8 nm varied from 117 to 367 nm depending on the culture medium used. This group showed no significant increase in Ti levels in various GIT tissues of rats after TiO_2 ingestion. There was a considerable amount of NPs found, however, in one section of Peyer's patches, which lead the authors to hypothesize that Caco-2 monolayer system is likely to underestimate the potential for oral absorption of NPs (Janer et al. 2014). This could also be due to the high likelihood of agglomeration experienced with NPs.

Since TiO_2 NPs are found in many foods and used in food manufacturing, it is critical to consider their possible absorption and accumulation after ingestion. Even with very low uptake from the GIT, human daily oral exposure can be expected to give rise to a very low but steady increase in titanium levels in tissues over time (Geraets et al. 2014). This is of particular interest due to older research that suggests accumulation of inorganic particles including, but not limited, to TiO_2 NPs in intestinal cells could play a part in inflammatory bowel diseases in humans (Powell et al. 2010; Ruiz et al. 2017). Earlier, it was shown that TiO_2 NPs can induce genotoxicity and inflammation in mice by upregulation of the pro-inflammatory cytokines tumor necrosis factor-α (TNF-α), interferon- γ (IFN- γ), and the mouse ortholog of interleukin-8 (IL-8) in peripheral blood (Trouiller et al. 2009). TiO_2 NPs can induce inflammation in the small intestine by inducing a Th-1-mediated inflammatory response in mice after 10 days at high dosage, as reported by Nogueira et al. 2012. This study demonstrated significant increases in the cytokines IL-12, IL-4, IL-23, TNF-α, IFN-γ and transforming growth factor-β (TGF-β) as well as $CD4^+$ T-cells. It is worth noticing, however, that treatment in the above study was given at a higher dose than the average daily ingestion rate in humans. A more recent study by Tada-Oikawa et al. (2016) had also demonstrated increases in IL-8 in an *in vitro* model. Another elegant study confirms the trend and shows that TiO_2 NPs significantly exacerbate existing colitis in WT (C57BL/6J) mice, as well as increase disruption of the mucosal epithelium (Ruiz et al. 2017). Mice deficient in the NLRP3 inflammasome ($Nlrp3^{-/-}$) and mice, which did not have pre-existing colitis before TiO_2 NP treatment, did not exhibit any significant changes in the GIT, indicating TiO_2 NPs may only cause damage when the condition is already present in an individual. In the same study, titanium levels from human subjects with Crohn's disease (CD) and ulcerative colitis (UC) were also analyzed, which indicated those individuals with active UC had significantly higher levels of titanium in their bloodstream. Origin and type of titanium, however, could not be determined, but these findings suggest it may be worthwhile to investigate TiO_2 NP-induced UC.

Urrutia-Ortega et al. (2016) has also confirmed similar effects of TiO_2 NPs increasing tumor formation and inflammation in colitis-associated cancer (CAC) BALB/c mice. Mice with pre-existing CAC treated with TiO_2 NPs exhibited a significant increase in tumor formation and tumor progression markers when compared to CAC mice without TiO_2 NP treatment. Goblet cells were also shown to decrease dramatically in CAC mice treated with TiO_2 NPs. It is interesting to compare the observations of Urrutia-Ortega et al. (2016) who had reported significant GIT changes in normal mice treated with only 5 mg/kg body weight (BW) TiO_2 NPs with those of Ruiz et al. (2017) who could not find changes even at higher dosage

(50–500 mg/kg BW). This could be due to the fact that Ruiz et al. (2017) only used mice that exhibited colitis, while Urrutia-Ortega et al. (2016) had used mice that were already diagnosed with colitis and cancer. Based on the comparison, it could be hypothesized that once the cancer stage is reached it is significantly more difficult to prevent deleterious effects of TiO_2 NPs related to gastrointestinal disease. In addition, Urrutia-Ortega et al. (2016) had also reported an upregulation of nuclear factor-kB (NF-κB) pathway. This signaling pathway consists of many protein products, which serve as transcription factors related to inflammation and plays significant roles in cancer pathogenesis. There may be promise in examining the effects of long-term, low-dose exposure of TiO_2 NPs on the gut inflammatory responses in light of these findings. Another critical aspect to consider when addressing TiO_2 NPs effects in the GIT is translocation to other organs via the gut-associated lymphoid tissues (GALT). The observations of TiO_2 NP uptake by the GIT attract deeper examinations to unravel the underlying connection for inflammatory diseases of the gut.

Another dimension of nano-Ti toxicity on the gut health is to look at its effect on the composition of gut microbiota. Investigators have started examining this interaction. Dudefoi et al. (2017) examined a defined human model of intestinal bacterial community *in vitro*. DNA profiles and phylogenetic distributions in this study confirmed limited effects on the bacterial community, with a modest decrease in the relative abundance of the dominant *Bacteroides ovatus* in favor of *Clostridium cocleatum* (−13% and +14%, respectively, $p < 0.05$). Such minor shifts in the treated consortia suggest that food grade and nano-TiO_2 particles do not have a major effect on human gut microbiota when tested *in vitro* at relevant low concentrations. However, the cumulative effects of chronic TiO_2 NP ingestion remain to be tested.

3 Liver Toxicity *In Vivo*

Table 1.2 shows the results of several experiments conducted in the past few years. Many *in vivo* toxicity studies on TiO_2 NPs have indicated the liver as a primary target for NP accumulation after oral, intravenous (IV), and intraperitoneal (IP) administration (Jeon et al. 2013; Bruno et al. 2014; Geraets et al. 2014; Shinohara et al. 2014a; Shrivastava et al. 2014; Silva et al. 2015; Hendrickson et al. 2016; Yao et al. 2016). Geraets et al. (2014) and Hendrickson et al. (2016) indicated low bioavailability from oral routes. This may be due to inability of TiO_2 NPs to fully penetrate the GIT, as discussed previously. Similar findings of no accumulation of TiO_2 NPs or genotoxicity in the liver of rats after exposure, most likely due to inability of being absorbed, have been demonstrated recently as well (Cho et al. 2013; MacNicoll et al. 2015; Asare et al. 2016; Donner et al. 2016). All of these with the exception of Asare et al. (2016) utilized oral uptake routes. Accumulation was shown to be influenced by the type of exposure route, with IV exposure leading to more TiO_2 NP accumulation in the liver, while oral exposure results in less accumulation in the

Table 1.2 Effects of TiO$_2$ NPs on liver *in vivo*

Model	NP primary size	NP hydrodynamic/ agglomerate size	NP structure	Treatment parameters	Route	Results	Reference
CD-1 mice	25 nm 80 nm	Not specified	Not specified	5 g/kg one time	Oral gavage	No toxicity observed, increased AST and ALT	Wang et al. (2007)
CD-1 mice	5 nm 10–15 μm (bulk)	Not specified	Anatase Rutile	0–150 mg/kg once a day for 14 days	IP	Accumulation in liver; Changes in ALT, ALP, AST, LDH; hepatocyte damage; elevated inflammatory cytokines	Ma et al. (2009)
C57B1/6Ip^{un}/p^{un} mice	21 nm	160±5 nm	Anatase/ rutile	0–600 μg/ml (3–7 ml) for 5 days	Oral	Oxidative DNA damage	Trouiller et al. (2009)
CD-1 mice	5 nm	Not specified	Anatase	0–50 mg/kg once a day for 60 days	IG	Accumulation in liver; histological changes; hepatocyte apoptosis; liver function damage	Cui et al. (2011)
CD-1 mice	5–6 nm	208–330 nm	Anatase	10 mg/kg once a day for 90 days	IG	Accumulation in the liver and aggregation in hepatocyte nuclei; increased inflammation; hepatocyte apoptosis; liver dysfunction.	Cui et al. (2012)
Wistar rats	10–20 nm	72 nm	Anatase	0–50 mg/kg once a week for 1 month	IV	Accumulation; increased apoptosis; mitochondria mediated cytotoxicity; oxidative stress induced genotoxicity; increased ALT, AST, ALP; decreased CAT, SOD, GSH	Meena and Paulraj (2012)
Wistar albino rats	Not specified	50.40 ± 5.60	Anatase	0–252 mg per rat one time	IP	Histologic changes; increased apoptosis; increased ALP and GOT indicate compromised liver function	Alarifi et al. (2013)
Sprague Dawley rats	21 nm	37.8 ± 0.4 nm	Anatase/ rutile	0–1041.5 mg/ kg once a day for 13 weeks	Oral	No significant accumulation	Cho et al. (2013)

Model	NP primary size	NP hydrodynamic/ agglomerate size	NP structure	Treatment parameters	Route	Results	Reference
ICR/CD mice	< 25 nm	429 nm	Anatase	1.5 mg per day for 7 days	IP	Histological changes; heavy accumulation in Kupffer cells; increased GOT, GPT, ALP enzymes indicating injury; decreased catalase, SOD, ALDH	Jeon et al. (2013)
Swiss albino mice	20–50 nm	281–301 nm	Anatase	0–100 mg/kg per day for 14 days	IP	Increased liver enzymes; histological changes; oxidative DNA damage (genotoxicity); ROS generation	Shukla et al. (2014)
ICR mice	42.30 ± 4.60 nm	100–500 nm	Anatase	0–1387 mg/kg one time	IV	no significant increase in ALT, TBIL, AST, ALP; vacuolar degeneration, inflammatory cells infiltration, hydropic degeneration, and spotty necrosis of hepatocytes; mice died on day 2 due to acute toxicity from high dose	Xu et al. (2013)
Wistar rats	MP150: 150 nm NP10: 10 nm NP5: 5 nm	Not specified	Anatase Anatase Anatase	1.6 g/100 g one time	IP	Accumulation consistently larger in liver, found in Kupffer cells and hepatocytes; increased CAT	Bruno et al. (2014)
Wistar rat	18 nm	Not specified	Anatase/ rutile	0–70 mg/kg once a day for 21 days	IP	increased ALP/ALT; several histological changes	Fatemeh and Mohammad (2014)

(continued)

Table 1.2 (continued)

Model	NP primary size	NP hydrodynamic/ agglomerate size	NP structure	Treatment parameters	Route	Results	Reference
CRL:WI (WU) Wistar rats	NM100: 42–90 nm NM101: 6 nm NM102: 20 nm NM103: 20 nm NM104: 20 nm	0–250 nm for all types NM102 showed additional mode 250–2500 nm	Anatase Anatase Anatase Rutile Rutile	8.4–14.1 mg/ kg one time (IV) Repeated (5) cumulative 42.3–71.9 mg/ kg (IV) 6.8–8.6 mg/kg one time (oral) Repeated (5) cumulative 34.1–59.9 mg/ kg (oral)	IV Oral	Low accumulation after oral doses; accumulation observed after IV doses; single IV treatment was recoverable after 24 hr	Geraets et al. (2014)
CD-1 (ICR) mice	5–6 nm	208–330 nm	Anatase	0–10 mg/kg per day for 6 months	IG	Accumulation which led to reductions in body weight, increases in liver indices, liver dysfunction, infiltration of inflammatory cells, and hepatocyte apoptosis/ necrosis	Hong et al. (2014)
SD rats	12–18 nm (diameter, 40–80 nm (length)	Not specified	Rutile	0–50 mg/kg one time	IP	Under non-oxidative stress conditions hepatocyte necrosis; under oxidative stress conditions hepatic changes; both conditions showed and increased AST, ALT, ALP	Sha et al. (2014)
F344/DuCrlCrlj rats	21 nm	62.5–65.7 nm	Anatase/ rutile	1 mg/kg one time	IV	Significant Ti burden in liver at 6 h post-exposure and did not decrease at 30 days after dose	Shinohara et al. (2014a)

Model	NP primary size	NP hydrodynamic/ agglomerate size	NP structure	Treatment parameters	Route	Results	Reference
F344 rats	21 nm	143–148 nm	Anatase/ rutile	0–6 mg/kg one time	II	Significant Ti burden at 3 d–26w in 6.0 mg group; translocation of NPs from lung to liver	Shinohara et al. (2014b)
Swiss albino mice	<75 nm	Not specified	Anatase/ rutile	500 mg/kg once a day for 21 days	Oral	Inhibition of CuZnSOD and MnSOD; NPs in endosomes and Kupffer cells; increase in ROS	Shrivastava et al. (2014)
Fisher F344 rats	21.5 ± 7.2 nm	In H_2O: 163.5 ± 12.6 In saline: 520.9 ± 41.7 nm	Anatase/ rutile	1 mg/kg one time	IV	Significant amount of Ti in liver 356 days post-exposure	Disdier et al. (2015)
Wistar rats	10–15 nm	Not specified	Anatase/ rutile	0–300 ppm once a day for 7 days	IP	Increased AST initially at 2 days, but then decreased at 7 and 14 days; histological changes	Doudi and Setorki (2015)
Wistar rats	Not specified	Not specified	Anatase	5 mg/kg one time	IV	Increased AST at 14 days, but decreased by day 28; slight increase in ALT; slight oxidative stress; increased GSR	Gonzalez-Esquivel et al. (2015)
ICR mice	21 nm	<50 nm	Not specified	0–0.5 mg once per week for 4 weeks	II	No changes in liver histology	Huang et al. (2015)
C57BL/6 mice	20.6 nm	Large range from 31.4–5407 nm depending on concentration and filters	Rutile	18 and 162 µg one time	II	Translocation of NPs from lung to livers observed up 28 days post-exposure (slow clearance); changes in expression of acute phase response genes	Husain et al. (2015)

(continued)

Table 1.2 (continued)

Model	NP primary size	NP hydrodynamic/agglomerate size	NP structure	Treatment parameters	Route	Results	Reference
Sprague-Dawley rats	Nanoamor: 40 nm Nanoamor:120 nm Sigma: 40–50 nm Sigma: up to 5 μm	Majority of particles 600–700 nm	Anatase Anatase Rutile Rutile	5 mg/kg one time	Oral	No significant uptake of Ti	MacNicoll et al. (2015)
Wistar rats	<100 nm	Not specified	Anatase/rutile	0–100 mg/kg once a day for 14 days	IG	Decreased ALP; no changes in ALT; increased AST	Vasantharaja et al. (2015)
Swiss albino mice	118 + 30 nm	Not specified	Anatase/rutile	2 mg/kg once a day for 10 days	IP	Accumulation of Ti; no change in ALT levels; slight increase in AST	Silva et al. (2015)
Wistar rats	20–30 nm	Not specified	Anatase	20 mg/kg every 2 days for 20 days	IP	Increased AST/ALT ratio; increased LDH activity; histological changes	Younes et al. (2015)
C57BL6 (WT) and $Ogg1^{-/-}$ KO mice	21 nm	245.8 ± 0.8 nm	Not specified	5 mg/kg one time	IV	No significant genotoxicity observed	Asare et al. (2016)
Crl:CD (SD) rats	Uf-1: 31 nm Uf-2: 20 nm Uf-3: 25 nm Pg-1: 191 nm Pg-2: 206 nm Pg-3: 88 nm	43 nm 42 nm 47 nm 153 nm 195 nm 213 nm	Anatase/rutile Anatase Rutile Anatase Rutile Rutile	500–2000 mg/kg one time	Oral	No presence of NPs in liver	Donner et al. (2016)
Male albino rats	21 nm	Not specified	Not specified	1200 mg/kg once a day for 12 weeks	Oral gavage	Toxic effects due to oxidative stress; increased AST and ALT; decreased GSH due to oxidative stress; genotoxicity observed; histological changes; TNF-α present	El-Zahed et al. (2016)

Model	NP primary size	NP hydrodynamic/ agglomerate size	NP structure	Treatment parameters	Route	Results	Reference
Sprague-Dawley rats	NM-101: 5–10 nm NM-25: 20–25 nm	Approx. 100 nm Larger than 100 nm	Anatase Anatase	250 mg/kg once a day for 28 days	IG	Accumulation of Ti highest in liver and spleen; no mortality or toxicity observed	Hendrickson et al. (2016)
ICR mice	25 nm 50 nm	Not specified	Anatase/rutile Rutile	40 mg/kg one time	IV, IP, IG	Accumulation in organs depended on route of exposure; liver one of highest accumulations in all treatments; little effect on liver function in all 3 routes; slight inflammation found in livers with longer exposure time	Yao et al. (2016)

liver (Yao et al. 2016). Shinohara et al. (2014a) demonstrated Ti burden did not decrease for up to 30 days after TiO_2 NP exposure. In an even longer study, Disdier et al. (2015) reported the persistence of Ti burden after a one-time IV dose of only 1 mg/kg BW in rats for up to 356 days post-exposure. This indicates the ability of TiO_2 NPs to accumulate and inability to clear TiO_2 NPs over a long period of time. Even though IV exposure is highly unlikely in a practical scenario, exposure to TiO_2 NPs through oral and inhalation routes occurs often. It has been shown that TiO_2 NPs possess the ability to translocate from the lungs to liver after inhalation in mice and rats after one-time exposure, producing significant Ti burden in the liver that does not decrease over long periods of time (Shinohara et al. 2014b; Husain et al. 2015). If these NPs are distributed through the bloodstream to the liver, and other organs, the possibility for toxicity is a threat. Although oral exposure seems to be non-threatening based on the available data, inhalation of TiO_2 NPs may prove to be much more detrimental. Therefore, this calls for further investigations on NP accumulation in the liver via inhalation routes for a full picture.

Liver enzymes such as aspartate aminotransferase (AST), alanine aminotransferase (ALT), and alkaline phosphatase (ALP) are often used as markers for hepatic injury. These are the key enzymes often monitored to assess liver toxicity. It has been reported many times that these biomarkers increase significantly after exposure to TiO_2 NPs (Meena and Paulraj 2012; Alarifi et al. 2013; Jeon et al. 2013; Shukla et al. 2014; Fatemeh and Mohammad 2014; Doudi and Setorki 2015; Gonzalez-Esquivel et al. 2015; El-Zahed et al. 2016). Only IV and IP exposure routes were explored in these studies, leaving oral effects of TiO_2 in question. It is particularly interesting that increases were seen at doses of as little as 1.5 mg/kg BW per day in short-term study, though TiO_2 NPs were administered intraperitoneally (Jeon et al. 2013). In addition, Doudi and Setorki (2015) demonstrated significant increases in AST, but then decreases at 7 and 14 days post exposure, indicating possible clearance of NPs from the kidneys.

One study has found surprising decreases in ALP and no changes in ALT at 50 and 100 mg/kg BW TiO_2 NPs (Vasantharaja et al. 2015). AST was still increased, however, in at both doses. Hydrodynamic diameter and agglomeration of NPs were not characterized in this study though, and so particle agglomeration could be responsible for inability of TiO_2 NPs to infiltrate liver tissue. Silva et al. (2015) also reported no changes in ALT levels with slight increases in AST after TiO_2 NP exposure. Xu et al. (2013) also demonstrated no increases in ALT, ALP, or AST even at very high doses. Both Xu et al. (2013) and Silva et al. (2015) had characterized NPs in solution as having aggregates of over 100 nm, which may have played a role in affecting enzyme levels. Unfortunately, oral studies monitoring liver enzyme activity after realistic daily are lacking. Previously, Wang et al. (2007) demonstrated no significant changes in AST and ALP after oral TiO_2 NP exposure in mice, but increased ALT levels were observed. AST/ALT ratio, which is typically a common indicator of hepatic injury, was also increased significantly. Interestingly, very large doses of 5 g/kg BW were used in this study and yet the expected elevation of most liver enzymes was not demonstrated. Since hydrodynamic diameter of NPs was not characterized in this study, NPs could have agglomerated given the very large amount given to the test animals. This could cause insufficient uptake into the liver if the NPs were not properly absorbed by the GIT. However, it should be noted that there also seem to be other factors affecting TiO_2 NPs behavior when in solution. Even though the previously mentioned studies which found little to no changes in liver enzyme levels had large hydrodynamic characterization (Xu et al. 2013; Vasantharaja et al. 2015; Silva et al. 2015), so did some of those which showed significant increases in enzyme levels (Jeon et al. 2013; Shukla et al. 2014). In addition, hydrodynamic size was also not characterized in many of the studies which showed increases in liver enzymes (Alarifi et al. 2013; Fatemeh and Mohammad 2014; Gonzalez-Esquivel et al. 2015). Based on this data, it is clear other factors are playing a role in TiO_2 NP liver toxicity concerning enzyme levels. Given the lack of studies monitoring liver enzyme levels after low-dose oral exposure, more investigation is needed in this area. This would most certainly be an avenue of research to consider, since TiO_2 NPs are often found in daily consumer products and foods.

Earlier *in vivo* studies on TiO_2 liver toxicity have demonstrated oxidative genotoxicity, accumulation of particles in the liver, hepatic injury due to increased liver enzyme activity hepatocyte structure damage, histological changes, and increased inflammatory cytokines (Trouiller et al. 2009; Ma et al. 2009; Cui et al. 2011, 2012; Meena and Paulraj 2012). Similar results have been demonstrated in many recent studies. Alarifi et al. (2013) demonstrated apoptosis in liver cells after TiO_2 NP exposure in albino Wistar rats. It was hypothesized that oxidative stress and generation of ROS lead to these changes and cytotoxicity, although this was not explored specifically. Genotoxicity due to oxidative stress and ROS generation have also been confirmed in an oral study by Shukla et al. (2014) using daily doses of 50 and 100 mg/kg BW in Swiss albino mice. Genotoxicity was also demonstrated by El-Zahed et al. (2016) after TiO_2 treatment, but at much higher levels of 1200 mg/kg BW per day. In this study however, genotoxicity was confirmed to be due to oxidative stress by Comet assay. It has been shown in many studies that decreases in specific antioxidant mediators such as GSH, catalase (CAT), superoxide dismutase (SOD) and its variants (CuZnSOD and MnSOD), and aldehyde dehydrogenase (ALDH) accompany increases in oxidative stress and ROS levels (Jeon et al. 2013; Shrivastava et al. 2014; Bruno et al. 2014; El-Zahed et al. 2016).

Oxidative stress already present in the cell can also contribute to the toxicity of TiO_2 NPs. The previously mentioned study by Sha et al. (2014) showed significant increase in liver enzyme activity while under the effects of oxidative stress in rats. Liver enzyme levels also increased without inducing oxidative stress (TiO_2 NP treatment only), but not as significantly as while under oxidative conditions. In addition, deleterious hepatocyte changes were observed at the lowest dosage of TiO_2 NPs (0.5 mg/kg) while under oxidative conditions, whereas TiO_2 NP treatment only resulted in hepatocyte changes at 5 mg/kg. These changes included intake of NPs and hepatocyte necrosis. Based on the wide range of data available, TiO_2 NP-induced ROS can lead to toxicity within liver cells. Therefore, a closer look at long-term exposure to TiO_2 NPs is warranted for public health safety reasons.

Physical changes have also been reported by many of the aforementioned groups investigating effects of TiO_2 NPs on the liver (Alarifi et al. 2013; Xu et al. 2013;

Shukla et al. 2014; Jeon et al. 2013; Fatemeh and Mohammad 2014; Doudi and Setorki 2015; El-Zahed et al. 2016). These include changes such as hydropic degeneration, congested and dilated central veins, destruction of nuclei, mononuclear cell accumulation, nuclear envelope swelling in Kupffer cells, spotty necrosis, and congestion of sinusoidal spaces. In addition, appearance of inflammatory cells and cytokines has been well documented over several years (Ma et al. 2009; Cui et al. 2011; Cui et al. 2012; Meena and Paulraj 2012; Alarifi et al. 2013; Xu et al. 2013; Hong et al. 2014). Hong et al. (2014) showed effects of chronic, low-dose TiO_2 NP exposure over the course of 6 months, which provides insight into realistic long-term effects. Yao et al. (2016) confirmed similar results, showing slight inflammation in livers with longer exposure time to TiO_2 NPs over the course of 120 days. El-Zahed et al. (2016) have also reported increased TNF-α in albino rats after TiO_2 NP treatment. Interestingly it has been shown that intratracheal instillation (II) of TiO_2 NPs does not induce histological changes in liver cells (Huang et al. 2015). Based on the number of studies in which inflammation and histological changes are demonstrated, the necessity to continue thorough investigation of TiO_2 NPs on liver function is apparent. The effects shown by Hong et al. (2014) in particular are very striking, considering the dosages used and length of experiment.

In vivo study of TiO_2 NP toxicity in the liver has shown that the liver is a possible primary target of NPs. However, many studies focus on IV and IP exposure, which is not a realistic analog for daily human exposure. Oral and intratracheal instillation exposures are by far the two most common routes for humans and other organisms to come into contact with TiO_2 NPs. These routes should be explored further in low-dose experiments to gauge practical toxicity effects. Liver enzymatic activity and tissue observation after NP exposure are also key to establish toxic effects. ROS, inflammation, and oxidative stress are also key factors that are present after TiO_2 NP exposure in liver tissue, as seen in many other tissue types.

4 Kidney Toxicity *In Vivo*

In vivo research on kidney toxicity after TiO_2 NP exposure has grown significantly in recent years, as shown in Table 1.3. A number of *in vivo* studies assessing accumulation of TiO_2 NPs in the kidney have demonstrated that uptake can occur across oral, IV, and IP routes of exposure in short- and long-term treatments (Gui et al. 2013; Bruno et al. 2014; Shinohara et al. 2014a; Geraets et al. 2014; Disdier et al. 2015; Hong et al. 2015; Huang et al. 2015; Silva et al. 2015; Hendrickson et al. 2016). In both short- and long-term oral and IV studies it was shown that Ti burden in the kidney may increase significantly initially, but decreases over time (Shinohara et al. 2014a; Geraets et al. 2014; Disdier et al. 2015). This suggests elimination of Ti in kidney tissue may occur more readily when compared to other organs. In addition, some studies which showed significant accumulation of TiO_2 NPs in the kidneys also indicated significant histological changes including infiltration of inflammatory cells, fatty degeneration, fibrosis, cell abscission, tubular dilation, brush border loss,

Table 1.3 Effects of TiO$_2$ NPs on kidney *in vivo*

Model	NP primary size	NP hydrodynamic/agglomerate size	NP structure	Treatment Parameters	Route	Results	Reference
Sprague Dawley rats	21 nm	37.8 ± 0.4 nm	Anatase/rutile	0–1041.5 mg/kg per day for 13 weeks	Oral	No significant accumulation	Cho et al. (2013)
CD-1 (ICR) mice	5–6 nm	209–330 nm	Anatase	0–10 mg/kg every day for 6 months	IG	Accumulation and ROS at all doses; renal dysfunction; apoptosis; increased NF-κB, CREA; decreased UA, BUN, GSH, Nrf2	Gui et al. (2013)
ICR mice	42.30±4.60	100–500 nm	Anatase	0–1387 mg/kg one time	IV	No significant increase in BUN, CREA; increased UA; swelling in renal glomerulus	Xu et al. (2013)
Wistar rats	5 nm 10 nm 150 nm	Not specified	Anatase Anatase Anatase	1.6 g/100 g one time	IP	Accumulation seen at 12 mo. post-exposure; lack of histological changes	Bruno et al. (2014)
CRL:WI (WU) Wistar rats	NM100: 42–90 nm NM101: 6 nm NM102: 20 nm NM103: 20 nm NM104: 20 nm	0–250 nm for all types NM102 showed additional mode 250–2500 nm	Anatase Anatase Anatase Rutile Rutile	8.4–14.1 mg/kg one time (IV) Repeated cumulative 42.3–71.9 mg/kg (IV) 6.8–8.6 mg/kg one time (oral) Repeated cumulative 34.1–59.9 mg/kg (oral)	IV Oral	Slight accumulation 24-h and 90-days post-exposure	Geraets et al. (2014)
F344/DuCrlCrlj rats	21 nm	62.5–65.7 nm	Anatase/rutile	1 mg/kg one time	IV	Ti burden high post-exposure, but decreased significantly over time over 30 days	Shinohara et al. (2014a)

(continued)

Table 1.3 (continued)

Model	NP primary size	NP hydrodynamic/agglomerate size	NP structure	Treatment Parameters	Route	Results	Reference
F344 rats	21 nm	143–148 nm	Anatase/rutile	0–6 mg/kg one time	II	No significant accumulation	Shinohara et al. (2014b)
Fisher F344 rats	21.5 ± 7.2 nm	In H$_2$O: 163.5 ± 12.6 In saline: 520.9 ± 41.7 nm	Anatase/rutile	1 mg/kg one time	IV	Ti burden high post-exposure, but decreased significantly over time over 7 days	Disdier et al. (2015)
Wistar rats	Not specified	Not specified	Anatase	5 mg/kg one time	IV	Increased renal oxidative stress marker MDA; GPx increased; renal damage and oxidative effects	Gonzalez-Esquivel et al. (2015)
CD-1 (ICR) mice	5–6 nm	208–330 nm	Anatase	0–5 mg/kg every day for 9 months	IG	Accumulation; increased nephrotoxicity markers; CREA and CREA/protein ratio increased; decreased UA and BUN; histological changes including fibrosis and necrosis; Wnt pathway activated	Hong et al. (2015)
ICR mice	21 nm	< 50 nm	Not specified	0–0.5 mg once per week for 4 weeks	II	Accumulation in cytoplasm and organelles; histological changes including fibrosis; increased BUN; increased ROS and TGF-β; increased mRNA for IL-1B and TNF-α	Huang et al. (2015)
Sprague-Dawley rats	Nanoamor, 40 nm Nanoamor,120 nm Sigma, 40–50 nm Sigma, up to 5 μm	Majority of particles 600–700 nm	Anatase Anatase Rutile Rutile	5 mg/kg one time	Oral	No significant accumulation	MacNicoll et al. (2015)
Wistar rats	< 100 nm	Not specified	Anatase/rutile	0–100 mg/kg once a day for 14 days	IG	Increased BUN and UA; no change in CREA	Vasantharaja et al. (2015)
Swiss albino mice	118 ± 30 nm	Not specified	Anatase/rutile	2 mg/kg once a day for 10 days	IP	Accumulation; decreased urea; no change in CREA	Silva et al. (2015)

Model	NP primary size	NP hydrodynamic/ agglomerate size	NP structure	Treatment Parameters	Route	Results	Reference
Wistar rats	20–30 nm	Not specified	Anatase	20 mg/kg every 2 days for 20 days	IP	No significant changes in accumulation, histology, or function markers observed	Younes et al. (2015)
Wistar rats	10–15 nm	Not specified	Anatase/ rutile	0–70 mg/kg every other day for 21 days	IV	Changes in urea and UA initially, but returned to control levels; histological changes but not morphological changes	Fartkhooni et al. (2016)
Sprague-Dawley rats	NM-101: 5–10 nm NM-25: 20–25 nm	Approx. 100 nm Larger than 100 nm	Anatase Anatase	250 mg/kg once a day for 28 days	IG	Slight accumulation; no toxicity or mortality	Hendrickson et al. (2016)
WT BALB/c mice, CAC BALB/c mice	Food grade—Not specified	50–600 nm	Not specified	5 mg/kg every day for 10 weeks	IG	No changes in function markers; significant change in kidney weight in CAC model	Urrutia-Ortega et al. (2016)
ICR mice	25 nm 50 nm	Not specified	Anatase/ rutile Rutile	40 mg/kg one time	IV, IP, IG	Little effect on function; no toxicity	Yao et al. (2016)

and tissue necrosis (Gui et al. 2013; Hong et al. 2015; Huang et al. 2015). The report from Huang et al. 2015 is particularly interesting because it utilized a weekly low-dose, intratracheal instillation over 4 weeks for the exposure route. In addition, it was also shown that renal fibrosis in TiO$_2$ NP-exposed mice was much worse when compared to controls, and ROS markers were significantly increased. If the inhalation route of TiO$_2$ NPs is capable of producing such effects, then workers exposed to these NPs daily may be at risk. TiO$_2$ NPs exposure is most often associated with oral and inhalation uptake routes, and so these effects should be investigated further. It should be noted that Shinohara et al. (2014b) showed no significant uptake in kidney tissues in rats also using intratracheal instillation, but this was a one-time dose study. Chronic, low-dose exposure as carried out by Huang et al. (2015) may be more accurate in assessing practical exposure limits for workers who are in contact with TiO$_2$ NPs daily. Cho et al. (2013) and MacNicoll et al. (2015) reported no uptake of Ti in the kidney after oral exposure at 13 weeks and 96 h, respectively. In the case of MacNicoll et al. (2015) this could be due to lack of absorption due to large hydrodynamic size (600–700 nm), or the possibility that the low, one-time dose was excreted before observation. However, Cho et al. (2013) subjected rats to oral exposure every day for 13 weeks, at considerably higher concentrations (260.4–1041.5 mg/kg BW), and still did not observe any Ti accumulation in the kidneys. Hydrodynamic size was also relatively close to primary particle size of 21 nm. The authors note that this was an unexpected result and suggested more research into methods for NP detection may be worthwhile.

Liver function markers such as creatinine (CREA), blood urea nitrogen (BUN), and uric acid (UA) are often used as indicators of nephrotoxicity. In humans, elevated CREA and BUN are associated with kidney dysfunction, while both decreased and increased UA can be associated with certain kidney diseases. Table 1.4 shows the results of several studies based on these markers. The results are overlapping, and it is difficult to ascertain a particular trend with respect to liver function markers and TiO$_2$ NP exposure. ROS and oxidative stress effects also are somewhat difficult to understand due to lack of many *in vitro* studies focusing on these aspects specifically in the kidney. Gui et al. (2013) demonstrated significant increases in ROS generation related to lipid, protein, and DNA peroxidation, which was also accompanied by decreased GSH. Gonzalez-Esquivel et al. (2015) also showed significant increases in the oxidative stress markers malondialdehyde (MDA) and GPx, and hypothesized the observed renal damage was due to oxidative effects on kidney cells.

As with other cell types and tissues, significant upregulation of nuclear factor-κB (NF-κB) pathway was observed *in vivo* in the kidney (Gui et al. 2013). NF-κB is often associated with activation of inflammatory pathways, and its increased presence suggests TiO$_2$ NPs can cause inflammation. In addition, this study indicated decreased Nrf2 activity, which, as discussed previously, has been shown to have protective effects during TiO$_2$ NP exposure. Huang et al. (2015) reported increases in the inflammatory cytokine TGF-β as well as increases in mRNA expression of TNF-α and IL-1B. Hong et al. (2015) demonstrated increased activation of the Wnt pathway after TiO$_2$ NP exposure, which is associated with poor prognosis in renal cell carcinoma patients and may promote renal inflammation and fibrosis. Yao et al.

Table 1.4 Effects of TiO₂ NP on kidney function markers *in vivo*

Author	Route	Exposure parameters	CREA increase	CREA decrease	CREA no change	UA increase	UA decrease	UA no change	BUN increase	BUN decrease	BUN no change
Gui et al. (2013)	Oral	2.5–10 mg/kg every day for 6 months	√	–	–	–	√	–	–	√	–
Xu et al. (2013)	IV	0–1387 mg/kg BW 1 time	–	–	√	√	–	–	–	–	√
Vasantharaja et al. (2015)	Oral	50/100 mg/kg BW per day for 14 days	–	–	–	√	–	–	√	–	–
Silva et al. (2015)	IP	2 mg/kg per day for 10 days	–	–	√	–	√	–	–	–	–
Huang et al. (2015)	II	0.1–0.5 mg/week for 4 weeks	√ (slight)	–	–	–	–	–	√	–	–
Urrutia-Ortega et al. (2016)	Oral	5 mg/kg for 10 weeks	–	–	√	–	–	–	–	–	√
Fartkhooni et al. (2016)	IV	30–70 mg/kg every other day for 21 days	–	–	√	√	–	–	–	–	–

(2016) has shown inflammation in kidneys persists up to 120 days regardless of exposure route, but mice in this study showed no significant changes in kidney coefficients when compared to controls and there was little effect on kidney function observed.

The results illustrated in Table 1.3 show some conflicting data, especially concerning accumulations. However, some low-dose oral and intratracheal instillation exposure studies have invariably shown that TiO_2 NPs can induce ROS and oxidative stress, as well as inflammation and changes in kidney function markers (Gui et al. 2013; Huang et al. 2015). Given that the primary role of the kidney is to filter out unnecessary substances it is not surprising that initial NP accumulation in the tissue decreases over time. Many studies indicate low accumulation and overall low deleterious effects after TiO_2 NP exposure in the kidneys. However, those studies which demonstrated negative effects warrant continued screening of TiO_2 NP effects in kidney tissues.

5   Lung Toxicity *In vivo*

Since inhalation is considered one of the leading ways for humans to come into contact with TiO_2 NPs, it is no surprise that there is a large pool of *in vivo* studies that have been conducted in recent years, as shown in Tables 1.5 and 1.6. An overwhelming majority of studies reviewed here found accumulation of TiO_2 NPs in lung tissues and cells after intratracheal instillation and inhalation exposure, often times at significant amounts (McKinney et al. 2012; Sun et al. 2012a; Sun et al. 2012b; Li et al. 2013; Tang et al. 2013; Shinohara et al. 2014b; Disdier et al. 2015; Hashiguchi et al. 2015; Yoshiura et al. 2015; Yao et al. 2016). It has been reported that TiO_2 NPs may initially infiltrate lung tissue and cells, but then decrease in accumulation over time, possibly due to elimination (Geraets et al. 2014; Shinohara et al. 2014b; Morimoto et al. 2016). This is particularly interesting because inhalation, intratracheal instillation, oral, and IV exposure routes were examined in these studies, indicating elimination from lung tissue may be possible across different routes of exposure. However, Disdier et al. (2015) demonstrated Ti persistence in lungs of rats up to 356 days after a single, low-dose IV exposure similar to those in the aforementioned studies. This may be attributed to experimental and NP differences. Hendrickson et al. (2016) also showed negligible Ti accumulation in lungs after TiO_2 NP exposure in rats, but this is most likely due to low bioavailability due to the oral exposure route used. As indicated by *in vitro* studies, phagosomes and uptake by phagocytization seem to play a role in uptake of Ti (Porter et al. 2013; Bruno et al. 2014; Morimoto et al. 2016). Given the large pool of data available regarding NP accumulation in the lungs, it is obvious that accumulation does occur and may persist according to NP characteristics.

Aggregation of TiO_2 NPs near inflammatory cells of severely affected regions in mice has also been demonstrated (Hashiguchi et al. 2015). It was also found that the inflammatory marker IFN-γ and chemokine CCL5, both markers of pneumonia,

Table 1.5 Effects of TiO$_2$ NPs on lung *in vivo*

Model	NP primary size	NP hydrodynamic/ agglomerate size	NP structure	Treatment parameters	Route	Results	Reference
Sprague-Dawley rats	110 nm	Not specified	Not specified	314, 826, or 3638 mg/m³ 2–4 times a day for 1–4 days	Inhalation	No significant pulmonary changes noted at low and middle dose levels; high dose caused significant increases in breathing rate, pulmonary inflammation, and lung cell injury	McKinney et al. (2012)
CD-1 (ICR) mice	5–6 nm	208–330 nm	Anatase	0–10 mg/kg every day for 90 days	II	Accumulation; edema, emphysema, and inflammation cell infiltration; increased O^{2-} and H$_2$O$_2$; increased Nrf2 expression	Sun et al. (2012a)
CD-1 (ICR) mice	5–6 nm	208–330 nm	Anatase	0–10 mg/kg every day for 90 days	II	Accumulation; edema, emphysema, inflammation cell infiltration, and blood vessel congestion; increased inflammatory cytokines (NF-κB, IL-6, HO-1); increased ROS and oxidative stress; decreased GSH\	Sun et al. (2012b)
Wistar rats	50 nm (diameter), 500 nm (length)	Not specified	Nanofiber	4 mg one time	II	Lower alveolar macrophage viability; increased multinucleated cells; decreased SOD	Hurbánková et al. (2013)

(continued)

Table 1.5 (continued)

Model	NP primary size	NP hydrodynamic/ agglomerate size	NP structure	Treatment parameters	Route	Results	Reference
BALB/c mice	21 nm	1.52±0.16 μm (largest fraction); 100–200 nm	Anatase/ rutile	Nebulizer particle flow rate: 7.8±0.2 l/min; up to 8 doses over 35 nm	Nebulizer	Increased inflammation in naïve models; single NP dose before OVA challenge (asthmatic model) enhanced allergic airway reaction	Jonasson et al. (2013)
CD-1 (ICR) mice	5–6 nm	208–330 nm	Anatase	0–10 mg/kg everyday for 90 days	Nasal instillation	Accumulation; inflammatory cell infiltration, edema, apoptosis, mitochondrial swelling; increased ROS leading to lipid, protein, and DNA peroxidation	Li et al. (2013)
Wistar rats	Not specified	22.82 ± 5.30 nm	Not specified	0, 3.5, or 17.5 mg/ kg every 2 days for 5 weeks	II	Decreased GSH, SOD, and NO; increased TNF-α decreased viability of AMs	Liu et al. (2013)
C57BL/6J mice	NS: 70–190 nm NB1: 1–5 μm long, 40–120 nm wide NB2: 6–12 μm long, 60–140 nm wide	354.2 nm Not specified Not specified	Anatase Anatase Anatase	0–30 μg one time	Pharyngeal aspiration	NP shape and length affect response; NB1 and NB2 induced increased PMNs and inflammation; NB2 group showed alveolitis and fibrosis	Porter et al. (2013)
Sprague-Dawley rats	P25: 24 nm TiO_2A: 28 nm TiO_2NB: 7000 (L) × 200 (W) × 10 (T) nm	209 ± 8 nm 292 ± 70 nm 2879 ± 117 nm	Anatase/ rutile spheres Anatase spheres Anatase nanobelts	0–200 μg one time	II	P25 and TiO_2A reached deep lung, but no toxicity observed; TiO_2NB group showed increased inflammation and histological changes	Silva et al. (2013)
Rats	20–50 nm	Not specified	Anatase	0–10 mg/ml one time	II	Thickened and ruptured alveolar walls; accumulation; increased lipid peroxidation	Tang et al. (2013)

Model	NP primary size	NP hydrodynamic/agglomerate size	NP structure	Treatment parameters	Route	Results	Reference
ICR mice	42.30 ± 4.60 nm	100–500 nm	Anatase	0–1387 mg/kg one time	IV	Infiltration of inflammatory cells; fibrosis and lesions	Xu et al. (2013)
Wistar rats	5 nm 10 nm 150 nm	Not specified	Anatase Anatase Anatase	1.6 g/100 g one time	IP	Accumulation in parenchyma; Increased SOD; NPs more persistent in lung than MPs	Bruno et al. (2014)
Sprague-Dawley rats	21 nm	195.2 nm	Anatase/rutile	0–32 mg/kg twice a week for 4 weeks	II	Disruption of alveolar septa; mitochondria dissolution; ER expansion	Chang et al. (2014)
Wistar rats (CRL:WI (WU))	NM100: 42–90 nm NM101: 6 nm NM102: 20 nm NM103: 20 nm NM104: 20 nm	0–250 nm for all types NM102 showed additional mode 250–2500 nm	Anatase Anatase Anatase Rutile Rutile	8.4–14.1 mg/kg one time (IV) Repeated cumulative 42.3–71.9 mg/kg (IV) 6.8–8.6 mg/kg one time (oral) Repeated cumulative 34.1–59.9 mg/kg (oral)	IV Oral	Lungs were one of the primary targets for NP accumulation initially, but levels returned to normal at 90 days post-exposure	Geraets et al. (2014)
BALB/c mice	35 nm	Not specified	Rutile	0–5 mg/kg one time	II	NPs exacerbated pneumonia in RSV-infected mice; Increased CCL5 chemokine and histological changes	Hashiguchi et al. (2014)
F344 rats	21 nm	143–148 nm	Anatase/rutile	0–6 mg/kg one time	II	Significant Ti burden in lung initially post-exposure, but decreased over time	Shinohara et al. (2014b)

(continued)

Table 1.5 (continued)

Model	NP primary size	NP hydrodynamic/agglomerate size	NP structure	Treatment parameters	Route	Results	Reference
WT and Nrf2 knockout C57bl6 mice	<25 nm	H_2O: 506±32 nm Saline: 2331 nm	Anatase	5 mg/kg 2 days/week for 4 weeks	Oropharyngeal aspiration	Nrf2 protects against NP-induced inflammation and oxidative stress; Nrf2 KO mice showed more inflammation and ROS	Delgado-Buenrostro et al. (2015)
Fisher F344 rats	21.5 ± 7.2 nm	In H_2O: 163.5 ± 12.6 In saline: 520.9 ± 41.7 nm	Anatase/rutile	1 mg/kg one time	IV	Ti burden in lungs significantly higher than controls at 356 days post exposure	Disdier et al. (2015)
Wistar rats	10–15 nm	Not specified	Anatase/rutile	0–300 ppm once a day for 7 days	IP	Alveolar wall thickening, vasculature hyperemia, alveolus destruction	Doudi and Setorki (2015)
C57BL/6 mice	$TiO_2NP^{10.5}$: 10.5 nm TiO_2NP^{38}: 38 nm TiO_2NP^{10}: 10 nm TiO_2NP^{10+}: 10 nm $TiO_2NP^{20.6}$: 20.6 nm SD–$TiO_2NP^{10.5+38}$: 12% and 24% by wt. SD–TiO_2NP^{38}: 36% by wt. SD–$TiO_2NP^{20.6}$: 10% by wt.	$TiO_2NP^{10.5}$: 130±3 nm TiO_2NP^{38}: 178 ± 86 nm TiO_2NP^{10}: 108 ±1 nm TiO_2NP^{10+}: 1978±153 nm $TiO_2NP^{20.6}$: 5224±832 nm SD–$TiO_2NP^{10.5+38}$: 280±16 nm SD–TiO_2NP^{38}: 517±60 nm SD–$TiO_2NP^{20.6}$: not specified	Rutile Rutile Rutile Rutile Not specified Not specified Not specified	18, 54, and 162 μg one time (NPs) 54, 162, and 486 μg one time (sanding dust)	II II	Free form NPs exhibited more differentially expressed genes when compared to sanding dust; primary particle size found to be strong predictor of pulmonary acute phase response; similar response pathways (NF-κB and Saa) activated across the range of NPs tested	Halappanavar et al. (2011)

Model	NP primary size	NP hydrodynamic/ agglomerate size	NP structure	Treatment parameters	Route	Results	Reference
Mice	35 nm	804–1022 nm	Rutile	0–2.5 mg/kg one time	Intranasal	Increased markers for pneumonia in RSV-infected mice post-exposure; aggregation near inflammatory cells; NPs exacerbated pulmonary condition	Hashiguchi et al. (2015)
ICR mice	21 nm	<50 nm	Not specified	0–0.5 mg once per week for 4 weeks	II	Granulocyte infiltration	Huang et al. (2015)
Crl:OF1 and BALB/c/ Sca mice	<5 μm	Not specified	Rutile	0–37 mg/m³ for 30 min; 16 mg/m³ repeated for a total of 16 hr	Aerosol	No pulmonary irritation, airflow limitation, or nasal/ pulmonary inflammation observed	Leppanen et al. (2015)
Wistar rats	18 nm	Not specified	Anatase/ rutile	0–70 mg/kg every other day for 21 days	IP	Increased ROS; chronic alveoli wall capillary congestion, bleeding, granulomas in parenchyma, hemorrhaging in alveoli walls	Mohammadi et al. (2015)
Swiss albino mice	118 ± 30 nm	Not specified	Anatase/ rutile	2 mg/kg once a day for 10 days	IP	No adverse histological effects observed	Silva et al. (2015)

(continued)

Table 1.5 (continued)

Model	NP primary size	NP hydrodynamic/ agglomerate size	NP structure	Treatment parameters	Route	Results	Reference
Wistar Hannover rats	15 nm	25 ± 5 nm	Not specified	0–1.0 mg one time	II	TiO_2 NP uptake by AMs; increased HO-1 initially; macrophage accumulation and lymphocyte infiltration in alveoli	Yoshiura et al. (2015)
Wistar rats	20–30 nm	Not specified	Anatase	20 mg/kg every 2 days for 20 days	IP	Significant accumulation in lung but no histological changes observed	Younes et al. (2015)
C57BL6 (WT) and Ogg1−/− KO mice	21 nm	245.8 ± 0.8 nm	Not specified	5 mg/kg one time	IV	Oxidative DNA damage in WT	Asare et al. (2016)
F344/DuCrlCrlj rats	AMT100: 6 nm MT1500AW: 28.8 × 7.6 nm TTO-S-3: 50–100 × 10–20 nm TTO-S-3 coated: 50–100 × 10–20 nm P25: 21 nm MP-100: 1000 nm FTL-100: 1680 × 130 nm	185 nm 58.5 nm 61.9 nm 241 nm 99.2 nm 531 nm Not specified	Anatase Rutile Rutile Rutile Anatase/ rutile Rutile Rutile	0–6 mg/kg one time	II	Inflammatory cell infiltration; hyperplasia of alveolar epithelium and thickening of alveolar walls; coated NPs showed more severe effects long term	Hashizume et al. (2016)

Model	NP primary size	NP hydrodynamic/ agglomerate size	NP structure	Treatment parameters	Route	Results	Reference
Sprague-Dawley rats	NM-101: 5–10 nm NM-25: 20–25 nm	Approx. 100 nm Larger than 100 nm	Anatase Anatase	250 mg/kg once a day for 28 days	IG	Low accumulation in lungs	Hendrickson et al. (2016)
BALB/c	4–8 nm	9.7 and 244 nm	Anatase	200 µg one time	IP	NPs exacerbated inflammatory responses via NF-κB in OVA-sensitized mice; increased IL-6 and TNF-α	Mishra et al. (2016)
F344 rats	12 $(w) \times 55$ (l) nm	II suspension: 50–100 nm Inhalation suspension: 20–300 nm	Rutile	0–1 mg one time 1.84 mg/m³ one time	II Inhalation	II group, but not inhalation group, showed increased neutrophils, HO-1, CINC-1, CINC-2; NPs found in AMs in both groups	Morimoto et al. (2016)
ICR mice	13.90 ± 1.80 nm	Not specified	Anatase/ rutile	0–1000 µg/kg one time	II	Increased IL-6 and TNF-α; increased LDH, indicating tissue damage	Pissuwan et al. (2016)
Mice	25 nm 50 nm	Not specified	Anatase/ rutile Rutile	40 mg/kg one time	IV, IP, IG	Lungs observed to be one of the highest accumulation target organs; inflammation present 120 days post-exposure	Yao et al. (2016)

(continued)

Table 1.6 Histological and morphological effects of TiO$_2$ NPs in murine lung tissue

Route	Reference	Results
Aerosol	McKinney et al. (2012)	Pulmonary inflammation, lung cell injury, significant increases in granulocytes in alveolar spaces
Intratracheal instillation	Sun et al. (2012a)	Pulmonary emphysema, edema, congestion, inflammatory cell infiltration; signs of apoptosis
Intratracheal instillation	Sun et al. (2012b)	Edema and emphysema; Inflammatory cell infiltration, blood vessel congestion
Intratracheal instillation	Hurbánková et al. (2013)	Lower alveolar macrophage viability; multinucleated cells
Nasal instillation	Li et al. (2013)	Infiltration of inflammatory cells, thickening of the pulmonary interstitium, and edema in lung; pneumonocytic ultrastructure showed apoptosis, including mitochondrial swelling, nuclear shrinkage, chromatin condensation, and evacuation of the pneumonocytic lamellar bodies
Pharyngeal aspiration	Porter et al. (2013)	Increase in granulocytes, alveolitis, significant interstitial fibrosis
Intratracheal instillation	Silva et al. (2013)	NPs found in macrophages; bronchiolitis/alveolitis near agglomerates
Intratracheal instillation	Tang et al. (2013)	Thickened alveolar walls, some ruptured
IV	Xu et al. (2013)	Perivascular infiltration of inflammatory and foamy cells; pulmonary fibrosis; granulomatous lesions
Intratracheal instillation	Chang et al. (2014)	Macrophage accumulation, extensive disruption of alveolar septa, slight alveolar thickness and expansion hyperemia, mitochondria dissolution, endoplasmic reticulum expansion
IP	Doudi and Setorki (2015)	Alveolar wall thickening, vasculature hyperemia, alveolus destruction
IP	Mohammadi et al. (2015)	Chronic alveoli wall capillary congestion, bleeding and granulomas in lung parenchyma, hemorrhage in alveoli walls, hemosiderin near bronchioles, vessels, and lymph follicles
Intratracheal instillation	Huang et al. (2015)	Granulocyte infiltration
Intratracheal instillation	Yoshiura et al. (2015)	macrophage accumulation, lymphocyte infiltration in alveoli
Intratracheal instillation	Hashizume et al. (2016)	Inflammation and inflammatory cell infiltration, hyperplasia of the alveolar epithelium, thickening of the alveolar wall
Intratracheal Instillation and inhalation	Morimoto et al. (2016)	Intra-alveolar infiltration of neutrophils; AMs with a pigment-like material in alveoli

were significantly increased in the bronchoalveolar lavage fluid (BALF) of mice infected with human respiratory syncytial virus (RSV) when compared to noninfected mice. Similar results of respiratory disease exacerbation have been reported using a mouse asthmatic model (Jonasson et al. 2013; Mishra et al. 2016). Jonasson et al. (2013) utilized aerosolized NPs, which is more similar to human exposure conditions when compared to instillation. This indicates workers with pre-existing respiratory diseases may be at risk if inhaling TiO_2 NPs on regular basis. In studies using non-disease murine models, inflammation seems to be a common side effect of NP exposure. Studies which have tracked inflammation in lungs have shown increases in various inflammatory cells and cytokines in BALF and lung tissues of rats and mice after TiO_2 intratracheal instillation (Sun et al. 2012b; Li et al. 2013; Liu et al. 2013; Porter et al. 2013; Hashizume et al. 2016; Pissuwan et al. 2016). These changes include increased neutrophils, eosinophils, TNF-α, and IL-6. Yao et al. (2016) observed inflammation in lung tissue 120 days after TiO_2 NP exposure, indicating ability to persist, but only IV, IP, and IG routes were investigated in this study. Unfortunately, none of these studies utilized inhalation routes, which is most related to human exposure. Yoshiura et al. (2015) showed no inflammation at low doses of 0.2 mg TiO_2 NPs by intratracheal instillation, but found inflammatory cells initially increased and then decreased at 1.0 mg for the duration of 1 month. Morimoto et al. (2016) demonstrated that intratracheal instillation of TiO_2 NPs led to increased inflammation, while inhalation did not induce any pulmonary inflammation in rats. This may be due to the fact that intratracheal instillation results in NPs being directly placed in the respiratory tract, whereas inhalation routes must also encounter several barriers before reaching respiratory tissues. Since humans are not subject to the intratracheal instillation routes, more focus is needed on inhalation studies *in vivo*. Studies focusing specifically on inflammatory markers after inhalation of TiO_2 NPs are somewhat lacking within recent years, and this would be an important avenue to investigate further.

ROS generation and oxidative stress are very apparent in lung toxicity studies that utilize TiO_2 NPs. Significant increases in superoxide anion (O_2^-) and hydrogen peroxide (H_2O_2) as well as decreases in GSH within lung tissue after TiO_2 NP exposure have been reported in low-dose, long-term studies on mice (Sun et al. 2012a; Sun et al. 2012b; Li et al. 2013). Liu et al. (2013) also reported decreased GSH, with lipid, protein, and DNA peroxidation. Lipid peroxidation and changes in SOD activity seem to be common among murine lung toxicity studies which monitor these factors (Sun et al. 2012a; Hurbánková et al. 2013; Tang et al. 2013; Bruno et al. 2014), and they are indicative of cellular oxidative stress. Asare et al. (2016) have reported oxidative DNA in lung tissues of mice, indicating TiO_2 NPs may cause genotoxicity. Interestingly, some studies have shown no TiO_2 effects on the oxidative markers HO-1 in BALF and lung tissue, or on the expression of the SOD gene *Sod1* (Yoshiura et al. 2015; Asare et al. 2016). Yoshiura et al. (2015) did find that HO-1 did significantly increase in BALF until the 1-month mark, but returned to levels close to control afterwards. Mohammadi et al. (2015) demonstrated TiO_2 NP dose-dependent increases in ROS initially, but they then decreased possibly due to aggregation and loss of reactivity. Aggregation as well as very low dosage concentrations may be responsible for studies, which show little effect on oxidative markers in lung tissues and BALF.

One area examined quite often in TiO_2 NP lung toxicity studies is that of cellular morphological and histological changes. In the vast majority of lung toxicity articles reviewed, an overwhelming majority reported some type of morphological and/or histological changes, ranging from mild to severe. Table 1.6 illustrates the wide variety of effects exhibited by lung cells *in vivo* after TiO_2 contact. Inflammatory cell infiltration and inflammation are two common side effects exhibited after TiO_2 NP exposure. In addition, increased granulocytes and granulomatous lesions are also somewhat common. It is important to note that these studies in Table 1.6 cover a wide range of NP characteristics, dosages, and exposure routes, and exposure times. These findings suggest that TiO_2 NP exposure should be continually assessed for occupational workers who may acquire through the inhalation exposure route. The *in vivo* uptake, persistence of Ti in lung tissues and associated histological and morphological changes caused by TiO_2 NPs warrant a closer look at toxicity risk assessment.

In vivo lung studies have shown that low-dose, long-term exposure to TiO_2 NPs via inhalation and intratracheal routes can cause a wide range of negative effects. As with other cell types, the NF-κB pathway was shown to be upregulated in lung studies, exacerbating inflammatory effects. Interestingly, Nrf2 has been shown to have a protective effect against TiO_2 NP exposure. As with other tissue types, ROS is also a hallmark of TiO_2 NP toxicity in the lung. In addition, TiO_2 NPs exacerbate respiratory conditions, which also contributes to their potential toxicity. Since inhalation is a main route of exposure, it is critical to continue monitoring its toxic effects.

6 Brain Toxicity *In vivo*

In vivo studies focusing on accumulation in the brain have shown differences based on the route of exposure (Table 1.7). IV exposure to TiO_2 NPs has generally showed low accumulation in brain tissues (Meena and Paulraj 2012; Geraets et al. 2014). There is evidence of some uptake initially in brain microvasculature endothelial cells (BECs), but this decreased over time and there was no translocation to the parenchyma of the brain (Disdier et al. 2015). In the same fashion, Meena et al. (2015) demonstrated a significant accumulation after IV exposure only at high doses (25 and 50 mg/kg), whereas at 5 mg there was no significant accumulation. Similar to IV routes, oral exposure to TiO_2 NPs has shown little to no accumulation in the brain (Cho et al. 2013; MacNicoll et al. 2015). Interestingly, a study using IP TiO_2 NP exposure demonstrated significant accumulation in the brain 1 day after the last injection (Younes et al. 2015). Observations were not reported for any other post-exposure times, however, and so it is difficult to determine if the NPs remained in brain tissues. Intranasal exposure has shown itself to be a key exposure route for accumulation in brain tissues. Ze et al. (2014a, b, c) demonstrated in several experiments the ability of TiO_2 NPs to accumulate consistently at 1.0 mg in brain tissue after intranasal exposure in mice. What is more, the accumulation was observed after long-term, low-dose exposure (2.5 and 5 mg/kg) similar to what workers may encounter. Su et al. (2015) demonstrated similar results in a long-term, low-dose study using even lower TiO_2 NP dosages (0.5 and 1.0 mg/kg BW) than Ze et al. (2014a, b, c). Overall, based on the reports of accumulation reviewed here oral exposure to TiO_2 NPs poses little threat to brain function, most likely due to low bioavailability and inability to cross the GIT to enter into the bloodstream. Accumulation after IV exposure does occur in some instances; however, no reports have pointed to significant accumulation long term. Continued investigation should be maintained though, since NP characteristics and dosages vary greatly. Intranasal exposure poses the most risk concerning NP accumulation in the brain. This type of exposure would allow particles to travel up the nasal cavity and make contact with the olfactory nerves, where they can then be distributed to the brain tissue. Indeed, this type of exposure has been documented previously in mice intranasally exposed to TiO_2 NPs (Wang et al. 2008a, b), which indicates risk of neurotoxicity.

ROS and oxidative stress are prevalent in many *in vivo* studies which observe TiO_2 NP effects in the brain. All neurotoxicity studies reviewed here which tracked ROS generation and oxidative stress indicated some level of change in these two factors, with many being very significant that affected brain function in some way. Ze et al. (2013, 2014b) observed in two separate long-term, low-dose studies that intranasal TiO_2 NPs induced ROS in the form of O_2^- and H_2O_2, as well as lipid, protein, and DNA peroxidation. Changes started at doses as low as 2.5 mg/kg, which indicates the ability of TiO_2 NPs to cause damage at a low concentration. Ze et al. (2013) also indicated oxidative stress was attributed to an activation of the p38-Nrf2 pathway at the protein level, which has also been confirmed in other studies focusing on different areas of toxicity. Krawczyńska et al. (2015) demonstrated decreases in GPx and GSR and significant reductions in GSH, as well as significant increases in GSSG and SOD activity after 1-time IV TiO_2 NP exposure of 10 mg/kg in rats. Although no effect on overall health was observed in this study, observation of mRNA expression in the brain showed TiO_2 NPs were strong modulators of a wide array of gene expressions. It is also worth noting these observations were made 28 days post-injection, which signifies the ability of TiO_2 NP effects to persist in effects well after exposure. Meena et al. (2015) similarly showed significant decreases in GPx after IV TiO_2 NP exposure, but only at higher concentrations of 25 and 50 mg/kg BW once a week for 4 weeks. Interestingly, SOD activity decreased in a dose-dependent manner, which is unexpected since TiO_2 NPs have shown to be upregulators of its activity as discussed in other sections. Significant oxidative stress as increased H_2O_2 release and genotoxicity were reported, but only at 25 and 50 mg/kg. However, lipid peroxidation was evident at the 5 mg dose after observation of significantly increased MDA levels. One particularly interesting oral study by Shrivastava et al. (2014) demonstrated significant increases in ROS and decreases in SOD and CAT activity in the brains of mice after TiO_2 NP exposure every day for 21 days, but the dosage was so large (500 mg/kg BW) that it is difficult to ascertain as to whether these effects are relatable to practical daily exposure limits. However, it is clear from the above account that there is significant risk to exposure if NPs are capable of reaching the brain. Therefore, continued study and observation is critical in maintaining proper risk assessment for workers who are exposed to TiO_2 NPs regularly.

Some rodent studies have pointed to neuro-inflammation after intranasal exposure to TiO_2 NPs. Y. Ze et al. (2014a) demonstrated significant activation in several inflammation mediators such as toll-like receptor 2 (TLR2), TLR4, TNF-α, and

Table 1.7 Effects of TiO₂ NPs on brain *in vivo*

Model	NP primary size	NP hydrodynamic/agglomerate size	NP structure	Treatment Parameters	Route	Results	Reference
Wistar rats	10–20 nm	72 nm	Anatase	0–50 mg/kg once a week for 1 month	IV	Mild accumulation of NPs	Meena and Paulraj (2012)
Sprague Dawley rats	21 nm	37.8 ± 0.4 nm	Anatase/rutile	260.4–1041.5 mg/kg per day for 13 weeks	Oral	No significant accumulation	Cho et al. (2013)
Sprague Dawley rats	9 nm	Not specified	Not specified	2 mg/ml one time	ICV	Chronic behavioral deterioration	Kim et al. (2013)
ICR mice	42.30±4.60	100–500 nm	Anatase	0–1387 mg/kg one time	IV	Neuronal cell degeneration; vacuoles in hippocampal neuronal cells	Xu et al. (2013)
CD-1 ICR mice	5–6 nm	208–330 nm	Anatase	0–10 mg/kg everyday for 90 days	Intra-nasal	Hemorrhaging observed; increased p38, c-JNK, NF-κB, Nrf-2, and HO-1 which led to increased ROS and lipid, protein, and DNA peroxidation	Ze et al. (2013)
Sprague-Dawley rats	5 nm	Not specified	Anatase	1 µg/µl 5 times during gestation	Sub-Q	Depressive like behavior in pups; oxidative DNA and lipid damage in pup brains	Cui et al. (2014)

Model	NP primary size	NP hydrodynamic/ agglomerate size	NP structure	Treatment Parameters	Route	Results	Reference
CRL:WI (WU) Wistar rats	NM100, 42–90 nm NM101, 6 nm NM102, 20 nm NM103, 20 nm NM104, 20 nm	0–250 nm for all types NM102 showed additional mode 250–2500 nm	Anatase Anatase Anatase Rutile Rutile	8.4–14.1 mg/ kg one time (IV) Repeated cumulative 42.3–71.9 mg/ kg (IV) 6.8–8.6 mg/kg one time (oral) Repeated cumulative 34.1–59.9 mg/ kg (oral)	IV Oral	Mild accumulation in brain after IV exposure	Geraets et al. (2014)
Lactating Wistar rats	10 nm	Not specified	Anatase	100 mg/kg for 21 days	Oral	Impaired memory and learning in offspring possibly due to timing of hippocampal development	Mohammadipour et al. (2016)
Swiss albino mice	<75 nm	Not specified	Anatase/ rutile	500 mg/kg once a day for 21 days	Oral	Increased ROS production; decreased SOD and CAT; increased NE and DA neuro-transmitters	Shrivastava et al. (2014)
CD-1 ICR mice	5–6 nm	208–330 nm	Anatase	0–10 mg/kg everyday for 90 days	Intra-nasal	Accumulation in hippocampus; impairment of spatial memory; increased expression of TLR2, TLR4, TNF-α, NF-κB, p52, and p65; suppressed IκB and IL-2; neuron necrosis and cellular degeneration	Ze et al. (2014a)
CD-1 ICR mice	5–6 nm	208–330 nm	Anatase	0–10 mg/kg everyday for 90 days	Intra-nasal	Accumulation; increased oxidative stress; overproliferation of glial cells; tissue necrosis and apoptosis	Ze et al. (2014b)

(continued)

Table 1.7 (continued)

Model	NP primary size	NP hydrodynamic/ agglomerate size	NP structure	Treatment Parameters	Route	Results	Reference
CD-1 ICR mice	5–6 nm	208–330 nm	Anatase	0–10 mg/kg everyday for 90 days	Intra-nasal	Accumulation; decreased spatial memory; nuclear degeneration, mitochondrial swelling, edema in glial cells, and apoptosis in hippocampal neurons	Ze et al. (2014c)
CD-1 ICR mice	5–6 nm	208–330 nm	Anatase	0–5 mg/kg once a day for 9 months	Intra-nasal	Altered glutamate metabolism; inhibition of NR1, NR2A, NR2B, and mGluR2 expression	Ze et al. (2014d)
Fisher F344 rats	21.5 ± 7.2 nm	In H_2O: 163.5 ± 12.6 In saline: 520.9 ± 41.7 nm	Anatase/ rutile	1 mg/kg one time	IV	Initial uptake in brain which decreased over time; no effects on BBB; increased IL-1β; Ti present in other locations may induce mediators transported by blood	Disdier et al. (2015)
Wistar rats	21 nm	233.7 ± 8.5	Not specified	0 or 10 mg/kg one time	IV	Decreased GPx, GSR, and GSH; increased GSSG and SOD; no effects on overall health	Krawczyńska et al. (2015)
Sprague-Dawley rats	Nanoamor, 40 nm Nanoamor, 120 nm Sigma, 40–50 nm Sigma, up to 5 µm	Majority of particles 600–700 nm	Anatase Anatase Rutile Rutile	5 mg/kg one time	Oral	No accumulation present	MacNicoll et al. (2015)
CD-1 ICR mice	5–6 nm	208–330 nm	Anatase	0–1 mg/kg every day for 9 months	Intra-nasal	Accumulation; over-proliferation of glial cells; hippocampal tissue necrosis; injury associated with reduced expression of many neurotrophin-mediated signaling pathways	Su et al. (2015)

Model	NP primary size	NP hydrodynamic/ agglomerate size	NP structure	Treatment Parameters	Route	Results	Reference
Wistar rats	20–30 nm	Not specified	Anatase	20 mg/kg every 2 days for 20 days	IP	Accumulation; normal histology; increased anxious index	Younes et al. (2015)
Wistar rats	21 nm	H_2O: 72.3 ± 9.6 nm PBS: 114 ± 17.8 nm	Anatase	0–50 mg/kg once a week for 4 weeks	IV	Accumulation; increased ROS, NF-κB (p65), HSP 60, p38, IFN-γ, TNF-α, caspase-3, and IFN; decreased SOD, melatonin, glutamate, and AChE; genotoxicity observed; apoptosis	Meena et al. (2015)
Swiss albino mice	118+30 nm	Not specified	Anatase/ rutile	2 mg/kg once a day for 10 days	IP	No changes in brain tissue	Silva et al. (2015)
Pregnant Sprague-Dawley rats	170.9 ± 6.4 nm (aerodynamic diameter)	Not specified	Not specified	10.4 ± 0.4 mg/ m^3 for 5 hr/day for 7.8 ± 0.5 days	Inhalation	In pups: significant working impairments; maternal NP exposure produces psychological deficits in pups that persist to adulthood	Engler-Chiurazzi et al. (2016)
Wistar rat	5–10 nm	Not specified	Anatase	0–200 mg/kg everyday for 60 days	IG	Decreased AChE; increased IL-6	Grissa et al. (2016)

several members of the NF-κB signal transduction pathway. Suppression of IκB, which acts as a suppressor of the NF-κB pathway, was observed as well. As discussed previously, this pathway has also been shown to be activated in other organs that experience inflammation after TiO_2 NPs, suggesting it plays central role in mediating toxicity via inflammatory responses. Meena et al. (2015) demonstrated involvement of the NF-κB signal transduction pathway by observing dose-dependent increases in p65 (part of NF-κB family) and p38 in brain tissue after TiO_2 NP IV exposure. As mentioned earlier, p38 acts along with Nrf-2 in oxidative stress mechanisms of TiO_2 NP-exposed brain tissue (Y. Ze et al. 2013). The evidence of modulation of these two pathways offers great insight into how TiO_2 NPs are able to induce toxicity via oxidative stress and inflammation. Meena et al. (2015) also observed significant increases in IFN-γ and TNF-α at 5 and 25 mg/kg BW, respectively. Other studies have shown increases in the inflammatory cytokines IL-1β and IL-6 levels in the brain after TiO_2 NP contact (Disdier et al. 2015; Grissa et al. 2016). Interestingly, Disdier et al. (2015) did not observe Ti uptake in the brain parenchyma where increased IL-1β levels were observed, and attributed the effect to distal TiO_2 NP uptake that was then circulated through the blood to the brain. This corresponds to other IV studies, which have demonstrated systemic uptake and circulation of TiO_2 NPs throughout the body of murine models. Grissa et al. (2016) demonstrated increases in IL-6 after oral TiO_2 NP exposure, however only at very high doses of 100 and 200 mg/kg BW every day for 60 days. Based on the repeated results of increased neuro-inflammation markers it is necessary to promote more study in this area.

Changes in brain cell morphology and histology are not surprising, given the data on oxidative stress and inflammation that has been reported (Table 1.7). Meena et al. (2015) reported apoptosis in brain cells, which was attributed to increased expression of p53, Bax, and caspase-3. This mechanism has been reported in other cell types as well, as discussed previously. Excessive proliferation of glial cells and edema after TiO_2 contact has been reported by multiple studies (Ze et al. 2014a; b; c; Su et al. 2015). Observation of neuronal cell degeneration, necrosis, and abscission of perikaryon has also been demonstrated after TiO_2 NP exposure (Xu et al. 2013; Ze et al. 2014a). In the hippocampus, several cellular changes have been documented as well. Appearance of vacuoles (indicative of fatty degeneration), cellular degeneration, apoptosis, tissue necrosis, mitochondrial swelling, and nuclear membrane collapse have all been reported in hippocampal cells (Xu et al. 2013; Ze et al. 2014a, b, c; Su et al. 2015). Pyramidal cells, which are a special neurons located in the hippocampus responsible for primary excitation of nerves, have shown dispersive replication, decreased size of cell volume, nuclear irregularity, degeneration, nuclear pyknosis, and appearance of cytoplasm feosin after TiO_2 NP exposure (Y. Ze et al. 2014c). Other TiO_2 NP effects identified in brain tissue include over-proliferation of spongiocytes, hemorrhage, and cell shrinkage (Ze et al. 2013, 2014a). What is particularly remarkable about all of these observed effects is that with the exception of the Xu et al. (2013) results, the rest all occurred after intranasal exposure. In fact, many of the studies reported significant changes at low-dose, long-term exposure rates. As mentioned previously, intranasal exposure is representative of practical TiO₂ NP exposure since particles inhaled may travel through the nasal cavity and make contact with olfactory nerves leading to the brain. Therefore, it is critical to continue to observe these effects at realistic concentration and exposure rates.

These morphological and histological changes in brain tissue due to TiO_2 NPs may also contribute to changes in brain function. The hippocampus plays a significant role in spatial memory. It has been shown that intranasal exposure to TiO_2 NPs resulted in deficiencies and impairment of spatial memory in mice (Ze et al. 2014a, c). Mohammadipour et al. (2016) demonstrated the passage of TiO_2 NPs through maternal milk to pups also resulted in impaired learning and memory in offspring, and hypothesized this was due to the timing of hippocampal development. Lesions have been documented in the hippocampus of mice after intranasal TiO_2 NP administration (Wang et al. 2008a), which could contribute to a loss of spatial memory. Another study utilizing prenatal exposure to TiO_2 NPs has revealed significant working impairments and psychological deficits in pups that persist into adulthood (Engler-Chiurazzi et al. 2016). Increases in neurotransmitters, which play a key role in memory, awareness, thought, and consciousness, such as norepinephrine and dopamine have also been documented in the cerebral cortex of mice after oral TiO_2 NP administration (Shrivastava et al. 2014). Although behavior was not analyzed in this study, the results do offer insight into other studies which have documented depressive-like behavior, behavioral deterioration, and increases of anxious index in rats (Kim et al. 2013; Cui et al. 2014; Younes et al. 2015). Changes in the neurotransmitter glutamate, a precursor to glutamine, have also been observed in the brain of rats and mice. Fluctuation in glutamate/glutamine levels, inhibition of glutamate receptor expression, and glutamine synthetase have been observed after TiO_2 NP administration (Ze et al. 2014; Meena et al. 2015). Changes to the levels of glutamate and its related mediators in the brain have been implicated in many neurodegenerative diseases, such as amyotrophic lateral sclerosis (ALS), Alzheimer's disease, Huntington's disease, and Parkinson's disease (Sheldon and Robinson 2007). Decreases in acetylcholine esterase (AChE), an important enzyme that metabolizes acetylcholine to choline and Acetyl-CoA, have also been observed in rats after TiO_2 NP administration (Meena et al. 2015; Grissa et al. 2016). Proper maintenance of the central cholinergic system is necessary for cognitive function, and decreases in AChE could play a role in studies which show abnormal behavior (Meena et al. 2015). Based on the data reviewed, monitoring neurotransmitters after TiO_2 NP exposure may reveal mechanisms for decreased brain function and behavioral abnormalities.

7 Conclusion

Studies that concentrate on oral and inhalation routes provide more insight into effects in humans, since these are the most common exposure routes. Oral toxicity has shown to be less severe than inhalation, most likely due to the inability of NPs to be absorbed by the GIT. There are still some cases, however, that show toxic effects after large oral doses. In addition, many experiments do not specify full details of NP preparation or experimental procedures, especially the case of hydrodynamic diameter. This is a key factor that should not be overlooked when assessing TiO_2 NP toxicity, since larger agglomerates are generally considered safer than smaller particles. Tests that track circulation of TiO_2 NPs systemically in the bloodstream have proven useful in the area of target organs for accumulation. ROS, oxidative stress, and inflammation have been shown to be toxicity markers in the organs or whole animals examined throughout this review. This is not surprising that inflammatory pathways such as NF-κB are often activated after TiO_2 NP exposure in multiple tissue types. Apoptotic pathways such as p53 and the caspase cascade are also commonly upregulated after TiO_2 NP exposure, increasing cytotoxic potential. These pathways are likely molecular mechanisms involved in toxicity. Mitochondrial damage, which is also associated with increased oxidative stress, has also been observed in many studies examining toxicity of TiO_2 NPs. Nevertheless, production of ROS and oxidative stress is common; it is not clear without shadows whether they are the cause or effect of above molecular events. Based on the studies reviewed here, it is evident there is a need to continue monitoring the toxicity of TiO_2 NPs. Many experiments have shown that low-dose, long-term exposure can result in toxic effects, particularly in the lung and brain. Since TiO_2 NPs are widely used in food, medicine, and other domestic products, it is critical to increase standardization for NP toxicity studies and monitor toxicity effects over extended periods.

References

Alarifi, S., Ali, D., Al-Doaiss, A., Ali, B., Ahmed, M., & Al-Khedhairy, A. (2013). Histologic and apoptotic changes induced by titanium dioxide nanoparticles in the livers of rats. *International Journal of Nanomedicine, 8*, 3937–3943. https://doi.org/10.2147/ijn.s47174.

Asare, N., Duale, N., Slagsvold, H., Lindeman, B., Olsen, A., Gromadzka-Ostrowska, J., Meczynska-Wielgosz, S., Kruszewski, M., Brunborg, G., & Instanes, C. (2016). Genotoxicity and gene expression modulation of silver and titanium dioxide nanoparticles in mice. *Nanotoxicology, 10*(3), 312–321. https://doi.org/10.3109/17435390.2015.1071443.

Baranowska-Wójcik, E., Szwajgier, D., Oleszczuk, P., et al. (2020). Effects of titanium dioxide nanoparticles exposure on human health—a review. *Biological Trace Element Research, 193*, 118–129. https://doi.org/10.1007/s12011-019-01706-6.

Brun, E., Barreau, F., Veronesi, G., Fayard, B., Sorieul, S., Chanéac, C., Carapito, C., Rabilloud, T., Mabondzo, A., Herlin-Boime, N., & Carrière, M. (2014). Titanium dioxide nanoparticle impact and translocation through ex vivo, in vivo and in vitro gut epithelia. *Particle and Fibre Toxicology, 11*, 16. https://doi.org/10.1186/1743-8977-11-13.

Bruno, M., Tasat, D., Ramos, E., Paparella, M., Evelson, P., Rebagliati, R., Cabrini, R., Guglielmotti, M., & Olmedo, D. (2014). Impact through time of different sized titanium dioxide particles on biochemical and histopathological parameters. *Journal of Biomedical Materials Research— Part A, 102*(5), 1439–1448. https://doi.org/10.1002/jbm.a.34822.

Chang X, Fu, Y., Zhang, Y., Tang, M., Wang, B. Effects of Th1 and Th2 cells balance in pulmonary injury induced by nano titanium dioxide. (2014) Environmental Toxicology and Pharmacology, 37 (1), 275–283 https://doi.org/10.1016/j.etap.2013.12.001

Cho, W., Kang, B., Lee, J., Jeong, J., Che, J., & Seok, S. (2013). Comparative absorption, distribution, and excretion of titanium dioxide and zinc oxide nanoparticles after repeated oral administration. *Particle and Fibre Toxicology, 10*(1). https://doi.org/10.1186/1743-8977-10-9.

Cui, Y., Liu, H., Ze, Y., Zhang, Z., Hu, Y., Cheng, Z., Hu, R., Gao, G., Wang, L., Tang, M., & Hong, F. (2012). Gene expression in liver injury caused by long-term exposure to titanium dioxide nanoparticles in mice. *Toxicological Sciences, 128*(1), 171–185. https://doi.org/10.1093/toxsci/kfs153.

Cui, Y., Liu, H., Zhou, M., Duan, Y., Li, N., Gong, X., Hu, R., Hong, M., & Hong, F. (2011). Signaling pathway of inflammatory responses in the mouse liver caused by TiO$_2$ nanoparticles. *Journal of Biomedical Materials Research—Part A, 96A*(1), 221–229.

Cui, Y., Zhou, Z., Lei, Y., Ma, M., Cao, R., Sun, T., Xu, J., Huo, M., Cao, R., Wen, C., & Che, Y. (2014). Prenatal exposure to nanoparticulate titanium dioxide enhances depressive-like behaviors in adult rats. *Chemosphere, 96*, 99–104. https://doi.org/10.1016/j.chemosphere.2013.07.051.

Delgado-Buenrostro, N., Medina-Reyes, E., Lastres-Becker, I., Freyre-Fonseca, V., Ji, Z., Hernandez-Pando, R., Marquina, B., Pedraza-Chaverri, J., Espada, S., Cuardrado, A., & Chirino, Y. (2015). Nrf2 protects the lung against inflammation induced by titanium dioxide nanoparticles: a positive regulator role of Nrf2 on cytokine release. *Environmental Toxicology, 30*(7), 782–792. https://doi.org/10.1002/tox.2195.

Disdier, C., Devoy, J., Cosnefroy, A., Chalansonnet, M., Herlin-Boime, N., Brun, E., Lund, A., & Mabondzo, A. (2015). Tissue biodistribution of intravenously administered titanium dioxide nanoparticles revealed blood-brain barrier clearance and brain inflammation in rat. *Particle and Fibre Toxicology, 12*(1). https://doi.org/10.1186/s12989-015-0102-8.

Donner, E., Myhre, A., Brown, S., Boatman, R., & Warheit, D. (2016). In vivo micronucleus studies with 6 titanium dioxide materials (3 pigment-grade & 3 nanoscale) in orally-exposed rats. *Regulatory Toxicology and Pharmacology, 74*, 64–74. https://doi.org/10.1016/j.yrtph.2015.11.003.

Doudi, M., & Setorki, M. (2015). Influence of titanium dioxide nanoparticles on oxidative stress and pulmonary dysfunction. *Zahedan Journal of Research in Medical Sciences, 17*(9). https://doi.org/10.17795/zjrms-1062.

Dudefoi, W., Moniz, K., Allen-Vercoe, E., Ropers, M.-H., & Walker, V. (2017). Impact of food grade and nano-TiO$_2$ particles on a human intestinal community. *Food and Chemical Toxicology, 106*(Part A), 242–249. https://doi.org/10.1016/j.fct.2017.05.050.

El-Zahed, E., El-Sayed, H., Ibraheem, O., & Omran, B. (2016). Hepatotoxic effects of titanium dioxide nanoparticles & the possible protective role of N-acetylcysteine in adult male albino rats (histological & biochemical study). *Journal of American Science, 11*(7), 79–91.

Engler-Chiurazzi, E., Stapleton, P., Stalnaker, J., Ren, X., Hu, H., Nurkiewicz, T., McBride, C., Yi, J., Engels, K., & Simpkins, J. (2016). Impacts of prenatal nanomaterial exposure on male adult sprague-dawley rat behavior and cognition. *Journal of Toxicology and Environmental Health, Part A, 79*(11), 447–452. https://doi.org/10.1080/15287394.2016.1164101.

Fartkhooni, F., Noori, A., & Mohammadi, A. (2016). Effects of Titanium Dioxide Nanoparticles Toxicity on the Kidney of Male Rats. *International Journal of Life Sciences, 10*(1), 65–69. https://doi.org/10.3126/ijls.v10i1.14513

Fatemeh, M., & Mohammad, F. (2014). The histological and biochemical effects of titanium dioxide nanoparticle (TiO 2) on the liver in Wistar rat. *International Research Journal of Biological Sciences, 3*(6), 1–5.

Faust, J. J., Doudrick, K., Yang, Y., Westerhoff, P., & Capco, D. G. (2014). Food grade titanium dioxide disrupts intestinal brush border microvilli in vitro independent of sedimentation. *Cell Biology and Toxicology, 30*(3), 169–188. https://doi.org/10.1007/s10565-014-9278-1

Geraets, L., Oomen, A., Krystek, P., Jacobsen, N., Wallin, H., Laurentie, M., Verharen, H., Brandon, E., & de Jong, W. (2014). Tissue distribution and elimination after oral and intravenous administration of different titanium dioxide nanoparticles in rats. *Particle and Fibre Toxicology, 11*, 30. https://doi.org/10.1186/1743-8977-11-30.

Gonzalez-Esquivel, A., Charles-Ninõ, C., Pacheco-Moises, F., Ortiz, G., Jaramillo-Juarez, F., & Rincon-Sanchez, A. (2015). Beneficial effects of quercetin on oxidative stress in liver and kidney induced by titanium dioxide (TiO$_2$) nanoparticles in rats. *Toxicology Mechanisms and Methods, 25*(3), 166–175. https://doi.org/10.3109/15376516.2015.1006491.

Grand, F., & Tucci, P. (2016). Titanium dioxide nanoparticles: a risk for human health? *Mini Reviews in Medicinal Chemistry, 25*(9), 762e769. https://doi.org/10.2174/138955751666616 0321114341.

Grissa, I., Guezguez, S., Ezzi, L., Chakroun, S., Sallem, A., Kerkeni, E., Elghoul, J., El Mir, L., Mehdi, M., Cheikh, H., & Haouas, Z. (2016). The effect of titanium dioxide nanoparticles on neuroinflammation response in rat brain. *Environmental Science and Pollution Research.*. Advance online publication. doi: https://doi.org/10.1007/s11356-016-7234-8

Gui, S., Li, B., Zhao, X., Sheng, L., Hong, J., Yu, X., Sang, X., Sun, Q., Ze, Y., Wang, L., & Hong, F. (2013). Renal injury and Nrf2 modulation in mouse kidney following chronic exposure to TiO$_2$ nanoparticles. *Journal of Agricultural and Food Chemistry, 61*(37), 8959–8968. https://doi.org/10.1021/jf402387e.

Halappanavar, S., Jackson, P., Williams, A., Jensen, K. A., Hougaard, K. S., Vogel, U., . . . Wallin, H. (2011). Pulmonary Response to Surface-Coated Nanotitanium Dioxide Particles Includes Induction of Acute Phase Response Genes, Inflammatory Cascades, and Changes in MicroRNAs: A Toxicogenomic Study. *Environmental and Molecular Mutagenesis, 52* (6), 425–439. https://doi.org/10.1002/em.20639

Hashiguchi, S., Yoshida, H., Akashi, T., Hirose, A., Kurokawa, M., & Watanabe, W. (2014). Effects of titanium dioxide nanoparticles on the pneumonia in respiratory syncytial virus-infected mice. *Toxicology Letters, 229*, S193. https://doi.org/10.1016/j.toxlet.2014.06.654.

Hashiguchi, S. et al. (2015). Titanium dioxide nanoparticles exacerbate pneumonia in respiratory syncytial virus (RSV)-infected mice. *Environmental Toxicology and Pharmacology, 39*, 879–886. https://doi.org/10.1016/j.etap.2015.02.017

Hashizume, N., Oshima, Y., Nakai, M., Kobayashi, T., Sasaki, T., Kawaguchi, K., Honda, K., Gamo, M., Yamamoto, K., Tsubokura, Y., Ajimi, S., Inoue, Y., & Imatanaka, N. (2016). Categorization of nano-structured titanium dioxide according to physicochemical characteristics and pulmonary toxicity. *Toxicology Reports, 3*, 490–500. https://doi.org/10.1016/j.toxrep.2016.05.005.

Hendrickson, O., Pridvorova, S., Zherdev, A., Klochkov, S., Novikova, O., Shevtsova, E., Bachurin, S., & Dzantiev, B. (2016). Size-dependent differences in biodistribution of titanium dioxide nanoparticles after sub-acute intragastric administrations to rats. *Current Nanoscience, 12*(2), 228–236. https://doi.org/10.2174/1573413711666151008013943.

Hong, F., Hong, J., Wang, L., Zhou, Y., Liu, D., Xu, B., Yu, X., & Sheng, L. (2015). Chronic exposure to nanoparticulate TiO$_2$ causes renal fibrosis involving activation of the wnt pathway in mouse kidney. *Journal of Agricultural and Food Chemistry, 63*(5), 1639–1647. https://doi.org/10.1021/jf5034834.

Hong, J., Wang, L., Zhao, X., Yu, X., Sheng, L., Xu, B., Liu, D., Zhu, Y., Long, Y., & Hong, F. (2014). Th2 factors may be involved in TiO$_2$ NP-induced hepatic inflammation. *Journal of Agricultural and Food Chemistry, 62*(28), 6871–6878. https://doi.org/10.1021/jf501428w.

Huang, K., Wu, C., Huang, K., Lin, W., Chen, C., Guan, S., Chiang, C., & Liu, S. (2015). Titanium nanoparticle inhalation induces renal fibrosis in mice via an oxidative stress upregulated transforming growth factor-β Pathway. *Chemical Research in Toxicology, 28*(3), 354–364. https://doi.org/10.1021/tx500287f.

Hurbánková, M., Černá, S., Kováčiková, Z., Wimmerová, S., Hrašková, D., Marcišiaková, J., & Moricová, Š. (2013). Effect of TIO2 nanofibres on selected bronchoalveolar parameters in acute and subacute phase - experimental study. *Central European Journal of Public Health, 21*(3), 165–170.

Husain, M., Wu, D., Saber, A., Decan, N., Jacobsen, N., Williams, A., Yauk, C., Wallin, H., Vogel, U., & Halappanavar, S. (2015). Intratracheally instilled titanium dioxide nanoparticles translocate to heart and liver and activate complement cascade in the heart of C57BL/6 mice. *Nanotoxicology, 9*(8), 1013–1022. https://doi.org/10.3109/17435390.2014.996192.

International Agency for Research on Cancer. (2010). Carbon Black, Titanium Dioxide, and Talc. *IARC Monographs on the Evaluation of Carcinogenic Risk to Humans, 93*, 193–275.

Janer, G., Mas del Molino, E., Fernandez-Rosas, E., Fernandez, A., Vazquez-Campos, S. (2014). Cell uptake and oral absorption of titanium dioxide nanoparticles. *Toxicology Letters, 228*(2), 103–110. https://doi.org/10.1016/j.toxlet.2014.04.014

Jeon, Y., Kim, W., & Lee, M. (2013). Studies on liver damage induced by nanosized-titanium dioxide in mouse. *Journal of Environmental Biology, 34*(2), 283–287.

Jonasson, S., Gustafsson, A., Koch, B., & Bucht, A. (2013). Inhalation exposure of nano-scaled titanium dioxide (TiO2) particles alters the inflammatory responses in asthmatic mice. *Inhalation Toxicology, 25*(4), 179–191. https://doi.org/10.3109/08958378.2013.770939.

Jones, K., Morton, J., Smith, I., Jurkschat, K., Harding Helen, A., & Evans, G. (2015). Human in vivo and in vitro studies on gastrointestinal absorption of titanium dioxide nanoparticles. *Toxicology Letters, 233*(2), 95–101. https://doi.org/10.1016/j.toxlet.2014.12.005.

Kawamata, H., & Manfredi, G. (2010). Mitochondrial dysfunction and intracellular calcium dysregulation in ALS. *Mechanisms of Ageing and Development, 131*(7-8), 517–526. https://doi.org/10.1016/j.mad.2010.05.003.

Kim, E., Palmer, P., Howard, V., Elsaesser, A., Taylor, A., Staats, G., & O'Hare, E. (2013). Effect of intracerebroventricular injection of TiO2 nanoparticles on complex behaviour in the rat. *Journal of Nanoscience and Nanotechnology, 13*(12), 8325–8330. https://doi.org/10.1166/jnn.2013.8217.

Krawczyńska, A., Dziendzikowska, K., Gromadzka-Ostrowska, J., Lankoff, A., Herman, A. P., Oczkowski, M., Królikowski, T., Wilczak, J., Wojewódzka, M., & Kruszewski, M. (2015). Silver and titanium dioxide nanoparticles alter oxidative/inflammatory response and renin-angiotensin system in brain. *Food and Chemical Toxicology, 85*, 96–105. https://doi.org/10.1016/j.fct.2015.08.005.

Leppanen M. et al. (2015). Negligible respiratory irritation and inflammation potency of pigmentary TiO2 in mice. *Inhalation Toxicology 27*(8) 10.3109/08958378.2015.1056890

Li, B., Ze, Y., Sun, Q., Zhang, T., Sang, X., Cui, Y., Wang, X., Gui, S., Tan, D., Zhu, M., Zhao, X., Sheng, L., Wang, L., Hong, F., & Tang, M. (2013). Molecular mechanisms of nanosized titanium dioxide-induced pulmonary injury in mice. *PLos ONE, 8*(2). https://doi.org/10.1371/journal.pone.0055563.

Liu, H., Yang, D., Yang, H., Zhang, H., Zhang, W., Fang, Y., Lin, Z., Tian, L., Lin, B., Yan, J., & Xi, Z. (2013). Comparative study of respiratory tract immune toxicity induced by three sterilisation nanoparticles: Silver, Zinc oxide and titanium dioxide. *Journal of Hazardous Materials, 248-249*(1), 478–486. https://doi.org/10.1016/j.jhazmat.2013.01.046.

Ma, L., Zhao, J., Wang, J., Liu, J., Duan, Y., Liu, H., Li, N., Yan, J., Ruan, J., Wang, H., & Hong, F. (2009). The acute liver injury in mice caused by nano-anatase TiO2. *Nanoscale Research Letters, 4*(11), 1275–1285. https://doi.org/10.1007/s11671-009-9393-8.

MacNicoll, A., Kelly, M., Aksoy, H., Kramer, E., Bouwmeester, H., & Chaudhry, Q. (2015). A study of the uptake and biodistribution of nano-titanium dioxide using in vitro and in vivo models of oral intake. *Journal of Nanoparticle Research, 17*(2), 20. https://doi.org/10.1007/s11051-015-2862-3.

McKinney, W., Jackson, M., Sager, T., Reynolds, J., Chen, B., Afshari, A., Krajnak, K., Waugh, S., Johnson, C., Mercer, R., Frazer, D., Thomas, T., & Castranova, V. (2012). Pulmonary and cardiovascular responses of rats to inhalation of a commercial antimicrobial spray containing titanium dioxide nanoparticles. *Inhalation Toxicology, 24*(7), 447–457. https://doi.org/10.3109/08958378.2012.685111.

Meena, R., Kumar, S., & Paulraj, R. (2015). Titanium oxide (TiO2) nanoparticles in induction of apoptosis and inflammatory response in brain. *Journal of Nanoparticle Research, 17*(1). https://doi.org/10.1007/s11051-015-2868-x.

Meena, R., & Paulraj, R. (2012). Oxidative stress mediated cytotoxicity of TiO 2 nano anatase in liver and kidney of Wistar rat. *Toxicological & Environmental Chemistry, 94*(1), 146–163. https://doi.org/10.1080/02772248.2011.638441.

Mishra, V., Baranwal, V., Mishra, R., Sharma, S., Paul, B., & Pandey, A. (2016). Titanium dioxide nanoparticles promote development of allergic airway inflammation and Socs3 expression via NF-κB pathway in murine model of asthma. *Biomaterials, 92*, 90–102. https://doi.org/10.1016/j.biomaterials.2016.03.016.

Mohammadi, F., Sadeghi, L., Mohammadi, A., Tanwir, F., Yousefi, B., & Izadnejad, M. (2015). The effects of Nano titanium dioxide (TiO2NPs) on lung tissue. *Bratislava Medical Journal, 116*(5), 296–301. https://doi.org/10.4149/BLL.

Mohammadipour, A., Hosseini, M., Fazel, A., Haghir, H., Rafatpanah, H., Pourganji, M., & Ebrahimzadeh Bideskan, A. (2016). The effects of exposure to titanium dioxide nanoparticles during lactation period on learning and memory of rat offspring. *Toxicology and Industrial Health, 32*(2), 221–228. https://doi.org/10.1177/0748233713498440.

Morimoto, Y., Izumi, H., Yoshiura, Y., Tomonaga, T., Lee, B., Okada, T., Oyabu, T., Myojo, T., Kawai, K., Yatera, K., Shimada, M., Kubo, M., Yamamoto, K., Kitajima, S., Kuroda, E., Horie, M., Kawaguchi, K., & Sasaki, T. (2016). Comparison of pulmonary inflammatory responses following intratracheal instillation and inhalation of nanoparticles. *Nanotoxicology, 10*(5), 607–618. https://doi.org/10.3109/17435390.2015.1104740.

National Institute for Occupational Safety and Health (NIOSH) (2011). Occupational exposure to titanium dioxide. *Current Intelligence Bulletin 63.* Retrieved February 9, 2015 from http://www.cdc.gov/niosh/docs/2011-160/pdfs/2011-160.pdf

Nogueira, C., de Azevedo, W., Dagli, M., Toma, S., Leite, A., Lordello, M., Nishitokukado, I., Ortiz-Agostinho, C., Duarte, M., Ferreira, M., & Sipahi, A. (2012). Titanium dioxide induced inflammation in the small intestine. *World Journal of Gastroenterology, 18*(34), 4729-4735. doi: https://doi.org/10.3748/wjg.v18.i34.4729

Pissuwan, D., Somkid, K., Kongseng, S., Sukwong, P., & Yoovathaworn, K. (2016). Respiratory tract toxicity of titanium dioxide nanoparticles and multi-walled carbon nanotubes on mice after intranasal exposure. *Micro & Nano Letters, 11*(4), 183–187. https://doi.org/10.1049/mnl.2015.0523.

Porter, D., Wu, N., Hubbs, A., Mercer, R., Funk, K., Meng, F., Li, J., Wolfarth, M., Battelli, L., Friend, S., Andrew, M., Hamilton, R., Sriram, K., Yang, F., Castranova, V., & Holian, A. (2013). Differential mouse pulmonary dose and time course responses to titanium dioxide nanospheres and nanobelts. *Toxicological Sciences, 131*(1), 179–193. https://doi.org/10.1093/toxsci/kfs261.

Powell, J., Faria, N., Thomas-McKay, E., & Pele, L. (2010). Origin and fate of dietary nanoparticles and microparticles in the gastrointestinal tract. *Journal of Autoimmunity, 34*(3), J226–J233. https://doi.org/10.1016/j.jaut.2009.11.006.

Powell, J., Harvey, R., Ashwood, P., Wolstencroft, R., Gershwin, M., & Thompson, R. (2000). Immune potentiation of ultrafine dietary particles in normal subjects and patients with inflammatory bowel disease. *Journal of Autoimmunity, 14*(1), 99–105. https://doi.org/10.1006/jaut.1999.0342.

Ruiz, P., Moron, B., Becker, H., Lang, S., Atrott, K., Spalinger, M., Scharl, M., Wojtal, K., Fischbeck-Terhalle, A., Frey-Wagner, Hausmann, M., Kraemer, T., & Rogler, G. (2017). Titanium dioxide nanoparticles exacerbate DSS-induced colitis: role of the NLRP3 inflammasome. *Gut.* 66:1216–1224

Sha, B., Gao, W., Wang, S., Gou, X., Li, W., Liang, X., Qu, Z., Xu, F., & Lu, T. (2014). Oxidative stress increased hepatotoxicity induced by nano-titanium dioxide in BRL-3A cells and Sprague—Dawley rats. *Journal of Applied Toxicology, 34*(4), 345–356. https://doi.org/10.1002/jat.2900.

Sheldon, A., & Robinson, M. (2007). The role of glutamate transporters in neurodegenerative diseases and potential opportunities for intervention. *Neurochemistry International, 51*(6-7), 333–355. https://doi.org/10.1016/j.neuint.2007.03.012.

Shinohara, N., Danno, N., Ichinose, T., Sasaki, T., Fukui, H., Honda, K., & Gamo, M. (2014a). Tissue distribution and clearance of intravenously administered titanium dioxide (TiO2) nanoparticles. *Nanotoxicology, 8*(2), 132–141. https://doi.org/10.3109/17435390.2012.763001.

Shinohara, N., Oshima, Y., Kobayashi, T., Imatanaka, N., Nakai, M., Ichinose, T., Sasaki, T., Zhang, G., Fukui, H., & Gamo, M. (2014b). Dose-dependent clearance kinetics of intratracheally administered titanium dioxide nanoparticles in rat lung. *Toxicology, 325*, 1–11. https://doi.org/10.1016/j.tox.2014.08.003.

Shrivastava, R., Raza, S., Yadav, A., Kushwaha, P., Flora, & Swaran, J. (2014). Effects of sub-acute exposure to TiO2, ZnO and Al2O3 nanoparticles on oxidative stress and histological changes in mouse liver and brain. *Drug and Chemical Toxicology, 37*(3), 336–347. https://doi.org/10.3109/01480545.2013.866134.

Shukla, R., Kumar, A., Vallabani, N., Pandey, A., & Dhawan, A. (2014). Titanium dioxide nanoparticle-induced oxidative stress triggers DNA damage and hepatic injury in mice. *Nanomedicine, 9*(9), 1423–1434. https://doi.org/10.2217/nnm.13.100.

Silva, M. R. et al. (2013). Biological Response to Nano-Scale Titanium Dioxide (TiO2): Role of Particle Dose, Shape, and Retention. *Journal of Toxicology and Environmental Health*, Part A: 76(16) https://doi.org/10.1080/15287394.2013.826567

Silva, A., Locatelli, C., Filho, U., Gomes, B., De Carvalho Jú Nior, R., De Gois, J., Borges, D., & Creczynski-Pasa, T. (2015). Visceral fat increase and signals of inflammation in adipose tissue after administration of titanium dioxide nanoparticles in mice. *Toxicology and Industrial Health*. Advance online publication. https://doi.org/10.1177/0748233715613224.

Su, M., Sheng, L., Zhao, X., Wang, L., Yu, X., Hong, J., Xu, B., Liu, D., Jiang, H., Ze, X., Zhu, Y., Long, Y., Zhou, J., Cui, J., Li, K., Ze, Y., & Hong, F. (2015). Involvement of neurotrophins and related signaling genes in TiO $_2$ nanoparticle—induced inflammation in the hippocampus of mice. *Toxicology Research, 4*(2), 344–350. https://doi.org/10.1039/C4TX00106K.

Sun, Q., Tan, D., Ze, Y., Sang, X., Liu, X., Gui, S., Cheng, Z., Cheng, J., Hu, R., Gao, G., Liu, G., Zhu, M., Zhao, X., Sheng, L., Wang, L., Tang, M., & Hong, F. (2012b). Pulmotoxicological effects caused by long-term titanium dioxide nanoparticles exposure in mice. *Journal of Hazardous Materials, 235-236*, 47–53. https://doi.org/10.1016/j.jhazmat.2012.05.072.

Sun, Q., Tan, D., Zhou, Q., Liu, X., Cheng, Z., Liu, G., Zhu, M., Sang, X., Gui, S., Cheng, J., Hu, R., Tang, M., & Hong, F. (2012a). Oxidative damage of lung and its protective mechanism in mice caused by long-term exposure to titanium dioxide nanoparticles. *Journal of Biomedical Materials Research—Part A, 100A*(10), 2554–2562. https://doi.org/10.1002/jbm.a.34190.

Tada-Oikawa, S., Ichihara, G., Fukatsu, H., Shimanuki Y., Tanaka, N., Watanabe, E., Suzuki, Y., Murakami, M., Izuoka, K., Chang, J., Wu, W., Yamada, Y., & Ichihara, S. (2016). Titanium Dioxide Particle Type and Concentration Influence the Inflammatory Response in Caco-2 Cells. *International Journal of Molecular Sciences, 17*(4). https://doi.org/10.3390/ijms17040576

Tang, Y., Wang, F., Jin, C., Liang, H., Zhong, X., & Yang, Y. (2013). Mitochondrial injury induced by nanosized titanium dioxide in A549 cells and rats. *Environmental Toxicology and Pharmacology, 36*(1), 66–72. https://doi.org/10.1016/j.etap.2013.03.006.

Tassinari, R., La Rocca, C., Stecca, L., Tait, S., De Berardis, B., Ammendolia, M., Iosi, F., Di Virgilio, A., Martinelli, A., & Maranghi, F. (2015). *In vivo and in vitro toxicological effects of titanium dioxide nanoparticles on small intestine. AIP Conference Proceedings*. Melville, NY: AIP Publishing. https://doi.org/10.1063/1.4922572.

Trouiller, B., Reliene, R., Westbrook, A., Solaimani, P., & Schiestl, R. H. (2009). Titanium dioxide nanoparticles induce DNA damage and genetic instability *in vivo* in mice. *Cancer Research, 69*(22). https://doi.org/10.1158/0008-5472.CAN-09-2496.

U.S. Geological Survey. (2016). *Mineral Commodity Summaries 2016*. Washington, D.C.: U.S. Government Printing Office. https://doi.org/10.3133/70140094.

Urrutia-Ortega, I., Garduño-Balderas, L., Delgado-Buenrostro, N., Freyre-Fonseca, V., Flores-Flores, J., González-Robles, A., Pedraza-Chaverri, J., Hernández-Pando, R., Rodríguez-Sosa, M., León-Cabrera, S., Terrazas, L., van Loveren, H., & Chirino, Y. (2016). Food-grade titanium dioxide exposure exacerbates tumor formation in colitis associated cancer model. *Food and Chemical Toxicology, 93*, 20–31. https://doi.org/10.1016/j.fct.2016.04.014.

Vance, M. E., Kuiken, T., Vejerano, E. P., et al. (2015). Nanotechnology in the real world: Redeveloping the nanomaterial consumer products inventory. *Beilstein J Nanotechnol., 6*, 1769–1780. https://doi.org/10.3762/bjnano.6.181.

Vasantharaja, D., Ramalingam, V., & Reddy, G. (2015). Oral toxic exposure of titanium dioxide nanoparticles on serum biochemical changes in adult male Wistar rats. *Nanomedicine Journal, 2*(1), 46–53.

Wang, J., Chen, C., Liu, Y., Jia, F., Li, W., Lao, F., Lia, Y., Lia, B., Ge, C., & Zhou, G. (2008a). Potential neurological lesion after nasal instillation of TiO_2 nanoparticles in the anatase and rutile crystal phases. *Toxicology Letters, 183*(1-3), 72–80. https://doi.org/10.1016/j.toxlet.2008.10.001.

Wang, J., Liu, Y., Jiao, F., Lao, F., Li, W., Gu, Y., Li, Y., Ge, C., Zhou, G., & Li, B. (2008b). Time-dependent translocation and potential impairment on central nervous system by intra-nasally instilled TiO(2) nanoparticles. *Toxicology, 254*(1-2), 82–90. https://doi.org/10.1016/j.tox.2008.09.014.

Wang, J., Zhou, G., Chen, C., Yu, H., Wang, T., Ma, Y., , Jia, G., Gao, Y., Li, B., Sun, J., Li, Y., Jiao, F., Zhao, Y., & Chai, Z. (2007). Acute toxicity and biodistribution of different sized titanium dioxide particles in mice after oral administration. *Toxicology Letters, 168*(2), 176-185. doi: https://doi.org/10.1016/j.toxlet.2006.12.001

Wang, Y., Chen, Z., Ye, Y., Wang, J., Wu, Y., Zhang, H., Wang, Y., Ba, T., Zhuo, L., Chen, S., & Jia, G. (2016). The safety and toxicity of food-related titanium dioxide nanoparticle. *Nanomedicine: Nanotechnology, Biology and Medicine, 12*(2), 464. https://doi.org/10.1016/j.nano.2015.12.054.

Weir, A., Westerhoff, P., Fabricius, L., Hristovski, K., & von Goetz, N. (2012). Titanium dioxide nanoparticles in food and personal care products. *Environmental Science and Technology, 46*(4), 2242-2250. doi: https://doi.org/10.1021/es204168d.

Xu, J., Shi, H., Ruth, M., Yu, H., Lazar, L., Zou, B., Yang, C., Wu, A., & Zhao, J. (2013). Acute toxicity of intravenously administered titanium dioxide nanoparticles in mice. *PLoS ONE, 8*(8), 1–6. https://doi.org/10.1371/journal.pone.0070618.

Yao, C., Li, C., Ding, L., Fang, J., Yuan, L., Hu, X., Wang, Y., & Wu, M. (2016). Effects of exposure routes on the bio-distribution and toxicity of titanium dioxide nanoparticles in mice. *Journal of Nanoscience and Nanotechnology, 16*(7), 7110–7117. https://doi.org/10.1166/jnn.2016.11349.

Yoshiura, Y., Izumi, H., Oyabu, T., Hashiba, M., Kambara, T., Mizuguchi, Y., Lee, B., Okada, T., Tomonaga, T., Myojo, T., Yamamoto, K., Kitajima, S., Horie, M., Kuroda, E., & Morimoto, Y. (2015). Pulmonary toxicity of well-dispersed titanium dioxide nanoparticles following intra-tracheal instillation. *Journal of Nanoparticle Research, 17*(6), 1–11. https://doi.org/10.1007/s11051-015-3054-x.

Younes, N., Amara, S., Mrad, I., Ben-Slama, I., Jeljeli, M., Omri, K., El Ghoul, J., El Mir, L., Rhouma, K., Abdelmelek, H., & Sakly, M. (2015). Subacute toxicity of titanium dioxide (TiO2) nanoparticles in male rats: emotional behavior and pathophysiological examination. *Environmental Science and Pollution Research International, 22*(11), 8728-8737. doi: https://doi.org/10.1007/s11356-014-4002-5

Ze, X., Su, M., Zhao, X., Jiang, H., Hong, J., Yu, X., Liu, D., Xu, B., Sheng, L., Zhou, Q., Zhou, J., Cui, J., Li, K., Wang, L., Ze, Y., & Hong, F. (2014d). TiO2 nanoparticle-induced neurotoxicity may be involved in dysfunction of glutamate metabolism and its receptor expression in mice. *Environmental Toxicology, 31*(6), 655–662. https://doi.org/10.1002/tox.22077.

Ze, Y., Hu, R., Wang, X., Sang, X., Ze, X., Li, B., Su, J., Wang, Y., Guan, N., Zhao, X., Gui, S., Cheng, Z., Cheng, J., Sheng, L., Sun, Q., Wang, L., & Hong, F. (2014b). Neurotoxicity and gene-expressed profile in brain-injured mice caused by exposure to titanium dioxide nanoparticles. *Journal of Biomedical Materials Research - Part A, 102*(2), 470–478. https://doi.org/10.1002/jbm.a.34705.

Ze, Y., Sheng, L., Zhao, X., Hong, J., Ze, Z., Yu, X., Pan, X., Lin, A., Zhao, Y., Zhang, C., Zhou, Q., Wang, L., & Hong, F. (2014a). TiO2 nanoparticles induced hippocampal neuroinflammation in mice. *PLoS One, 9*(3), e92230. https://doi.org/10.1371/journal.pone.0092230.

Ze, Y., Sheng, L., Zhao, X., Ze, X., Wang, X., Zhou, Q., Liu, J., Yuan, Y., Gui, S., Sang, X., Sun, Q., Hong, J., Yu, X., Wang, L., Li, B., & Hong, F. (2014c). Neurotoxic characteristics of spatial recognition damage of the hippocampus in mice following subchronic peroral exposure to TiO2 nanoparticles. *Journal of Hazardous Materials, 264*, 219–229. https://doi.org/10.1016/j.jhazmat.2013.10.072.

Ze, Y., Zheng, L., Zhao, X., Gui, S., Sang, X., Su, J., Guan, N., Zhu, L., Sheng, L., Hu, R., Cheng, J., Cheng, Z., Sun, Q., Wang, L., & Hong, F. (2013). Molecular mechanism of titanium dioxide nanoparticles-induced oxidative injury in the brain of mice. *Chemosphere, 92*(9), 1183–1189. https://doi.org/10.1016/j.chemosphere.2013.01.094.

Chapter 2
Zebrafish Models of Nanotoxicity: A Comprehensive Account

M. d'Amora and S. Giordani

Contents

1 Introduction

Due to unique the physicochemical and electrical properties of materials at the nanoscale, nanomaterials are increasingly utilized these days in various fields of applications, including manufacturing, biotechnology, nanomedicine, and electronics. All these fields, particularly biomedical applications, explicitly demand nanomaterials that are bio safe. However, data concerning the possible toxic effects of the different nanomaterials are still scarce. To this end, different biological models can be used to assess nanomaterials toxicity. Well-established in vitro models are usually used to evaluate the nanomaterials toxicity. They are preferred in comparison to other biological systems, thanks to their low cost and easy maintenance.

M. d'Amora
Nano Carbon Materials, Istituto Italiano di Tecnologia (IIT), Genoa, Italy

S. Giordani (✉)
School of Chemical Sciences, Dublin City University, Glasnevin, Dublin, Ireland
e-mail: silvia.giordani@dcu.ie

© Springer Nature Switzerland AG 2021
N. Sharma, S. Sahi (eds.), *Nanomaterial Biointeractions at the Cellular, Organismal and System Levels*, Nanotechnology in the Life Sciences, https://doi.org/10.1007/978-3-030-65792-5_2

However, after a preliminary investigation obtained in cells, the possible nanomaterials toxicity needs to be further investigated by using an in vivo system, significantly more complex than cultured cells. Indeed, there are several in vivo toxicological models including *Drosophila Melanogaster*, zebrafish, and mouse. Each system presents several advantages and limitations for toxicology screening. In particular, the main limitation of *Drosophila melanogaster* is that it possesses only four chromosome pairs. In contrast, mice are good candidates, but the main problem is that they are expensive and time-consuming. In this framework, zebrafish represents alternative and complementary model organisms, with several peculiarities, making them established systems for toxicity screening, in comparison to other species (Kalueff et al. 2016; Wiley et al. 2017).

In this chapter, we underline the different peculiarities of zebrafish that make them excellent candidates as model systems for toxicological screening. Moreover, we describe the different parameters used as toxicological endpoints. In particular, we focus on the developmental toxicity, describing the mortality and hatching rates, cardiac and swimming activities, immunotoxicity, genotoxicity, and neurotoxicity. In addition, we give a brief overview of previous toxicity studies on different classes of nanomaterials performed using zebrafish as model. These include carbon-based nanomaterials (fullerenes, carbon nanotubes, carbon dots, nanodiamonds, carbon nano-onions, carbon nano-horns, and graphene oxide), metallic nanoparticles (gold and silver nanoparticles), and semiconductor nanoparticles (silicon-based nanoparticles).

2 Unique Characteristics of Zebrafish

Nowadays, zebrafish are employed in different studies to assess the toxicity of nanomaterials, involving both embryos and adult organisms. The increasing use of zebrafish in the nanotoxicology field is due to several powerful peculiarities they possess. Zebrafish are vertebrates and therefore share a high homology degree with mammals, including humans (Howe et al. 2013; Kalueff et al. 2014). The cardiovascular, nervous, and digestive systems of these model animals are similar to the mammal's ones. Thanks to the similarity in the cellular and developmental mechanisms with the other vertebrates, studies performed in zebrafish can give insight into human mechanisms. Other advantages of employing zebrafish in a toxicological screening over other vertebrates are their small size, their low cost, and easy maintenance. Adults are 3 cm long, reducing the housing space and costs. In addition, the minute size of embryos allows performing toxicity experiments by placing together a high number of samples in 96 multi-well plates. Embryos are able to absorb the nanomaterials dissolved in their medium directly by soaking (d'Amora et al. 2017, 2018a, b). This provides several replicates at one time and reduces the amount of nanomaterials used and therefore the cost of the toxicological screening. In addition, all these peculiarities give the possibility to create high-throughput screens for toxicity screening in which embryos and larvae can develop in testing plates

(Horzmann and Freeman 2018; Hill et al. 2005). In this way, a large number of nanomaterials can be screened contemporary and rapidly. Moreover, zebrafish have a large number of offspring in each generation. The females produce with external fertilization around 200–300 eggs per week. The organogenesis occurs quickly and the major organs are formed within 5 days post fertilization (dpf). In addition, embryos are transparent, and this allows to easily identifying the developmental staging and assessing the toxicological endpoints during a complete toxicity screening. Their transparency is very useful when immunohistochemistry (IHC) and in situ hybridization (ISH) are employed.

3   Toxicological Screening in Zebrafish

Zebrafish are powerful platforms to test the effects of different nanomaterials. The toxicity of nanomaterials is assessed by examining different toxicological endpoints during the zebrafish development (Heiden et al. 2007), based on both external phenotypic and internal organs changes (Pham et al. 2016). The mortality (or survival rate) and hatching rates are the first endpoints analyzed during a toxicological screening. Subsequently, other biological parameters, such as the cardiac (heartbeat rate) and swimming activities, could be evaluated (Fig. 2.1). In addition, the possible presence of malformations in embryos and larva exposed to different concentrations of nanomaterials is an important endpoint. Moreover, zebrafish are employed to assess immunotoxicity, genotoxicity, and neurotoxicity in a complete toxicity testing (Fig. 2.1).

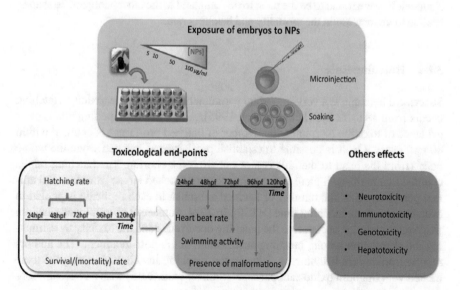

Fig. 2.1 A schematic representation of the complete toxicological screening in *C. elegans*

3.1 Developmental Toxicity

The visualization of the different toxicological endpoints in zebrafish is done at specific stages that correspond to crucial points of the development. All the observations are carried out on in vivo embryos and larvae.

3.1.1 Mortality/Survival Rate

The mortality rate of embryos and larvae is evaluated and noted at different concentrations of nanomaterials and throughout the whole exposure period. Since different studies reported the presence of mortality in the early life stage (Fraysse et al. 2006), the evaluation of this parameter starts at 4 h post fertilization (hpf). The mortality rate of zebrafish at this stage is calculated by determining the eggs blocked at the blastomeric stage and the unfertilized eggs, using a dissecting stereomicroscope. After 4 hpf, zebrafish embryos or larvae are established as dead if there is no more heartbeat rate or no longer moving or the appearance of the tissue changed from the normal transparency to the opacity (Ali et al. 2011). The mortality rate is subsequently recorded every 24 h (at 24, 48, 72, 96 and 120 hpf). Treatment with acute doses of nanomaterials enhances the increase in the mortality rate. For instance, the increase of the exposure concentrations of silica nanoparticles induced an increase in mortality and cell death (Duan et al. 2013). Moreover, different studies indicated a relationship between the incidences of the mortality and the different shapes of nanomaterials. Hence, the toxicity induced by zinc oxide nanoparticles (ZnO NPs) with a different shape, including nanospheres, nano-sticks, and cuboidal submicron particles, was investigated (Hua et al. 2014). Nano-sticks were found to be the most toxic compared to the other nanoparticle shapes, leading to an increase in the mortality and hatching rate.

3.1.2 Hatching Rate

Successful hatching is a very important parameter to conceive the toxicity. Hatching occurs from 48 to 72 hpf (Kimmel et al. 1995). A delay in the hatching rate is clear evidence of toxicity. Normally, the number of hatched pro-larvae is counted within 80 hpf, every 2 h. It is possible to establish pro-larvae as hatched when the whole body (from the head to the tail) comes out from the chorion. The hatching rate is counted in a multi-well plate as the percentage of hatched larvae from the total animal per plate. After the number of hatched zebrafish in each replicate is pooled to calculate the mean hatching time (HT50). In 2014, Samaee et al. employed titanium oxide nanoparticles to conceive the hatching occurrence and the toxicity by estimating the correlation among hatching success and hours post-treatment. The authors reported premature hatching or delay in the hatching in zebrafish embryos/larvae treated with titanium oxide nanoparticles (Samaee et al. 2015) with a concentration-dependent toxicity.

3.1.3 Possible Presence of Malformations

The presence of malformations in zebrafish embryos and larvae exposed to nanomaterials is another endpoint frequently assessed during toxicological testing. Since development has been well characterized (Kimmel 1989; Kimmel et al. 1995) and embryos and larvae are transparent, it is possible to easily observe the main zebrafish developmental defects and abnormalities using little magnification. The discrimination between normal and anomalous development is generally made using the organogenesis description of Kimmel et al. (Kimmel et al. 1995). Zebrafish are placed in a multi-well under a common stereomicroscope at different developmental stages and the malformations are noted. These involve incomplete organ development, or defects in different body parts, including the heart, notochord, and brain. In this framework, silica nanoparticles were found to induce typically zebrafish malformations, including yolk sac edema, tail malformations, pericardial edema, and head malformation. The pericardial edema was the most incident defect caused by silicon-based nanoparticles (Duan et al. 2013).

3.1.4 Neurotoxicity

Zebrafish have emerged as powerful model to assess the neurotoxicity (Giordani and d'Amora 2018). It is well known that the developmental processes of the central nervous system of zebrafish and other vertebrates are highly conserved (Belousov 2011). This homology comprises also the blood brain barrier (BBB) development (Eliceiri et al. 2011). Moreover, many brain subdivisions of mammals during the development have a counterpart in zebrafish (Wullimann 2009). In the neurotoxicity assessment, thanks to the transparency of zebrafish, it is possible to visualize specific neurons or axonal tracts in vivo, by means of different biomarkers and dyes or in fixed samples using in situ hybridization and immunohistochemistry techniques. Different endpoints can be investigated in order to investigate the possible neurotoxicity induce by nanomaterials, including neurobehavioral profiling and neural morphogenesis (Kalueff et al. 2014; Truong et al. 2014). Different studies evaluated the neuronal apoptosis by using specific staining and in situ hybridization. Kim et al. showed that treatment of zebrafish with small gold nanoparticles leads to neuronal damage. Moreover, using in situ hybridization, the expression of several transcription factors involved in the eye development, including *sox10, pax6a,* and *pax6b,* were analyzed and found to be re-pressed (Kim et al. 2013).

Nowadays, another crucial parameter to understand and evaluate the neurotoxicity is represented by the zebrafish behavioral response to the nanomaterials exposure (Locomotion and Behavioral Toxicity in Larval Zebrafish: Background, Methods, and Data). It is possible to assess the swimming activities in terms of spontaneous movements or number of movements by means of video recording tools. Normally this measurement is performed on larvae at 72 hpf or after 72 hpf. Before this developmental stage, the zebrafish rolled up in the chorion and this membrane disturbs their movements. Treatment of zebrafish with different carbon-based nanomaterials revealed that the frequency of movements of larvae at 96 hpf had no significant reduction in the case of carbon nano-onions and carbon nano-horns, while a significant reduction was found in the case of graphene oxide exposure (d'Amora et al. 2017). In contrast, the number of spontaneous movements was not affected in zebrafish embryos treated with single-walled carbon nanotubes (Ong et al. 2014). Duan et al. evaluated the total distance of swimming of zebrafish treated with silicon-based nanoparticles (Duan et al. 2013). The results reported a decrease on the swimming distance concentrations dependent. In addition, zebrafish exposed to graphene quantum dots (GQDs) presented a decrease in the total swimming distance and speed. In both studies, the total distance of swimming was measured by means of a visible light test (Wang et al. 2015).

3.1.5 Cardiotoxicity

The heart is the first organ to develop in zebrafish and the heartbeat starts around 22 hpf. The cardiovascular system is formed and completely functional within 48 hpf (Thisse and Zon 2002). Even if there are physiological differences between the heart of zebrafish and mammalian, zebrafish can be considered a prominent candidate to assess the cardiotoxicity. In fact, the mechanisms of heart ontogenesis are well conserved between vertebrates (Staudt and Stainier 2012) and the electrical properties are highly homolog to the human's one (Arnaout et al. 2007; Sedmera et al. 2003). Moreover, the optical transparency of embryos allows in vivo real-time imaging. A study on zebrafish cardiotoxicity assays reported that embryos can live for days in the presence of abnormalities affecting the circulatory system, validating zebrafish as good tools in this field (Chen et al. 1996). Moreover, it was found that tyrosine kinases are conserved expressed during the early developmental stage (Challa and Chatti 2013). All these peculiarities make zebrafish suitable and powerful systems to assess the cardiotoxic effects (Cheng et al. 2011). In particular, the embryos have been employed to assess the effects of different nanomaterials on cardiovascular development, while adults have been used to investigate acute and chronic effects on cardiac function (Sarmah and Marrs 2016). Thanks to the optical transparency of the zebrafish, the cardiac function, including heartbeat rate, the presence of malformations can be easily evaluated by using different in vivo biomarkers.

During a cardiotoxicity screening, the effects of nanomaterials on zebrafish are assessed at 48hph by using a common stereomicroscope. Zebrafish were used to investigate the possible cardiotoxicity induced by several nanomaterials. The possible cardiotoxicity of silica nanoparticles (Si NPs) during the development was deeply evaluated via intravenous microinjection. Silica nanoparticles caused bradycardia and pericardial edema. Moreover, treated embryos presented oxidative stress and neutrophil-mediated cardiac inflammation. Histology techniques on the heart of embryos and larvae treated with silica nanoparticles allowed observing the presence of inflammatory cells in the atria (Duan et al. 2016).

3.2   Immunotoxicity

The immune system is known to be highly sensitive to nanomaterials exposure in particular, in terms of inflammation induction and activation of macrophages and neutrophils (Johnston et al. 2018). Different studies reported that nanomaterials modulate cytokines production by generating free radicals. Moreover, exposure to nanomaterials can induce allergic sensitization and asthma (Di Gioacchino et al. 2011). Zebrafish treated with small gold nanoparticles (1.5 nm core) functionalized with three different ligands presented a perturbation in the inflammation and immune response mechanisms (Truong et al. 2013). On the other side, silver nanoparticles (Ag NPs) induced immunotoxicity in adult zebrafish (Krishnaraj et al. 2016).

3.3   Genotoxicity

An important component in nanotoxicity is the evaluation of genotoxicity. Nanomaterials can induce genotoxicity, leading to DNA damage and gene mutations. Genotoxicity is a crucial risk determinant for long-term toxicity, including tumorigenesis. In the past, mice have been widely used to assess the genotoxicity of nanomaterials, by using micronucleus assays and gene profiling techniques (Manjanatha et al. 2014). Recently, zebrafish come out as powerful genotoxic tools. The genotoxicity induced by several nanomaterials was studied and assessed in zebrafish by using different techniques. Dedeh et al. and Geffroy et al. evaluated the effects of gold nanoparticles (AuNPs) using RAPD coupled with PCR genotoxicity test (Geffroy et al. 2012; Dedeh et al. 2015). After gold nanoparticles exposure, an alteration of genome composition was found (Geffroy et al. 2012). Subsequently, other techniques were employed to assess the genotoxicity of compounds in zebrafish (Villacis et al. 2017) (Rocco et al. 2015). Adult animals were treated with different doses of iron oxide nanoparticles (IONPs). The results demonstrated significant concentration-dependent genotoxic effects of IONPs. In particular, a high number of transcripts of liver samples were found to have a different expression in comparison to the controls, by using microarray analysis (Villacis et al. 2017). In addition, the potential genotoxicity of titanium dioxide nanoparticles (TiO$_2$ NPs) was analyzed in zebrafish by means of diffusion assay, RAPD-PCR technique, and comet assay (Rocco et al. 2015). The maximum concentrations of nanoparticles tested caused the highest genotoxicity.

3.4 Hepatotoxicity

The organogenesis of the liver begins at 72 hpf and is completely functional within 120 hpf (Chu and Sadler 2009). Several reports indicated that compounds are metabolized by zebrafish during the development with mechanisms similar to those of humans (Quinlivan and Farber 2017; Vliegenthart et al. 2014). Different nanomaterials can cause liver injury. It is possible to evaluate the hepatotoxicity in zebrafish by means of different tests, including enzymes assays and histology techniques. The hepatotoxicity can be easily evaluated visually on the liver tissue. Another technique consists in use biomarkers for liver injury. The levels of these biomarkers are measured in the circulation of treated animals (Vliegenthart et al. 2014). The visualization of the liver damage can be performed by using transgenic lines with a liver-dye expression. The size and the number of hepatocytes can be calculated by analyzing the intensity of the fluorescence before and after the treatment (Zhang et al. 2014).

4 Toxicity Studies in Zebrafish

4.1 Carbon-Based Nanomaterials

Carbon nanomaterials (CNMs) have gained increased interest in different fields, thanks to their unique electronic, optical, and physical characteristics (d'Amora and Giordani 2018). They include fullerenes (C_{60}) (Kroto et al. 1985), nanodiamonds (NDs) (Greiner et al. 1988), carbon nanotubes (CNTs) (Iijima 1991), carbon nano-onions (CNOs) (Ugarte 1992), carbon nano-horns (CNHs) (Iijima et al. 1999), carbon dots (CDs) (Xu et al. 2004), and graphene (Novoselov et al. 2004). Their biocompatibility plays a crucial role in their different applications, including nanomedicine and bioimaging. In the last few years, a careful evaluation of the possible toxic effects of different CNMs in zebrafish during the development has been carried out, reporting their in vivo biosafety (d'Amora et al. 2017; Nicholas et al. 2018).

4.1.1 Fullerenes

Fullerenes are employed in several biomedical and biological applications, including imaging and drug delivery, thanks to their intrinsic photoluminescence, nano-meter size, and hollow cavity (Levi et al. 2006). Using zebrafish as model system, the toxicity of different fullerenes was tested. The effects of the dendro [C_{60}]fullerene DF-1, with antioxidant properties, were assessed, by monitoring the survival rate and the possible presence of malformations (Daroczi et al. 2006). DF-1 exerted no detectable toxicity on zebrafish at the tested concentration. Usenko et al. exposed zebrafish to C_{60}, its hydroxylated derivative $C_{60}(OH)_{24}$, and C_{70} (Usenko et al. 2007, 2008). All these fullerenes caused a high percentage of developmental abnormalities and mortality, with $C_{60}(OH)_{24}$ less toxic than C_{60}. In another study, dendrofullerenes (mono-adducts of C_{60}) and e,e,e-trismalonic acid-like fullerenes (C3-like fullerenes), anionic water-soluble fullerenes were found to be more toxic than oxo-amino fullerenes, anionic fullerenes with similar structures (Beuerle et al. 2007). In addition, it was observed that the toxicity of anionic fullerenes varied from very low to moderate depending on the structures. The biological consequences of different fullerenes on zebrafish were also studied in terms of effects on the proteomic profiles. For instance, the comparison of proteomic profiles between the phosphatidylcholine-based phospholipid nanoparticles containing fullerene C_{60} and the control reported low toxicity of the nanoparticles on zebrafish (Kuznetsova et al. 2014).

4.1.2   Carbon Nanotubes

Since their discovery (Iijima 1991; Iijima and Ichihashi 1993), CNTs have raised increasing interest from different fields for their unique chemical, optical, electrical, and thermal (Bachilo et al. 2002; Ruoff and Lorents 1995) properties. Thanks to these properties, they are employed in different applications. CNTs comprised sp2 carbon atoms organized in single or multiple coaxial tubes of graphitic sheets resulting in single-walled carbon nanotubes (SWCNTs) and multiple-walled carbon nanotubes (MWCNTs), respectively. In 2009, Cheng et al. studied the biological consequences and in vivo biodistribution of fluorescent-labelled MWCNT (FITC-BSA-MWCNTs) in zebrafish at the different developmental stages (Cheng et al. 2009). No lethal effects and no developmental defects were observed after FITC-BSA-MWCNTs injection. Moreover, the data suggested that purification and functionalization of carbon nanotubes improved their biosafety. Subsequently, the same group evaluated the effects of BSA-MWCNTs sonicated in nitric acid for 24 h (MWCNTs-24 h) and 48 h (MWCNTs-48 h). The sonication time affected the length of the MWCNTs. MWCNTs-24 h presented a length of 0.8 ± 0.5 µm, while MWCNTs-48 h had a length of 0.2 ± 0.1 µm. Zebrafish embryos were microinjected with MWCNTs-24 h and MWCNTs-48 h to check their effects. MWCNTs-24 h did not affect the embryos, while the MWCNTs-48 h caused significant toxic effects (Cheng and Cheng 2012). The authors suggested that shorter BSA-MWCNTs were more toxic in zebrafish embryos after injection. Perhaps another factor could be the production of carbonaceous fragments during the nitric acid treatment (Del Canto et al. 2011; Salzmann et al. 2007).

4.1.3   Carbon Dots

Carbon dots, also known as carbon quantum dots (C-dots) or graphene quantum dots (GQDs), are interesting materials to be used in imaging application as they have high photo-stability and exhibit an intrinsic fluorescence (Zheng et al. 2015). The toxicity of C-dots has been investigated in zebrafish in terms of liability after soaking or microinjection of C-dots in embryos (Kang et al. 2015). Zebrafish grow normally, with a low percentage of developmental abnormalities. C-dots demonstrated good biocompatibility. Recently, Khajuria et al. reported similar results for carbon dots doped with nitrogen (N-CDs). Embryos soaked in N-CDs solutions with different concentrations presented viabilities of more than 75%, with no malformations. These data confirmed the biosafety of C-dots after soaking (Khajuria et al. 2017). On the other hand, GQDs exhibited high biocompatibility, without affecting zebrafish at a concentration lower than 2 mg mL^{-1} (Roy et al. 2015).

4.1.4 Nanodiamonds

Nanodiamonds (NDs) have been employed in several biomedical applications, including drug delivery and imaging, thanks to good optical and biological properties (Mochalin et al. 2011). Lin et al. evaluated the possible adverse effects as well as the persistent effects on larval behavior of nanodiamonds. After microinjection in the yolk, only high concentrations (5 mg/ml) of NDs affected the zebrafish growth, inducing body axis curvature (Lin et al. 2016). Recently, we assessed the possible toxicological effects induced by small carbon dot decorated nanodiamonds (CD-DNDs) on zebrafish (Nicholas et al. 2018). CD-DNDs caused no significant effect on the survival, hatching, and heartbeat rates, and the zebrafish organogenesis. Our results clearly demonstrated the biosafety of CD-DNDs.

4.1.5 Carbon Nano-Onions

Multi-shell fullerenes, known as carbon nano-onions (CNOs), are promising CNMs for imaging (Bartelmess et al. 2015; Lettieri et al. 2017a) and diagnostic applications (Lettieri et al. 2017b; Giordani et al. 2014). Small CNOs (average diameters of 5 nm) show high cellular uptake and weak inflammatory potential (Yang et al. 2013).

Our group has been investigating the toxicity of benzoic acid functionalized CNOs (benz-CNOs) and fluorescent boron dipyrromethene (BODIPY) tagged CNOs (BODIPY-CNOs) in zebrafish during the development. We evaluated the survival and hatching rates, cardiac activity, frequency of movements, and possible morphological abnormalities of zebrafish embryo and larvae treated with 5, 10, 50, and 100 µg mL^{-1} of benz-CNOs and BODIPY CNOs for 120 hpf. We observed no considerable changes in all the toxicological endpoints analyzed in treated embryos and larvae. In particular, the survival and hatching rates of treated zebrafish were found to be similar to the untreated control. Moreover, no reduction in the total frequency of movements and no cardiac effects were observed and the total percentages of abnormalities during the organogenesis was less than 4%. Our result clearly revealed that benz-CNOs and BODIPY-CNOs presented non-toxicity and good biocompatibility in zebrafish (d' Amora et al. 2016).

Furthermore, we reported that oxi-CNOs possessed higher biocompatibility than other classes of CNMs such as oxi-carbon nano-horns (CNHs) (d'Amora et al. 2017).

4.1.6 Carbon Nano-Horns

Carbon nano-horns (CNHs) are conical carbon nanostructures, suitable for biomedical applications, such as drug delivery (Xu et al. 2008).

Our group assessed for the first time the in vivo biological consequences of carbon nano-horns in zebrafish during the development. We exposed the embryos to different concentrations of oxidized CNHs (oxi-CNHs) of 5–8 nm in horn diameter and 30–50 nm in length. Oxi-CNHs induced no significant differences in survival/hatching rates and heartbeat rate of treated embryos and larvae. Moreover, no reduction in the cardiac and swimming activities was observed in the larvae treated with the different concentrations of CNHs. Our results demonstrated that oxi carbonhorns presented no toxicity.

4.1.7 Graphene Oxide

Graphene oxide (GO) presents different properties such as high surface area, layer number, and lateral dimensions that make them able to transport drugs, genes, and proteins in certain cell types or specific body regions. Thanks to these properties, GO is employed in cancer treatment, biological imaging, and drug delivery. Several groups have assessed the in vivo toxicity of graphene oxide. In 2014, Liu et al. (Liu et al. 2014) treated zebrafish eggs with different concentration of GO (1, 5, 10, 50, 100 mg/l) and analyzed different biological parameters. GO (average size 512 nm) resulted to be toxic, inducing a disturbance in the hatching and larvae length. Subsequently, Chen and his group reported that, after exposure to zebrafish, part of the GO adhered to the chorion of the embryo, occluding the pore and consequently causing hypoxia and hatching delay (Hu et al. 2016). Moreover, the amount of GO up taken by the embryos induced damage in the mitochondria, a reduction of the heartbeat rate and different developmental abnormalities affecting the eye, the heart, and the tail.

Our group has investigated the toxicity of commercially available GO (lateral size 15 μm) in zebrafish (d'Amora et al. 2017). GO caused adverse effects on zebrafish development at high concentrations (50 and 100 μg ml^{-1}). The treated embryos/larvae presented a developmental delay. The heartbeat rates and the frequency of movements of treated larvae were reduced. Moreover, different developmental abnormalities in zebrafish, including pericardial and yolk sac edema, fold flexure and tail flexure have been found. The percentage of malformations reached 13.5% and 11% in the case of pericardial and yolk sac edema, respectively.

4.2 Metallic Nanoparticles

Metallic nanoparticles have a wide range of applications in different fields, including nanomedicine. Among metallic nanoparticles, the gold and silver NPs are mainly employed; therefore, an accurate assessment of their toxicity is needed.

4.2.1 Gold Nanoparticles

Gold nanoparticles (Au NPs) are mainly employed in nanomedicine applications, as diagnostic agents (Huang et al. 2006) or drugs carriers (Dykman and Khlebtsov 2011). Since gold nanoparticles can induce cytotoxicity (Gerber et al. 2013), their possible effects are further investigated in zebrafish.

Adult zebrafish treated with gold nanoparticles of two sizes presented genome alterations and different dysfunctions (Geffroy et al. 2012) in several tissues. It has been reported the surface functionalization of gold nanoparticles can influence their toxicity. In particular, gold nanoparticles functionalized with positively charged N,N,N-triethylammoniumethanol (TMAT) caused a high mortality rate in zebrafish, without a significant presence of developmental abnormalities. On the other hand, gold nanoparticles functionalized with 2-mercaptoethanatesulfonate (MES) have completely different behavior. In fact, zebrafish treated with these nanoparticles have no significant percentage of mortality while presented a high incidence of developmental defects. Other studies confirmed the dependence of the gold nanoparticles toxicity from the functionalization (Truong et al. 2012). Small gold nanoparticles caused a disruption in eye development with consequent neuronal damage and changes in the behavioral profile (Kim et al. 2013). One factor mediating the toxicity of gold nanoparticles is represented by the different shapes. Gold nanoparticles of different shapes were exposed to adult zebrafish (Sangabathuni et al. 2017). Rod-shaped Au NPs presented higher uptake and faster clearance compared to spherical gold nanoparticles and stars NPs.

4.2.2 Silver Nanoparticles

Silver nanoparticles (Ag NPs) are extensively applied in different biomedical applications as antimicrobial agents (Kőrösi et al. 2016), drug delivery systems (Jin and Ye 2007), therapeutic agents (Czupryna and Tsourkas 2006), and biosensors (Jin and Ye 2007). Because of the widespread use and the increased exposure to Ag NPs (Benn et al. 2010), it is important to access the toxic effects related to their exposure. In 2008, Asharani et al. reported for the first time toxic effects and biodistribution of Ag NPs on zebrafish during the development (Asharani et al. 2008). Embryo treated with different concentrations of Ag NPs presented high mortality, a hatching delay, and different malformations, including pericardial edema and tail flexure. In addition, Ag NPs localized preferentially in the yolk, heart, and brain. The toxicity of silver nanoparticles was found to be dependent on their size (Bar-Ilan et al. 2009). In the last decade, many studies further investigated the toxicity of Ag NPs of different sizes and with different surface coatings. In order to evaluate the toxicity dependence from the different coating surfaces, Lee et al. synthesized Ag NPs functionalized with three biocompatible peptides (CALNNK, CALNNS, CALNNE). They investigated the toxic effects of Ag-CALNNK NPs$^{+\zeta}$, Ag-CALNNS NPs$^{-2\zeta}$, and Ag-CALNNE NPs$^{-4\zeta}$ and demonstrated charge-dependent toxicity. Ag-CALNNK NPs$^{+\zeta}$ and Ag-CALNNE NPs$^{-4\zeta}$

were the most and less biocompatible nanoparticles, respectively (Lee et al. 2013). Recently, the different behavior of flat and spherical Ag NPs was investigated in zebrafish. Silver nanoplates were found to be more toxic than Ag nanospheres (Abramenko et al. 2018). The effects of silver nanoparticles on zebrafish larvae were also investigated in terms of bio-interactions with subcellular structures (d'Amora et al. 2015), by evaluating the possible effects of small-sized NPs on the cytoskeletal architecture.

4.3 Semiconductors Nanoparticles

Over the past few years, semiconductor nanomaterials such as silicon and germanium became attractive materials for bio photonics and personalized medicine, i.e., imaging and therapeutic agents, applications (Hashim et al. 2014) (Maji et al. 2014; Li et al. 2014). Nevertheless, principal trouble that can restrict their use in these applications are the toxicity due to heavy metal (Ambrosone et al. 2012).

4.3.1 Silicon-Based Nanoparticles

During the last decade, the possible toxicity of silicon-based nanomaterial has been deeply investigated in zebrafish. Duan et al. reported that silicon-based nanoparticles caused high mortality and a hatching delay concentration-dependent (Duan et al. 2013). Moreover, Si NPs induced different types of developmental defects, such as head malformation and yolk sac edema, and a decrease in the total swimming distance. Another study reported that Si NPs did not internalize in the embryos and were mostly accumulated in the chorion surface (Fent et al. 2010).

The effects of silicon-based nanoparticles produced by laser ablation have been studied in zebrafish during the development (d'Amora et al. 2018a, b). The results showed that these NPs did not affect any biological parameters in the zebrafish embryos and larvae, demonstrating their biosafety.

5 Conclusions

As the use of nanomaterials in daily life, and in particular in nanomedicine applications, constantly increases, their possible adverse effects need to be carefully evaluated. In this chapter, we report that zebrafish have become excellent in vivo systems for toxicological screening at the whole animal level. These models are cheaper, quicker, and more efficient than other vertebrates, including mice. We have highlighted how their use gives the opportunity to investigate specific physiological impacts at the different stage of the zebrafish growth. Notwithstanding the toxicity studies of different nanomaterials based on vertebrate models have been reported in the recent past, the use of zebrafish in nanotoxicity is relatively new. In the near future, zebrafish may become an alternative to other mammalian organisms in evaluation of the toxicity of different nanomaterials.

References

Abramenko, N. B., Demidova, T. B., Abkhalimov, E. V., Ershov, B. G., Krysanov, E. Y., & Kustov, L. M. (2018). Ecotoxicity of different-shaped silver nanoparticles: Case of zebrafish embryos. *Journal of Hazardous Materials, 347*, 89–94. https://doi.org/10.1016/j.jhazmat.2017.12.060.

Ali, S., HGJV, M., & Richardson, M. K. (2011). Large-scale assessment of the zebrafish embryo as a possible predictive model in toxicity testing. *PLoS One, 6*(6), e21076. https://doi.org/10.1371/journal.pone.0021076.

Ambrosone, A., Mattera, L., Marchesano, V., Quarta, A., Susha, A. S., Tino, A., Rogach, A. L., & Tortiglione, C. (2012). Mechanisms underlying toxicity induced by CdTe quantum dots determined in an invertebrate model organism. *Biomaterials, 33*(7), 1991–2000. https://doi.org/10.1016/j.biomaterials.2011.11.041.

Arnaout, R., Ferrer, T., Huisken, J., Spitzer, K., Stainier, D. Y., Tristani-Firouzi, M., & Chi, N. C. (2007). Zebrafish model for human long QT syndrome. *Proceedings of the National Academy of Sciences of the United States of America, 104*(27), 11316–11321. https://doi.org/10.1073/pnas.0702724104.

Asharani, P. V., Lian Wu, Y., Gong, Z., & Valiyaveettil, S. (2008). Toxicity of silver nanoparticles in zebrafish models. *Nanotechnology, 19*(25), 255102. https://doi.org/10.1088/0957-4484/19/25/255102.

Bachilo, S. M., Strano, M. S., Kittrell, C., Hauge, R. H., Smalley, R. E., & Weisman, R. B. (2002). Structure-assigned optical spectra of single-walled carbon nanotubes. *Science, 298*(5602), 2361–2366. https://doi.org/10.1126/science.1078727.

Bar-Ilan, O., Albrecht, R. M., Fako, V. E., & Furgeson, D. Y. (2009). Toxicity assessments of multi-sized gold and silver nanoparticles in zebrafish embryos. *Small (Weinheim an der Bergstrasse, Germany), 5*(16), 1897–1910. https://doi.org/10.1002/smll.200801716.

Bartelmess, J., Quinn, S. J., & Giordani, S. (2015). Carbon nanomaterials: Multi-functional agents for biomedical fluorescence and Raman imaging. *Chemical Society Reviews, 44*(14), 4672–4698. https://doi.org/10.1039/C4CS00306C.

Belousov, L. V. (2011). Scott F. Gilbert—Developmental biology, 2010, Sinauer associates, Inc., Sunderland, MA ninth edition. *Russian Journal of Developmental Biology, 42*(5), 349. https://doi.org/10.1134/s1062360411050043.

Benn, T., Cavanagh, B., Hristovski, K., Posner, J. D., & Westerhoff, P. (2010). The release of nanosilver from consumer products used in the home. *Journal of Environmental Quality, 39*(6), 1875–1882.

Beuerle, F., Witte, P., Hartnagel, U., Lebovitz, R., Parng, C., & Hirsch, A. (2007). Cytoprotective activities of water-soluble fullerenes in zebrafish models. *Journal of Experimental Nanoscience, 2*(3), 147–170. https://doi.org/10.1080/17458080701502091.

Challa, A. K., & Chatti, K. (2013). Conservation and early expression of zebrafish tyrosine kinases support the utility of zebrafish as a model for tyrosine kinase biology. *Zebrafish, 10*(3), 264–274. https://doi.org/10.1089/zeb.2012.0781.

Chen, J. N., Haffter, P., Odenthal, J., Vogelsang, E., Brand, M., van Eeden, F. J., Furutani-Seiki, M., Granato, M., Hammerschmidt, M., Heisenberg, C. P., Jiang, Y. J., Kane, D. A., Kelsh, R. N., Mullins, M. C., & Nusslein-Volhard, C. (1996). Mutations affecting the cardiovascular system and other internal organs in zebrafish. *Development (Cambridge, England), 123*, 293–302.

Cheng, H., Kari, G., Dicker, A. P., Rodeck, U., Koch, W. J., & Force, T. (2011). A novel pre-clinical strategy for identifying cardiotoxic kinase inhibitors and mechanisms of cardiotoxicity. *Circulation Research, 109*(12), 1401–1409. https://doi.org/10.1161/CIRCRESAHA.111.255695.

Cheng, J., Chan, C. M., Veca, L. M., Poon, W. L., Chan, P. K., Qu, L., Sun, Y. P., & Cheng, S. H. (2009). Acute and long-term effects after single loading of functionalized multi-walled carbon nanotubes into zebrafish (Danio rerio). *Toxicology and Applied Pharmacology, 235*(2), 216–225. https://doi.org/10.1016/j.taap.2008.12.006.

Cheng, J., & Cheng, S. H. (2012). Influence of carbon nanotube length on toxicity to zebrafish embryos. *International Journal of Nanomedicine, 7*, 3731–3739. https://doi.org/10.2147/IJN.S30459.

Chu, J., & Sadler, K. C. (2009). New school in liver development: Lessons from zebrafish. *Hepatology (Baltimore, Md), 50*(5), 1656–1663. https://doi.org/10.1002/hep.23157.

Czupryna, J., & Tsourkas, A. (2006). Suicide gene delivery by calcium phosphate nanoparticles: A novel method of targeted therapy for gastric cancer. *Cancer Biology & Therapy, 5*(12), 1691–1692.

d' Amora, M., Rodio, M., Bartelmess, J., Sancataldo, G., Brescia, R., Cella Zanacchi, F., Diaspro, A., & Giordani, S. (2016). Biocompatibility and biodistribution of functionalized carbon nanoonions (f-CNOs) in a vertebrate model. *Scientific Reports, 6*, 33923. https://doi.org/10.1038/srep33923. https://www.nature.com/articles/srep33923#supplementary-information.

d' Amora, M., Camisasca, A., Lettieri, S., & Giordani, S. (2017). Toxicity assessment of carbon nanomaterials in zebrafish during development. *Nanomaterials, 7*(12), 414.

d' Amora, M., Cassano, D., Pocoví-Martínez, S., Giordani, S., & Voliani, V. (2018a). Biodistribution and biocompatibility of passion fruit-like nano-architectures in zebrafish. *Nanotoxicology*, 1–9. https://doi.org/10.1080/17435390.2018.1498551.

d' Amora, M., Gaser, A. N., Lavagnino, Z., Sancataldo, G., Zanacchi, F. C., & Diaspro, A. (2015). Zebrafish larvae as model system to study possible toxicity of silver nanoparticles at cytoskeletal level by means of advanced microscopy. *Biophysical Journal, 108*(2, Supplement 1), 217a. https://doi.org/10.1016/j.bpj.2014.11.1201.

d' Amora, M., & Giordani, S. (2018). In G. Ciofani (Ed.), *7 - carbon nanomaterials for nanomedicine, Smart nanoparticles for biomedicine* (pp. 103–113). Elsevier. https://doi.org/10.1016/B978-0-12-814156-4.00007-0.

d' Amora, M., Rodio, M., Sancataldo, G., Diaspro, A., & Intartaglia, R. (2018b). Laser-fabricated fluorescent, ligand-free silicon nanoparticles: Scale-up, Bio-safety, and 3D live imaging of zebrafish under development. *ACS Applied Bio Materials*. Just Accepted Manuscript. https://doi.org/10.1021/acsabm.8b00609.

Daroczi, B., Kari, G., McAleer, M. F., Wolf, J. C., Rodeck, U., & Dicker, A. P. (2006). In vivo radioprotection by the fullerene nanoparticle DF-1 as assessed in a zebrafish model. *Clinical Cancer Research, 12*(23), 7086–7091. https://doi.org/10.1158/1078-0432.ccr-06-0514.

Dedeh, A., Ciutat, A., Treguer-Delapierre, M., & Bourdineaud, J. P. (2015). Impact of gold nanoparticles on zebrafish exposed to a spiked sediment. *Nanotoxicology, 9*(1), 71–80. https://doi.org/10.3109/17435390.2014.889238.

Del Canto, E., Flavin, K., Movia, D., Navio, C., Bittencourt, C., & Giordani, S. (2011). Critical investigation of defect site functionalization on single-walled carbon nanotubes. *Chemistry of Materials, 23*(1), 67–74. https://doi.org/10.1021/cm101978m.

Di Gioacchino, M., Petrarca, C., Lazzarin, F., Di Giampaolo, L., Sabbioni, E., Boscolo, P., Mariani-Costantini, R., & Bernardini, G. (2011). Immunotoxicity of nanoparticles. *International Journal of Immunopathology and Pharmacology, 24*(1 Suppl), 65s–71s.

Duan, J., Yu, Y., Li, Y., Li, Y., Liu, H., Jing, L., Yang, M., Wang, J., Li, C., & Sun, Z. (2016). Low-dose exposure of silica nanoparticles induces cardiac dysfunction via neutrophil-mediated inflammation and cardiac contraction in zebrafish embryos. *Nanotoxicology, 10*(5), 575–585. https://doi.org/10.3109/17435390.2015.1102981.

Duan, J., Yu, Y., Shi, H., Tian, L., Guo, C., Huang, P., Zhou, X., Peng, S., & Sun, Z. (2013). Toxic effects of silica nanoparticles on zebrafish embryos and larvae. *PLoS One, 8*(9), e74606. https://doi.org/10.1371/journal.pone.0074606.

Dykman, L. A., & Khlebtsov, N. G. (2011). Gold nanoparticles in biology and medicine: Recent advances and prospects. *Acta Naturae, 3*(2), 34–55.

Eliceiri, B. P., Gonzalez, A. M., & Baird, A. (2011). Zebrafish model of the blood-brain barrier: Morphological and permeability studies. *Methods in Molecular Biology (Clifton, NJ), 686*, 371–378. https://doi.org/10.1007/978-1-60761-938-3_18.

Fent, K., Weisbrod, C. J., Wirth-Heller, A., & Pieles, U. (2010). Assessment of uptake and toxicity of fluorescent silica nanoparticles in zebrafish (Danio rerio) early life stages. *Aquatic Toxicology (Amsterdam, Netherlands), 100*(2), 218–228. https://doi.org/10.1016/j.aquatox.2010.02.019.

Fraysse, B., Mons, R., & Garric, J. (2006). Development of a zebrafish 4-day embryo-larval bioassay to assess toxicity of chemicals. *Ecotoxicology and Environmental Safety, 63*(2), 253–267. https://doi.org/10.1016/j.ecoenv.2004.10.015.

Geffroy, B., Ladhar, C., Cambier, S., Treguer-Delapierre, M., Brèthes, D., & Bourdineaud, J.-P. (2012). Impact of dietary gold nanoparticles in zebrafish at very low contamination pressure: The role of size, concentration and exposure time. *Nanotoxicology, 6*(2), 144–160. https://doi.org/10.3109/17435390.2011.562328.

Gerber, A., Bundschuh, M., Klingelhofer, D., & Groneberg, D. A. (2013). Gold nanoparticles: Recent aspects for human toxicology. *Journal of Occupational Medicine and Toxicology (London, England), 8*(1), 32. https://doi.org/10.1186/1745-6673-8-32.

Giordani, S., Bartelmess, J., Frasconi, M., Biondi, I., Cheung, S., Grossi, M., Wu, D., Echegoyen, L., & O'Shea, D. F. (2014). NIR fluorescence labelled carbon nano-onions: Synthesis, analysis and cellular imaging. *Journal of Materials Chemistry B, 2*(42), 7459–7463. https://doi.org/10.1039/C4TB01087F.

Giordani, S., & d'Amora, M. (2018). The utility of zebrafish as a model for screening developmental neurotoxicity. *Frontiers in Neuroscience, 12*, 976.

Greiner, N. R., Phillips, D. S., Johnson, J. D., & Volk, F. (1988). Diamonds in detonation soot. *Nature, 333*, 440. https://doi.org/10.1038/333440a0.

Hashim, Z., Green, M., Chung, P. H., Suhling, K., Protti, A., Phinikaridou, A., Botnar, R., Khanbeigi, R. A., Thanou, M., Dailey, L. A., Commander, N. J., Rowland, C., Scott, J., & Jenner, D. (2014). Gd-containing conjugated polymer nanoparticles: Bimodal nanoparticles for fluorescence and MRI imaging. *Nanoscale, 6*(14), 8376–8386. https://doi.org/10.1039/C4NR01491J.

Heiden, T. C., Dengler, E., Kao, W. J., Heideman, W., & Peterson, R. E. (2007). Developmental toxicity of low generation PAMAM dendrimers in zebrafish. *Toxicology and Applied Pharmacology, 225*(1), 70–79. https://doi.org/10.1016/j.taap.2007.07.009.

Hill, A. J., Teraoka, H., Heideman, W., & Peterson, R. E. (2005). Zebrafish as a model vertebrate for investigating chemical toxicity. *Toxicological Sciences, 86*(1), 6–19. https://doi.org/10.1093/toxsci/kfi110.

Horzmann, K. A., & Freeman, J. L. (2018). Making waves: New developments in toxicology with the zebrafish. *Toxicological Sciences, 163*(1), 5–12. https://doi.org/10.1093/toxsci/kfy044.

Howe, K., Clark, M. D., Torroja, C. F., Torrance, J., et al. (2013). The zebrafish reference genome sequence and its relationship to the human genome. *Nature, 496*(7446), 498–503. https://doi.org/10.1038/nature12111.

Hu, X., Sun, J., & Zhou, Q. (2016). Specific nanotoxicity of graphene oxide during zebrafish embryogenesis AU – Chen, Yuming. *Nanotoxicology, 10*(1), 42–52. https://doi.org/10.3109/17435390.2015.1005032.

Hua, J., Vijver, M. G., Richardson, M. K., Ahmad, F., & Peijnenburg, W. J. (2014). Particle-specific toxic effects of differently shaped zinc oxide nanoparticles to zebrafish embryos (Danio rerio). *Environmental Toxicology and Chemistry, 33*(12), 2859–2868. https://doi.org/10.1002/etc.2758.

Huang, X., El-Sayed, I. H., Qian, W., & El-Sayed, M. A. (2006). Cancer cell imaging and Photothermal therapy in the near-infrared region by using gold Nanorods. *Journal of the American Chemical Society, 128*(6), 2115–2120. https://doi.org/10.1021/ja057254a.

Iijima, S. (1991). Helical microtubules of graphitic carbon. *Nature, 354*, 56. https://doi. org/10.1038/354056a0.

Iijima, S., & Ichihashi, T. (1993). Single-shell carbon nanotubes of 1-nm diameter. *Nature, 363*, 603. https://doi.org/10.1038/363603a0.

Iijima, S., Yudasaka, M., Yamada, R., Bandow, S., Suenaga, K., Kokai, F., & Takahashi, K. (1999). Nano-aggregates of single-walled graphitic carbon nano-horns. *Chemical Physics Letters, 309*(3), 165–170. https://doi.org/10.1016/S0009-2614(99)00642-9.

Jin, S., & Ye, K. (2007). Nanoparticle-mediated drug delivery and gene therapy. *Biotechnology Progress, 23*(1), 32–41. https://doi.org/10.1021/bp060348j.

Johnston, H. J., Verdon, R., Gillies, S., Brown, D. M., Fernandes, T. F., & Henry, T. B. (2018). Adoption of in vitro systems and zebrafish embryos as alternative models for reducing rodent use in assessments of immunological and oxidative stress responses to. *Nanomaterials, 48*(3), 252–271. https://doi.org/10.1080/10408444.2017.1404965.

Kalueff, A. V., Echevarria, D. J., Homechaudhuri, S., Stewart, A. M., Collier, A. D., Kaluyeva, A. A., Li, S., Liu, Y., Chen, P., Wang, J., Yang, L., Mitra, A., Pal, S., Chaudhuri, A., Roy, A., Biswas, M., Roy, D., Podder, A., Poudel, M. K., Katare, D. P., Mani, R. J., Kyzar, E. J., Gaikwad, S., Nguyen, M., & Song, C. (2016). Zebrafish neurobehavioral phenomics for aquatic neuropharmacology and toxicology research. *Aquatic Toxicology, 170*, 297–309. https://doi. org/10.1016/j.aquatox.2015.08.007.

Kalueff, A. V., Echevarria, D. J., & Stewart, A. M. (2014). Gaining translational momentum: More zebrafish models for neuroscience research. *Progress in Neuro-Psychopharmacology & Biological Psychiatry, 55*, 1–6. https://doi.org/10.1016/j.pnpbp.2014.01.022.

Kang, Y.-F., Li, Y.-H., Fang, Y.-W., Xu, Y., Wei, X.-M., & Yin, X.-B. (2015). Carbon quantum dots for zebrafish fluorescence imaging. *Scientific Reports, 5*, 11835. https://doi.org/10.1038/ srep11835. https://www.nature.com/articles/srep11835#supplementary-information.

Khajuria, D. K., Kumar, V. B., Karasik, D., & Gedanken, A. (2017). Fluorescent nanoparticles with tissue-dependent affinity for live zebrafish imaging. *ACS Applied Materials & Interfaces, 9*(22), 18557–18565. https://doi.org/10.1021/acsami.7b04668.

Kim, K.-T., Zaikova, T., Hutchison, J. E., & Tanguay, R. L. (2013). Gold nanoparticles disrupt zebrafish eye development and pigmentation. *Toxicological Sciences, 133*(2), 275–288. https:// doi.org/10.1093/toxsci/kft081.

Kimmel, C. B. (1989). Genetics and early development of zebrafish. *Trends in Genetics, 5*(8), 283–288.

Kimmel, C. B., Ballard, W. W., Kimmel, S. R., Ullmann, B., & Schilling, T. F. (1995). Stages of embryonic development of the zebrafish. *Developmental Dynamics, 203*(3), 253–310. https:// doi.org/10.1002/aja.1002030302.

Kőrösi, L., Rodio, M., Dömötör, D., Kovács, T., Papp, S., Diaspro, A., Intartaglia, R., & Beke, S. (2016). Ultrasmall, ligand-free ag nanoparticles with high antibacterial activity prepared by pulsed laser ablation in liquid. *Journal of Chemistry, 2016*, 4143560.

Krishnaraj, C., Harper, S. L., & Yun, S. I. (2016). In vivo toxicological assessment of biologi- cally synthesized silver nanoparticles in adult zebrafish (Danio rerio). *Journal of Hazardous Materials, 301*, 480–491. https://doi.org/10.1016/j.jhazmat.2015.09.022.

Kroto, H. W., Heath, J. R., O'Brien, S. C., Curl, R. F., & Smalley, R. E. (1985). C60: Buckminsterfullerene. *Nature, 318*, 162. https://doi.org/10.1038/318162a0.

Kuznetsova, G. P., Larina, O. V., Petushkova, N. A., Kisrieva, Y. S., Samenkova, N. F., Trifonova, O. P., Karuzina, I. I., Ipatova, O. M., Zolotaryov, K. V., Romashova, Y. A., & Lisitsa, A. V. (2014). Effects of fullerene C60 on proteomic profile of Danio rerio fish embryos. *Bulletin of Experimental Biology and Medicine, 156*(5), 694–698. https://doi.org/10.1007/ s10517-014-2427-y.

Lee, K. J., Browning, L. M., Nallathamby, P. D., & X-HN, X. (2013). Study of charge-dependent transport and toxicity of peptide-functionalized silver nanoparticles using zebrafish embryos and single nanoparticle Plasmonic spectroscopy. *Chemical Research in Toxicology, 26*(6), 904– 917. https://doi.org/10.1021/tx400087d.

Lettieri, S., Camisasca, A., d'Amora, M., Diaspro, A., Uchida, T., Nakajima, Y., Yanagisawa, K., Maekawa, T., & Giordani, S. (2017a). Far-red fluorescent carbon nano-onions as a biocompatible platform for cellular imaging. *RSC Advances, 7*(72), 45676–45681. https://doi.org/10.1039/C7RA09442F.

Lettieri, S., d'Amora, M., Camisasca, A., Diaspro, A., & Giordani, S. (2017b). Carbon nano-onions as fluorescent on/off modulated nanoprobes for diagnostics. *Beilstein Journal of Nanotechnology, 8*, 1878–1888. https://doi.org/10.3762/bjnano.8.188.

Levi, N., Hantgan, R. R., Lively, M. O., Carroll, D. L., & Prasad, G. L. (2006). C60-fullerenes: Detection of intracellular photoluminescence and lack of cytotoxic effects. *Journal of Nanobiotechnology, 4*, 14. https://doi.org/10.1186/1477-3155-4-14.

Li, Z., Sun, Q., Zhu, Y., Tan, B., Xu, Z. P., & Dou, S. X. (2014). Ultra-small fluorescent inorganic nanoparticles for bioimaging. *Journal of Materials Chemistry B, 2*(19), 2793–2818. https://doi.org/10.1039/C3TB21760D.

Lin, Y. C., Wu, K. T., Lin, Z. R., Perevedentseva, E., Karmenyan, A., Lin, M. D., & Cheng, C. L. (2016). Nanodiamond for biolabelling and toxicity evaluation in the zebrafish embryo in vivo. *Journal of Biophotonics, 9*(8), 827–836. https://doi.org/10.1002/jbio.201500304.

Liu, X. T., Mu, X. Y., Wu, X. L., Meng, L. X., Guan, W. B., Ma, Y. Q., Sun, H., Wang, C. J., & Li, X. F. (2014). Toxicity of multi-walled carbon nanotubes, graphene oxide, and reduced graphene oxide to zebrafish embryos. *Biomedical and Environmental Sciences, 27*(9), 676–683. https://doi.org/10.3967/bes2014.103.

Maji, S. K., Sreejith, S., Joseph, J., Lin, M., He, T., Tong, Y., Sun, H., Yu, S. W.-K., & Zhao, Y. (2014). Upconversion nanoparticles as a contrast agent for photoacoustic imaging in live mice. *Advanced Materials, 26*(32), 5633–5638. https://doi.org/10.1002/adma.201400831.

Manjanatha, M. G., Bishop, M. E., Pearce, M. G., Kulkarni, R., Lyn-Cook, L. E., & Ding, W. (2014). Genotoxicity of doxorubicin in F344 rats by combining the comet assay, flow-cytometric peripheral blood micronucleus test, and pathway-focused gene expression profiling. *Environmental and Molecular Mutagenesis, 55*(1), 24–34. https://doi.org/10.1002/em.21822.

Mochalin, V. N., Shenderova, O., Ho, D., & Gogotsi, Y. (2011). The properties and applications of nanodiamonds. *Nature Nanotechnology, 7*, 11. https://doi.org/10.1038/nnano.2011.209.

Nicholas, N., Marta, D. A., Neeraj, P., Alexander, M. P., Natalya, F., Marco, D. T., Igor, V., Philipp, R., Brant, G., Jessica, M. R., Silvia, G., & Olga, S. (2018). Fluorescent single-digit detonation nanodiamond for biomedical applications. *Methods and Applications in Fluorescence, 6*(3), 035010.

Novoselov, K. S., Geim, A. K., Morozov, S. V., Jiang, D., Zhang, Y., Dubonos, S. V., Grigorieva, I. V., & Firsov, A. A. (2004). Electric field effect in atomically thin carbon films. *Science, 306*(5696), 666–669. https://doi.org/10.1126/science.1102896.

Ong, K. J., Zhao, X., Thistle, M. E., MacCormack, T. J., Clark, R. J., Ma, G., Martinez-Rubi, Y., Simard, B., Loo, J. S. C., Veinot, J. G. C., & Goss, G. G. (2014). Mechanistic insights into the effect of nanoparticles on zebrafish hatch. *Nanotoxicology, 8*(3), 295–304. https://doi.org/10.3109/17435390.2013.778345.

Pham, D.-H., De Roo, B., Nguyen, X.-B., Vervaele, M., Kecskés, A., Ny, A., Copmans, D., Vriens, H., Locquet, J.-P., Hoet, P., & de Witte, P. A. M. (2016). Use of zebrafish larvae as a multi-endpoint platform to characterize the toxicity profile of silica nanoparticles. *Scientific Reports, 6*, 37145–37145. https://doi.org/10.1038/srep37145.

Quinlivan, V. H., & Farber, S. A. (2017). Lipid uptake, metabolism, and transport in the larval zebrafish. *Frontiers in Endocrinology, 8*, 319. https://doi.org/10.3389/fendo.2017.00319.

Rocco, L., Santonastaso, M., Mottola, F., Costagliola, D., Suero, T., Pacifico, S., & Stingo, V. (2015). Genotoxicity assessment of TiO2 nanoparticles in the teleost Danio rerio. *Ecotoxicology and Environmental Safety, 113*, 223–230. https://doi.org/10.1016/j.ecoenv.2014.12.012.

Roy, P., Periasamy, A. P., Lin, C.-Y., Her, G.-M., Chiu, W.-J., Li, C.-L., Shu, C.-L., Huang, C.-C., Liang, C.-T., & Chang, H.-T. (2015). Photoluminescent graphene quantum dots for in vivo imaging of apoptotic cells. *Nanoscale, 7*(6), 2504–2510. https://doi.org/10.1039/C4NR07005D.

Ruoff, R. S., & Lorents, D. C. (1995). Mechanical and thermal properties of carbon nanotubes. *Carbon, 33*(7), 925–930. https://doi.org/10.1016/0008-6223(95)00021-5.

Salzmann, C. G., Llewellyn, S. A., Tobias, G., Ward, M. A. H., Huh, Y., & Green, M. L. H. (2007). The role of carboxylated carbonaceous fragments in the functionalization and spectroscopy of a single-walled carbon-nanotube material. *Advanced Materials, 19*(6), 883–887. https://doi.org/10.1002/adma.200601310.

Samaee, S. M., Rabbani, S., Jovanovic, B., Mohajeri-Tehrani, M. R., & Haghpanah, V. (2015). Efficacy of the hatching event in assessing the embryo toxicity of the nano-sized TiO(2) particles in zebrafish: A comparison between two different classes of hatching-derived variables. *Ecotoxicology and Environmental Safety, 116*, 121–128. https://doi.org/10.1016/j.ecoenv.2015.03.012.

Sangabathuni, S., Murthy, R. V., Chaudhary, P. M., Subramani, B., Toraskar, S., & Kikkeri, R. (2017). Mapping the Glyco-gold nanoparticles of different shapes toxicity, biodistribution and sequestration in adult zebrafish. *Scientific Reports, 7*(1), 4239. https://doi.org/10.1038/s41598-017-03350-3.

Sarmah, S., & Marrs, J. A. (2016). Zebrafish as a vertebrate model system to evaluate effects of environmental toxicants on cardiac development and function. *International Journal of Molecular Sciences, 17*(12), 2123. https://doi.org/10.3390/ijms17122123.

Sedmera, D., Reckova, M., deAlmeida, A., Sedmerova, M., Biermann, M., Volejnik, J., Sarre, A., Raddatz, E., McCarthy, R. A., Gourdie, R. G., & Thompson, R. P. (2003). Functional and morphological evidence for a ventricular conduction system in zebrafish and Xenopus hearts. *American Journal of Physiology Heart and Circulatory Physiology, 284*(4), H1152–H1160. https://doi.org/10.1152/ajpheart.00870.2002.

Staudt, D., & Stainier, D. (2012). Uncovering the molecular and cellular mechanisms of heart development using the zebrafish. *Annual Review of Genetics, 46*, 397–418. https://doi.org/10.1146/annurev-genet-110711-155646.

Thisse, C., & Zon, L. I. (2002). Organogenesis--heart and blood formation from the zebrafish point of view. *Science, 295*(5554), 457–462. https://doi.org/10.1126/science.1063654.

Truong, L., Reif, D. M., St Mary, L., Geier, M. C., Truong, H. D., & Tanguay, R. L. (2014). Multidimensional in vivo hazard assessment using zebrafish. *Toxicological Sciences, 137*(1), 212–233. https://doi.org/10.1093/toxsci/kft235.

Truong, L., Saili, K. S., Miller, J. M., Hutchison, J. E., & Tanguay, R. L. (2012). Persistent adult zebrafish behavioral deficits results from acute embryonic exposure to gold nanoparticles. *Comparative Biochemistry and Physiology Toxicology & Pharmacology, 155*(2), 269–274. https://doi.org/10.1016/j.cbpc.2011.09.006.

Truong, L., Tilton, S. C., Zaikova, T., Richman, E., Waters, K. M., Hutchison, J. E., & Tanguay, R. L. (2013). Surface functionalities of gold nanoparticles impact embryonic gene expression responses. *Nanotoxicology, 7*(2), 192–201. https://doi.org/10.3109/17435390.2011.648225.

Ugarte, D. (1992). Curling and closure of graphitic networks under electron-beam irradiation. *Nature, 359*(6397), 707–709. https://doi.org/10.1038/359707a0.

Usenko, C. Y., Harper, S. L., & Tanguay, R. L. (2007). In vivo evaluation of carbon fullerene toxicity using embryonic zebrafish. *Carbon N Y, 45*(9), 1891–1898. https://doi.org/10.1016/j.carbon.2007.04.021.

Usenko, C. Y., Harper, S. L., & Tanguay, R. L. (2008). Fullerene C60 exposure elicits an oxidative stress response in embryonic zebrafish. *Toxicology and Applied Pharmacology, 229*(1), 44–55. https://doi.org/10.1016/j.taap.2007.12.030.

Villacis, R. A. R., Filho, J. S., Pina, B., Azevedo, R. B., Pic-Taylor, A., Mazzeu, J. F., & Grisolia, C. K. (2017). Integrated assessment of toxic effects of maghemite (gamma-Fe2O3) nanoparticles in zebrafish. *Aquatic Toxicology (Amsterdam Netherlands), 191*, 219–225. https://doi.org/10.1016/j.aquatox.2017.08.004.

Vliegenthart, A. D., Tucker, C. S., Del Pozo, J., & Dear, J. W. (2014). Zebrafish as model organisms for studying drug-induced liver injury. *British Journal of Clinical Pharmacology, 78*(6), 1217–1227. https://doi.org/10.1111/bcp.12408.

Wang, Z. G., Zhou, R., Jiang, D., Song, J. E., Xu, Q., Si, J., Chen, Y. P., Zhou, X., Gan, L., Li, J. Z., Zhang, H., & Liu, B. (2015). Toxicity of graphene quantum dots in zebrafish embryo. *Biomedical and Environmental Sciences, 28*(5), 341–351. https://doi.org/10.3967/bes2015.048.

Wiley, D. S., Redfield, S. E., & Zon, L. I. (2017). Chemical screening in zebrafish for novel biological and therapeutic discovery. *Methods in Cell Biology, 138*, 651–679. https://doi.org/10.1016/bs.mcb.2016.10.004.

Wullimann, M. F. (2009). Secondary neurogenesis and telencephalic organization in zebrafish and mice: A brief review. *Integrative Zoology, 4*(1), 123–133. https://doi.org/10.1111/j.1749-4877.2008.00140.x.

Xu, J., Yudasaka, M., Kouraba, S., Sekido, M., Yamamoto, Y., & Iijima, S. (2008). Single wall carbon nanohorn as a drug carrier for controlled release. *Chemical Physics Letters, 461*(4), 189–192. https://doi.org/10.1016/j.cplett.2008.06.077.

Xu, X., Ray, R., Gu, Y., Ploehn, H. J., Gearheart, L., Raker, K., & Scrivens, W. A. (2004). Electrophoretic analysis and purification of fluorescent single-walled carbon nanotube fragments. *Journal of the American Chemical Society, 126*(40), 12736–12737. https://doi.org/10.1021/ja040082h.

Yang, M., Flavin, K., Kopf, I., Radics, G., Hearnden, C. H., McManus, G. J., Moran, B., Villalta-Cerdas, A., Echegoyen, L. A., Giordani, S., & Lavelle, E. C. (2013). Functionalization of carbon nanoparticles modulates inflammatory cell recruitment and NLRP3 inflammasome activation. *Small (Weinheim an der Bergstrasse, Germany), 9*(24), 4194–4206. https://doi.org/10.1002/smll.201300481.

Zhang, X., Li, C., & Gong, Z. (2014). Development of a convenient in vivo Hepatotoxin assay using a transgenic zebrafish line with liver-specific DsRed expression. *PLoS One, 9*(3), e91874. https://doi.org/10.1371/journal.pone.0091874.

Zheng, X. T., Ananthanarayanan, A., Luo, K. Q., & Chen, P. (2015). Glowing graphene quantum dots and carbon dots: Properties, syntheses, and biological applications. *Small (Weinheim an der Bergstrasse, Germany), 11*(14), 1620–1636. https://doi.org/10.1002/smll.201402648.

Chapter 3
Caenorhabditis elegans: A Unique Animal Model to Study Soil–Nanoparticles–Organism Interactions

Ashley Cox and Nilesh Sharma

Contents

1 Introduction

Caenorhabditis elegans (*C. elegans*) is a free-living nematode that is bacterivorous. Nematodes are abundant and ubiquitous multicellular animals that inhabit the terrestrial, freshwater, and marine environments. They have been successfully used as

A. Cox
Department of Biomedical Sciences, Marshall University School of Medicine, Huntington, WV, USA

N. Sharma (✉)
Department of Biology, Western Kentucky University, Bowling Green, KY, USA
e-mail: Nilesh.sharma@wku.edu

© Springer Nature Switzerland AG 2021
N. Sharma, S. Sahi (eds.), *Nanomaterial Biointeractions at the Cellular, Organismal and System Levels*, Nanotechnology in the Life Sciences, https://doi.org/10.1007/978-3-030-65792-5_3

environmental indicators because of their abundance and unique roles in ecosystems. Because of its short life cycle and ease in maintaining in lab cultures, it has become a favorite model of nanotoxicological investigations, specifically for metal oxide nanomaterials. The importance of *C. elegans* in toxicology studies stems from its small size, 1 mm-long (convenient to maintain on agar plates), life cycle of 72 h from larva to adult developmental stages (Fig. 3.1), a fully sequenced genome, and easy availability of a large number of mutants advantageous for genetic studies (Hu et al. 2018b). Changes in a worm's phenotype such as body length, moving behavior (locomotion), or reproduction (changes in population) are critical endpoints in toxicology. Recently, the use of various biological models in biosafety assessment of nanomaterials has been increasing, especially ecological risk assessment, and the nematode *C. elegans* is one of them. The evaluation of nano-bio-interactions in *C. elegans* can provide the fate and toxicity of NPs in a multicellular organism. Many studies have demonstrated that toxicity of NPs to *C. elegans* is highly dependent on the uptake and exposure time, both of which are related to surface chemistry and particle size (Eom and Choi 2019). These studies suggest that understanding uptake mechanism is particularly important to explain toxicity and risk of nanomaterials.

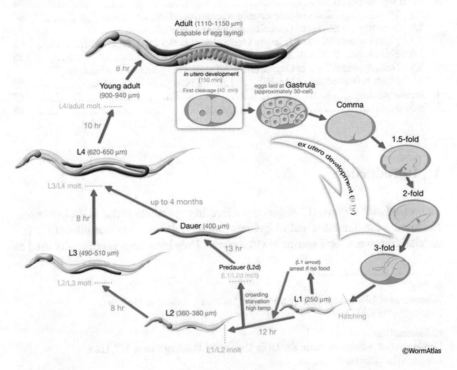

Fig. 3.1 Life cycle of *C. elegans* at 22 degree C: Redrawn from *WormAtlas*. https://doi.org/10.3908/wormatlas.1.1

Following the availability of whole-genome microarrays, *C. elegans* has served as a model organism for studies in neurobiology, developmental biology, genetics, and the environmental impact of xenobiotics. Approximately 20.000 genes encode for the nematode's proteins and the majority of human genes (60–80%) including the genes involved in human disease have a counterpart/homolog in the worm. An adult hermaphrodite worm consists of 959 cells, which include 302 neurons. Neural circuits have been mapped that may consist of just a few muscle and neural cells. However, these simple neural circuits drive complex neuromuscular behaviors; for example, two hermaphrodite-specific motor neurons (HSNs) are known to control egg laying (Scharf et al. 2015). This worm is transparent which enables imaging of labeled compounds including reporter proteins in single neural cells of living worms. The ease of worm cultivation in solid or liquid culture additionally makes *C. elegans* amenable to biochemical methods and medium to high-throughput screening.

Furthermore, changes in worm locomotion behavioral assays can be related to altered motor neuronal function, which is important for understanding neurological diseases, such as amyotrophic lateral sclerosis (ALS) and Parkinson's disease. The increasing use of nanotechnology in our daily life can have many unintended effects affecting human health, environment, and ecosystems.

2 Growth, Reproduction, and Mortality Effects

2.1 Effects of Silver Nanoparticles (AgNPs)

Silver nanoparticles (AgNPs) are the most widespread metallic nanomaterials applied in numerous consumer products with the production up to 500 t/year worldwide (Nanotech Project 2016). Widespread applications of Ag nanomaterials include textiles, plastics, and medical devices principally for their antimicrobial properties (Starnes et al. 2015). During manufacture, use, and product disposal, these nanomaterials can be released to wastewater streams, where they have been shown to partition largely to sewage sludge during the wastewater treatment process. Therefore, land application of biosolids is the primary route of Ag-MNP introduction to terrestrial ecosystems heightening ecological concerns (Starnes et al. 2015). Further, the behavior of silver nanoparticles (AgNPs) is influenced by environmental factors that affect their bioaccumulation and toxicity potentials. Understanding the relationships between the physicochemical properties of AgNPs and their toxicity is critical for environmental and health risk analysis. Table 3.1 summarizes the studies on AgNPs.

As sulfidation is a major transformation product for manufactured silver nanoparticles (Ag-MNPs) in the wastewater treatment process, Starnes et al. (2015) compared effects of pristine AgNPs versus sulfidized AgNPs (sAgNPs). They convincingly demonstrated that sulfidation decreases solubility of AgNPs and

Table 3.1 Effects of different forms of silver nanoparticles (AgNPs) on *C. elegans* life cycle

Year	Authors	NPs type	NP sizes used	Hydrodynamic diameter	Concentrations used	Length of treatment	Significant findings
2015	Starnes et al.	Pristine-AgNP; sulfidized-sAgNP	AgNP 58 nm sAgNP 64 nm	AgNP 79 nm sAgNP 88 nm	5–10,000 µg/L	24 for growth and mortality assay 48 h for reproduction	Sulfidation decreases solubility of AgNPs and reduce bioavailability of intact sAgNPs, thus reducing the toxicity of sAgNPs in comparison to pristine AgNP. Also shows reproduction was the most sensitive endpoint tested for both Ag-MNPs and sAg-MNPs
2016	Starnes et al.	PVP-coated AgNP (pristine) Sulfidized AgNP		79.6 ± 0.5 nm a 88.5 ± 0.5	350 ug Ag/L for PVP-AgNP 1500 ug Ag/L for s-AgNP	48 h	Distinct toxicity mechanism between pristine and sulfidized AgNP: Processes most affected by Ag-MNP relate to metabolism, while those affected by sAg-MNP relate to molting and the cuticle, and the most impacted processes for AgNO3 exposed nematodes was stress related
2016	Schultz et al.	PVP-coated AgNP Sulfidized AgNP	$58.3 + 12.9$ $64.5 + 19.4$	79–105 nm 213–298 nm	0–14.4 mg Ag/L 0–240 mg Ag/L	96 h per generation, 10 generations total	Ag exposure significantly reduced reproduction for Ag-PVP concentrations above 9.6 mg Ag/L and s-AgNP concentrations above 15 mg Ag/L Sensitivity to Ag was retained over generations, suggesting epigenetic mechanisms Increased time to egg laying as generations increased. No significant changes to life span, but stronger effects on life span in earlier generations were lost in later generations

Year	Authors	NPs type	NP sizes used	Hydrodynamic diameter	Concentrations used	Length of treatment	Significant findings
2017	Yang et al.	Citrate coated; PVP-coated AgNP Rhodamine-coated AgNP	26 and 83 nm 23 and 70 nm Not given	Not given	26 nm: 0.01–1 mg/L 23, 70, and 83 nm: 0.01–20 mg/L Uptake: 5 mg/L	Uptake tests: 24–48 h	Citrate-coated AgNPs posed higher toxicity than PVP-coated AgNPs in their corresponding range of sizes. IC50 are in the order of 26 nm citrate-coated <23 nm PVP-coated <83 nm citrate-coated <70 nm PVP-coated AgNPs, suggesting that both sizes and coatings of AgNPs have significant influence on toxicity
2018	Yang et al.	PVP AgNPs (25 and 75 nm)	25.73 ± 4.763 77.43 ± 6.433	50.21 ± 2.763 139.23 ± 3.564 nm	0–5 mg/L (*E. coli* exposed to AgNPs, then plated for worm uptake)	12 h (*E. coli* exposure) 24 h (worm exposure to *E. coli*)	Ionic strength significantly enhanced the reproductive toxicity (germ cell corpses, brood size, and life span) and neurotoxicity Higher ionic strength increased bioaccumulation of AgNPs in *E. coli* and the resulting Ag body burden in worms increased ROS significantly increased in worms after AgNP-*E. coli* exposure
2019	Moon et al.	Ag NP (<100 nm) Ag nanowires (41–42 nm) Ag nano plates (36 nm)	10 μm-nanowires 20 μm nanowires and nano plates		1–5 mg Ag/kg soil	L1 larvae exposed for 40 h	The shape of silver nanomaterials significantly affects their toxicity. Ag nanoparticles and nanoplates were found to inhibit the growth and reproduction of *C. elegans*, whereas Ag nanowires had a negligible effect

reduces bioavailability of intact sAgNPs. Both decreased solubility and decreased uptake of sAgNPs contribute to reductions in toxicity. These results also indicated that reproduction was the most sensitive endpoint tested for both Ag-MNPs and sAg-MNPs.

Toxicity of pristine Ag-MNPs has been shown to be driven by both ion- or particle-specific effects which are greater than in equivalent ionic (metal salt) concentrations. The above observation of greatly reduced toxicity of sulfidized AgNPs has thus significant environmental implications and suggest that hazards of Ag-MNP exposure are likely greater in applications where Ag-MNPs are introduced directly to soil rather than through biosolids application.

In another related study, Starnes et al. (2016) examined toxicogenomic responses of pristine Ag-MNP, sulfidized Ag-MNP (sAg-MNP), and AgNO3 on *C. elegans* reproduction. Transcriptomic profiling revealed only 11% of differentially expressed genes were common among the three treatments. This study further identified expression of four genes (numr-1, rol-8, col-158, and grl-20) using qRT-PCR. Gene ontology enrichment analysis also revealed that Ag-MNP and sAg-MNP had distinct toxicity mechanisms and did not share any of the biological processes. The processes most affected by Ag-MNP relate to metabolism, while those processes most affected by sAg-MNP relate to molting and the cuticle. AgNO3-related toxicity was mostly stress related. Based on these findings, investigators finally suggested that cuticle damage may be the primary mechanism of toxicity of sAg-MNP to *C. elegans*. While the decreased solubility of sAg-MNP relative to Ag-MNP is correlated with reduced toxicity, release of free ions is not the primary mechanism of reproductive toxicity for fully sulfidized sAg-MNP.

Comparative effects of pristine (PVP-coated) AgENPs and sulfidized AgENPs were examined in a multigenerational study by Schultz et al. (2016). They examined how exposure over ten generations affects the sensitivity of the nematode *C. elegans* to pristine, sulfidized Ag ENPs, and AgNO3. In line with the findings of Starnes et al. (2015), toxicity of the different silver forms decreased in the order AgNO3, Ag ENPs, and Ag2S ENPs. In early generations, life span was shown to shorten, but these effects were not seen in later generations. Reduced body size was also observed for AgNO3 and Ag-PVP exposed nematodes after ten generations. Output of parent (P) generation decreased in a concentration-dependent manner with increasing concentration. Continuous multigenerational exposure to Ag (ionic and particulate) gradually increased time to first egg laying. By the F9 generation, egg-laying period had been extended from 96 to 120 h in silver exposed populations. This was not observed in reference population or recovery populations.

The most important observation made by these investigators are that continuous exposure to Ag ENPs and AgNO3 caused pronounced sensitization (approximately tenfold) in the F2 generation, lasting until F10. These sensitization effects were manifest less for Ag2S ENP exposures, indicating different toxicity mechanisms. The sensitivity to Ag ENPs and AgNO3 resulting from the initial multigenerational exposure persisted in the recovery populations as well. These investigators suggest that the transgenerational transfer of sensitivity may be controlled through the epigenome regulation.

Yang et al. (2017) developed *C. elegans*-based biomarkers as critical risk indicators on assessing the impact of silver nanoparticles on soil ecosystems. This approach uses methods for rapidly screening and assessing the potential toxicity risk of AgNPs in general and sludge-treated soils in particular. Investigators assessed the risk for soils exceeding a threshold of *C. elegans* locomotion related neurotoxicity based on the statistical models. Results indicate that locomotion inhibition of *C. elegans* was dependent on surface properties, diameter, and exposure time of AgNPs. Further, this modeling study shows that the soil contamination risks by sewage sludge-released AgNPs are significantly low in the countries of Europe, the USA, and Switzerland. However, large production and widespread applications of AgNPs are highly likely to pose long-term ecotoxicity risk on sludge-treated or untreated soils, particularly for 26 nm citrate-coated AgNPs. The approach of integrating probabilistic risk model and *C. elegans*-based ecological indicator provides an effective tool to rapidly screen and assess the impacts of STPs-released AgNPs on soil environment (Fig. 3.2). Yang's (2017) model based on predictive risk threshold and risk characterization protocol, as displayed in Fig. 3.2, can be extended to environmental studies related to other nanomaterial releases. Investigators suggest that *C. elegans* can be considered as a proxy for estimating soil risk metrics, which can help develop methods of management for mitigating the metal NPs-induced toxicity on terrestrial ecosystems.

In another study, Yang et al. (2018) investigated the ecotoxicity of AgNPs by using three different exposure media (deionized water as control, EPA water as a low ionic strength exposure medium, and KM as a high ionic strength exposure medium) to pretreat AgNPs. *E. coli* was then exposed to these transformed AgNPs before they were fed to *C. elegans*. Results in this investigation indicated that ionic strength significantly enhanced the reproductive toxicity (germ cell corpses, brood size, and life span) of AgNPs in *C. elegans*. Authors suggest that the ionic strength-treated AgNPs played a major role in the bioaccumulation of nanoparticles and ROS production leading to the increased toxicity. Finally, this study confirms that potential ecotoxicity of AgNPs is associated with the environmental factors.

An excellent comparative analysis of the shape-dependent toxicity of Ag nanomaterials (AgNMs) in *C. elegans* was conducted by Moon et al. (2019). In this study, AgNMs of different shapes and sizes (silver nanoparticles, AgNPs; 10 µm silver nanowires, 10-AgNWs; 20 µm silver nanowires, 20-AgNWs; silver nanoplates, AgPLs) were added to natural soil and their effects on the growth and reproduction of elegans were measured. Ag nanoparticles and nanoplates were found to inhibit the growth and reproduction of *C. elegans*, whereas Ag nanowires had a negligible effect. Among these nanomaterials, AgNPs were found to be the most toxic. This study clearly confirms that the shape of AgNPs plays a significant role in their toxicity level.

Notwithstanding detected concentrations of AgNMs were higher in the above study than typical environmental concentrations which are in the range of 1.2 ng to 0.1 mg/kg soil per year, results of this study suggest that AgNMs concentrations in a range of 5 mg/kg in soil may inhibit growth or induce reproductive defects in nematodes. Several factors, such as temperature, agglomeration, surface charge,

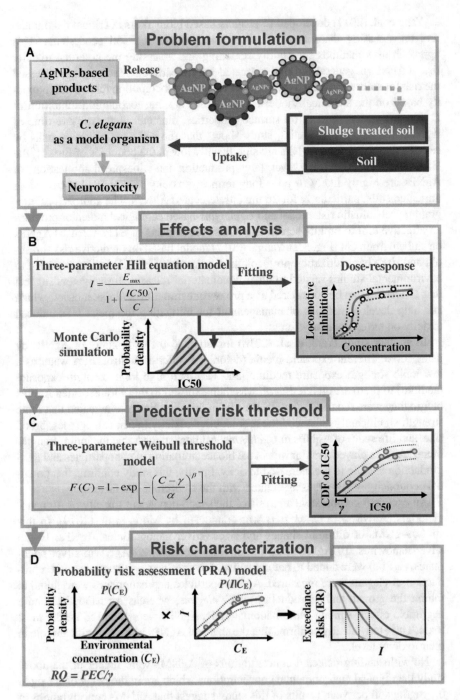

Fig. 3.2 Proposed *C. elegans*-based probabilistic risk assessment framework (reproduced from Yang et al. 2017)

and coating agent, are known to affect the toxicity of AgNMs, but the role of nano-material shapes in influencing toxic fate is new. Furthermore, soil-based nano-Ag toxicity investigations are underrepresented compared to water-based AgNM toxicity studies and need further explorations.

2.2 Effects of Gold Nanoparticles (AuNPs)

The use of gold nanomaterials has rapidly increased in various industrial sectors and commercial products including food, facial cosmetic products, as imaging contrast agents, semiconductors, and for water-hygiene management or water remediation. The recent applications of gold nanomaterials particularly in biomedical fields such as therapeutics, diagnostics, and biosensors have added greatly to its potential for release into the environment (Gonzalez-Moragas et al. 2017). The predicted concentrations of AuNPs in water and soil ecosystems in the United Kingdom are 0.14 and 5.9 parts per billion, respectively (Moon et al. 2017). Hence, there is a need to develop new platforms and approaches to evaluate AuNPs-containing products, in particular to expedite the early stages of safe nanoparticle development in the cosmetic, food, and pharmaceutical industries. Having several unique physical and chemical properties, gold (Au) is one of the most studied metals for the synthesis of nanoparticles. Table 3.2 summarizes various studies focused on AuNPs interactions.

The multigenerational exposure of AuNPs from F0 to F4 was first examined by Kim et al. (2013). This investigation used food exposure method where *E. coli* populations are treated with various doses of nanoparticles before their transfer to the nematode growth medium. No significant changes on survival rate in all generations were recorded in this study. However, reproduction rate was significantly affected in F2, and then gradually recovered in the subsequent generations. Reproductive system abnormalities exhibited a close relationship with low reproduction rates. Abnormalities increased as reproduction rate decreased in F2. Frequency of these abnormalities decreased in F3 and F4 with rising rates of reproduction. Reproductive abnormalities observed included bag of worms (BOW), abnormal uterus, abnormal vulva, and no eggs. Authors suggests the germ-line transfer of toxicity as the cause of transgenerational effects in abnormalities as well as lower reproduction rates.

Moon et al. (2017) also investigated multigenerational effects of AuNPs on *C. elegans* but in two different settings. The design of this experiment included continuous or intermittent food exposure (*E. coli* populations were treated with AuNPs before their transfer to nematodes) methods. Exposure was applied at the rate of $0–10^{10}$ particles/mL for 48 h. This study also recorded negligible effects of AuNP on survival when compared with controls. After continuous exposure, F2-F4 generations exhibited significant reductions in populations accompanied by reproductive abnormalities. Intermittent exposures, given at the P0 and F3 only, had more pronounced effects on F3 worms, possibly due to damage during convalescence period up through F2. Authors suggested that intermittent exposures had greater

Table 3.2 Effects of different forms of gold nanoparticles (AuNPs) on *C. elegans* life cycle

Year	Authors	NPs type	NP sizes used	Hydrodynamic diameter	Concentrations used	Length of treatment	Significant findings
2013	Kim et al.	Au NPs	10 nm		AuNPs 1% v/v	Food exposure of 48 h	No effect on survival from F0 to F4 generations, but transgenerational effects were observed on progeny population because of germ line transfer of toxicity
2017	Gonzalez-Moragas et al.	Citrate-capped Au	11 nm 150 nm	16 ± 3 nm 221 ± 49 nm	100 ug/mL	Feeding experiments—24 h Survival and brood size experiments—24 h	NPs were not able to cross intestinal and dermal barriers. 11 nm more toxic—Significant decrease in survival and brood size (reproductive output); larger NPs did not show toxicity for survival or brood size
2017	Moon et al.	Au nanocolloid	8.5–12.0 nm	In distilled water: 0–300 nm In media: 350–950 nm	0, 5, 25, and 50 × 10^10 particles/ mL	48 h	Continuous exposure showed reproductive abnormalities in F2-F4 Intermittent exposure in p0 and F3 reproductive system abnormalities and inhibited reproduction rates were observed in F2 and F3. Although shorter in exposure time, this group experienced greater effects Survival rates unaffected in both intermittent and continuous exposure

Year	Authors	NPs type	NP sizes used	Hydrodynamic diameter	Concentrations used	Length of treatment	Significant findings
2018	Hu et al.	AuNPs with and without 11-MUA coat; Au-to-MUA ratios of 0.5, 1, 3	AuNP: 6.45 nm 0.5 ratio: 1.83 nm 1 ratio: 1.26 nm 3 ratio: 0.80 nm	Not given	Maintenance and exposure: Uptake: 100 µL of 0.1 mg/mL Endpoint toxicity, body length, locomotion, brood size: Neuron primary culture: 100 µL of 0.1 mg/mL Microarray: 0.1 mg/mL bare NPs or ratios of MUA-NPs	72 h 12 h 72 h 3 day	Decreased locomotion; shorter neuron axon growth Gene expression changes relating to cell defense, morphogenesis, neuron function, and metal detoxification/homeostasis MUA-to-Au ratios of 0.5, 1 and 3 reduced the worm population, body size, thrashing, and brood size. Bare NPs significantly reduced worm population, body length, and brood size (not locomotion, though) Bare NPs and MUA-AU NPs significantly reduced axonal length; increasing the MUA-to-Au ratios increasingly reduced neuronal survival
2019	Gallud et al.	Ammonium-modified AuNPs	20 nm	Au-20-COOH—9–28 nm AU-20-NR3+ 10–54 nm	0–500 µg/mL	24 h	Lethality observed when using Au-20-NR3+ NPs but not with AU-20-COOH NPs. Key necrosis protease *clp1* mediated.

effects on reproduction and abnormalities than continuous exposure. Findings in this study point to the occurrence of multigenerational effects following different exposure patterns, exposure levels, and recovery periods.

Effects of 11 nm and 150 nm citrate-capped AuNPs on *C. elegans* survival, reproduction performance, and biodistribution were elegantly studied by Gonzalez-Moragas et al. (2017). Throughout 0–500 µg/mL treatments, researchers did not observe crossing of AuNPs through intestinal and dermal barriers of the nematode, which marks a departure from other metal oxide treatments in *C. elegans* and other animal models—indicating an internalization of nanoparticles by the intestinal cells. AuNPs at 100 µg/mL significantly reduced survival rate and brood size with respect to controls in 11 nm AuNP treated worms. AuNPs (11 nm) exhibited higher toxicity. In case of treatment with larger NPs, brood size was also not significantly different from control. By employing a combination of microscopy techniques, chemical analysis, absorbance microspectroscopy, this study revealed the biodistribution of 11-nm and 150-nm citrate-capped AuNPs, which were largely accumulated in the intestinal lumen. This study confirmed that the nanoparticles were not able to cross the intestinal and dermal barriers, based on the absence of endocytosis and adsorption, respectively. It is important to mention that transcription of selected markers of endocytosis and intestinal barrier integrity (elt-2, eps-8, act-5, chc-1, and dyn-1) were not altered in worms exposed to AuNPs.

Size-tunable gold nanoparticles coated with 11-mercaptoundecanoic acid (MUA-Au NPs) were used in this study to investigate the toxicogenomic responses in *C. elegans* (Hu et al. 2018b). By controlling the MUA-to-Au ratio, they varied the size of Au NPs up to <1 nm, the size which is largely used in bioimaging. This investigation demonstrated the adverse effects of bare Au and MUA-Au NPs on the population, body length, and locomotion of worms as well as on axonal neuron growth. MUA played a critical role in tuning the size of Au NPs, leading to differentially observed toxic effects. The internalization and absorption of bare Au and MUA-Au NPs into worm's tissues and body cavities were confirmed by X-ray and confocal microscopy. Changes in the gene expression of some critical genes involved in cellular defense, body morphogenesis, embryonic neuron development, and metal detoxification were recorded (Hu et al. 2018b). These gene expressions accurately reflected changes in the observed worm phenotypes after AU NP exposure.

To examine the role of surface chemistry for the toxicity of Au-NPs, Gallud et al. (2019) tested two groups of gold nanoparticles (20 nm) functionalized with different surface chemical moieties: alkyl ammonium bromide (Au-NR3+ NPs) or alkyl sodium carboxylate (Au-COOH NPs) capped Au-NPs in *C. elegans*. Toxicity was evaluated by studying the survival rate in young adult worms (24 h post L4 larvae stage) exposed to Au-COOH NPs and Au-NR3+ NPs in the concentration range 0–500 µg/mL. Results show that the Au-20-NR3+ NPs induced a significant, dose-dependent decrease in the survival of worms, compared to worms treated with the same concentrations of Au-20-COOH NPs. In particular, 500 µg/mL of Au-NR3+ resulted in more than 75% death of young adults, while exposure to the same concentration of Au-COOH NPs did not cause any lethality. This investigation assessed further if the Au-NPs caused lethality in *C. elegans* through the apoptotic, necrotic, or autophagic cell death pathway. Toward this end, mutant worms defective in the key proteases involved in apoptosis (*ced-3*) and necrosis (*clp-1*), or the autophagy gene (*lgg-1*), were exposed to 500 µg/mL of Au-NR3+ NPs or Au-COOH NPs. RNA sequencing combined with proteomics and functional assays confirmed that the ammonium-modified Au-NPs elicited mitochondrial dysfunction (Gallud et al. 2019). The loss of *clp-1* reduced lethality induced by Au-NR3+ NPs. These results confirm that surface functionalization plays a role for the toxicity of Au-NPs.

2.3 Effects of Titanium Oxide Nanoparticles (TiO₂ NPs)

Titanium nanoparticles (TiO$_2$ NPs: E171) have rapidly gained applications in a range of industrial and consumer products including food, confectionaries, pastries, toothpastes, sunscreen, cosmetics, paints, and medicines with the assumption that they are biocompatible and safe to human health (Baranowska-Wójcik et al. 2020). Multiple sources of nanoscale TiO$_2$ can thus result in human exposure through water, air, and soil sediments. Weir et al. (2012) have shown the simulated exposure to TiO$_2$ for the US population, with an average of 1–2 mg TiO$_2$/kg bodyweight/day for children under the age of 10 years and approximately 0.2–0.7 mg TiO$_2$/kg bodyweight/day for the other consumer age groups. Industrial applications add enormously to the list of on-the-shelve grocery products. Table 3.3 presents an overview of studies on TiO$_2$ NPs.

In a toxicogenomic study, Rocheleau et al. (2015) used anatase and rutile nanoparticles (TiO$_2$ NPs) to compare with their bulk counterpart TiO$_2$ particles. These results indicated that selected nano-TiO2 particles had no significant effect on *C. elegans* survival at concentrations up to 500 mg/L. In contrast, reproduction was affected at much lower concentrations (at concentrations of 5, 10, and 20 mg/L). Nano-rutile NAM30 was significantly more toxic than bulk rutile at all three concentrations tested. Nano-anatase particles were less toxic than bulk anatase particles. Although little toxicity differences were seen between nano-anatase and nano-rutile, bulk anatase had a greater toxic effect than bulk rutile. Researchers also observed that anatase agglomerate sizes were inversely proportional to its nominal size, whereas rutile agglomerate sizes were generally proportional to their nominal size. This interesting observation thus indicates that agglomerate size can be a better predictor of toxicity than particle size (Rocheleau et al. 2015).

Investigation by Rocheleau et al. (2015) is probably the first study comparing the effects of nano-TiO$_2$ with bulk-TiO$_2$ particles using a whole-genome *C. elegans* microarray. Microarray analysis in this study indicates that each type of nanoparticle triggers a different spectrum of effects in the worm; for example, anatase had a greater effect on metabolic pathways than rutile, whereas rutile exhibited a greater effect on developmental processes than anatase. Whole-genome microarray analysis by these researchers indicated further that the regulation of glutathione-S-transferase *gst-3*, cytochrome P450 *cypp33-c11*, stress resistance regulator *scl-1*, oxidoreductase *wah-1*, and embryonic development *pod-2* genes was significantly affected by nano-sized and bulk-TiO2 particles.

Table 3.3 Effects of different forms of titanium dioxide nanoparticles (TiO$_2$ NPs) on *C. elegans* life cycle

Year	Authors	NPs type	NP sizes used	Hydrodynamic diameter	Concentrations Used	Length of treatment	Significant findings
2014	Rocheleau et al.	Anatase TiO$_2$ NP, Rutile TiO$_2$ NP TiO$_2$ bulk	Anatase 6–16 nm Rutile 69–164 nm	Anatase 881–928 nm Rutile 729–1317 nm	5–500 mg/L	Adult worms exposed for 24–72 h	Nano-rutile NAM30 was significantly more toxic than bulk rutile at all three concentrations tested. Anatase had a greater effect on metabolic pathways than rutile, whereas rutile exhibited a greater effect on developmental processes than anatase.
2015	Ratnasekhar et al.	TiO$_2$ NP TiO$_2$ bulk	11 nm 124 nm	239 nm 451 nm	7.7 and 38.5 µg/mL	6–24 h	Progeny population significantly decreased in worms exposed to even 7.7 mg/mL of TiO2 NPs. Metabolomics analysis suggest disturbances mainly occurred in TCA cycle metabolism, glyoxalate and tricarboxylate metabolism; depletion of arachidonic acid as an important biomarker for oxidative stress
2018a	Hu et al.	Anatase TiO$_2$ NPs; Rutile TiO$_2$	32 nm 30 nm	206 nm 291 nm	100–500 µg/mL	42–72 h	Rutile TiO$_2$ NPs negatively affect both the body size and worm population was more toxic than anatase form; exposure adversely affects these neurons' growth. DNA microarray assays reflect altered gene expression patterns
2018	Wang et al.	TiO$_2$ NPs	5 and 15 nm	230–259 for 5 nm nano-Ti; 715–903 for 15 nm nano-Ti	Life span and germ apoptosis: 2 and 4 µg/mL TiO$_2$ NPs +10 µM heavy metals	Life span: Continuous until death Apoptosis: 12 h	TiO$_2$ NPs significantly increased and enhanced reproductive and developmental toxicity of heavy metals (cadmium, arsenate, nickel)—Induced apoptosis in meiotic phases of gonads. Size of 5 nm had greatest effects

Year	Authors	NPs type	NP sizes used	Hydrodynamic diameter	Concentrations Used	Length of treatment	Significant findings
2019	Ma et al.	Food-grade TiO_2 Bulk TiO2 TiO2 NP	53–308 nm 64–259 nm 11–52 nm	Not given	1–10 mg/L	Photocatalytic activity: 24 h Reproduction: 4–5 days Uptake: 3 h	All three TiO_2 (1–10 mg/L) induced concentration-dependent effects on reproduction, with a reduction in brood size; reproductive toxicity was independent of particle size, all causing a reduction of worm life span, accompanied by an increased frequency of age-associated vulval integrity defects, TiO_2 NPs were most toxic for these endpoints. Photocatalytic activity tests showed TiO_2 NPs as most toxic, but only when worms were exposed for 3 h to UV light

Ratnasekhar et al. (2015) used an excellent metabolomics approach for the evaluation of nanotoxicity caused by nanotitania (<25 nm) at sublethal 7.7 and 38.5 µg/mL concentrations. TiO_2 NP exposure significantly affected the metabolome profile of *C. elegans* (Fig. 3.3). This study clearly demonstrated that most of the significant perturbations occurred in organic acids and amino acids. Differential marker metabolites identified from the metabolomics analysis suggest that the disturbances mainly occurred in TCA cycle metabolism, glyoxalate and tricarboxylate metabolism, glutamine and glutamate metabolism, and inositol phosphate metabolism. A correlation between metabolite changes and functional characteristic such as reproduction was shown in this study; the progeny population significantly decreased in worms exposed to even 7.7 mg/mL concentration of TiO2 NPs. Further, the investigators identified depletion of arachidonic acid as an important biomarker for oxidative stress in nanotitania exposure. These results indicate that metabolomics could be used as a reliable and sensitive tool for health risk assessment of nanomaterials (Ratnasekhar et al. 2015).

Hu et al. (2018a) tested the higher concentrations of nano-Ti with the rationale for understanding the implication of chronic release or accidental corrosion of nanoparticles from the human bone/body Ti-implants. In this investigation, anatase TiO_2 had no effect on body length of the worm; however, there was a reduction in offspring population by 50% after 72 h at 500 µg/L concentration. Rutile TiO_2 NPs severely reduced both body length and population size. This study also demonstrated that the isolated neurons when exposed to TiO_2 NPs grow shorter axons, which are likely the cause of impeded worm locomotion behavior. Furthermore, investigators in this study also conducted DNA microarray assays to determine the change in gene expression. Microarray assays revealed the changes in the expression levels: three genes (mtl-2, C45B2.2, and nhr-247) involved in metal binding and/or detoxification and five genes (C41G6.13, C45B2.2, srr-6, K08C9.7, and C38C3.7) involved in neuronal functions (Hu et al. 2018a).

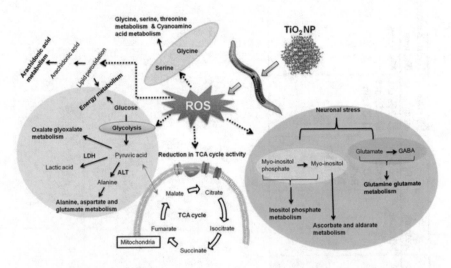

Fig. 3.3 Schematic representation of biological pathways affected in *C. elegans* on exposure to TiO2 NP (Ratnasekhar et al. 2015: Nanotoxicology, 2015; 9(8): 994–1004)

Departing from the direct interaction between a nanomaterial and organism, Wang et al. (2018) examined the interactions of TiO_2 NPs (5 and 15 nm) with heavy metals like cadmium (Cd), arsenate (As(III)), and nickel (Ni) to influence heavy metal bioaccumulations and toxicities in *C. elegans* during the process of sedimentation in aquatic environment. Results showed that heavy metals accelerated the aggregation of nano-TiO_2. The rapid aggregation and sedimentation of Ti nanoparticles resulted in the vertical distribution of heavy metals through adsorption and induced increased and prolonged exposure to benthic species. TiO_2 NPs at nontoxic concentrations enhanced the heavy metal bioaccumulation and reproductive toxicity to *C. elegans* in a dose- and size-dependent manner. Study by Wang et al. (2018) convincingly suggests that the interaction and fate of both TiO_2 NPs and heavy metals on their shared contributions to toxicity—during the sedimentation—should be considered as an integral part of risk assessment in the ecological system (Wang et al. 2018).

Ma et al. (2019) compared toxicity potentials of three sources of TiO_2: food additive TiO_2, bulk TiO_2, and nano P25 TiO_2 with primary particle size of 149 nm, 129 nm, and 26 nm, respectively. All these sources of Ti (at 1–10 mg/l) induced concentration-dependent effects on the worm's reproduction, with a reduction in brood size by 8.5–34%. However, food TiO_2 exhibited greater phototoxicity than bulk TiO_2 despite the two had similar primary particle size and size distribution, suggesting factors other than particle size or size distribution are important in affecting toxicity. Under long-term exposure without UV radiation, all three TiO_2 materials caused chronic toxicity in *C. elegans* including reduced reproduction, shortened life span, and increased incidence of age-associated vulval integrity defects. The impact on reproduction was independent of primary particle size, as the three categories of Ti had comparable effects. However, the impact on life span and vulval integrity was more pronounced for P25 than for food additive and bulk TiO2, indicating the impact of primary particle size on these long-term toxicity effects (Ma et al. 2019).

2.4 Effects of Zinc Oxide Nanoparticles (ZnO NPs)

Zinc oxide engineered nanoparticles (ZnO-ENPs) possess unique physicochemical properties that allow for a wide array of applications: biosensors, optical devices, and personal care products. Recent estimates for global production of ZnO-ENPs exceed 550 tons annually the lion's share of which are applied in cosmetics, sunscreens, and paints (Spisni et al. 2016). Given the widespread applications in consumer products, the primary pathway that facilitates entry of these nanomaterials into the environment is through the land application of biosolids from wastewater treatment plants. After entering the wastewaters, nano-ZnO partitions to sewage sludge, which is popularly used as a soil amendment (Starnes et al. 2019). Table 3.4 presents an overview of studies on ZNO NPs.

The prolonged exposure to ZnO NPs 50 µg/L significantly reduced body length when compared to controls (Wu et al. 2013). The exposure to ZnO NPs at 0.5–50 µg/L also significantly reduced brood size compared to controls. No effects on the brood size were recorded under the lower doses tested. ZnO NPs exhibited greatest reproductive toxicity when compared to TiO_2 and SiO_2 NPs in the same study. ROS production significantly correlated to toxicity effects on reproduction. However, these markers of oxidative stress were neutralized by the antioxidant treatment.

Engineered nanomaterials like ZnO-NPs accumulate in the sediment because of aggregation, agglomeration, and sedimentation processes after being released into aquatic environments. The toxicity and bioaccumulation of ENMs in soil invertebrates through different exposure routes—soil, sediment, or water—have been reported. To evaluate the chronic toxicity of ZnO-NPs in sediments, Huang et al. (2019) conducted ZnO-NPs- or $ZnCl_2$-spiked sediment exposure study using *C. elegans*. Long-term exposure (96 h) was not recorded to cause growth and reproduction toxicity in *C. elegans*. However, neurological defects affecting locomotion behavior (frequency of body bends and head thrashes) were observed. Metabolic toxicity was also determined based on changes in bioluminescence, reflecting the metabolic ATP level in the transgenic strain PE254. The results indicated that the ATP level decreased by about 30–40% upon sediment exposure to ZnO-NPs at the examined concentrations (600 and 2000 mg/kg) compared to the 0 mg/kg control.

In addition, significant increases in intracellular reactive oxygen species and lipid peroxidation were induced by the long-term sediment exposure to ZnO-NPs. This exposure also led to mobilization of the transcription factor DAF-16/FOXO from the cytoplasm to the nuclei targeting expression of the corresponding stress-responsive genes (mtl-1 and sod-3). This study demonstrates that sediment exposure to ZnO-NPs results in oxidative stress in the benthic organism *C. elegans*, mediated by the transcription factor DAF-16 triggering stress-responsive gene activation (Huang et al. 2019).

As in the case of AgNPs described above, Starnes et al. (2019) compared pristine and transformed ZnO manufactured NPs for their effects on *C. elegans* mortality using microarray analysis and qRT-PCR confirmation. The transformed nanomaterials were either phosphatized (pZnO-NPs) or sulfidized (sZnO-NPs)—expected to be present in biosolids that are applied to agricultural fields. Transformations of ZnO-NPs decreased toxicity by approximately tenfold. No differences in toxicity between ZnO NPs and ZnSO4 were recorded. Authors suggest dissolution of Zn was the main driver of toxicity. Microarray data and pathway analysis exhibited that there are distinct responses to MNP exposures and ZnSO4 exposure. However, all nanoparticles treatments induced responses of the three genes associated with metal ion toxicity and 50% of the genes responding to ZnSO4 were shared among all forms of ZnO-MNP treatments. It is also noteworthy that unique pathways and GO Terms in both transformed treatments suggest that the responses to pZnO-NPs and sZnO-NPs are distinct from those to the pristine ZnO-MNPs (Starnes et al. 2019).

Table 3.4 Effects of different forms of zinc oxide nanoparticles (ZnO NPs) on *C. elegans* life cycle

Year	Authors	NPs type	NP sizes used	Hydrodynamic diameter	Concentrations Used	Length of treatment	Significant findings
2013	Wu al.	ZnO NPs TiO$_2$ NPs SiO$_2$ NPs	30 nm each		0.05–5 µg/L	Exposure of L1–larvae to adult	Toxicity order for the examined metal oxide NPs was: ZnO NPs > TiO$_2$ NPs > SiO$_2$NPs affecting growth, locomotion behavior, reproduction, and ROS production as endpoints
2019	Huang et al.	ZnO NPs	<50 nm		600 and 2000 mg/kg of sediment	Long-term, chronic exposure-96 h	Reproduction not affected but frequency of body bends and head thrashes and ATP levels adversely affected; intracellular reactive oxygen species and lipid peroxidation triggered by the transcription factor DAF-16
2019	Starnes et al.	Pristine ZnO NPs & transformed ZnO NP	30 nm	Pristine ZnO NPs: 265 nm.	5–200 mg/L for mortality assay; 0.5–20 mg/L for reproduction assay	L1 exposed for 24–48 h	Transformation of ZnO-MNPs reduced their toxicity by nearly tenfold; when comparing pristine and transformed ZnOMNPs, 66% and 40% of genes were shared between ZnO-MNPs and sZnO-MNPs or pZnO-MNPs, respectively

2.5 Effects of Copper Oxide (CuO NPs); Silicon (Si NPs) and Nickel Nanoparticles (Ni NPs)

Table 3.5 presents an overview of studies on copper, silicon, and nickel nanoparticles. Mashock et al. (2016) attempted to delineate the effects of two categories of copper: copper oxide NPs and coper sulfate (a source of Cu ions) on *C. elegans* toxicity. Although the toxic effects of CuO NPs on nematodes has not been well studied, copper ions from copper salts have been shown to be toxic to several species, including *C. elegans*. In this background, this study described the physiological effects of CuO NPs to *C. elegans* several wild strains as well as Bristol N2 laboratory strain. CuO NPs produced greater inhibitory effects in all strains with respect to feeding, reproduction, and development. Reproduction was significantly reduced only at the highest copper dose; however, more pronounced effects were observed with copper oxide nanoparticles compared to copper sulfate treatment. The results support an increased sensitivity to CuO NPs compared to copper sulfate in a genetically broad group of *C. elegans* strains. The nanoparticle-specific effects particularly included the toxicity in *C. elegans* growth (reduced body length), reproduction, and feeding possible from nanoparticles affecting only the gut-proximal cells (Mashock et al. 2016).

As the mechanism of Si NPs uptake is not fully understood, Eom and Choi (2019) investigated the exposure of nano Si to *C. elegans* using microarray and pathway analysis focusing the uptake mechanisms and related toxicity effects. No mortality was observed after 24 h exposure to Si NPs. However, reproductive ability was significantly reduced at the same concentrations. To determine a global mechanism of toxicity, microarray was conducted on the worms exposed to Si NPs at 10 mg/l. Microarray results indicated that genes involved in reproduction, such as msp (Major Sperm Protein) genes, were significantly downregulated in the nematode exposed to NPs. Pathway analyses on differentially expressed genes (DEGs) revealed that endocytic pathway was factoring into the acquisition of Si nanomaterials. The inhibitor assay indicated that an internalization process facilitated by clathrin-mediated endocytosis is involved in the uptake of Si NPs. Functional analysis using endocytosis defective mutants (cav-1, cup-2, and chc-1) confirmed the role of endocytosis on the reproductive toxicity of Si NPs (Eom and Choi 2019). The findings from *C. elegans* provide information to help understand potential risks of Si NPs.

To determine the effect of nickel nanoparticles (Ni NPs) on *C. elegans* reproduction, worms were treated with 1.0, 2.5, and 5.0 $\mu g/cm^2$ of Ni NPs or nickel microparticles (Ni MPs). Reproductive toxicities observed in the treated worms included a decrease in brood size, fertilized egg, and spermatide activation (Kong et al. 2017). Results indicated that Ni NPs induced higher reproductive toxicity to the nematode than Ni MPs under the same treatment doses. This study does not show how Ni nanoparticles affected the growth and viability of the exposed populations.

Table 3.5 Effects of silica nanoparticles (Si NPs), nickel nanoparticles (Ni NPs), and copper oxide nanoparticles (CuO NPs) on *C. elegans* life cycle

Year	Authors	NPs type	NP sizes used	Hydrodynamic diameter	Concentrations used	Length of treatment	Significant findings
2015	Scharf, Guhrs, & Mikecz	Silica NP and bulk	50 nm 500 nm		2.5 mg/mL	24 h	Neurotoxic such as accelerated amyloid protein aggregation; neuromuscular defects of the egg laying apparatus
2019	Hyun-Jeong Eom & Jinhee Choi	SiNP	60–100 nm— Aggregation in liquid and K-medium		qRT-PCR— survival—50–500 mg/L Reproduction—20–200 mg/L	qRT-PCR—24 h Survival—24 and 48 h Reproduction—72 h	No mortality after 24 h exposure, but reproductive ability significantly reduced at same concentrations. Reproductive genes significantly downregulated. Clathrin-mediated endocytosis key for reproductive toxicity
2016	Kong et al.	Ni NP	90 nm	5 µg/mL: 260–725 nm 12.5 µg/mL: 400–879 nm	1.0, 2.5, and 5.0 µg/cm^2	24–72 h	Reproductive toxicities included a decrease in brood size, fertilized egg, and spermatid activation; increase in generation time and out-of-round spermatids Significantly reduced GSH activity at 5.0 µg Increased pro-apoptotic genes at all concentrations. Oxidative stress and apoptosis play an important role in negatively regulating reproductive processes
2016	Mashock et al.	CuO$_2$	28 nm	Less than 300 nm— Tested at 0, 24, 96 h at various concentrations	1.8, 3.8, 7.9, and 15.9 mg Cu/L	96 h	Reduction of average body length starting at 7.9 mg/L; reduced feeding behavior starting at 3.8 mg/L; reduced reproduction starting at 7.9 mg/L; increased stress response (HSP16) starting at 7.9 mg/L; degeneration of dopaminergic neurons

3 Neurological Effects

Most importantly, this model organism has a well-conserved and fully described nervous system that makes it ideal for use in neurotoxicity assessment of nanoparticles. A typical adult organism has 302 neurons and 7000 synapses that compose two independent nervous systems, a large somatic system with 282 neurons and a small pharyngeal system with 20 neurons (Sinis et al. 2019). The connectivity of the neuronal network has been fully shown and majority of gene families relevant to neurotransmission in mammals have been identified in this worm. Furthermore, the transparency of this animal allows visualization of neuron damage and regeneration with the assistance of fluorescent markers (Hisamoto and Matsumoto 2017). Relevant to nanotoxicity assessment protocol, the worm behavior comprises a number of nonspecific phenotypes connected to aberrant nerve functions. The major phenotypes used in NP screening are: (a) locomotion (frequently evaluated by head thrashing and body bends); (b) pharyngeal pumping; (c) defecation; and (d) egg laying. Well-characterized neural circuits control each biological process, dysfunction of which could be a plausible explanation of abnormal outcomes in otherwise anatomically and functionally intact worms (Sinis et al. 2019). Some pressing examples include the damage of neurons, which innervate muscles of a fully developed vulva, may result in perturbation of egg laying, reduction of progeny and bag of worms (BOW) phenomenon (hatching of larvae in the paternal body) or defect in the motor neurons that affects locomotion. Tables 3.1–3.5 summarize the neurological effects of different nanomaterials reviewed in this chapter.

3.1 Neurological Effects of Silver Nanoparticles (Ag NPs)

To evaluate the fate of pristine nano-Ag after being released into the environment, Yang et al. (2018) used three different exposure media (deionized water as control, EPA water as a low ionic strength exposure medium, and KM as a high ionic strength exposure medium) to pretreat AgNPs. E. coli cells were exposed to these transformed nanomaterials before they were fed to the nematode. The results indicated that ionic strength significantly enhanced the neurotoxicity (reducing head thrash and body bend) besides reproductive toxicity (reducing germ cell corpses, brood size, and life span) in C. elegans. After 24 h of treatment with the transformed Ag NPs (25 and 75 nm), locomotion behavior was significantly affected at both endpoints of evaluation: body bend and head thrash (Yang et al. 2018).

3.2   Neurological Effects of Gold Nanoparticles (au NPs)

As described above in A. II, Hu et al. (2018b) utilized size-tunable gold nanoparticles coated with 11-mercaptoundecanoic acid (MUA-Au NPs) to investigate the toxicogenomic responses in *C. elegans*. Exposure of Au NPs significantly affected locomotion as a result of impeded motor neuron function. When primary neurons isolated from *C. elegans* embryos were exposed to Au NPs, the axonal growth was significantly impeded. Interestingly, while this effect was independent of the MUA-to-Au ratio, the effect on cell viability (using the MTT assay) was dependent on this ratio, indicating that increasing the MUA-to-Au ratio also decreases the neuronal survival rate. Interestingly, cell viability was decreased at only Au NP particle sizes equal to or smaller than 1.26 ± 0.25. Smaller particles are likely capable of triggering higher oxidative stress on cells than larger particles because of their extensive surface areas. Changes in gene expression patterns correlated well to the changes in the phenotypes observed. For example, dpy-14 (expressed in embryonic neurons), clec-174 (involved in cellular defense), and cut-3 (involved in body morphogenesis) were upregulated while mtl-1 (functions in metal detoxification and homeostasis) was downregulated (Hu et al. 2018b).

3.3   Neurological Effects of Titanium Oxide Nanoparticles (TiO$_2$ NPs)

As outlined in A. III, Hu et al. (2018a) compared the effects of rutile and anatase TiO$_2$ NPs on *C. elegans* axonal growth and locomotion behavior. Differential effects of anatase and rutile nanotitania were observed on the locomotion; for anatase, thrashing frequency decreased starting at 500 µg/mL, while for rutile, it decreased at 100–500 µg/mL. Primary cultured neurons exposed to anatase at 200 and 300 µg/mL and rutile at 300 µg/mL showed significant decreases in axonal length. DNA microarray assays revealed changes in the gene expression levels of many genes involved in neuronal function [C41G6.13, C45B2.2, srr-6, K08C9.7, and C38C3.7] (Hu et al. 2018a).

3.4   Neurological Effects of Zinc Oxide Nanoparticles (Zn NPs)

Wu et al. (2013) conducted a prolonged exposure (from L1 larvae to adult worms) of the nematode *C. elegans* to TiO2 NPs and ZnO NPs (30 nm each) at the equal concentrations of 0.05 µg/L and observed significant ($P < 0.01$) decrease in both body bends and head thrashes of nematodes compared with control. The neurotoxicity of ZnO–NPs was greater than in TiO$_2$ NPs-exposed nematodes. Moreover, treatment with antioxidants of ascorbate or NAC effectively inhibited the buildup of oxidative stress and retrieved the adverse effects of TiO_2 NPs or ZnO NPs.

As described in A. IV, the long-term sediment exposure to ZnO NPs did not affect growth and reproduction of *C. elegans*, but locomotive behaviors (frequency of body bends and head thrashes) and ATP levels were adversely affected (Huang et al. 2019). In addition, significant increases in intracellular reactive oxygen species and lipid peroxidation were induced by long-term sediment exposure to ZnO-NPs. Huang et al. (2019) demonstrated that *mtl-1* and *sod-3* gene expressions were upregulated upon long-term sediment exposure to ZnO NPs (2000 mg/kg) compared to control ($P < 0.001$). The most significant induction was observed in *mtl-1* gene expression, with a 70% increase upon ZnO NPs sediment exposure. This study further indicated that long-term sediment exposure to ZnO NPs induced DAF-16-targeted stress-responsive genes expression (*mtl-1 and sod-3*).

3.5   Neurological Effects of Copper Oxide Nanoparticles (CuO NPs), Aluminum Oxide Nanoparticles (Al_2O_3 NPs), and Silicon Nanoparticles (Si NPs)

The effects of copper oxide nanoparticles and copper sulfate on neurons—cells with known vulnerability to heavy metal toxicity—were investigated by Mashock et al. (2016). About 10% of the exposed population showed degeneration of dopaminergic neurons after copper oxide nanoparticle exposure. Additionally, this study showed an increased tolerance in knockout mutants (for divalent-metal transporters: *smf-1* or *smf*-2) to copper exposure, implicating the role of both transporters in copper acquisition and consequent neurodegeneration. These results highlight the complex nature of CuO nanoparticle toxicity, in which a nanoparticle-specific effect was observed in some traits (average body length, feeding behavior) as described above and a copper ion specific effect was observed for other traits (neurodegeneration, response to stress). Neuronal deformation occurred in a similar portion of the *C. elegans* population after exposure with either CuO NPs or copper sulfate at equal amounts of copper. This indicates further to the released copper ions from CuO NPs as a major factor contributing to the observed NPs effect on neuronal health and organismal stress response (Mashock et al. 2016).

To determine the role of an oxidant in amelioration of aluminum nanoparticles-led neurotoxicity, three transgenic strains of *C. elegans*—producing the fluorescent signals labeling the GABAergic neurons, AFD sensory neurons, or AIY interneurons—were used along with a wild-type N2 in an investigation by Yu et al. (2015). Exposing the worms to Al_2O_3 NPs (1–10 mg/L) led to neurodegeneration-related phenotypes including neuronal loss, abnormality of axon development, and gap formation on nerve cords in the GABAergic nervous system and some behavioral deficits in the transgenic strains. Al2O3-NPs at 10 mg/L decreased thermotaxis learning (TL) in exposed animals compared to the control. Wild-type N2 nematodes exposed to 10 mg/L of Al_2O_3 NPs also showed a significant decrease in body bend and head thrash. Interestingly, the pretreatment of the worm with 200 mg/L of vitamin E prevented the neurotoxicity of Al_2O_3-NPs by reducing both the neurodegeneration and behavioral deficits. Investigators showed the pretreatment with vitamin E prevented the induction of oxidative stress, and sustained the normal intestinal permeability and development in Al_2O_3-NPs exposed population (Yu et al. 2015). Vitamin E-pretreated animals also showed less abnormality in the development of neurons involved in behavioral control and expression pattern of genes regulating cell identity of the corresponding neurons. These results if reliably reproduced may be helpful in designing strategies for nanotoxicity amelioration in humans.

Significant decreases in body bends and head thrashes at 5–50 µg/L concentrations of Si NPs were recorded while comparing its effects with those of ZnO_2 or TiO_2 NPs (Wu et al. 2013). Investigators connected toxicity effects with the production of intracellular ROS, which were reversed after an antioxidative treatment.

Scharf et al. (2015) conducted a well-designed, mechanistic study to demonstrate the neurological effects of Si NP in *C. elegans*. In this study, adult worms were exposed to Si for 24 h after which observations were made with respect to two neuromuscular behaviors: locomotion (body bends) and egg laying defect (bag-of-worms-assay or BOW-assay). Both of these phenotypes demonstrated significant adverse effects of nano-Si. The defective phenotypes carried the signatures of fibrillation of proteins to an SDS-insoluble aggregome. Analysis of the proteomic data revealed that the silica NP-specific aggregome was enriched in proteins containing beta strand structures in comparison with the total proteome, consistent with the notion that beta-strands possess an intrinsic propensity for amyloid protein fibrillation. The results suggest that silica NPs promote a cascade of events including disturbance of protein homeostasis, widespread protein aggregation, and inhibition of serotonergic neurotransmission, which can be intervened by pharmacological compounds preventing amyloid fibrillation (Scharf et al. 2015).

4 Immunotoxicity

Nanomaterials have been shown to interact with the immune system cells/components either by stimulating or suppressing immune functions through the mediation of oxidative stress or inflammation in in vitro and in vivo models (Roy et al. 2015). *C. elegans* does not have an adaptive immune system or mobile immune cells, such as professional phagocytes. Intestinal epithelial cells provide an essential line of defense for *C. elegans* against ingested pathogens. It is important to note that *C. elegans* intestinal epithelial cells bear a striking resemblance to human intestinal cells. The nematode anatomy thus facilitates study of host–pathogen interactions in the intestine (Pukkila-Worley and Ausubel 2012). Principal among the immune regulators in *C. elegans* is the NSY-1/SEK-1/PMK-1 MAP. This pathway is orthologous to the ASK1 (MAP kinase kinase kinase)/MKK3/6 (MAP kinase kinase)/p38 (MAP kinase) pathway in mammals kinase pathway (Pukkila-Worley and Ausubel 2012). The p38 MAP kinase pathway coordinates the basal and infection-induced regulation of immune effectors required for defense against most nematode pathogens, but it does not act alone. The transcription factor ZIP-2 controls an immune signaling pathway that acts independently of PMK-1 and is induced only by virulent strains of *P. aeruginosa*. There are few studies to throw light on how different nanomaterials released into the environment affect immunity and immune regulators of this soil model animal.

It is known that the p38 MAPK pathway is required for the activation of transcription factor SKN-1/Nrf following *P. aeruginosa* PA14 infection in *C. elegans*, suggesting that SKN-1 plays an important role in pathogenic stress and environmental oxidative stress signaling (Papp et al. 2012). In one excellent study, Li et al. (2020) investigated the effects of long-term ZnO-NPs exposure (from L1 larvae to adults) on innate immunity and its underlying mechanisms using a wild-type, N_2, and mutant strains of *C. elegans* (defective in *pmk-1, sek-1, and nsy-1*, which encode for the p38 MAPK pathway) as hosts and *Pseudomonas aeruginosa* PA14 as a pathogen. The results showed that the uptaken ZnO-NPs mainly accumulated in the worm intestine and the chronic exposure of ZnO-NPs at environmentally relevant concentrations (50 and 500 mg/L) decreased the survival of wild-type worms that were infected with *P. aeruginosa* PA14. ZnO-NPs at 500 mg/L also increased the accumulation of live *P. aeruginosa* PA14 colonies in the nematode intestine by threefold (Li et al. 2020). The above effects correlated well with inhibition of the intestinal nuclear translocation of SKN-1, along with the downregulation of its target gene gcs-1. These findings also indicate that ZnO-NPs and $ZnCl_2$ exert different modes of action on innate immunity with early-life, long-term exposure in *C. elegans*. This suggests that the immunotoxicity is mostly due to the particulate effects of the ZnO-NPs. Because SKN-1/Nrf and the p38 MAPK pathways are highly conserved evolutionarily in eukaryotes, the negative impact on innate immunity caused by the chronic ZnO-NPs exposure may also apply to other organisms, including humans.

One potentially interesting application of *C. elegans* pathogenesis assays involves their use in large-scale screens to identify novel antimicrobials (Peterson and Pukkila-Worley 2018). Antimicrobial peptides play an important role in the innate immune response of nematodes; alteration in expression patterns of genes encoding antimicrobial peptides could be used to assess the immunotoxicity of NPs.

References

Baranowska-Wójcik, E., Szwajgier, D., Oleszczuk, P., et al. (2020). Effects of titanium dioxide nanoparticles exposure on human health—A review. *Biological Trace Element Research, 193*, 118–129. https://doi.org/10.1007/s12011-019-01706-6.

Eom, H. J., & Choi, J. (2019). Clathrin-mediated endocytosis is involved in uptake and toxicity of silica nanoparticles in Caenohabditis elegans. *Chemico-Biological Interactions, 311*, 108774. https://doi.org/10.1016/j.cbi.2019.108774.

Gallud, A., Klöditz, K., Ytterberg, J., et al. (2019). Cationic gold nanoparticles elicit mitochondrial dysfunction: A multi-omics study. *Scientific Reports, 9*, 4366. https://doi.org/10.1038/s41598-019-40579-6.

Gonzalez-Moragas, L., Berto, P., Vilches, C., Quidant, R., Kolovou, A., Santarella-Mellwig, R., Schwab, Y., Stürzenbaum, S., Roig, A., & Laromaine, A. (2017). In vivo testing of gold nanoparticles using the Caenorhabditis elegans model organism. *Acta Biomaterialia, 53*, 598–609. https://doi.org/10.1016/j.actbio.2017.01.080.

Hisamoto, N., & Matsumoto, K. (2017). Signal transduction cascades in axon regeneration: insights from C. Elegans. *Current Opinion in Genetics & Development, 44*, 54–60. https://doi.org/10.1016/J.GDE.2017.01.010.

Hu, C. C., Wu, G. H., Hua, T. E., Wagner, O. I., & Yen, T. J. (2018a). Uptake of TiO_2 nanoparticles into C. elegans neurons negatively affects axonal growth and worm locomotion behavior. *ACS Applied Materials and Interfaces, 10*(10), 8485–8495. https://doi.org/10.1021/acsami.7b18818.

Hu, C. C., Wu, G. H., Lai, S. F., Muthaiyan Shanmugam, M., Hwu, Y., Wagner, O. I., & Yen, T. J. (2018b). Toxic effects of size-tunable gold nanoparticles on Caenorhabditis elegans development and gene regulation. *Scientific Reports, 8*, 15245. https://doi.org/10.1038/s41598-018-33585-7.

Huang, C. W., Li, S. W., & Liao, V. H. C. (2019). Long-term sediment exposure to ZnO nanoparticles induces oxidative stress in: Caenorhabditis elegans. *Environmental Science: Nano, 6*(8), 2602–2614. https://doi.org/10.1039/c9en00039a.

Kim, S. W., Kwak, J. I., & An, Y. J. (2013). Multigenerational study of gold nanoparticles in Caenorhabditis elegans: Transgenerational effect of maternal exposure. *Environmental Science and Technology, 47*(10), 5393–5399. https://doi.org/10.1021/es304511z.

Kong, L., Gao, X., Zhu, J., Zhang, T., Xue, Y., & Tang, M. (2017). Reproductive toxicity induced by nickel nanoparticles in Caenorhabditis elegans. *Environmental Toxicology, 32*(5), 1530–1538. https://doi.org/10.1002/tox.22373.

Li, S.-W., Huang, C.-W., & Liao, V. H.-C. (2020). Early-life long-term exposure to ZnO nanoparticles suppresses innate immunity regulated by SKN-1/Nrf and the p38 MAPK signaling pathway in *Caenorhabditis elegans*. *Environmental Pollution, 256*, 11338. https://doi.org/10.1016/j.envpol.2019.113382.

Ma, H., Lenz, K. A., Gao, X., Li, S., & Wallis, L. K. (2019). Comparative toxicity of a food additive TiO 2, a bulk TiO 2, and a nano-sized P25 to a model organism the nematode C. elegans. *Environmental Science and Pollution Research, 26*(4), 3556–3568. https://doi.org/10.1007/s11356-018-3810-4.

Mashock, M. J., Zanon, T., Kappell, A. D., Petrella, L. N., Andersen, E. C., & Hristova, K. R. (2016). Copper oxide nanoparticles impact several toxicological endpoints and cause neurodegeneration in caenorhabditis elegans. *PLoS One, 11*(12), 1–19. https://doi.org/10.1371/journal.pone.0167613.

Moon, J., Kwak, J., II, & An, Y. J. (2019). The effects of silver nanomaterial shape and size on toxicity to Caenorhabditis elegans in soil media. *Chemosphere, 215*, 50–56. https://doi.org/10.1016/j.chemosphere.2018.09.177.

Moon, J., Kwak, J. I., Kim, S. W., & An, Y. J. (2017). Multigenerational effects of gold nanoparticles in Caenorhabditis elegans: Continuous versus intermittent exposures. *Environmental Pollution, 220*, 46–52. https://doi.org/10.1016/j.envpol.2016.09.021.

Nanotech Project (2016). Project on emerging nanotechnologies. consumer products inventory, http://www.nanotechproject.org/inventories/consumer/analysis draft.

Papp, D., Csermely, P., & Soti, C. (2012). A role for SKN-1/Nrf in pathogen resistance and immunosenescence in Caenorhabditis elegans. *PLoS Pathogens, 8*, e1002673.

Peterson, N. D., & Pukkila-Worley, R. (2018). Caenorhabditis elegans in high-throughput screens for anti-infective compounds. *Current Opinion in Immunology, 2018*(54), 59–65.

Pukkila-Worley, R., & Ausubel, F. (2012). Immune defense mechanisms in the *Caenorhabditis elegans* intestinal epithelium. *Current Opinion in Immunology, 24*(1), 3–9. https://doi.org/10.1016/j.coi.2011.10.004.

Ratnasekhar, C., Sonane, M., Satish, A., & Mudiam, M. K. R. (2015). Metabolomics reveals the perturbations in the metabolome of Caenorhabditis elegans exposed to titanium dioxide nanoparticles. *Nanotoxicology, 9*(8), 994–1004. https://doi.org/10.3109/17435390.2014.993345.

Rocheleau, S., Arbour, M., Elias, M., Sunahara, G. I., & Masson, L. (2015). Toxicogenomic effects of nano- and bulk-TiO2 particles in the soil nematode Caenorhabditis elegans. *Nanotoxicology, 9*(4), 502–512. https://doi.org/10.3109/17435390.2014.948941.

Roy, R., Das, M., & Dwivedi, P. D. (2015). Toxicological mode of action of ZnO nanoparticles: Impact on immune cells. *Molecular Immunology, 63*(2), 184–192. https://doi.org/10.1016/j.molimm.2014.08.001.

Scharf, A., Guhrs, K.-H., & Mikecz, A. (2015). Anti-amyloid compounds protect from silica nanoparticle-induced neurotoxicity in the nematode C. elegans. *Nanotoxicology*, 1–10. https://doi.org/10.3109/17435390.2015.1073399. Early Online.

Schultz, C. L., Wamucho, A., Tsyusko, O. V., Unrine, J. M., Crossley, A., Svendsen, C., & Spurgeon, D. J. (2016). Multigenerational exposure to silver ions and silver nanoparticles reveals heightened sensitivity and epigenetic memory in Caenorhabditis elegans. *Proceedings of the Royal Society B: Biological Sciences, 283*(1832). https://doi.org/10.1098/rspb.2015.2911.

Sinis, S. I., Gourgoulianisb, K. I., Hatzogloua, C., & Zarogiannisa, S. G. (2019). Mechanisms of engineered nanoparticle induced neurotoxicity in Caenorhabditis elegans. *Environmental Toxicology and Pharmacology, 67*, 29–34. https://doi.org/10.1016/j.etap.2019.01.010.

Spisni, E., Seo, S., Joo, S. H., & Su, C. (2016). Release and toxicity comparison between industrial- and sunscreen-derived nano-ZnO particles. *International journal of Environmental Science and Technology, 13*, 2485–2494.

Starnes, D., Unrine, J., Chen, C., Lichtenberg, S., Starnes, C., Svendsen, C., Kille, P., Morgan, J., Baddar, Z. E., Spear, A., Bertsch, P., Chen, K. C., & Tsyusko, O. (2019). Toxicogenomic responses of Caenorhabditis elegans to pristine and transformed zinc oxide nanoparticles. *Environmental Pollution, 247*, 917–926. https://doi.org/10.1016/j.envpol.2019.01.077.

Starnes, D. L., Lichtenberg, S. S., Unrine, J. M., Starnes, C. P., Oostveen, E. K., Lowry, G. V., Bertsch, P. M., & Tsyusko, O. V. (2016). Distinct transcriptomic responses of Caenorhabditis elegans to pristine and sulfidized silver nanoparticles. *Environmental Pollution, 213*, 314–321. https://doi.org/10.1016/j.envpol.2016.01.020.

Starnes, D. L., Unrine, J. M., Starnes, C. P., Collin, B. E., Oostveen, E. K., Ma, R., Lowry, G. V., Bertsch, P. M., & Tsyusko, O. V. (2015). Impact of sulfidation on the bioavailability and toxicity of silver nanoparticles to Caenorhabditis elegans. *Environmental Pollution, 196*, 239–246. https://doi.org/10.1016/j.envpol.2014.10.009.

Wang, J., Dai, H., Nie, Y., Wang, M., Yang, Z., Cheng, L., Liu, Y., Chen, S., Zhao, G., Wu, L., Guang, S., & Xu, A. (2018). TiO2 nanoparticles enhance bioaccumulation and toxicity of heavy metals in Caenorhabditis elegans via modification of local concentrations during the sedimentation process. *Ecotoxicology and Environmental Safety, 162*(July), 160–169. 10.1016/j.ecoenv.2018.06.051.

Weir, A., Westerhoff, P., Fabricius, L., Hristovski, K., & von Goetz, N. (2012). Titanium dioxide nanoparticles in food and personal care products. *Environmental Science and Technology, 46*(4), 2242–2250. https://doi.org/10.1021/es204168d

Wu, Q., Nouara, A., Li, Y., Zhang, M., Wang, W., Tang, M., Ye, B., Ding, J., & Wang, D. (2013). Comparison of toxicities from three metal oxide nanoparticles at environmental relevant concentrations in nematode Caenorhabditis elegans. *Chemosphere, 90*(3), 1123–1131. https://doi.org/10.1016/j.chemosphere.2012.09.019.

Yang, Y., Cheng, Y., & Liao, C. (2017). Nematode-based biomarkers as critical risk indicators on assessing the impact of silver nanoparticles on soil ecosystems. *Ecological Indicators, 75*, 340–351. https://doi.org/10.1016/j.ecolind.2016.12.051.

Yang, Y., Xu, G., Xu, S., Chen, S., Xu, A., & Wu, L. (2018). Effect of ionic strength on bioaccumulation and toxicity of silver nanoparticles in Caenorhabditis elegans. *Ecotoxicology and Environmental Safety, 165*, 291–298. https://doi.org/10.1016/j.ecoenv.2018.09.008.

Yu, X., Guan, X., Wu, Q., Zhao, Y., & Wang, D. (2015). Vitamin E ameliorates neurodegeneration related phenotypes caused by neurotoxicity of Al2O$_3$-nanoparticles in C. elegans. *Toxicology Research, 4*(5), 1269–1281. https://doi.org/10.1039/c5tx00029g.

Chapter 4
Response to Engineered Nanomaterials in *Arabidopsis thaliana*, a Model Plant

Sinilal Bhaskaran and Shivendra Sahi

Contents

1   Introduction

Nanomaterials are diverse in the chemical composition, size, shape, surface characteristics, purity, stability and optical, thermal and electrical properties and have gained manifold applications in the modern industrial society. Their interactions with living organisms are significant because of the increased permeability conferred by the small size. The property of target-specific, controlled release of nanomaterials is utilized to deliver a variety of molecules into plant and animal cells. In plants, nano fertilizers have been advocated for enhancement of nutrient-use efficiency by the controlled release in meeting the plant's demands (Zulfiqar et al. 2019). This application in turn prevents nutrient loss through runaway water and transformation to chemical forms that are not consumable to plants. Pesticide application based on nanotechnology helps to reduce the dosage and frequency of application (Hayles et al. 2017; Ojha et al. 2018). The resultant drastic reduction in the usage of fertilizers and plant protectants anticipates lower investment and harm to the environment (Adisa et al. 2019). Biosensor technology using nanomaterials is underway for detection of plant pathogens and plant metabolic flux (Chaudhry et al. 2018). Some nanomaterials are used for targeted delivery of DNA, but their share is negligible compared to other uses (Riley and

S. Bhaskaran (✉)
Fatima Mata National College, Kollam, Kerala, India
e-mail: sinilab@fatimacollege.net

S. Sahi
Department of Biological Sciences, University of Sciences, Philadelphia, PA, USA
e-mail: s.sahi@usciences.edu

© Springer Nature Switzerland AG 2021
N. Sharma, S. Sahi (eds.), *Nanomaterial Biointeractions at the Cellular, Organismal and System Levels*, Nanotechnology in the Life Sciences, https://doi.org/10.1007/978-3-030-65792-5_4

Vermerris 2017). Because of the desirable physical and chemical properties mentioned above, new applications are on way every day; at the same time, concern over the uncontrolled release of nanomaterials into the environment is soaring in scientific community. Before a novel nanomaterial enters the human supply chain, their entry into plants from air, water and soil and how they interact with plant cells in vivo need to be scrutinized as they can be transferred to next trophic levels affecting the ecosystem. Unfortunately, most of the applications of nanomaterials have not undergone such scrutinizes.

Attempts to understand the interaction of nanomaterials tested in vitro and *in vivo* narrate both stimulatory and inhibitory effects. In plants, the changes occurring in morphology, anatomy, physiology and gene expression as a result of exposure to nanomaterials are usually observed to assess the impact. Crop species have been studied mostly to explore the intended outcomes of specific applications. However, the limited information on physiology and gene interactions of nanomaterials in most of the species poses a question about the depth and reliability of results. On the other hand, results of the studies conducted in experimental/ model organisms are more informative because of several reasons. This includes the ease in conducting experiments in controlled conditions, analysis using standard protocols, reproducibility in results and the possibilities for exploring molecular mechanisms using the techniques of genomics and metabolomics due to the availability of complete genome sequence (Montes et al. 2017). *Arabidopsis thaliana* is preferred over other model plants because of its small size enabling the growth of large number of plants in a small area, short life cycle, self-fertile nature and high potential for producing mass of seeds. The easier transformation using *Agrobacterium tumefaciens*, availability of gene overexpression and mutant lines, small diploid genome and the public availability of its resources convenient for genetics and genomics analyses make Arabidopsis system ideal for nanointeractions. Realizing the merits of the system, *A. thaliana* has been exposed to a range of nanomaterials for an in-depth characterization of the effects, which are discussed below. For convenience, the impact of different nanomaterials by virtue of their chemical compositions is discussed here, giving emphasis to the morphological/anatomical, physiological and genetic effects.

Gold: Influence of gold nanoparticles (AuNPs) on growth of Arabidopsis has been characterized by seed germination and its percentage, measuring fresh weight and dry weight, root growth by increase in root length, branching, etc. Both inhibitory and promotive effects have been reported. Low concentrations promoted growth, exhibiting about twofold increase in the percentage of seed germination (Kumar et al. 2013). Growth promotion was also observed in terms of shoot, root length, number of lateral roots and rosette leaves bringing an increase in fresh weight ranging from three- to sixfold. Increase in water content was also noticed in the treated plants, which imparted the increase in fresh weight. The overall growth promotive effects were found persistent throughout the life of the plants evidenced by early flowering and increase in yield even after retraction from exposure for 15 days. In a similar study, AuNP treatment (100 mg/L) significantly reduced the number and length of lateral roots at a high concentration and the effect was irrespective of the particle shapes (Siegel et al. 2018). However, size influenced growth as inhibition of primary root and promotion of root hair growth in the experimental group treated with AuNPs of 10 nm. Surface charge of the particles was also observed to be influencing growth (Hendel et al. 2019). Compared to neutral AuNPs, the charged ones reduced growth of root meristematic region. Negatively charged AuNPs induced root hair and lateral root growth.

AuNPs enter plants through the root system and get transported to other parts of the plant via vascular system (Koo et al. 2015a, b). Analysis of the surface charge of AuNPs and its role in absorption indicated the preference of negatively charged AuNPs over those having positive charge (Avellan et al. 2017). Particles were lodged in the border cells and the mucilage secreted by the cells at the root tip. Majority of the positively charged particles were found accumulated outside and a few spotted inside were considered internalized by the process of endocytosis. In addition to the surface charge, size also influenced uptake and translocation. Surprisingly, a similar study conducted using positively, negatively and neutral AuNPs came up with contradictory results (Hendel et al. 2019). Neutral AuNPs induced vacuolization in the rhizodermal and cortical cells at the root tip which is a typical response to heavy metals. Regardless of the surface charge, AuNPs were found accumulated in the vicinity to the root surface. Treatment with neutral and positively charged particles stimulated the detachment of plasma membrane and the space formed was found filled with a secretion which normally occurs in case of abiotic stress. Even though negatively charged AuNPs did not impose membrane detachment, increase in cell wall thickness was observed which is to be considered as a mechanism to prevent their entry. Charge on the particles facilitated their entry into protoplasts. Positively charged AuNPs were favoured than the latter. Both endocytotic and non-endocytotic modes of entries were observed in this study. Thus, it is clear that the studies conducted so far in Arabidopsis are focused mainly on the morphological and anatomical changes. Few of them tried to understand the mechanism of entry, transport and accumulation. Monitoring physiological changes can give some idea on the metabolic pathways interacting with AuNPs. Analyses of few antioxidant enzyme activities and the expression of few microRNAs conclude that the changes observed were triggered by the related micro RNAs (Kumar et al. 2013). AuNPs accumulated inside plant tissues generate heat according to an analysis using thermal imaging. High induction of heat shock proteins associated with the condition is an indication of that (Koo et al. 2015a, b). Stress symptoms like reduction of chlorophyll and formation of anthocyanin were accompanied with the changes in test groups fed with higher AuNP concentrations (Wang et al. 2013). So, it is evident that a comprehensive study of the mechanism of interaction of AuNPs in Arabidopsis is yet to come. Arabidopsis can be effectively utilized to unravel the molecular interactions of nanomaterials based on its available genomic information.

Apart from being utilized as an agent for molecular delivery in plants, AuNPs themselves can be synthesized by plants and plant products (Shankar et al. 2004). Green engineering approach for nanofabrication is receiving attention in light of the nanoparticles produced with novel surface coatings and the environment-friendly nature of the process. Another application of such nanomaterials could be directed to phytoextraction of soil pollutants or in the water filtration systems. In plant-inspired process, gold is fed to growing plants as water-soluble salts that transform to gold nanoparticles after reacting with plant chemicals in vivo. The regulatory points of the redox reactions involved are yet to be deciphered for a smooth production cycle. Further, the ions that are potentially reactive may be translocated to different parts of the plant by metal transporter proteins and get converted to the metallic form by certain other reducing agents to make them unreactive (Taylor et al. 2014; Jain et al. 2014). Tiwari et al. (2016) reported majority of metal responsive and binding genes were upregulated when plants were exposed to gold salts (KAuCl$_4$) forming a bulk of AuNPs in their root and shoot tissues. Among differentially expressed genes, 70 genes were upregulated up to twofold in root (Fig. 4.1; Table 4.1). The classification of upregulated genes based on metal responses indicates that 12.46% genes were associated with cation binding. The expression of ferric reduction oxidase 5 (FRO5) was highest (17.53-fold) among upregulated genes (Fig. 4.1; Table 4.1). The induced loci encode different types of transporters such as copper transporter, nitrate transporter, ABC transporter, heavy-metal-associated protein (HMA), zinc (Zn) transporter, malate transporter and phosphate transporter. These genes are responsible for uptake of essential elements and nutrients such as Fe, Cu, Zn, NO3 and PO4.

Silver: Silver nanoparticles (AgNPs) occupy a major part of the commercially utilized nanoparticles. They are available as coated and uncoated forms. Their interaction when studied by supplying 0.01 to 100 mg/L to Arabidopsis demonstrated dose-dependent effects (Wang et al. 2013). Like gold nanoparticles, lower concentrations of AgNPs enhanced growth while the higher concentrations were found inhibitory. Parameters of growth like size of rosette leaves, root length and shoot weight exhibited the same distinct patterns across the different concentrations tested. Smaller nanoparticles were found to be more toxic in this investigation. The results obtained were comparable to those of AgNO$_3$ that produced AgNPs inside plant tissues, largely in leaves. In a similar study conducted with AgNPs and Ag+, more or less similar patterns in the growth parameters were observed in the concentrations treated (Kaveh et al. 2013). Silver concentration was found higher in plants exposed to AgNO$_3$ attesting the higher permeability of silver ions than their particles. AgNPs and Ag$^+$ were inhibitory to root growth and were of the same toxic responses (Baghkheirati and Lee 2015). In another study, a comparison between the effects of AgNPs and Ag$^+$ demonstrated higher interferences of AgNPs in growth, reproduction and metabolism (Ke et al. 2018). Shoot and root growth were inhibited. Reduction in chlorophyll and the production of anthocyanin in leaves indicated stress. Decrease in reproductive efficiency was observed which was marked by the reduction in bolting height, bud number, number of pods, pod length and their biomass. Notwithstanding that sugar, phenylpropanoid and amino acid pathways were affected by both AgNPs and Ag+, enhancement of galactose metabolism and the reduction in the levels of amino acids valine, serine and aspartate were characteristic to AgNPs. Stretching of the vegetative growth is another effect induced which resulted in shortening of the reproductive phase followed by a poor regeneration capacity of the seeds (Geisler-Lee et al. 2014). Particles were found accumulated inside the root cells, vascular tissues, cotyledons and stomata at different days after the treatment. Toxicity affected meristematic cells at the root tips preventing the

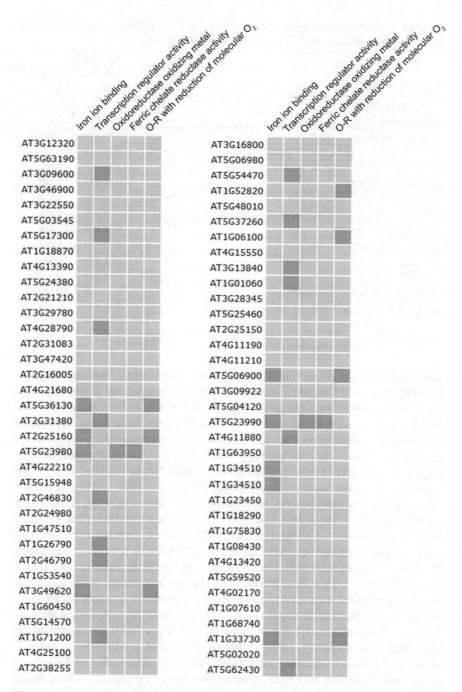

Fig. 4.1 Plant gene set enrichment analysis of significantly upregulated genes when Arabidopsis seedlings were exposed to 10 ppm KAuCl$_4$ (Source: Tiwari et al. (2016): *Scientific Reports*, 6:21733, DOI: https://doi.org/10.1038/srep21733)

Table 4.1 Comparison of microarray data by quantitative RT-PCR (Source: Tiwari et al. (2016): Scientific Reports, 6:21733, DOI: https://doi.org/10.1038/srep21733)

| | | | qRT-PCR | |
| | | | 10 ppm | 25 ppm |
Locus ID	Annotation	Microarray	Au	Au
AT5G23990	FRO5	17.54	7.269	1.855
AT2G46830	CCA1	3.88	4.645	0.472
AT1G63950	HMA3	3.833	26.459	0.025
AT3G28345	ABC transporter B family member	3.102	2.215	0.526
AT5G14570	High affinity nitrate transporter	3.4	2.250	2.406
AT3G46900	COPT2	4.22	4.835	1.192
AT1G01060	LHY	3.59	2.978	1.638
AT3G12320	Light-inducible and clock-regulated 3, LNK3	4.81	8.889	0.498
AT1G33730	CYP76C5	3.77	1.685	2.308
AT5G17300	RVE1	2.919	2.689	1.986
AT5G24380	YSL2	3	1.332	0.386
AT3G09600	Myb	2.37	2.087	2.288
AT4G10530	Subtilase family protein	−6.85	0.073	0.450
AT5G49770	Leucine-rich repeat receptor-like protein kinase	−2.79	0.183	0.385
AT2G44840	ERF13	−2.01	0.179	1.116

growth of primary and lateral roots. Localization using isotopic labelling and single particle Inductively Coupled Plasma (ICP) analysis detected the particles in roots and their transport to the other parts (Nath et al. 2018). On analysis using single-cell ICP, the particles were found predominantly at the middle lamella and cell walls of the root and a smaller portion found translocated to other parts of the plant; however those detected inside the cells were aggregated (Bao et al. 2016). Exposure of seedlings to 0.2–1 mg/L AgNPs for a maximum period of 72 hrs resulted in detecting silver inside the plant tissues with the help of ICP analysis (Nair and Chung 2014a). Since toxicity has been reported in other plant and animal systems, AgNP usage cannot be considered to be safe. Changes in physiological parameters like reduction in chlorophyll content, increase in anthocyanin, etc. were observed when seedlings were exposed to the above concentrations of AgNPs for 14 days (Nair and Chung 2014b) as observed in an earlier study (Wang et al. 2013). Increase in lipid peroxidation, Reactive Oxygen Species (ROS) production and change in mitochondrial membrane potential were recorded in roots of the seedlings exposed. Shape of the particles also influences the responses in plants (Syu et al. 2014). Spherical particles induced anthocyanin production and high level of Super Oxide Dismutase (SOD) indicating their inhibitory nature. Whereas decahedral particles promoted root growth and were recognized as the one produced the lowest level of SOD. Expression of proteins associated with ROS accumulation and cell proliferation were found common in all types. Stress induced by AgNPs in plants when compared to that of cold, salt, drought and heat stresses were found different and milder (Baghkheirati and Lee 2015). ROS generation was found common in all. The stress induced by

AgNPs was observed to be having highest similarity with cold stress compared to the others. AgNPs above 300 mg/L were found to be interfering with potassium (K+) efflux and calcium (Ca_2+) influx impairing the transport through plasma membrane (Sosan et al. 2016). This study also noticed reduction of root growth, photosynthesis rate and the formation of ROS as reported in several other studies conducted in Arabidopsis. Oxidation of apoplastic L-ascorbic acid was also observed and annotated as an effect of AgNPs.

Significant upregulation of the genes involved in glutathione (GSH) biosynthesis on AgNP- exposure clearly demonstrated the toxic interferences of AgNPs in plants (Nair and Chung 2014b). Gene expression patterns in response to AgNPs were found similar to those in response to fungal infection and anion transport. The two major categories of genes responded were of oxidative stress and cell proliferation. An overlap in gene expression pattern was visible in response to $AgNO_3$ and AgNPs (Kaveh et al. 2013). Cell cycle regulating genes *AtPCNA1* and *AtPCNA2* were upregulated up to 72 hrs and observed to be downregulated after that. DNA mismatch repair genes *AtMSH2, AtMSH3, AtMSH6* and *AtMSH7* showed downregulation in the plants exposed (Nair and Chung 2014a, b, c). Gene expression results also revealed the prominent role of systemic signalling in toxic responses exhibited by the plant. Most of the comparative studies using AgNPs and Ag + performed for distinguishing their effects separately concluded that the toxic effects of AgNPs are because of the Ag^+ ions released from the particles. However, in a study using equivalent concentrations of the two, both were found producing similar effects denying the chances for Ag^+ ions to interfere (Zhang et al. 2019). Both induced ROS accumulation, reduction in efficiency of photosynthesis, and showed similarity in gene expression pattern. Genes associated with photosynthesis, oxidative stress, signal perception and response, etc. were found differentially expressed. Genes involved in the synthesis of Glucosinolates, the group of secondary metabolites were identified as specifically regulated by AgNPs, designating the triggering of defence mechanisms.

Nanoparticles are often used for delivery of other molecules that surface-bond with the particles, and their individual effects have to be differentiated from those of the combinations. Such an evaluation conducted for the conjugate of herbicide Imazethapyr with AgNPs demonstrates enhancement of the toxicity (Wen et al. 2016). AgNPs alone at 25 μM and 50 μM enhanced plant growth, but decreased chlorophyll content. However, the treatment did not increase the free amino acids as observed in the responses associated with heavy metal exposure. When leaves of the AgNP-treated plants were examined, amino acid content increase was noticed, and it was later confirmed due to the release of Ag^+ ions from the AgNPs. Outer surface of the roots was covered with AgNPs and their concentration was much less inside attesting their formation in conjunction with the Ag^+ moved inside. As observed in several other studies, ROS was not detected upon staining in the AgNP-treated group, but increased activities of detoxifying enzymes SOD and Catalase (CAT) were recorded. Toxicity observed was higher in the experimental group subjected to the herbicide alone and was still higher with the AgNP-herbicide conjugate. Inhibition of geotropic root growth was observed in response to high concentration of AgNP exposure (300 ng/L) and was identified to be because of the inhibition in auxin synthesis (Sun et al. 2017). Since expression of the concerned receptors was found downregulated and *AFB4*, a negative regulator of auxin signalling, as upregulated, the chances for blockade of auxin signalling were predicted. It might be occurring through the physical blockade created by the nanoparticles lodged at the intercellular spaces and inside the cells. The conjugate of herbicide Diclofop-methyl with AgNPs when used in Arabidopsis showed less inhibitory effects than AgNPs used alone (Li et al. 2018). Growth inhibition, increase in anthocyanin, accumulation of H_2O_2 and decrease in the rate of photosynthesis were observed in the experimental group with AgNP alone. However, the conjugate had reduced values in these parameters, which is presumed to occur due to the low stability of the Ag^+ ions released from the AgNPs. When particles of two sizes (10 and 60 nm) were exposed to the plants, most of them found aggregated on root surface and a very small percentage distributed inside (Wang et al. 2019).

Copper: Copper oxide nanoparticles (CuONPs) are not directly being utilized for applications related to agriculture. However, they have several other industrial applications because of their catalytic activity. Most of the studies conducted in Arabidopsis to assess the interaction of CuONPs were carried out along with Cu^+ for differentiating individual effects. Since the release of metal ions from nanoparticles observed in the case of few others, studying the two in parallel and comparing the effects can identify the specific effects of CuONPs. These particles are observed to be inhibiting growth, effected by reduction in biomass and inhibition of root growth (Tang et al. 2016; Wang et al. 2016a, b; Yuan et al. 2016; Landa et al. 2017; Zhao et al. 2018). On the contrary, the particles were found less toxic than Cu^+ in a similar investigation done in Arabidopsis (Ke et al. 2017). Loss of root gravitropism also has reported in a study conducted only with CuONPs (Nair and Chung 2014c). CuONPs were mainly found accumulated on the root surface and cell walls (Yuan et al. 2016) and induced changes in anatomy of the root, mainly manifested as lignin deposition, increase in width of cells and swelling at the root elongation zone (Wang et al. 2016a, b). Cells at root tips died due to the toxicity (Nair and Chung 2014c; Tang et al. 2016). A small portion of the particles were localized in aerial parts (Soria et al. 2019), which induced vacuole shrinkage and cell death (Yuan et al. 2016). Major physiological changes were generation of reactive oxygen species (Nair and Chung 2014c; Tang et al. 2016; Wang et al. 2016a, b; Yuan et al. 2016) and increase in anthocyanin (Nair and Chung 2014c; Ke et al. 2017). Increase in saturated fatty acids and decrease in unsaturated fatty acids lead to collapse of membranes (Yuan et al. 2016). CuONPs triggered the differential expression of a subset of genes regulating cell division and stress response. SOD, CAT and Peroxidase (PRX) were expressed in accordance with the ROS generated (Nair and Chung 2014c; Tang et al. 2016; Wang et al. 2016a, b; Landa et al. 2017). Auxin responsive genes (Wang et al. 2016a, b) and lignin biosynthesis related genes (Tang et al. 2016) were found upregulated. Stress induction and response was found triggered because of the changes in genes involved in proline biosynthesis and sulphur assimilation (Nair and Chung 2014c), heat shock proteins, methionine synthesis (Tang et al. 2016), metallo chaperonins and water deficiency response (Landa et al. 2017). Metabolite profiling identified the increase in isothiocyanates, scopoletin and jasmonates which are functioning in defence signalling and response (Soria et al. 2019).

Zinc: Zinc oxide nanoparticles (ZnONPs) are extensively used in cosmetics, especially because of its UV-reflective nature. Responses in Arabidopsis have been studied after supplementing into soil and artificial media of liquid and solid nature at different concentrations. Majority of the works reported reduction in growth expressed as low percentage of seed germination and the reduction in biomass contributed by the decrease in number of leaves, rosette size and the length of primary and the lateral roots (Lee et al. 2010; Landa et al. 2012; Landa et al. 2015; Wang et al. 2016a, b; Nair and Chung 2017; Vankova et al. 2017; Yang et al. 2018). But these interferences are reported to be not due to their internalization, rather than the adsorption on to root surface. Anthocyanin induction and reduction in chlorophyll content were two visible changes induced, which are characterized as stress indicators (Wang et al. 2016a, b; Vankova et al. 2017). However, a lateral increase in the level of carotenoids was also noticed along with that (Wang et al. 2016a, b). Reduction in the rate of photosynthesis, transpiration and conductance of leaf stomata were few other notable physiological changes observed. Increase in the concentration of Zn was experienced in tissues, which in turn affected the nutrient homeostasis resulting in reduction of total concentrations of K, S and Cu (Nair & Chung, 2017). Hormonal changes including increase of cytokinin in roots, ABA in leaves and apex, reduction of Zeatin and IAA in apex, etc. have been induced by ZnONPs (Vankova et al. 2017). This hormonal imbalance was accompanied with the reduction of jasmonic acid and jasmonate isoleucine in apex. Gene expression changes observed were somewhat similar to the other types of nanoparticles discussed earlier. Genes related to stress response, especially oxidative stress response, signal transduction, hypoxia, detoxification, wound response and defence, metal ion transport and homeostasis were found expressed in excess (Landa et al. 2012; Landa et al. 2015; Wang et al. 2016a, b; Nair and Chung 2017). Reduction of the rate of photosynthesis observed in few studies were later proved as due to the downregulation of genes involved in chlorophyll synthesis, photosystem I and electron transport (Landa et al. 2012; Wang et al. 2016a, b). Growth inhibition observed especially at the apices was also effected by the downregulation of the genes functioning in microtubule synthesis and protein translation in addition to the above-mentioned ones (Landa et al. 2012). ZnONPs also increased the frequency of homologous recombination and induced epigenetic changes which was tested and confirmed using transgenic Arabidopsis (Yang et al. 2018). Since entry of ZnONPs into the plant cells is prevented, it cannot be considered that the particles can induce the wide variety of responses we have seen (Nath et al. 2018). Based on the results of the studies conducted along with Zn^+ ions, it can be concluded that the responses are of the Zn^+ ions liberated into the growth media from the particles (Yang et al. 2018). However, it has been shown that the adsorption of the particles to root surface can induce signal transduction pathways which can trigger an array of changes in the plant (Landa et al. 2012 & Landa et al. 2015).

Iron: Iron oxide nanoparticles (FeONPs) are used in drug delivery, magnetic resonance imaging, groundwater treatments, photocatalytic reactions, environmental remediation, etc. Interaction with Arabidopsis demonstrated inhibitory effects at high concentrations expressing morphologically as reduction in growth rate (Marusenko et al. 2013; Bombin et al. 2015). As observed in the case of some other nanoparticles, lower concentrations produced promotive effects (Kim et al. 2014). Native particles were not observed to be transported into the plant, but their charged counterparts were found distributed in roots and different parts of the shoot including stem, leaves, flowers and seeds (Bombin et al. 2015). Reduction in chlorophyll (Marusenko et al. 2013), pollen viability, pollen tube length and number of seeds (Bombin et al. 2015) was also observed. Growth promotion at lower concentrations is observed to be happening in different ways. H_2O_2 formed on FeONP exposure induces loosening of cell wall, resulting in reduction in cell wall thickening, reorientation of microfibrils and increased incidence of endocytosis (Kim et al. 2014). Cell wall loosening has also increased leaf surface area (Kim et al. 2015). Increase in the activity of the plasma membrane H^+ ATPase activity due to the reduction of apoplastic pH increased stomatal opening facilitating increased intake of CO_2 without encountering excess water loss (Kim et al. 2015). Notwithstanding that *A. thaliana* offers the possibility of an in-depth investigation, especially for understanding the interactions at genetic level, none of the studies have attempted that.

Carbon: Carbon nanomaterials are available in different forms as fullerenes, nano-onions, nano-cones, nano-horns, carbon dots, carbon nanotubes, nano-beads, nano-fibres, nano-diamonds, and graphene. They are different in structure, size and shape and hence find applications in diverse fields like electronics, optics, nano-medicine, biosensors, renewable energy production, environmental remediation and as carriers for delivering molecules, metals, atoms, etc. Their responses in Arabidopsis have not been investigated in detail so far. Those that were studied in the plant are of carbon nanotubes (CNTs) examined at the cellular level. Single-walled carbon nanotubes (SWCNTs) were observed to be entering protoplasts and mesophyll cells (Shen et al. 2010; Yuan et al. 2012). They induce ROS, chromatin condensation and DNA breakage and result in cell death. ROS evolution confirmed by staining and expression analyses of the concerned genes is assumed to be inducing apoptosis (Shen et al. 2010). Multiwalled nanotubes (MWCNTs) also have somewhat similar effects, denoted by the reduction in cell viability, chlorophyll content, etc. (Lin et al. 2009). Both SWCNTs and MWCNTs form aggregates in media and inside the cells. Smaller aggregates of MWCNTs were found more inhibitory in effect than the larger ones (Lin et al. 2009).

Cerium: Cerium dioxide nanoparticles (CeO₂NPs) are used as a polishing material, additive in glass and ceramic, fuel cell material, in agricultural products and automotive industry. Their interaction with Arabidopsis has been investigated to some extent. CeO₂NPs at high concentrations inhibit growth, observed as redox ion in biomass of shoot and root (Yang et al. 2017). However, concentrations below 500 mg/L showed increase in biomass (Tumburu et al. 2017; Wu et al. 2018). Particles were found distributed in leaves and even inside chloroplast (Wu et al. 2017; Yang et al. 2017; Wu et al. 2018). Negatively charged particles enter easily through their interaction with the positively charged plasma membrane (Wu et al. 2017 & Wu et al. 2018). Increase in H_2O_2, MDA and the associated reduction in chlorophyll with respect to CeO_2NPs treatment designates toxicity development in the plant (Yang et al. 2017). On the other hand, negatively charged CeO_2NPs have opposite effects. They are capable of scavenging free radicals and thereby support to survive under stress (Wu et al. 2017 & Wu et al. 2018). These particles induce upregulation of the genes involved in transcription, ageing, H_2O_2 regulation, cell cycle, stress responsive genes in shoot which is accompanied by downregulation of auxin stimulus and cell wall modification (Tumburu et al. 2017). In roots, genes functioning in transcription, phenyl propanoid metabolism, seed maturation, and response to GA were upregulated and those of cell wall formation, syncytium formation, cell signalling, cell cycle and polysaccharide catabolism were downregulated. In essence, CeO_2NPs are identified as toxic. Experiments conducted with bulk CeO_2 along with CeO_2NPs proved the ineffectiveness of CeO_2 ions in inducing toxic changes equivalent to the CeO_2NPs (Yang et al. 2017). Hence it cannot be assumed that the Ce ions released from disintegrating nanoparticles in suspension are responsible for the changes.

Titanium: Titanium dioxide is the ingredient imparting white tint to almost all products used in industry, medicine, cosmetics, etc. Their range of applications widened as the size of the particles narrowed down. As a result, titanium dioxide nanoparticles (TiO_2NPs) emerged as the one produced in the largest quantity worldwide and has resulted in its release in large amount into the environment. Arabidopsis has been used to characterize the effects in plants brought about by the particles. Almost all studies done up to date are carried out either in solid or liquid growth media. Conjugation with the dye alizarin deduced their entry into almost all parts of the plant body (Kurepa et al. 2010). Lower concentrations promoted growth evidenced as increase in germination rate and biomass (García-Sánchez et al. 2015; Liu et al. 2017). Whereas the effect was opposite in higher concentrations bringing about reduction in biomass and yield (Liu et al. 2017). Prominent physiological impacts noticed were increase in chlorophyll and total protein at lower concentrations and their reversal upon treatment with higher concentrations. Higher concentrations elevated the activity of antioxidant enzymes SOD, CAT, POD and APX depicting oxidative stress (Liu et al. 2017). Reduction in chlorophyll, increase in oxygen and vitamin E production were also observed in case of higher concentrations. Chlorophyll breakdown results in phytol formation, which is being utilized for vitamin E production by *Vte5* gene product (Szymańska et al. 2016). Experiments with Alizarin conjugated particles identified that the particles disrupt microtubule network of the cell, which was characterized by the reduction in alpha tubulin, beta tubulin and ubiquitin with a corresponding increase in ubiquinated proteins (Wang et al. 2011). Increase in light absorption, fluorescence quantum yield of chloroplast, electron transfer, photolysis, oxygen evolution and hill reaction on exposure to low concentrations enhance the efficiency of photosynthesis and hence can be suggested for enhancement of the process of carbohydrate fixation (Ze et al. 2011). Expression profiling techniques have identified the genes functioning in light absorption, photosynthesis (Ze et al. 2011; Tumburu et al. 2015, 2017), nutrient transport (García-Sánchez et al. 2015; Tumburu et al. 2017), root development and growth (García-Sánchez et al. 2015; Tumburu et al. 2015, 2017), stress response and hormone response (Tumburu et al. 2015, 2017) gets differentially regulated during exposure to TiO_2NPs. Gene expression changes accompanied with morphological and physiological changes were upregulation of tocochromanol biosynthesis gene which utilizes phytol formed during the degradation of chlorophyll (Szymańska et al. 2016), downregulation of GST and GR with a corresponding increase in the activities of SOD, CAT, POD and Ascorbate Peroxidase (APX) in roots (Liu et al. 2017). Overall, TiO_2NPs enhance the light capturing capability and increase the efficiency of photosynthesis and hence beneficial to plants. However, we cannot conclude as beneficial in all respects because of the demonstration of breakdown of chlorophyll and the rupture of microtubule network associated.

Quantum Dots: These particles are mostly employed in electronic industry because of their electrical and luminescence properties. Some of them are also used for labelling or staining applications in biology. They are prepared as binary compounds such as lead sulphide, lead selenide, cadmium selenide, cadmium sulphide, cadmium telluride, iridium arsenide and iridium phosphite. Even though they are identified as toxic to animal system because of their small size, photolytic activity, high surface reactivity and mechanical stability, very few of them have been investigated for phytotoxicity, especially in Arabidopsis. Exposure of the plant to cadmium quantum dots formed in combination with selenium, sulphur and zinc produced differential responses according to the concentration, type of treatment and its duration. Cadmium sulphide quantum dots (CdS QDs) in high concentration inhibited seed germination and root growth (Marmiroli et al. 2014; 2020). Inhibition of the root growth was preceded by swelling and bursting which is assumed to be due to the clogging of the vasculature due to aggregation of the particles. Appearance of chlorosis, necrotic regions, increased wax deposition and reduction in stomatal and trichome density on leaves clearly indicated toxicity. Overall reduction in respiration was noticed. Increase in the biosynthesis of phenolic compounds, glutathione, antioxidant activity and lipid peroxidation indicated evolution of ROS. Predominant increase in the expression of genes associated with heat shock, temperature stress regulation, ROS metabolism and ion transport reinforced the assumption. Comparison of the effects with that of $CdSO_4$ helped to differentiate the specific effects of CdS QDs (Marmiroli et al. 2014, 2020). For understanding the role of surface charge in uptake and transport, cadmium selenide QDs (CdSe QDs) were coated with charged molecules and supplied to plant through liquid growth medium (Koo et al. 2015a, b). Charged particles observed to be moving fast through the vascular tissue and getting distributed into almost all parts of the plant. On the other hand, much of the neutral QDs formed aggregates and found lodged at short distances. Finally, anionic particles were localized in the apoplast and the cationic ones intracellular. Anionic particles induced chlorosis and their cationic counterparts produced browning of the leaves. Hence experimental evidences indicate that surface properties have determinant role in the uptake and transport of QDs. Experiments with CdSe QDs coupled with salicylic acid binding moiety were found successful in internalization and proved the utility of modifying surface properties for intracellular targeting (Liu et al. 2015).

Zinc selenide QDs (ZnSe QDs) when used at two different concentrations (100 & 250 µM) exhibited different responses (Kolackova et al. 2019). Increase in gallic acid, phytochelatin, phenols, H_2O_2 scavenging and antioxidant activity were induced by the lower dose, and the higher one was characterized by a reduction in membrane lipidemic compounds, H_2O_2 scavenging, and antioxidant activity. Expression of genes in phytochelatin biosynthesis found upregulated in case of 100 µM and the genes in glutathione synthesis and ROS scavenging in 250 µM group. Overall, quantum dots at higher concentrations are inhibitory to plant growth and development.

2 Conclusion and the Future Perspectives

Seed germination and the changes in growth pattern of root and shoot are considered for the assessment of the effects due to exposure to nanomaterials. The parameters used are not reliable because of the lack of phenotypes for most of the changes occurring at molecular level. Results vary depending upon concentration, size, shape, surface charge and the chemical composition of the nanomaterials tested. Hence, it is important to carry out nanoparticle testing with the parameters for assessing changes in physiology at molecular level. Nanomaterials at low concentrations show growth promotion. In case of FeONPs, the changes associated with growth enhancement were persuaded by the H_2O_2 generated inside. H_2O_2 induces loosening of cell wall and reduction of cell wall thickening, reorientation of microfibrils and increased incidence of endocytosis (Kim et al. 2014). Cell wall loosening also results in an increased leaf surface area (Kim et al. 2015). Reduction in apoplastic pH increases the activity of the plasma membrane H + ATPase activity which, in turn, augment stomatal opening facilitating increased intake of CO_2 without encountering excess water loss (Kim et al. 2015). Growth enhancement can also happen through the improvement of the efficiency of carbohydrate fixation as observed in case of TiO_2NPs (Ze et al. 2011). Thus, the mode of operation of growth enhancement is different in case of different nanomaterials and demands more extensive studies needed to figure out the details. On the contrary, exposure to high concentration of nanoparticles induces stress characterized by the production of anthocyanin and inhibition of chlorophyll biosynthesis (Nair and Chung 2014b; Wang et al. 2016a, b; Vankova et al. 2017). Stress signalling is also mediated through ROS (Baghkheirati and Lee 2015; Wang et al. 2016a, b; Landa et al. 2017; Liu et al. 2017). Chlorophyll degradation is activated by the ROS developed as byproducts inside the chloroplast (Rogers and Munné-Bosch 2016). ROS can also damage a large variety of cellular biomolecules, including carbohydrates, nucleic acids, lipids, proteins, etc., and alter their functions. Antioxidant enzymes at this point prevent the burst in ROS level and prevent sudden cell death. This is accompanied with the changes in the GSSG/GSH and MDA pool (Hasanuzzaman et al. 2019). Anthocyanin having antioxidant activity is also produced in excess to act against ROS (Zhang et al. 2012). Changes in gene expressions are known in case of some nanomaterials which facilitate to monitor the effects at molecular level and to establish the link between the pathways of signal perception and response.

Nanomaterials can enter the plant body through different routes because of their presence in air, water and soil. In majority of the studies, the nanomaterials tested have been applied into the growth medium owing to their absorption through the root system. Uptake through aerial parts, mainly through openings like stomata need to be considered and application methods including foliar spray should be examined (Kolackova et al. 2019). As the mode of translocation will be different in this case, its impact with respect to the root to shoot translocation should be compared. Arabidopsis being a plant with small, slender vasculature, the possibilities for in vivo localization or accumulation remain unexplored for majority of nanomaterials. Another important limitation of the studies conducted so far is the short period of exposure, mostly done for 2–3 weeks. That kind of an experimental set-up cannot mimic the actual conditions prevalent in any ecosystems. The short life cycle of Arabidopsis is ideal for characterizing the responses at different stages of growth and few studies have explored these possibilities (Ke et al. 2018). The type of nanomaterials and their volume released into the environment is different according to their cost of production and usage. For instance, the production of TiO_2NPs is higher with respect to AuNPs. Accordingly, we can expect a higher percentage of TiO_2NPs in the environment than the latter. Stability of nanomaterials in the environment should also be verified. They can interact with other chemical moieties in air, water and soil and get transformed to higher reactive or toxic forms. This can also happen once they are inside the plant. Many nanomaterials form aggregates in plants and it will be interesting to know whether the conditions for aggregate formation are induced by plant molecules to prevent their movement and reactivity. The important advantages of Arabidopsis as an experimental plant are the availability of its sequenced genome and of gene overexpression and mutant lines. Nevertheless, gene expression studies at the transcriptomic level have been conducted in response to nanomaterial exposure; few attempts have been made to testify the effects using gene overexpression and mutant lines (Yang et al. 2018). Hence, we can expect more realistic studies in Arabidopsis to characterize the effect of nanomaterials in future.

References

Adisa, I. O., Pullagurala, V. L. R., Peralta-Videa, J. R., Dimkpa, C. O., Elmer, W. H., Gardea-Torresdey, J. L., & White, J. C. (2019). Recent advances in nano-enabled fertilizers and pesticides: A critical review of mechanisms of action. *Environmental Science: Nano, 6*, 2002–2030. https://doi.org/10.1039/C9EN00265K.

Avellan, A., Schwab, F., Masion, A., Chaurand, P., Borschneck, D., Vidal, V., Rose, J., Santaella, C., & Levard, C. (2017). Nanoparticle uptake in plants: Gold nanomaterial localized in roots of *Arabidopsis thaliana* by X-ray computed nanotomography and hyperspectral imaging. *Environmental Science and Technology, 51*, 8682––8691.

Baghkheirati, E. K., & Lee, J. G. (2015). Gene expression, protein function and pathways of *Arabidopsis thaliana* responding to silver nanoparticles in comparison to silver ions, cold, salt, drought, and heat. *Nanomaterials, 5*, 436–467. https://doi.org/10.3390/nano5020436.

Bao, D., Oh, Z. G., & Chen, Z. (2016). Characterization of silver nanoparticles internalized by Arabidopsis plants using single particle ICP-MS analysis. *Frontiers in Plant Science, 7*, 32. https://doi.org/10.3389/fpls.2016.00032.

Bombin, S., LeFebvre, M., Sherwood, J., Xu, Y., Bao, Y., & Ramonell, K. M. (2015). Developmental and reproductive effects of iron oxide nanoparticles in Arabidopsis thaliana. *International Journal of Molecular Sciences, 16*, 24174–24193.

Chaudhry, N., Dwivedi, S., Chaudhry, V., Singh, A., Saquib, Q., Azam, A., & Musarrat, J. (2018). Bio-inspired nanomaterials in agriculture and food: Current status, foreseen applications and challenges. *Microbial Pathogenesis, 123*, 196–200. https://doi.org/10.1016/j.micpath.2018.07.013.

García-Sánchez, S., Bernales, I., & Cristobal, S. (2015). Early response to nanoparticles in the Arabidopsis transcriptome compromises plant defence and root-hair development through salicylic acid signalling. *BMC Genomics, 16*, 341. https://doi.org/10.1186/s12864-015-1530-4.

Geisler-Lee, J., Brooks, M., Gerfen, J. R., Wang, Q., Fotis, C., Sparer, A., Ma, X., Berg, R. H., & Geisler, M. (2014). Reproductive toxicity and life history dtudy of silver nanoparticle effect, uptake and transport in *Arabidopsis thaliana. Nanomaterials, 4*, 301–318. https://doi.org/10.3390/nano4020301.

Hasanuzzaman, M., Bhuyan, M. H. M. B., Anee, T. I., Parvin, K., Nahar, K., Mahmud, J. A. I., & Fujita, M. (2019). Regulation of ascorbate-glutathione pathway in mitigating oxidative damage in plants under abiotic stress. *Antioxidants, 8*, 384. https://doi.org/10.3390/antiox8090384.

Hayles, J., Johnson, L., Worthley, C., & Losic, D. (2017). *Nanopesticides: A review of current research and perspectives In new pesticides and soil sensors* (pp. 193–225). Academic Press. https://doi.org/10.1016/B978-0-12-804299-1.00006-0.

Hendel, A. M., Zubko, M., Stróz, D., & Kurczynska, E. U. (2019). Effect of nanoparticles surface charge on the *Arabidopsis thaliana* (L.) roots development and their movement into the root cells and protoplasts. *International Journal of Molecular Science, 20*, 1650. https://doi.org/10.3390/ijms20071650.

Jain, A., Sinilal, B., Starnes, D. L., Sanagala, R., Krishnamurthy, S., & Sahi, S. V. (2014). Role of Fe-responsive genes in bioreduction and transport of ionic gold to roots of *Arabidopsis thaliana* during synthesis of gold nanoparticles. *Plant Physiology and Biochemistry, 84*, 189–196.

Kaveh, R., Li, Y. S., Ranjbar, S., Tehrani, R., Brueck, C. L., & Aken, B. V. (2013). Changes in Arabidopsis thaliana gene expression in response to silver nanoparticles and silver ions. *Environmental Science & Technology, 47*, 10637–10644.

Ke, M., Qu, Q., Peijnenburg, W. J. G. M., Li, X., Zhang, M., Zhang, Z., Lu, T., Pan, X., & Qian, H. (2018). Phytotoxic effects of silver nanoparticles and silver ions to Arabidopsis thaliana as revealed by analysis of molecular responses and of metabolic pathways. *Science of the Total Environment, 644*, 1070–1079.

Ke, M., Zhu, Y., Zhang, M., Gumai, H., Zhang, Z., Xu, J., & Qian, H. (2017). Physiological and molecular response of *Arabidopsis thaliana* to CuO nanoparticle (nCuO) exposure. *Bulletin of Environmental Contamination and Toxicology, 99*, 713–718.

Kim, J. H., Lee, Y., Kim, E. J., Gu, S., Sohn, E. J., Seo, Y. S., An, H. J., & Chang, Y. S. (2014). Exposure of iron nanoparticles to Arabidopsis thaliana enhances root elongation by triggering cell wall loosening. *Environmental Science and Technology, 48*, 3477–3485.

Kim, J. H., Oh, Y., Yoon, H., Hwang, I., & Chang, Y. S. (2015). Iron nanoparticle-induced activation of plasma membrane H+-ATPase promotes stomatal opening in Arabidopsis thaliana. *Environmental Science and Technology, 49*, 1113–1119.

Kolackova, M., Moulick, A., Kopel, P., Dvorak, M., Adam, V., Klejdus, B., & Huska, D. (2019). Antioxidant, gene expression and metabolomics fingerprint analysis of *Arabidopsis thaliana* treated by foliar spraying of ZnSe quantum dots and their growth inhibition of agrobacterium tumefaciens. *Journal of Hazardous Materials, 365*, 932–941.

Koo, Y., Ekaterina, Y., Lukianova-Hleb, E. Y., Pan, J., Thompson, S. M., Lapotko, D. M., & Braam, J. (2015b). *In planta* response of Arabidopsis to photothermal impact mediated by gold nanoparticles. *Small, 12*, 623–630.

Koo, Y., Wang, J., Zhang, Q., Zhu, H., Chehab, E. W., Colvin, V. L., Alvarez, P. J. J., & Braam, J. (2015a). Fluorescence reports intact quantum dot uptake into roots and translocation to leaves of *Arabidopsis thaliana* and subsequent ingestion by insect herbivores. *Environmental Science and Technology, 49*, 626–632.

Kumar, V., Guleria, P., Kumar, V., & Yadav, S. K. (2013). Gold nanoparticle exposure induces growth and yield enhancement in Arabidopsis thaliana. *Science of the Total Environment, 461–462*, 462–468.

Kurepa, J., Paunesku, T., Vogt, S., Arora, H., Rabatic, B. M., Lu, J., Wanzer, M. B., Woloschak, G. E., & Smalle, J. A. (2010). Uptake and distribution of ultrasmall anatase TiO2 alizarin red S nanoconjugates in Arabidopsis thaliana. *Nano Letters, 10*, 2296–2302.

Landa, P., Dytrych, P., Prerostova, S., Petrova, S., Vankova, R., & Vanek, T. (2017). Transcriptomic response of Arabidopsis thaliana exposed to CuO nanoparticles, bulk material, and ionic copper. *Environmental Science and Technology, 51*, 10814–10824.

Landa, P., Prerostova, S., Petrova, S., Knirsch, V., Vankova, R., & Vanek, T. (2015). The transcriptomic response of *Arabidopsis thaliana* to zinc oxide: A comparison of the impact of nanoparticle, bulk, and ionic zinc. *Environmental Science and Technology, 49*, 14537–14545.

Landa, P., Vankova, R., Andrlova, J., Hodek, J., Marsik, P., Storchova, H., White, J. C., & Vanek, T. (2012). Nanoparticle-specific changes in Arabidopsis thaliana gene expression after exposure to ZnO, TiO2, and fullerene soot. *Journal of Hazardous Materials, 241-242*, 55–62.

Lee, C. W., Mahendra, S., Katherine, Z., Li, D., Tsai, Y. C., Braam, J., & Pedro, J. J. A. (2010). Developmental phytotoxicity of metal oxide nanoparticles to Arabidopsis thaliana. *Environmental Toxicology and Chemistry, 29*, 669–675.

Li, X., Ke, M., Zhang, M., Peijnenburg, W. J. G. M., Fan, X., Xu, J., Zhang, Z., Lu, T., Fu, Z., & Qian, H. (2018). The interactive effects of diclofop-methyl and silver nanoparticles on Arabidopsis thaliana: Growth, photosynthesis and antioxidant system. *Environmental Pollution, 232*, 212–219.

Lin, C., Fugetsu, B., Su, Y., & Watari, F. (2009). Studies on toxicity of multi-walled carbon nanotubes on Arabidopsis T87 suspension cells. *Journal of Hazardous Materials, 170*, 578–583.

Liu, H., Ma, C., Chen, G., White, J. C., Wang, Z., Xing, B., & Dhankher, O. P. (2017). Titanium dioxide nanoparticles alleviate tetracycline toxicity to *Arabidopsis thaliana* (L.). *ACS Sustainable Chemistry and Engineering, 5*, 3204–3213.

Liu, J. W., Deng, D. Y., Yu, Y., Liu, F. F., Lin, B. X., Cao, Y. J., Hu, X. G., & Wu, J. Z. (2015). In situ detection of salicylic acid binding sites in plant tissues. *Luminescence, 30*, 18–25.

Marmiroli, M., Mussi, F., Pagano, L., Imperiale, D., Lencioni, G., Villani, M., Zappettini, A., White, J. C., & Marmiroli, N. (2020). Cadmium sulfide quantum dots impact on Arabidopsis Thaliana physiology and morphology. *Chemosphere, 240*, 124856.

Marmiroli, M., Pagano, L., Savo, S. M. L., Villani, M., & Marmiroli, N. (2014). Genome-wide approach in Arabidopsis thaliana to assess the toxicity of cadmium sulfide quantum dots. *Environmental Science and Technology, 48*, 5902–5909.

Marusenko, Y., Jessie Shipp, J., Hamilton, G. A., Morgan, J. L. L., Keebaugh, M., Hill, H., Dutta, A., Zhuo, X., Upadhyay, N., Hutchings, J., Herckes, P., Anbar, A. D., Shock, E., & Hartnett, H. E. (2013). Bioavailability of nanoparticulate hematite to Arabidopsis thaliana. *Environmental Pollution, 174*, 150–156.

Montes, A., Bisson, M. A., Gardella, J. A., & Aga, D. S. (2017). Uptake and transformations of engineered nanomaterials: Critical responses observed in terrestrial plants and the model plant Arabidopsis thaliana. *Science of the Total Environment, 607-608*, 1497–1596. https://doi.org/10.1016/j.scitotenv.2017.06.190.

Nair, P. M. G., & Chung, I. M. (2014a). Cell cycle and mismatch repair genes as potential biomarkers in Arabidopsis thaliana seedlings exposed to silver nanoparticles. *Bulletin of Environmental Contamination and Toxicology, 92*, 719–725. https://doi.org/10.1007/s00128-014-1254-1.

Nair, P. M. G., & Chung, I. M. (2014b). Assessment of silver nanoparticle-induced physiological and molecular changes in Arabidopsis thaliana. *Environmental Science and Pollution Research, 21*, 8858–8869. https://doi.org/10.1007/s11356-014-2822-y.

Nair, P. M. G., & Chung, I. M. (2014c). Impact of copper oxide nanoparticles exposure on Arabidopsis thaliana growth, root system development, root lignification, and molecular level changes. *Environmental Science and Pollution Research, 21*, 12709–12722.

Nair, P. M. G., & Chung, I. M. (2017). Regulation of morphological, molecular and nutrient status in Arabidopsis thaliana seedlings in response to ZnO nanoparticles and Zn ion exposure. *Science of the Total Environment, 575*, 187–198.

Nath, J., Dror, I., Landa, P., Vanek, T., Ashiri, I. K., & Berkowitz, B. (2018). Synthesis and characterization of isotopically-labeled silver, copper and zinc oxide nanoparticles for tracing studies in plants. *Environmental Pollution, 242*, 1827–1837.

Ojha, S., Singh, D., Sett, A., Chetia, H., Kabiraj, D., & Bora, U. (2018). Nanotechnology in crop protection *In* Nanomaterials in plants, algae, and microorganisms: Concepts and controversies., *1*, 345–391. https://doi.org/10.1016/B978-0-12-811487-2.00016-5.

Riley, M. K., & Vermerris, W. (2017). Recent advances in nanomaterials for gene delivery—A review. *Nanomaterials, 7*, 94. https://doi.org/10.3390/nano7050094.

Rogers, H., & Munné-Bosch, S. (2016). Production and scavenging of reactive oxygen species and redox signaling during leaf and flower senescence: Similar but different. *Plant Physiology, 171*, 1560–1568.

Shankar, S. S., Rai, A., Ankamwar, B., Singh, A., Ahmad, A., & Sastry, M. (2004). Biological synthesis of triangular gold nano prisms. *Nature Materials, 3*, 482–488.

Shen, C. X., Zhang, Q. F., Li, J., Bi, F. C., & Yao, N. (2010). Induction of programmed cell death in Arabidopsis and rice by single-wall carbon nanotubes. *American Journal of Botany, 97*, 1602–1609.

Siegel, J., Záruba, K., Švorčík, V., Kroumanová, K., Burketová, L., & Martinec, J. (2018). Round-shape gold nanoparticles: Effect of particle size and concentration on Arabidopsis thaliana root growth. *Nanoscale Research Letters, 13*(95). https://doi.org/10.1186/s11671-018-2510-9.

Soria, N. G. C., Bisson, M. A., Gokcumen, G. E. A., & Aga, D. S. (2019). High-resolution mass spectrometry-based metabolomics reveal the disruption of jasmonic pathway in Arabidopsis thaliana upon copper oxide nanoparticle exposure. *Science of the Total Environment, 693*, 133443.

Sosan, A., Svistunenko, D., Straltsova, D., Tsiurkina, K., Smolich, I., Lawson, T., Subramaniam, S., Golovko, V., Anderson, D., Sokolik, A., Colbeck, I., & Demidchik, V. (2016). Engineered silver nanoparticles are sensed at the plasma membrane and dramatically modify the physiology of *Arabidopsis thaliana* plants. *The Plant Journal, 85*, 245–257. https://doi.org/10.1111/tpj.13105.

Sun, J., Wang, L., Li, S., Yin, L., Huang, J., & Chen, C. (2017). Toxicity of silver nanoparticles to Arabidopsis: Inhibition of root gravitropism by interfering with auxin pathway. *Environmental Toxicology and Chemistry, 36*, 2773–2780.

Syu, Y., Hung, J. H., Chen, J. C., & Chuang, H. (2014). Impacts of size and shape of silver nanoparticles on Arabidopsis plant growth and gene expression. *Plant Physiology and Biochemistry, 83*, 57–64.

Szymańska, R., Kołodziej, K., Ślesak, I., Zimak-Piekarczyk, P., Orzechowska, A., Gabruk, M., Zadło, A., Habina, I., Knap, W., Burda, K., & Kruk, J. (2016). Titanium dioxide nanoparticles (100-1000 mg/l) can affect vitamin E response in Arabidopsis thaliana. *Environmental Pollution, 213*, 957–965.

Tang, Y., He, R., Zhao, J., Nie, G., Xu, L., & Xing, B. (2016). Oxidative stress-induced toxicity of CuO nanoparticles and related toxicogenomic responses in Arabidopsis thaliana. *Environmental Pollution, 212*, 605–614.

Taylor, A. F., Rylott, E. L., Anderson, C. W. N., & Bruce, N. C. (2014). Investigating the toxicity, uptake, nanoparticle formation and genetic response of plants to gold. *PLoS One, 9*(4), e93793. https://doi.org/10.1371/journal.pone.0093793.

Tiwari, M., Krishnamurthy, S., Shukla, D., Kiiskila, J., Jain, A., Datta, R., Sharma, N., & Sahi, S. V. (2016). Comparative transcriptome and proteome analysis to reveal the biosynthesis of gold nanoparticles in Arabidopsis. *Scientific Reports, 6*, 21733. https://doi.org/10.1038/srep21733.

Tumburu, L., Andersen, C. P., Rygiewicz, P. T., & Reichman, J. R. (2015). Phenotypic and genomic responses to titanium dioxide and cerium oxide nanoparticles in Arabidopsis germinants. *Environmental Toxicology and Chemistry, 34*, 70–83.

Tumburu, L., Andersen, C. P., Rygiewicz, P. T., & Reichman, J. R. (2017). Molecular and physiological responses to titanium dioxide and cerium oxide nanoparticles in Arabidopsis. *Environmental Toxicology and Chemistry, 36*, 71–82. https://doi.org/10.1002/etc.3500.

Vankova, R., Landa, P., Podlipna, R., Dobrev, P. I., Prerostova, S., Langhansova, L., Gaudinova, A., Motkova, K., Knirsch, V., & Vanek, T. (2017). ZnO nanoparticle effects on hormonal pools in Arabidopsis thaliana. *Science of the Total Environment, 593-594*, 535–542.

Wang, J., Koo, Y., Alexander, A., Yang, Y., Westerhof, S., & Zhang, Q. (2013). Phytostimulation of Poplars and Arabidopsis exposed to silver nanoparticles and Ag+ at sublethal concentrations. *Environmental Science and Technology, 47*, 5442–5449. https://doi.org/10.1021/es4004334.

Wang, S., Kurepa, J., & Smalle, J. A. (2011). Ultra-small TiO2 nanoparticles disrupt microtubular networks in Arabidopsis thaliana. *Plant Cell and Environment, 34*, 811–820.

Wang, T., Wu, J., Xu, S., Deng, C., Wu, L., Wu, Y., & Bian, P. (2019). A potential involvement of plant systemic response in initiating genotoxicity of ag-nanoparticles in Arabidopsis thaliana. *Ecotoxicology and Environmental Safety, 170*, 324–330.

Wang, X., Yang, X., Chen, S., Li, Q., Wang, W., Hou, C., Gao, X., Wang, L., & Wang, S. (2016a). Zinc oxide nanoparticles affect biomass accumulation and photosynthesis in Arabidopsis. *Frontiers in Plant Science, 6*, 1–9.

Wang, Z., Xu, L., Zhao, J., Wang, X., White, J. C., & Xing, B. (2016b). CuO nanoparticle interaction with Arabidopsis thaliana: Toxicity, parent-progeny transfer, and gene expression. *Environmental Science and Technology, 50*, 6008--6016.

Wen, Y., Zhang, L., Chen, Z., Sheng, X., Qiu, J., & Xu, D. (2016). Co-exposure of silver nanoparticles and chiral herbicide imazethapyr to Arabidopsis thaliana: Enantioselective effects. *Chemosphere, 145*, 207–214.

Wu, H., Shabala, L., Shabala, S., & Giraldo, J. P. (2018). Hydroxyl radical scavenging by cerium oxide nanoparticles improves Arabidopsis salinity tolerance by enhancing leaf mesophyll potassium retention. *Environmental Science: Nano, 5*, 1567–1583.

Wu, H., Tito, N., & Giraldo, J. P. (2017). Anionic cerium oxide nanoparticles protect plant photosynthesis from abiotic stress by scavenging reactive oxygen species. *ACS Nano, 11*, 11283–11297.

Yang, A., Wu, J., Deng, C., Wang, T., & Bian, P. (2018). Genotoxicity of zinc oxide nanoparticles in plants demonstrated using transgenic Arabidopsis thaliana. *Bulletin of Environmental Contamination and Toxicology, 101*, 514–520.

Yang, X., Pan, H., Wang, P., & Zhao, F. J. (2017). Particle-specific toxicity and bioavailability of cerium oxide (CeO2) nanoparticles to Arabidopsis thaliana. *Journal of Hazardous Materials, 322*, 292–300.

Yuan, H., Hu, S., Huang, P., Song, H., Wang, K., Ruan, J., He, R., & Cui, D. (2012). Single walled carbon nanotubes exhibit dual-phase regulation to exposed Arabidopsis mesophyll cells. *Nanoscale Research Letters, 7*, 1–9.

Yuan, J., He, A., Huang, S., Hua, J., & Sheng, G. D. (2016). Internalization and phytotoxic effects of CuO nanoparticles in Arabidopsis thaliana as revealed by fatty acid profiles. *Environmental Science and Technology, 50*, 10437--10447.

Ze, Y., Liu, C., Wang, L., Hong, M., & Hong, F. (2011). The regulation of TiO2 nanoparticles on the expression of light-harvesting complex II and photosynthesis of chloroplasts of Arabidopsis thaliana. *Biological Trace Element Research, 143*, 1131–1141.

Zhang, C. L., Jiang, H. S., Gu, S. P., Zhou, X. H., Lu, Z. W., Kang, X. H., Yin, L., & Huang, J. (2019). Combination analysis of the physiology and transcriptome provides insights into the mechanism of silver nanoparticles phytotoxicity. *Environmental Pollution, 252*, 1539–1549.

Zhang, Q., Su, L. J., Chen, J. W., Zeng, X. Q., Sun, B. Y., & Peng, C. L. (2012). The antioxidative role of anthocyanins in Arabidopsis under high-irradiance. *Biologia Plantarum, 56*, 97–104.

Zhao, S., Dai, Y., & Xu, L. (2018). Toxicity and transfer of CuO nanoparticles on Arabidopsis thaliana. *IOP Conference series: Earth and Environmental Science, 113*, 012021. https://doi.org/10.1088/1755-1315/113/1/012021.

Zulfiqar, F., Navarro, M., Ashraf, M., Akram, N. A., & Munné-Bosch, S. (2019). Nano fertilizer use for sustainable agriculture: Advantages and limitations. *Plant Science, 289*, 110270. https://doi.org/10.1016/j.plantsci.2019.110270.

Chapter 5
Metal Oxide Nanoparticle Toxicity in Aquatic Organisms: An Overview of Methods and Mechanisms

J. G. Parsons, Mataz Alcoutlabi, and R. K. Dearth

Contents

J. G. Parsons (✉)
Department of Chemistry, University of Texas, Brownsville, TX, USA
e-mail: jason.parsons@utrgv.edu

M. Alcoutlabi
Department of Mechanical Engineering, University of Texas, Rio Grande Valley,
Edinburg, TX, USA

R. K. Dearth
Department of Biology, University of Texas, Rio Grande Valley, Edinburg, TX, USA

© Springer Nature Switzerland AG 2021
N. Sharma, S. Sahi (eds.), *Nanomaterial Biointeractions at the Cellular,
Organismal and System Levels*, Nanotechnology in the Life Sciences,
https://doi.org/10.1007/978-3-030-65792-5_5

1 Introduction and Background

Nanostructured materials have been widely used in different applications including energy storage, catalysis, agriculture systems, and biomedicine. Nanomaterials (NMs) are an anomalous class of compounds that are known for their high toxicity when exposed to living organisms. The investigation of the behavior of toxic nano-materials for human exposure (e.g., nanoparticles, nanoflakes, and nanorods) has attracted much attention in recent years due to the fact that there is no quantitative evaluation to address its effects on human health and the environment. The anoma-lous behavior of nanomaterials such as nanoparticles (NP) is originated by the fact that these materials have strained surfaces, and generally have an oxidized layer of atoms at the surface (Zhang et al. 2014; Strasser et al. 2010; Casaletto et al. 2006). NMs have both natural and anthropogenic sources (Griffin et al. 2018; Rogers 2016; Jeevanandam et al. 2018; Currie and Silica in Plants 2007; Hough et al. 2008). NPs are commonly found in the environment from natural sources, for example, nanopar-ticles such as SiO_2, and Fe_3O_4 are commonly found in soils and plants (Griffin et al. 2018; Rogers 2016; Jeevanandam et al. 2018; Currie and Silica in Plants 2007; Hough et al. 2008). However, concerns of toxicity can raise awareness when using nanoparticles of anthropogenic origins in consumer products (Vance et al. 2015; Fröhlich and Roblegg 2012; Calzolai et al. 2012; Tulve et al. 2015). For example, currently there are more than 1800 consumer products based on NPs used in the market (Vance et al. 2015). Some of the applications of these products are intended for direct use in human health such as skin protection or treatment; for example, new cosmetics are being produced compounds based on NPs such as zinc oxide (ZnO) nanoparticles used in sunscreens (Vance et al. 2015; Fröhlich and Roblegg 2012; Calzolai et al. 2012; Tulve et al. 2015). Health drinks, antibacterial, antimi-crobial products are also being produced from noncompounds containing metallic NPs such as silver and copper (Vance et al. 2015; Fröhlich and Roblegg 2012; Calzolai et al. 2012; Tulve et al. 2015). A further example of the invasive application of NPs in consumer products is the generation of clothing impregnated with metal-lic NPs in the fabric in order to reduce the odor-producing bacteria (Tulve et al. 2015). However, the actual human and environmental health effects of NPs are not well understood. The literature on the toxicity of NPs is lacking and in some cases contradictory in nature. As time progresses, the risk of exposure to anthropogenic synthesized nanoparticles has been constantly increasing and may have unknown possible health effects. The toxic effect of these nanomaterials with respect to the effects on human health has not been systematically investigated.

Based on the interaction studies of a large number of species, the main toxicity effects or routs of toxicity can be summarized as in Fig. 5.1.

Most of the investigations reported in this area discuss the toxicity effects of NPs on living organisms, which have been focused for the most part on bacterial studies. However, studies have been reported on the toxicity of NPs in eukaryotic and com-plex organisms such as algae and multicellular aquatic organisms, but a clear mech-anism and the mode of nanomaterial actions are lacking. The generation of ROS

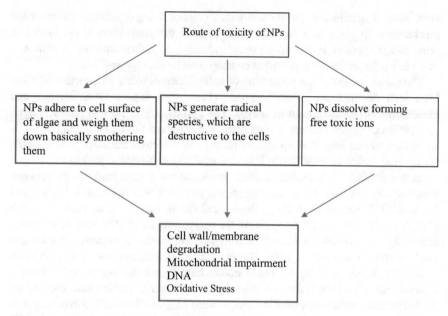

Fig. 5.1 Diagrammatic summary of the toxic effects of various NPs on cells and organisms found in the literature

Table 5.1 Common radical species generated in natural waters using sunlight and common compounds found in natural waters and the possible influence of photoactive NPs on radical generation (Stumm and Morgan 1996; Parke and Sapota 1996; Nosaka and Generation 2017; Sharma et al. 2019)

Name	Chemical Formula	Possible RXN
Singlet oxygen	$^1O_2$	Light interaction with DO, catalyzed with NPs
Super oxide	$O_2^{\bullet-}$	Deprotonation of $OH2^{\bullet}$ photolysis of metal complex in solution, photolytic product of Fenton like rxn of TiO_2 NPs
Hydrogen peroxyl	$OH_2^{\bullet}$	Found developed from atmosphere, protonation of the superoxide radical
Hydroxyl radical	$Oh^{\bullet}$	Photolysis of hydroxo or Fenton type reaction, surface oxidation on TiO_2 NPs, NO_3^-, NO_2^-, ozone decomposition, photolysis of peroxide
Organic peroxyl radical	$ROO^{\bullet}$	DO photolysis, electron transfer from surface of semiconductor NPs such as TiO_2 and ZnO

RXN is short for reaction

species has been shown to occur in natural waters using photoactive nanoparticles; the generation of ROS has been shown to occur in algal cells (Li et al. 2003). The generation of ROS species produced from photoactive NPs can be exemplified in Table 5.1. The generation of radical species including ROS is the process that occurs in natural water systems in the presence of sunlight. Interestingly, the reactive species generated from these chemical reactions may be amplified in the presence of

NPs. Radical species are also formed within cells of living organisms and there are mechanisms in place to reduce the damage from the generation of the ROS and other radial species. However, an overproduction of radical species in cells can overwhelm the defense mechanisms causing cellular damage and death.

The enhancement of the generation of radical species using photoactive NPs has been well documented in the literature (Parke and Sapota 1996; Nosaka and Generation 2017; Sharma et al. 2019). The application of NPs in photocatalytic destruction of organic chemicals has also been achieved through applying the ability of NPs to generate ROS species or through the electron transfers from the NPs (Deng et al. 2015; Le et al. 2015; Haldorai et al. 2014; Kumar et al. 2014).

In the studies on bacterial species, results show a high correlation between nanoparticles and their differential toxicity to different bacterial strains (Prabhu and Poulose 2012; Hwang et al. 2008; Durán et al. 2010; Fabrega et al. 2009; Greulich et al. 2012; Durán et al. 2016). Perhaps, the best example of NP toxicity in the literature is Ag-NP toxicity on *E. coli* as well as other common bacteria (Prabhu and Poulose 2012; Hwang et al. 2008; Durán et al. 2010; Fabrega et al. 2009; Greulich et al. 2012; Durán et al. 2016). These studies have shown that Ag in the NP form is typically more toxic than the dissolved Ag^+ ion. Bacterial studies have focused on the bactericidal effects of metallic nanoparticles such as silver NPs. Silver nanoparticles have shown to be toxic to both gram-negative and gram-positive bacteria. It is interesting that nanoparticles affect both the gram-positive and -negative bacteria, and NPs have been shown to transport across the cell walls of bacteria. Figs. 5.2 and 5.3 show *E. coli* and *Bacillus subtilis* after treatment of sodium citrate (control group), citrate-stabilized Ag NPs, and silver ions, respectively (the TEM size bars represent 500 nm). As can be seen in the NP-treated bacteria, the cells look very different from both the $AgNO_3$-treated and citrate-treated bacteria. The cell walls/cell membranes of the Ag NP-treated cells are difficult to see and are ill-defined.

The oxidative and reductive dissolution of NPs is well discussed in the literature (Schnippering et al. 2008; Misra et al. 2012; Wang et al. 2013; Ho et al. 2010). The dissolution of NPs in solution is especially well documented in metal oxide NPs, where the average size of NPs has been shown to change with time due to the reaction with dissolved species (Schnippering et al. 2008; Misra et al. 2012; Wang et al. 2013; Ho et al. 2010). The redox-based dissolution of NPs is due to the transfer of electrons from the surface of the NP to the dissolve species, which can be either organic or ionic in nature. Note that dissolution of metal oxides in solution has been well documented for different naturally occurring minerals such as hematite and magnetite as well as Iron (oxy)hydroxides (Sulzberger and Laubscher 1995; Borghi et al. 1991; Postma 1993). For example, the dissolution of NPs in solution has been shown to be enhanced in the presence of light generating ionic species, which may be more toxic to particular organisms in the water column (Shibata et al. 2004; Dasari et al. 2013).

However, there is a problem with determining the concentration of NPs in solution and making it comparable to the dissolved ion concentration. To further explore the differences, one must look at the traditional understanding of colloids in solution. A colloid has been traditionally defined as a particle ranging in size from 0.1 to 1000 nm,

Fig. 5.2 TEM images of cells from *E. Coli* treated with sodium citrate (**a**), Citrate stabilized Ag NPs (**b**), and cells treated with Ag ions (**c**) in a LB growth medium

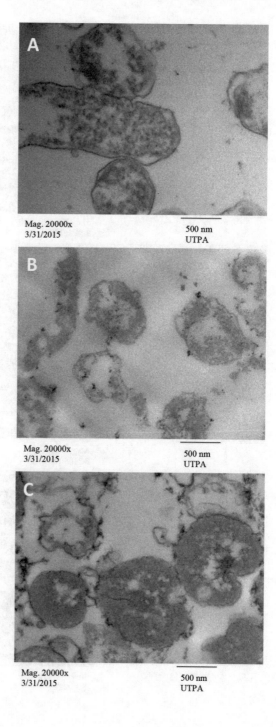

Fig. 5.3 TEM images of cells from *Bacillus subtilis* treated with sodium citrate (**a**), Citrate stabilized Ag NPs (**b**), and cells treated with Ag ions (**c**) in a LB growth medium

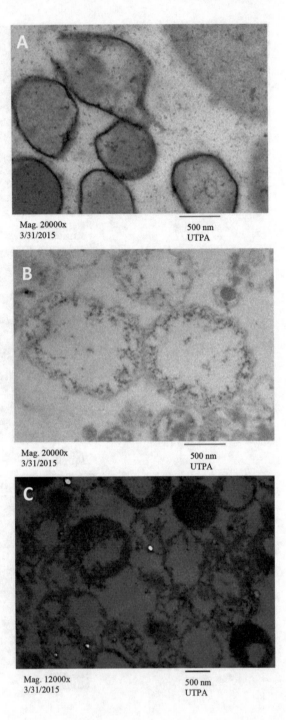

which is suspended in a medium, such as a solid in liquid (Gregory 2005). Interestingly, colloids have charged surfaces depending on their composition (e.g., structure and morphology) and can either be positive or negative (McDonogh et al. 1984; Zhong et al. 2004; Fafarman et al. 2011). Alternatively, using surfactants, NPs can have zero charge and very low dispersibility in water. The toxicity of chemical compounds can be linked to a few different factors; however, toxicity can be linked to dose, availability, and duration of exposure (Zhong et al. 2004; Fafarman et al. 2011).

2   Effect of Surfactant on Toxicity

Surfactants can play a vital role in the toxicity of nanoparticle and dispersion within the environment and subsequently within organisms. The earlier works on the synthesis of nanoparticles have reported the use of reducing agents such as citric acid, ascorbic acid, and sodium borohydride (Wuithschick et al. 2015; Kimling et al. 2006; Song et al. 2009). These reducing agents have been used to generate polydispersed mixtures of nanoparticles, and in some cases the reducing agents become surface coating agents. The geometry of nanoparticles can also affect their dispersion in a solution. Fig. 5.4 exhibits formation of geometrically regular nanoparticles using citrate. However, these geometries extend past the metallic NPs into the metal oxide as well as metal sulfide NPs (Grzelczak et al. 2008; Lim et al. 2007; Kitchens et al. 2005; Scarabelli et al. 2013; Rodriguez et al. 1996; Luis & Liz-Marzán 2002; Personick et al. 2011).

However, more recently the shift in nanoscience has been for the addition of surfactants to a solution to control the geometry of the resulting nanoparticle (Grzelczak et al. 2008; Lim et al. 2007; Kitchens et al. 2005; Scarabelli et al. 2013; Rodriguez et al. 1996; Luis & Liz-Marzán 2002; Personick et al. 2011). The desire to control the shape of nanoparticles can be related to the desired function of the nanoparticles. Nevertheless, the use of surfactants has one major drawback to their use, which is the coating of the nanoparticle surface with a chemical that may inhibit the way the particle interacts with the environment. Examples of a surface coated Au-NP with cetyltrimethylammonium bromide (CTAB) surfactant and citrate are shown in Fig. 5.5. The CTAB-extracted nanoparticles show a definite layer around the edge and the citrate-stabilized nanoparticle show a halo effect around the edge of the particle.

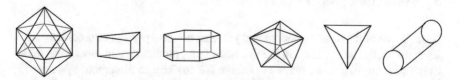

Fig. 5.4 Regular platonic geometric-shaped nanoparticles generated without surfactants in solution

Fig. 5.5 Examples of
surfactants surrounding
and encapsulating
nanoparticles after
synthesis

The development of surfactant-stabilized NPs has raised the following interesting question: What causes the toxicity of nanoparticles? In addition, the development of "biological methods" of NP synthesis for better biocompatibility of NPs raises even more questions that need to be answered about NP toxicity, the distribution of NPs in organisms, and the toxicity of the surfactant used for nanoparticle stabilization. As surfactants coat NPs, their ability to cross the cellular membrane increases, which may change the toxicity of the NP, or may cause carcinogenic effects. In addition, NP toxicity may be related to the stress of the organism and inability for the immune system to respond to pathogens.

The various nanomaterials currently synthesized or developed for human health-related applications include metal oxide nanoparticles, metallic nanoparticles, organic nanoparticles, nanofibers, and carbon nanoparticles. Moreover, there are many questions about NP toxicity and the validity of the toxicity, some particles show toxic effects in one organism but do not show the same in another organism. Current literature shows that the toxicity of NPs has been linked to the oxidation ability of the NPs, the ability to generate radical species such as the ROS and reactive nitrogen species (RNS) (Parke and Sapota 1996; Kimura et al. 2005; Hurst and Lymar 1997). Another important question about NP toxicity is the toxicity due to the NP itself or to dissolved ions in solution, or the generation of secondary species. NP toxicity has also been linked in some cases to the ability to attach to the cell walls in some organisms (Chwalibog et al. 2010; Hajipour et al. 2012). The focus of this book chapter is to highlight the toxicity of metal and metal oxide nanoparticles in different biological systems. The discussion of studies that have focused primarily on the accumulation or purely on the uptake of NPs by organisms is beyond the scope of this work.

3 Methodologies of Study

Perhaps one of the most important and yet least uniform parts of nanotoxicity studies is the methodology used in the synthesis (e.g., predicting and measuring toxicity). There are two basic ways to observe NP toxicity in biological systems: (1) using chemical techniques of analysis or (2) biological/cellular techniques. Each set of techniques has advantages and disadvantages when used to observe NP toxicity.

Testing the toxicity of NPs can be broken down into in vitro and/or in vivo testing. The results of toxicity testing performed in vivo and in vitro do not necessarily align with each other. However, this may be due to the fact that there is no direct effect (rather, in direct effect) of NP toxicity on the organismal development or behavior. In fact, some of the earlier studies on NP-plant interactions have shown positive effects of NPs on plant growth (Gardea-Torresdey et al. 2002; Rico et al. 2014; Arora et al. 2012; Kumar et al. 2013). For example, the treatment of *Medicago* sp. with potassium tetrachloroaurate salt resulted in the generation of gold NPs inside the live plants (Gardea-Torresdey et al. 2002). However, the treated plants showed enhanced growth when compared to the control plants. Similarly, this effect was shown in wheat and *Arabidopsis thaliana* (Rico et al. 2014; Arora et al. 2012; Kumar et al. 2013). Alternatively, the generation of reactive oxygen species (ROS) and other radical species have been shown to be important in investigating the toxicity of some NP species. Perhaps the dissolved ions, stabilizing agents, and secondary generated ROS are more toxic than the NPs are to the organism (Lee et al. 2009; Bozich et al. 2014; Yasun et al. 2015).

4 Chemical and Physical Methods to Observe NPs in Biological Systems

Taking a broad overview of the characterization methods used for NPs, one can summarize them as follows: dynamic light scattering (DLS), UV-Vis, X-ray diffraction (XRD), transmission electron microscopy (TEM), atomic force microscopy (AFM), selected area electron diffraction (SAED), atomic emission spectroscopy (AES), electron energy loss spectroscopy (EELS), and X-ray photoemission spectroscopy (XPS). These techniques are utilized for the characterization of the NPs themselves, at the same time also used for the evaluation of changes in NPs after uptake. In addition, ICP-MS, ICP-AES, fluorescence spectroscopy, flow cytometry, and Neutron activation analysis (NAA) have been used to determine NPs uptake as well (Jansa and Huo 2012; Pauksch et al. 2014; Govindaraju et al. 2008; Sun et al. 2005; Raffi et al. 2008; Akhtar et al. 2012; Asharani et al. 2008; Tetard et al. 2008; Kerkmann et al. 2004; Saif et al. 2008; Absar et al. 2005; Hu et al. 2007; Abdel-Mohsen et al. 2013; Heike et al. 2014; Merrifield et al. 2018).

As technology advances, the ability to evaluate the effects of NPs on organisms also gets refined. In recent years, there have been tremendous instrumental advances specifically in the area of Inductively Coupled Plasma Mass Spectroscopy (ICP-MS) for NP analysis (Heike et al. 2014; Merrifield et al. 2018; Love et al. 2012; Marquis et al. 2009). The development of the Single Particle (SP) ICP-MS has given investigators the ability to characterize true NP solution concentrations. Furthermore, as SP-ICP-MS is available, it can be utilized to detect the difference between NPs and dissolved ions (Heike et al. 2014; Merrifield et al. 2018). In addition, the improvement in TEM techniques has also allowed for better imaging of NPs within cells

(Love et al. 2012; Marquis et al. 2009). As well as TEM technology has advanced the ability to image and determine the phase of NPs in cells. The attachment of Selected SAED to TEM may allow, in the future, capturing the diffraction analysis of NPs in individual cells.

Fluorescence spectroscopy has also been utilized for the quantitative and qualitative determination of NPs in cells (Love et al. 2012; Marquis et al. 2009). The downside of the fluorescence microscopy method is the limitation of the quantification of the NPs that depends on the fluorescence properties of the NPs. The limitation can further exasperate by adding a fluorescence molecule to the particle, which may alter the uptake and NP behavior. To overcome this problem, spinning disk confocal microscopy was developed and has been used to monitor the movement of quantum dots within cells (Love et al. 2012; Marquis et al. 2009).

One of the most important issue to address NP toxicity is the reproducibility of the data reported on the same material by different research groups. The reproducibility problems sometimes can be related to the intrinsic properties of the NPs. For example, the attachment of a surfactant to a NP surface will change the global behavior of the NP such as the point of zero charge (PZC), and the interaction with the cellular walls and lipid membrane in cells (Love et al. 2012; Marquis et al. 2009). In addition, the surfactants can also change the agglomeration of the nanoparticles in solution, which can directly affect the transport of the NPs. Surfactants are commonly used as a stabilizer during the synthesis of the NPs to generate specific size range and geometrical shape.

5 Biochemical Assays for the Determination of NPs

Warheit et al. have developed a base set of toxicity tests using ultrafine TiO_2 NPs (Warheit et al. 2007). The crystal size, chemical composition, and surface reactivity of the NPs were determined using different experimental techniques such as hazard tests: pulmonary assay, skin irritation, skin sensitization, oral toxicity, and eye irritation. Additional genotoxicity and aquatic screening experiments were also performed on the NPs to study their NPs behavior in organisms. However, when strictly discussing methodologies from the biochemical/ molecular biological point of view, one delves into cellular assays, which have been developed to assess the vitality of cells. Cellular assays include, but are not limited to, the assays of the following: metabolic activity, hemolysis, apoptosis and necrosis, exocytosis, cell proliferation, oxidative stress, immunogenicity, gene expression, and DNA damage (Love et al. 2012; Marquis et al. 2009).

Proliferation can be determined using assays through assessing different cell properties and/or processes including metabolic shifts, DNA synthesis, and the ability to form colonies. Metabolic processes and properties are assessed using either MMT, XXT, or WST-1 assays; membrane integrity can be measured using dye uptake such as Alamar blue or Thymidine incorporation. Necrosis can be determined using LDH to determine cellular material leakage or uptake of dyes such as trypan blue, neutral red, and propidium iodide. Apoptosis can be assayed through the uptake and transport of Annexin-V and DNA laddering among other techniques. The determination of the mechanisms or mode of action of NPs is generally done using DNA damage assays such as the studying of the fragmentation, and breakage of DNA (Love et al. 2012; Marquis et al. 2009). This can be assessed through Comet, CSE, and TUNEL (which can be utilized to study cellular apoptosis) techniques (Love et al. 2012; Marquis et al. 2009). The other major mechanism of NP toxicity, oxidative stress is determined through the determination of ROS, lipid peroxidation, lipid hydroperoxides, depletion of antioxidants, SOD activity, and the expression of SOD. These cellular properties are usually determined using the interaction with dyes, or fluorescence probes, or the formation of stable radical species (Love et al. 2012; Marquis et al. 2009).

For in vivo studies of whole organisms, more techniques become available to investigators, which allow for the determination of the effects of "long-term" exposure to NPs. The in vivo studies allow for the determination of EC_{50} or LD_{50} for treatment of animal with NPs. Furthermore, in vivo studies allow for the determination of the movement of NPs in an organism between organisms and allow for the determination of the toxicity of a given NP and the route of exposure through the three common routs of exposure inhalation, dermal contact, or through ingestion. To determine the effects of NPs on living organisms, researchers have utilized studies such as histopathology/histology to observe morphology changes in the organs of the test animal. Hematology and serum chemistry were used to investigate at the composition of blood and serum as well as the changes occurring after treatment with NPs. The distribution of the nanoparticles in the body is another aspect that has been studied, which can be achieved using X-ray or MRI imaging as well as the extraction of the organs and determination of the effects of the NPs. All the techniques utilized previously allow for a better or clearer picture of the distribution and transport of NPS in living organisms. These studies also provide valuable information of the effect of biological systems on the toxicity of NPs.

6 Metal Oxide NP Toxicity: Case Studies

6.1 TiO₂ NP Toxicity

The toxicity of TiO_2 NPs has been investigated extensively in many different organisms including algae, plants human cells, and fish as outlined in the following sections. The interest in TiO_2 NP toxicity has been developed through its use and the application in consumer products such as sunscreens (Vance et al. 2015). TiO_2 is found in three different crystal structures, anatase, brookite, and rutile, of which, rutile is the most thermodynamically stable form, followed by anatase, and then brookite. In sunscreens, there are commonly two forms of TiO_2 present in the anatase and the rutile forms and are added for their UV absorption properties. Due to

Table 5.2 Commonly observed toxic effects in the treatment of multiple types of live organisms and cells with of TiO_2 NPs

Effect/impairment of biological process	Organisms
Oxidative stress/ROS	Algae, plankton/ diatoms, fish, human epithelial cells
Immune response	Bivalve, fish
Viability	Algae, fish, human epithelial cells

the low solubility of TiO_2, the toxicity of the NPs appears to be a function of the NPs and not to solubilized ions in solution. The commonly observed toxic effects of TiO_2 NPs on the different organisms are summarized below in Table 5.2.

7 TiO_2 Toxicity in the Treatment of Algae

Algae, particularly microalgae, are simple organisms that are abundantly found in lakes, rivers, or on lands. The toxicity of nanoparticles in these organisms can be studied where the whole organism can be exposed to nanoparticles in the growth medium. When considering the toxicity of nanoparticles to algal species, it is a whole organism treatment being studied, whose treatment with TiO_2 NPs appear to induce oxidative stress in the organisms and subsequently affect cell viability. Multiple studies on the toxicity of TiO_2 NPs have been performed and are summarized in the following examples extracted from the literature.

Chen et al. investigated the toxic effects of TiO_2 NP on unicellular green alga *Chlamydomonas reinhardtii* (Chen et al. 2012). The authors investigated TiO_2 NPs with an average size range of 21 nm at varying concentrations from 0.1 mg/L to 100 mg/L. The exposure of the algae to the TiO_2 showed inhibition of photosynthetic efficiency and cellular growth. However, the amount of chlorophyll (a) in the algae was not different than in the control samples but the carotenoids and chlorophyll (b) were observed to increase with increasing concentration. The authors also showed that as the concentration of TiO_2 NPs was increased, the cells were damaged while the chloroplasts were degraded, and other organelles were reduced. It was noted that the TiO_2 NPs were found inside the cells, coating the cell walls, and coating the cellular membranes. Along the same lines, Aruoja et al. investigated the toxicity of TiO_2 nanoparticles on a microalgae *Pseudokirchneriella subcapitata* (Aruoja et al. 2009). The authors showed that bulk TiO_2 had an EC_{50} of 35.9 mg Ti/l whereas the TiO_2 NP had an EC_{50} equal to 5.83 mg Ti/L (Aruoja et al. 2009). The data showed almost a sixfold increase in the EC_{50} for the NP compared with the bulk crystal phase. The authors also showed that the TiO_2 NPs formed characteristic aggregates that entrapped the algal cells. It was suggested that the toxicity of the TiO_2 NPs occurred through either the generation of reactive oxygen species (ROS), the inactivation of the cells through a combination of TiO_2 and visible light, or through the destruction of the cellular surface. However, the toxicity of the TiO_2 NPs was caused by the generation of the ROS-mediated toxicity through the hydroxyl radical species (Aruoja et al. 2009). Lee and An studied TiO_2 NP toxicity on green algae using irradiation from UV-A and UV-B light sources, visible light, as well as without irradiation (Lee and An 2013). The authors investigated the toxic effects of 21 nm TiO_2 NPs on the green alga *Pseudokirchneriella subcapitata*, which is found in fresh water. It was found that as the concentration of NPs was increased, the algal growth rate was observed to decrease. These effects were not observed in the pre-irradiation conditions and attributed to the effects of the photo catalytic ability of the TiO_2 NPs. Under visible light, the TiO_2 nanoparticles showed EC_{50} of 2.53 under visible light; however, under UV-A light, the EC_{50} increased slightly to 3.00 and was 2.95 using UV-B irradiation (Lee and An 2013). The authors tied the toxic effects of the TiO_2 nanoparticles to the formation of the superoxide ion by the TiO_2 which caused observed effects in the algae. Sadiq et al. have investigated the toxic effects of TiO_2 anatase NPs on *Scenedesmus* sp. and *Chlorella* sp. microalgal species (Sadiq et al. 2011). Inhibition of the growth for both microalgal species was observed. The alga species were treated with concentrations of 3–196 mg/L. The observed EC_{50} at 72 h were 16.12 for *Chlorella* sp. and 21.2 mg/L for *Scenedesum* sp. Through FTIR analysis of the samples, the authors observed the attachment of the nanoparticles to the cellular membrane during the study. Photocatalytic TiO_2 Degausa P-25 (a mixture of anatase and rutile) mixtures have also been studied for their toxic effects on the marine algae *Pseudokirchneriella subcapitata* (Sadiq et al. 2011). The authors investigated dose response experiments, where lipid peroxidation, chlorophyll A concentration, as well as direct cell counts were measured for toxicity. The study showed an EC_{50} of 6.5 particles per cell, and it was discovered that the critical particle size ranged from 4 to 30 nm for detrimental effects to be observed. Further analysis of the samples by the authors using SEM showed agglomeration of the NPs on the surface of the cells in layers. Metzler et al. also indicated the NPs showing lipid peroxidation in conjunction with the surface coverage may have been the cause of the negative effects of the NPs (Metzler et al. 2011). Fu et al. investigated the effects of TiO_2 nanoparticle on ROS production and inhibition of growth in freshwater algae (Fu et al. 2015). The authors used *Pseudokirchneriella subcapitata* (new name, *Raphidocelis subcapitata*) as the algal species and exposed these algae to TiO_2 NPs with UV irradiation for 3 h with and without a UV filter. The effects of TiO_2 NPs on algae pre-exposed to UV light were also investigated. The data showed that exposure to TiO_2 NPs with and without UV filters decreased algal growth. It was also determined that the EC_{50} were 8.7 and 6.3 mg/L for the algae exposed to UV with and without filters, respectively (Fu et al. 2015). Finally, it was concluded from the study that the exposure to the TiO_2 NPs the ROS species were not directly involved in the sublethal effects in the algae. Moreover, Li et al. studied the effects of TiO_2 on algal species and the production of ROS species (Li et al. 2015). In that work, two algal species *Karenia brevis* and *Skeletonema costatum* were used where the algal species were exposed to TiO_2 NPs with diameters in the range of 5–10 nm in the anatase phase. The authors observed that the TiO_2 NPs were transported inside the cells. The growth of both algal species was inhibited with an EC_{50} of 10.69 and 7.37 for *Karenia brevis* and *Skeletonema costatum* after 72 h treatment, respectively. The effects on the algal species were attributed to the oxidative stress caused by ROS generation inside the algal cells. Inhibition of electron transport showed the ROS generation site was the chloroplast for the *K. brevis* (Li et al. 2015). Clement et al. investigated the toxicity of anatase and rutile TiO_2 NPs on caldocerans, algae, and rotifers (Clément et al. 2013). The authors exposed the organisms to TiO_2 nanoparticles with crystallite sizes of 15, 32, and 25 nm as determined using XRD. The authors also exposed the organisms to microparticles of anatase and rutile. The treatments consisted of exposing the organisms to 100 mg/L of a specific material for up to 72 hrs (Clément et al. 2013). *Chlorella vulgaris* at a TiO_2 concentration of 100 mg/L 25 nm anatase and 1 um rutile with the average of 5.7% and 4.53% were observed, respectively. With *D. magna*, the authors performed tests at 48 and 72 h after exposure. Acute toxicity was observed and was directly related to the increase in particle concentrations (Clément et al. 2013). As the exposure time was extended to 72 h, it was determined that the EC_{50} was 1.30 mg/L, which was observed for the 15 nm anatase NPs. EC_{50} were greater when concentrations of 100 mg/L of the other nanoparticles were tested. More importantly, the anatase form was more toxic than the rutile form of the particles (Clément et al. 2013).

8 TiO_2 Toxicity in the Treatment of Diatoms

Similar to other algae, plankton and diatoms show toxicity to TiO_2 exposures. As diatoms are single-celled algae, they are also treated as a whole organism in the environmental toxicity assays. As in other forms of organisms, TiO_2 NPs induce oxidative stress in diatoms affecting their survival and growth. Clement et al. extended his nanotoxicity study to include *Phaeodactylum tricornutum* (Clément et al. 2013). All forms of TiO_2 used for exposure caused toxicity in this diatom, except the nanoparticles and microparticles. From the EC_{50} values, it was indicated that the 15 nm nanoparticles were more toxic than the other nanoparticles. In fact, the nanoparticles were toxic to the diatoms as is $K_2Cr_2O_7$, a well-known carcinogen. The authors also investigated *Brachionus plicatilis*, a rotifer species, and demonstrated that the micro-particles were much less toxic than $K_2Cr_2O_7$ while the 15 nm Anatase NPs were the most toxic (Clément et al. 2013). Plankton *Daphnia magna* were studied by Dabrunz et al. for the toxic effects of TiO_2 NPs (Dabrunz et al. 2011). The authors studied the plankton stressed with TiO_2 NPs with an approximate diameter of ~100 nm. It was found that concentrations of 3.8 and 0.73 ppm lead to toxic effects in the plankton (Dabrunz et al. 2011). It was also reported that increasing the particle size to ~200 nm, the toxicity was greatly reduced. A mechanism of action for the toxicity of the nanoparticles was proposed, which required the development of a surface coating of the TiO_2 NPs over the outside of the plankton, which lead to a molting disruption. Additionally, the treatment of the plankton with 2 ppm of TiO_2 NPs lead to the development of the surface coating; however after 36 h, the coating was not present in the treated samples (Dabrunz et al. 2011). However, after 96 h of treatment with surface coating, the mortality rate of the plankton was found to be 90%. In a similar study by Zhu et al., the toxicity and bioaccumulation of TiO_2 NPs in *Daphnia magna* was investigated (Zhu et al. 2010). A minimal toxicity of the nanoparticles was observed within 48 h of treatment. However, upon longer treatment, internalization of the nanoparticle was occurred while increased toxicity was observed. Mansfield et al. investigated the toxicity of TiO_2 NPs in *D. magna* under natural light (Mansfield et al. 2015). The *D. magna* were exposed to anatase phase of the TiO_2 NPs at concentrations of 20 and 200 ppm. The authors showed that under varying UV irradiation, the LC_{50} of the TiO_2 NPs was 778 ppb in the 50% irradiation samples. Whereas in the 100% UV irradiation samples, the LC_{50} was 139 ppb (Mansfield et al. 2015).

9 TiO_2 Toxicity in the Treatment of Bivalves

Moving into more complex organisms, studies have been performed on the toxicity of bivalve species using TiO_2 NPs. The environment for the treatment of bivalves is more complex than algae and plankton/diatom treatments. The sediment introduces a secondary medium aside from the treatment of the organisms with suspended NPs in solution. There are transport questions on the movement of NPs through the sediments. However, ultimately oxidative stress appears to be one of the larger mechanisms involved in the interaction of TiO_2 and bivalves. The effects of TiO_2 on bivalve species are summarized in the following section. Marisa et al. investigated the effects of TiO_2 NPs on clam *Ruditapes philippinarum* (Marisa et al. 2015). The authors investigated the effect of a mixture of anatase and rutile (70:30) with a focus on studying the hemocyte phagocytic activity. The authors showed that the effect on the hemocyte function was mediated by the internalization of the nanoparticles within the hemocytes. The TEM images taken of treated cells with the NPs were noted to be on the cell wall as well as inside the cell (Marisa et al. 2015). Along the same lines, Shi et al. investigated the effects of TiO_2 NPs on a commercial clam *Tegillarca granosa* (Shi et al. 2017). The authors exposed the clams to environmentally relevant concentrations of TiO_2 (no phase specified) at 10–100 ug/L. The authors also exposed the clams for 30 days where they studied the expression of genes, the immune-related molecules expressed in the clams after treatment. The clams showed a highly downregulated immune-related response, which indicates the TiO_2 NPs suppressed the immune system of the clams. Furthermore, the gene expression suggested pattern recognition receptors may be receptors for NPs in marine invertebrates. The authors also showed data suggesting LMS reduction, phagocytosis degrease, and NO production increases and lysosome release, the reduction of ROS production, and increase in HM in other marine organisms after TiO_2 treatment.

10   TiO$_2$ Toxicity in the Treatment of Fish

The effects of TiO$_2$ NPs on zebrafish have been investigated from different aspects, which include the toxicity, but also behavioral changes observed (Chakraborty et al. 2016). TiO$_2$ NPs have been shown to induce pre-mature hatching and have been found to be toxic to zebrafish embryos, which was determined to be concentration dependent. Exposure of zebrafish embryos to TiO$_2$ NPs has been shown to cause death (Chakraborty et al. 2016). Further studies with TiO$_2$ NPs and zebra fish have shown neurotoxicity. In fish studies, the translocation of TiO$_2$ NPs to different cell types and organs begins to be noticed. The treatment of zebrafish with TiO$_2$ NPs shows the enhanced expression of particular genes and the suppression of other genes resulting in brain damage. Stressing of the zebra fish has also shown to reduce the reproductive rates of both male and female fish and embryo development (Chakraborty et al. 2016). Chen et al. investigated the effects of TiO$_2$ NPs on the behavioral of zebrafish (*Danio rerio*) (Chen et al. 2011). The authors exposed embryos of zebrafish to P25 TiO$_2$ NPs, which is a well-known photocatalytic nanomaterial and is a mixture of the anatase and rutile TiO$_2$ phases. The TiO$_2$ NP were found to range in size from 25 to 70 nm. The authors noted that there were no changes in the survivability, hatchability, or morphology in the larvae at low concentrations. However, the mobility and the percentage of time inactive were lower in the 0.1, 0.5, and 1.0 treatments but were unchanged in 5 and 10 mg/L treatments (Chen et al. 2011). However, it was not indicated that the changes in behavior were due to physiological damage from the exposure to the TiO$_2$ NPs. It was also observed that the behavioral changes may be due to increase in antioxidant enzymes. The enzymes included in the discussion were superoxide dismutase, catalase, and peroxidase, which has been shown in other fish species exposed to TiO$_2$ NPs (Chen et al. 2011). Lammel and Struve have investigated the toxicity of TiO$_2$ NP on rainbow trout cell lines (Lammel and Sturve 2018). The authors investigated cell lines from liver and gills. In that work, TiO$_2$ nanoparticles with 21 nm diameter were used in the P25 mixture, which as previously mentioned, it is a photocatalytic mixture of the anatase and rutile phases of TiO$_2$. The cells were treated with NP concentrations from 1 to 100 ug/mL (or 1–100 ppm) for a 72 h period. The authors noted that there was a little cytotoxicity from the TiO$_2$ NPs after the 72 h exposure (Lammel and Sturve 2018). However, the nanoparticles were present as agglomerates in the intracellular vesicles of the liver and gill cells. Furthermore, the TiO$_2$ adsorbed to the plasma membrane were internalized in the cells (Marisa et al. 2015). Zhu et al. investigated the toxicity of TiO$_2$ NPs on zebrafish (Zhu et al. 2010). The zebrafish was exposed to nanoscale non-stabilized TiO$_2$ in the anatase phase with an average diameter of less than 20 nm. The studies were performed for 96 h using concentration of 1–500 mg/L. The zebrafish embryos had a hatching rate of approximately 100% even at concentrations of 500 mg/L. The TiO$_2$ NPs did not appear to have adverse effects on the zebrafish in the study (Zhu et al. 2010). In another study, Yang et al. investigated the toxicity of TiO$_2$ NPs on developing zebrafish (Yang et al. 2013). The authors examined the impact of humic acid with TiO_2 NPs on the growth and development of zebrafish embryos under simulated sunlight (Yang et al. 2013). P25 TiO_2 nanoparticles (3:1 anatase: rutile) with an average diameter of 21 nm were used in that study. The authors employed the TIO_2 NPs as the control sample to investigate the effects of the HA/TiO_2 NP treatment, which were TiO_2 NPs with concentrations of 0–1000 mg/L and HA 0–30 mg/L. The authors showed significantly decreased survival rate at 5 dpf as the concentration of the TiO_2 NPs and HA was increased as inherent toxicity without simulated sunlight (Yang et al. 2013). Similarly, in the presence of simulated sunlight, the survivability of the fish was reduced. However, the concentration showed to have a larger effect. Without simulated sunlight, there were approximately 50% of the embryos survived at 1000 ppm with HA and TiO_2 treatment and approximately an 80% survival rate was observed with the TiO_2 alone; however, under simulated sunlight, there were no survivors above 500 mg/L with and without the HA treatment. In developing zebrafish, the presence of the HA and TiO_2 showed increased oxidative stress in the fish. It was determined through lipid peroxidation that increased oxidation damage and DNA damage was present in the fish with simulated sunlight (Yang et al. 2013). Jovanovic et al. investigated the effects of TiO_2 on Fathead minnows (*pimephales promelas*) (Jovanovic et al. 2015). However, the authors investigated the stress with respect to the fish's ability to respond to pathogens. The TiO_2 NPs were observed to accumulate in the kidney, followed by the spleen, then the liver. The concentration in the liver was approximately equivalent to the concentration of NPs found in the whole fish. Additionally, It was observed that minnows treated at 2 ng/g and 10 ug/g (based on body weight) exposure to the TiO_2 NPs followed by exposure to the pathogens either *aeromonas hydrophila* or *Edwardsiella ictaluri* showed inhibited response in the immune system and increased mortality. Further study of the minnows injected with the TiO_2 NPs showed histopathology (Jovanovic et al. 2015). There appeared to be an interplay between the histopathology and the immune systems of organisms treated with NPs. Reeves et al. investigated the hydroxy radicals generated from TiO_2 on cytotoxicity and oxidative damage to DNA in fish (Reeves et al. 2008). The authors used 5 nm anatase NPs at concentrations from 0.1 to 100 ug/mL (ppm) with and without UV irradiation using goldfish cell. The authors observed that all dose levels showed significant increase in oxidative damage to DNA (Reeves et al. 2008). The analysis of the data showed there was no dose-dependent response in the oxidative damage to the nanoparticles at low concentrations of NPs. For the goldfish cells exposed to the TiO_2 NPs for 24 h in the absence of UV irradiation, a small decrease in cell viability was observed (~80% viability). When the cells were treated with 1000 ppm TiO_2 nanoparticles and were subsequently co-exposed to UVA irradiation, the cell viability decreased to ~40%. The authors concluded that with and without photosensitization with UVA light, TiO_2 NPs were potentially genotoxic to fish cells. In addition, the authors concluded that the major cause of toxicity was the generation of the hydroxyl radical (Reeves et al. 2008).

11   TiO$_2$ Toxicity in the Treatment of Mammals and Mammalian Cells

The treatment of mammals and cell lines show mixed results with TiO$_2$ NP toxicity. Ultimately, the toxicity manifests itself largely at the level of oxidative stress affecting cellular viability, similar to the other organisms mentioned earlier treated with TiO$_2$ NPs. Similar to the fish TiO$_2$ NP toxicity studies, the potential for translocation of nanomaterials to different organs exists, which may cause multisystem toxicity within the organism. However, for the most part, the observed effects on the mammals, excluding cell lines, appear to have an initial effect, which reduces with the exposure time. Wu et al. have investigated the toxicity and penetration of TiO$_2$ NPs in hairless mice and porcine skin after dermal exposure (Wu et al. 2009). The authors investigated Anatase NPs with sizes of 4 and 10 nm, rutile NPs with sizes of 25, 60, and 90 nm, and Degaus P25 (anatase/rutile mixture 70:30), a commercial photocatalytic material. The TiO$_2$ NPs were tested for their ability to transport across porcine skin in vitro and in vivo. The in vitro studies showed not penetration of the TiO$_2$ NPs inside the tissue. On the other hand, the in vivo studies indicated that TiO$_2$ NPs were in the stratum corneum, stratum granulosum, prickle cell layer, and basal cell layer; however, there was no deep penetration into the dermis (Wu et al. 2009). The author's study on the hairless mice showed that mice treated with the 10 nm, 25 nm, and P25 NPs had decreased weight after treatment. The liver and spleen coefficients were also significantly higher. The nanoparticles were found to accumulate in liver, spleen, and heart, but no transport was observed into the blood. In addition, the particles were observed in the liver; the bile ducts close to the nanoparticle aggregates were shown to be swollen indicating tissue damage (Wu et al. 2009). The effects of TiO$_2$ NPs after intraperitoneal injection to mice were investigated by Chen et al. (Chen et al. 2009). The mice were injected with TiO$_2$ NPS at concentrations ranging from 324 to 2592 mg/kg using the anatase phase of TiO$_2$, which they were synthesized utilizing a sol gel method. The average grain size of the NPs was approximately 3.6 nm. TiO$_2$ NPs were observed to be accumulated in the spleen after 24 h treatment. The authors observed a dose-dependent relationship as the concentration of the TiO$_2$ NPs was increased the accumulation of the NPs in organs increased. The spleen concentrations of the TiO$_2$ NPs decreased as time progressed. As the time increased, the lung, kidney, and liver showed the presence of TiO$_2$ NPs. The authors concluded that some of the NPs were excreted from the kidney and the toxicity of the NPs to the liver than kidney. The study also showed that TiO$_2$ NPs could transport to different tissues after injection. All the organs that contained TiO$_2$ NPs showed adverse effects to some degree but were minor in the extent of damage (Chen et al. 2009). Warheit et al. investigated the effects of TiO$_2$ NPs on pulmonary instillation in rats to determine if toxicity was a function of NP size (Warheit et al. 2006). The authors used rutile NPs with an average diameter of 300 nm, anatase nano-rods (up to 233 nm long), and nanodots which were in the anatase phase. The rats were treated intratracheally with TiO$_2$ NPs dispersed in phosphate-buffered saline (PBS) solution. Exposure to the nanoparticles produced a trainset short-lived pulmonary inflammation response within the first 24 h of treatment. However, the NPs administered at a high dosage of 5 mg/kg did not induce any long-lasting effects in the rats (Warheit et al. 2006). Gerloff et al. have investigated the toxicity of TiO_2 mixed anatase/rutile phase on Caco-2 cells from human intestine (Gerloff et al. 2012). The authors noted that TiO_2 has been a common food additive for many years. The authors studied pure anatase, and a mixture of anatase/rutile from various vendors. The average crystallite size was determined to range from 6.7 nm up to 215 nm. It was concluded from the study that specific surface area and crystallinity of the NPs was important for the toxicity to the intestinal cells. However, it was noted that there was no correlation or evidence for the TiO_2 nanoparticles playing a role in the ROS-mediated stress of the Caco-2 cells (Gerloff et al. 2012). Hsiao and Huan studied the effects of TiO_2 on human lung epithelial cells (Hsiao and Huang 2011). The authors purchased phase pure anatase nanoparticles and anatase/rutile mixed phase nanoparticles and subjected to treatments of the TiO_2 NPs from 50 to 1.56 µg/mL from time intervals of 12, 24, or 72 hrs. The authors followed the cell morphology and performed MMT assay for cell viability, followed the production of IL-8 for the concentration treatments. This exposure caused a decrease in toxicity as the average grain size of the nanoparticles increased for 12 and 72 h treatments. The amorphous particles induced more pro-inflammatory factor in human cells than the larger NPs. It was also noted that the anatase phase of the TiO_2 NP induced a higher cytotoxicity to the cells. The authors also indicted that point of zero charge (PZC) plays a role in the transport of the nanoparticles across the cells and plays a role in the toxic effects of the nanoparticles (Gerloff et al. 2012). Jeng and Swanson, on the other hand, studied the toxicity of metal oxide NPs on mammalian cells from mouse (Jeng and Swanson 2006). The authors studied TiO_2 among other metal oxide NPs with a mean size range from 50 to 70 nm. The authors used the MMT assay to investigate the cell viability of TiO_2 where no effect was observed on cell viability. Furthermore, it was shown that TiO_2 did not affect the mitochondrial function of the cells. The authors concluded that the TiO_2 NPs were only lightly toxic to the cells (Jeng and Swanson 2006). Karlsson et al. exposed human epithelial cells to TiO_2 NPs with an average diameter of 63 nm, which showed no reduced cell viability (Karlsson et al. 2008). These nanoparticles showed the ability to damage DNA. The TiO_2 NPS showed an increase in the ROS species generated within the cells over the concentrations in the control cells (Karlsson et al. 2008).

12 ZnO NP Toxicity

ZnO NPs have been investigated for their toxic effects in different biological systems. Like TiO_2 NPs ZnO NPs are a commercially important Type of NP. One of the common uses of ZnO has been traditionally in high SPF sunscreens. The application of ZnO in sunscreens especially with NPs raises several questions about the toxicity, considering the direct dermal application of sunscreen. Also, the high use

Table 5.3 Major commonly observed toxic effects in the treatment of multiple types of live organisms and cells with of ZnO NPs

Effect/impairment of biological process	Organisms
Oxidative stress/ROS	Algae, bivalves, fish, mammals/ mammalian cells
Immune response	Fish, mammals/ mammalian cells
Viability	Algae, fish, mammalian cells

of sunscreens in and around aquatic environments raises questions about the environmental transfer and potentially elevated concentrations of ZnO NPs. Unlike TiO_2 NPS ZnO NPs can be dissolved at physiological pHs and the Zn^{2+} ions are able to cause similar toxic effects as the NPs. The toxicity of ZnO NPs becomes a question of the solubility and organismal response to dissolve Zn^{2+} ions in solution as well as the ZnO NPs effects. The most commonly observed toxicity effects of ZnO NPs on the different organisms are summarized in Table 5.3 and in the following sections.

13   ZnO Toxicity in the Treatment of Algae

The toxicity of ZnO NPs is similar to that of TiO_2 NP in algal species, specifically with respect to the increase in ROS generation and decrease in cell viability. However, unlike TiO2, NPs ZnO NPs are soluble and can form ions. The formation of Zn^{2+} ions in solution complicates the study of ZnO toxicity. Zn^{2+} ions can be toxic in high enough concentration, and thus the question becomes: is the observed toxicity caused by the ZnO or the Zn^{2+} ion. Bhuvaneshwari et al. have investigated the cytotoxicity of ZnO NPs on *Scenedesmus obliquus* under low concentration, VU-light, as well as dark and visible light conditions (Bhuvaneshwari et al. 2015). The authors used ZnO NPs with sizes less than 100 nm and a second set of ZnO NPs with an average size of 40 nm. The authors used three concentrations of 0.25, 0.5, and 1.0 mg/L and used a 72 h treatment, with either UV-C, visible light, or dark conditions. Subsequent to treatment, the authors determined the following: oxidative stress, cellular membrane integrity, total organic carbon (TOC), and the surface interaction of the NPs with the cells. As well, the authors investigated the internalization and uptake of the NPs. The authors showed that cell viability had decreased for both NPs under all illumination techniques and increased with increasing concentration of the ZnO NPs. Under UV-C irradiation the amount of ROS species generated were increased, as well as the membrane integrity decreased with increasing concentration. The internalization of the NPs into the algal cells was also observed (Bhuvaneshwari et al. 2015). Ji et al. studied the toxicity of ZnO NPs on the green algae *Chlorella* sp. (Ji et al. 2011). The authors observed a nano-Zn particles-led inhibition of algal growth with a 6-day EC_{50} of 20 mg/L. The authors used 20 nm ZnO particles and showed that as the concentration of the NP increased, the toxicity was increased. It was also noted that the amount of dissolved Zn in solution increased with an increase in the concentration. The data suggested that the toxicity of the ZnO was caused by the NP dissolution (Bhuvaneshwari et al. 2015). Manzo et al. investigated the toxic effects of ZnO nanoparticles on a marine alga *Dunaliella tertiolecta* (Manzo et al. 2013). The authors used ZnO nanoparticles uncoated with an average diameter of 100 nm, as well the authors used $ZnCl_2$ to assay the Zn^{2+} ion toxicity. It was reported that the dissolution of the ZnO was almost complete after 24 h of exposure. The data from the study showed an EC_{50} of 1.94 for the ZnO NPs, and EC_{50} of 3.57 for bulk ZnO NPs, and an EC_{50} of 0.65 for the $ZnCl_2$. The authors concluded that the ZnO was more toxic than the ionic form (Ji et al. 2011). Peng et al. have investigated the toxicity of and the effect of morphology of ZnO NPs on marine algae (Peng et al. 2011). In that work, spherical NPs, nanoplatelets (2-D structures), and nanoneedles/nanorods were investigated (Peng et al. 2011). The algae *T. pseudonana* and *P. tricornutum* were exposed to 10-80 mg/mL of the ZnO NPs for up to 72 h. It was observed that the dissolution of the NPs occurred within hours of exposing the ZnO NPs to the growth medium releasing Zn^{2+} ions. The growth of both the *T. pseudonana* and *C. gracilis* occurred and was thought to be a result of the NP dissolution due to the acute toxicity of the Zn^{2+} to the organisms. However, the *P. tricornutum* growth was not as greatly affected by the Zn^{2+} ions and observed effects could be correlated to particle concertation and morphology (Peng et al. 2011). The 1-D structures showed greater toxicity than the 3-d morphologies. The study suggested that both the ZnO and Zn^{2+} were the cause of the toxicity in the algae(Peng et al. 2011). Miller et al. have studied the effects of ZnO nanoparticle on marine phytoplankton *S. marinoi, T. pseudonana, D. tertiolecta,* and *I. galbana* (Miller et al. 2010). The authors used zincite nanoparticles with an average diameter of 20-30 nm and exposure concentrations of 10–1000 ug/L for 96 h. The growth of the phytoplankton was significantly reduced after exposure to the ZnO NPs. More importantly, the ZnO NPs were quickly dissolved in the saltwater medium 32% of the NPs mass and was dissolved in the first 12 h and still observed to be dissolving at the 96 h time period.

14 ZnO Toxicity in the Treatment of Bivalves and Aquatic Organisms

Similar effects with respect to the toxicity have been observed within bivalve species that have been studied as have been observed in algal and plankton species. Similarities in toxicity have been observed even though bivalve spices are somewhat more complex than algal and plankton species. In addition, the route of exposure requiring the movement of the particle through a sediment is also different. Buffet et al. have studied the effects of ZnO NPs on clams and ragworms (Buffet et al. 2012). The authors exposed the organisms to 3 mg/kg concentrations based on the sediment, which they considered to be an environmental relevant concentration. The NPs were sized between 21 and 34 nm with a positive surface charge from diethylene glycol. The organisms were treated under three difference conditions, seawater alone, DEG alone, and ZnO NPs in sediment (3 mg/kg). It was concluded that there were no consistent changes in biochemical markers (Buffet et al. 2012). The effects appeared to be related to the DGE presence on the NPs while the behavior of burrowing for the DEG-treated and ZnO-treated NPs were not significantly different. The authors noted that ZnO NP toxicity was generally attributed to the presence of Zn^{2+} ions; however it was difficult to differentiate between the natural background Zn^{2+} and that accumulated from the NPs and NP dissolution (Buffet et al. 2012). Ali et al. observed the oxidative stress and genotoxic effect of ZnO NPs in a freshwater snail (Ali et al. 2012). The authors exposed the freshwater snails to ZnO NPs with an average diameter of 50 nm for 96 h at below lethal concentrations. It was reported that at 32 ppm, a reduction of glutathione was observed with increases in the malondialdehyde level and catalyze levels in the digestive glands. The authors also observed significant DNA damage with the organisms treated for 24 and 96 h. The authors concluded that the ZnO NPs induced genotoxicity in digestive gland due to induced oxidative stress (Ali et al. 2012).

15 ZnO Toxicity in the Treatment of Fish

Similar to fish studies with TiO_2 NPs, ZnO NPs have shown to be toxic. Again, the pertinent question is whether the observed toxicity is due to the interaction of NPs or Zn^{2+} ions. It is also observed in fish like algae, ZnO NPs caused oxidative stress as well as immune system impairment and affected viability. Zhu et al. investigated the effects of ZnO NPs on zebrafish (Zhu et al. 2008). The authors treated the fish with ZnO NPs (20 nm) and Bulk ZnO particles at concentrations from 1 to 50 for up to 96 h. At the 96th hour of treatment, the zebrafish survival rate dropped from approximately 98% in the controls to 0 in the 50-ppm treatment. Similar results were observed in the Bulk studies. As the ZnO concentrations were increased, the survivability of the zebrafish decreased. Similar results were observed for the hatching rate with a decrease of 0% for the 50 ppm treated fish. The EC_{50} for both the Bulk ZnO and NP ZnO were approximately 2. It was noted that the tissue ulceration when the hatched zebrafish embryos after 72 h treatment sand increased with increasing particle concentration (Zhu et al. 2008). Zhao et al. studied the acute effects of ZnO NPs exposure on the development of toxicity, oxidative stress, and DNA damage in zebrafish (Zhao et al. 2013). The authors suspended ZnO less than 100 nm NPs in zebrafish culture media with concertation variation from 1 to 100 ppm. The authors analyzed the concertation of free zinc in solution after treatment and prepared Zn solutions from $ZnSO_4$ to determine the toxic effects of the Zn^{2+} ions. The results of the study showed the embryo hatch was significantly slow and malformation increased at 96 h of exposure (Zhao et al. 2013). The ZnO NPs also induced DNA damage through ROS generation. In addition, the activities of defense enzymes were changed while MDA concentration in the larvae were increased. It was also indicated that the Zn^{2+} ions were less responsible for the damage than the ZnO NPs. The data from the study also showed mRNA was damaged, which can be used to encode response enzymes and responses for oxidative stress (Zhao et al. 2013).

16   ZnO Toxicity in the Treatment of Mammals and Mammalian Cells

ZnO NPs have been shown to be toxic to mammals and mammal cells. However, the toxic effects of the ZnO NPs seem to correlate strongly to the concentration of free Zn^{2+} ions. In addition, in in vivo studies using mice or other mammalian models, one begins to see multisystem effects with the treatment of whole organisms with ZnO NPs. In addition, some studies begin to show secondary effects or causal effects due to direct effect on a different cellular system. Song et al. have investigated the role of Zn^{2+} in ZnO NP toxicity in mouse macrophage Ana-1 (Song et al. 2010). The authors focused on the toxicity from the generation and damage caused by ROS. The study focused on nanoparticles of four different sizes: one set of particles was on the order of microns in size, the second set was less than 100 nm, the third set were 30 nm, and the fourth set were 10–30 nm in diameter. The cells were treated with NP concentrations of 2.5–100 ppm for 24 h. In that work, a dose-dependent toxic effect was observed (Song et al. 2010). In addition, there was no effect observed due to the size of the nanoparticles. But a Zn^{2+} concentration of 10 ppm induced a 50% death which was close to the NP toxicity. The cellular viability and membrane integrity were observed to have high correlation with the concentrations of the Zn^{2+} in the supernatant of the culture. Finally, the authors concluded that ROS may be generated by the ZnO and dissolved Zn^{2+} but had a minor role in cytotoxicity but was a cytotoxic response to ZnO NP and Zn^{2+} (Song et al. 2010). Roy et al. investigated the therapeutic effects of ZnO NPs in balb/c mice (Roy et al. 2014). Concentrations of 1–12 mg/mL of the ZnO NPs with average diameters below 50 nm were used in the study with a treatment time of 30 days. The ZnO NPs effectively boosted the immune response leading to inflammatory response, which was determined by the expression of Cox2, MMP, and PGE2. The authors concluded that the ZnO NPs could be potentially utilized in therapeutics (Roy et al. 2014). Esmaeillou et al. investigated the oral toxicity of ZnO NPs in adult mice (Esmaeillou et al. 2013). The authors used 20–30 nm ZnO NPs and were given to the mice using oral gavage at a concentration of 333.33 mg/kg/day. Loss of appetite, ever lethargy, and vomiting in some mice were observed over the exposure period. Within 3 days of treatment, one mouse was deceased. The authors performed a series of biochemical assays and showed lower HDL and LDL in the treated mice but the triglycerides in the serum were increased. The results of the assays also showed potential liver and kidney damage in the mice. The authors also showed that lung damage was induced by the ZnO NPs. ZnO NPs (50-70 nm) have shown to be toxic in mammalian cells (Esmaeillou et al. 2013). Jeng and Swanson showed nanoparticle-exposed Neuro-2A cells exposed to doses of 100 ppm changed in morphology, and size, as well became detached (Hsiao and Huang 2011). Furthermore, the mitochondrial function degreased at treatments from 50 to 100 ppm. The authors also noted that the cellular membrane integrity was affected as indicated by LDL leakage. Apoptosis occurred at under ZnO NP treatment and cells became necrotic with the increase in concentrations. Buerki-Thurnherr et al. have investigated the mechanisms of ZnO NP toxicity using Jurkat cell or human T cells (Buerki-Thurnherr et al. 2013). The authors used four different ZnO materials in the study, which included commercially available 10 nm ZnO NP, a mandelic modified 18.3 nm ZnO, a 29 nm ZnO NP with Silica shell, and lab made methoxy-coated NPs 8 nm. In that study, it was determined the toxicity induced in the T cells by the NPs was due to the dissolution of the ZnO NPs. The presence of the Zn^{2+} ions in the growth media, which resulted in the storage of the zincosomes and the expression of MTs was upregulated. Above the threshold concentration, apoptosis occurred and eventually cellular death. Interestingly, the authors noted that ROS were not a major contributor to the studied mechanism (Buerki-Thurnherr et al. 2013). Hsiao and Huang have studied the toxicity of ZnO NPs on lung cells (Hsiao and Huang 2011). This study was designed to test a common premise that the physiochemical properties of NPs are critical players in the toxicity of NPs. The authors prepared rod-shaped, spherical, and sphere-like ZnO NPs. The sphere ZnO NPs had diameter of approximately 5 nm and lengths of 16-48 nm. Whereas the spherical NPs had diameters in the range of 5–10 nm. The sphere like NPS had sizes from 36 to 60 and 50 to 122 nm. The cells were treated with 1.56-50 ppm concentrations for times of 12, 24, and 72 hrs. The authors demonstrated convincingly that the nanorods were more toxic than the nanospheres with EC_{50} values of 8.5 and 12.1 ug/mL, respectively. The shape appeared to control the toxicity of the NPs when size and surface area were comparable. The authors concluded that both size and shape of the ZnO NPs influence the mitochondrial activity and chemokine productivity (Hsiao and Huang 2011). Karlsson exposed epithelial cells to 100 nm ZnO NPs which showed a decrease in the cell viability of 38% (Jeng and Swanson 2006). The ZnO showed almost no damage to the DNA of the cells. The ZnO NPs showed increase in the ROS generation in the cells after treatment compared to the control cells (Jeng and Swanson 2006).

17   CuO NP Toxicity

Numerous studies were reported on the toxicity of CuO NPs, which are perhaps the third most studied NPs for their toxicity. Copper oxide NPs find commercial use in products such as antifouling paints, antimicrobials, and cosmetics (Khan et al. 2019; Grigore et al. 2016; Adeleye et al. 2016; Khashan et al. 2016). These commercial and personal care applications have the potential to increase the presence of CuO NPs in the environment and human exposure. Like ZnO NPs, CuO NPs can readily go through dissolution at physiological pHs and cause increased copper concentrations

Table 5.4 Major commonly observed toxic effects in the treatment of multiple types of live organisms and cells with of CuO NPs

Effect/ Impairment of Biological Process	Organisms
Oxidative stress/ROS	Algae, plankton, bivalves, fish, mammalian cells
Immune system downregulation	Bivalves,
Viability	Algae, plankton, bivalves, fish, mammalian cells

in cells. Elevated copper concentrations in humans have been linked to several different health conditions. The copper-related health conditions due to elevated copper concentrations can be summarized as follows: liver damage, gastrointestinal symptoms, and at extremely high concentration death (Institute of Medicine 1998; National Research Council Committee on Copper in Drinking Water 2000). It should also be noted that copper ions are essential to the healthy growth and development of organisms. However, the toxicity of CuO NPs may be a combination of the dissolved copper species and the NPs, or caused by either, as in nano zinc treatments. Mixed results have been observed when authors attempt to determine CuO NP toxicity. The most commonly observed toxic effects observed across the different organisms are summarized in Table 5.4.

18   CuO Toxicity in the Treatment of Algae

Similarly, effects with respect to the toxicity of ZnO NPs have been observed with CuO NPs. Ultimately, oxidative stress and cellular viability are the most common toxic effects of CuO NPs in algal species. Similar to ZnO NPs, CuO NPs are somewhat soluble in aqueous solution releasing Cu^{2+} ion, which are toxic to some algal species. In fact, $CuSO_4$ is a well-known algicide, which generates free Cu^{2+} ions in solution. Thus, the study of CuO NP toxicity is more than likely a combination of the particles and the dissolved ions in solution. Melegari et al. have investigated the toxicity of CuO NPs on *Chlamydomonas reinhardtii* a green algal species (Melegari et al. 2013). The authors treated the algal species with 20–30 nm CuO NPs, which were approximately spherical in shape. The study used NP concentrations of 0.1–1000 ppm CuO NPs with treatment times up to 72 h. It was reported that the NPs induced an increase in the reactive species within the algae as well as lipid peroxidation of cellular membranes. The study also showed that the NPs were able to transport across the cell wall and were internalized within the cells. The toxic effects of the NPs were noticed after 72 h of treatment with a concentration of 0.1 mg/L. Aruoja et al. investigated the effects of CuO NPs *Pseudokirchneriella subcapitata*, a microalga (Aruoja et al. 2009). The authors used a CuO NP with an average size of 30 nm for a treatment period of 72 hrs., with concentration of 100 ppm. The authors determined that the EC_{50} for the CuO NPs was 0.71 mg/L, which was much more toxic than the comparable CuO bulk EC_{50} of 11.55 mg. The data showed a decrease in the cell growth and complete inhibition of growth at 6.4 mg Cu/L using the CuO NPs. In addition, it was concluded that the bioavailability of the CuO in the NPs was much higher than the bulk phase (Aruoja et al. 2009). Suppi et al. investigated CuO NP toxicity to algae (Suppi et al. 2015). The authors used CuO NPs with average diameters of 30 nm and treatment concentration of up to 100 ppm. The results showed minimum biocidal effects starting at CuO 10 ppm NP (Suppi et al. 2015). Cheloni et al. investigated the effects of CuO NPs on *Chlamydomonas reinhardtii* using different irradiation compositions (Cheloni et al. 2016). The authors studied natural light, UV B, and light from an incubator which is a mixture of UV A, UV B, and visible light. The microalgae were treated with CuO NPs with a size range of 30-50 nm for up to 24 hrs. at two concentrations of 8 ug/mL and 0.8 mg/mL. It was found that the hydrodynamic diameter of the NPS changes from 800 nm to 600 nm within 2 h of dispersing and remained constant thereafter. The light was observed to affect the algal cells, under incubator light the cells showed a 1.7% growth inhibition at 0.8 ppm CuO, the 8 ug/L, the 0.8 mg/L treatment showed a 67.7% growth inhibition. As well under the natural sunlight, the 8 ug/L treatment showed a 2.3% growth inhibition whereas the 0.8 mg/L treatment showed approximately 6% growth inhibition. Under the UV B irradiation, the inhibition increased further for the 8 ug/L CuO NP treatment and showed approximately 82% inhibition of growth and the 0.8 mg/L treatment showed approximately 94% growth inhibition. Similarly, it was also noted that as the algae were treated with the CuO NPs with increasing concentration and changing light condition, the membrane permeability was altered with increasing Cu concentration especially under UV B. The authors concluded that the UV B in conjunction with the UV B irradiation had a synergistic effect on the algae growth and health (Cheloni et al. 2016). Sankar et al. have investigated the effect of larger particles 270 nm on *Microcystis aeruginosa* to investigate the growth inhibition of bloom formation (Sankar et al. 2014). The algal species were treated with CuO NP concentrations from 12.5 to 50 mg/L for up to 96 h. The authors showed that cell density in the growth media decreased with treatment time and with increasing concentration of the CuO NPs. At 50 ppm of the CuO NPs an inhibition rate of approximately 90% was observed. All NP treatments showed a decrease in the chlorophyll A and B as well as the total amount of carotenoids produced by the algae. The authors concluded that the CuO NPs were potentially a biocide to the algae to prevent the formation of algal blooms (Sankar et al. 2014).

19   CuO Toxicity in the Treatment of Plankton

Similar to algae species, plankton are affected by the toxicity of CuO NPs in solution. The major toxic effect observed in plankton species appears to be the generation of ROS species causing a decrease in the cellular viability. Heinlaan et al. investigated the effects of CuO NPs with an average dimeter of 30 nm on *Daphnia magna* and *T. platyurus* (Heinlaan et al. 2008). The authors treated the plankton and showed a EC_{50} of approximately 165 mg/L and 95 for *D. magna* and *T. platyurus*, respectively. The EC_{50} of the CuO NPs were much higher values than those observed for free Cu ions from $CuSO_4$. The *T. platyurus* was more sensitive to the CuO NPs than the *D. magna*; the authors also showed that the NPs did not have to enter the cells of the organisms to cause toxicity (Heinlaan et al. 2008). In another study, Heinlaan et al. investigated the effects of CuO NPs with an average dimeter of 30–60 nm on *Daphnia magna* (Heinlaan et al. 2011). The authors investigated the changes to the midgut of the organism using TEM analysis. At 48 h of treatment, the authors showed an EC_{50} of 175 mg/L for the CuO NPs. The concertation of solubilized copper in solution from the CuO NPs was 0.05 mg/L and the determined EC_{50} was similar to that observed for free copper ions. The authors noted that in the presence of CuO NPs in the midgut lumen, internalization of the CuO NPs was not evident. In addition, the CuO nPs after 48 hrs were no longer in the peritrophic membrane but had moved to the midgut epithelium microvilli (Heinlaan et al. 2011). Mwaanga et al. investigated the effects of CuO NPs with an average dimeter of less than 50 nm on *Daphnia magna* (Mwaanga et al. 2014). The authors exposed 5-day-old *D. magna* for 72 h to sublethal concentrations of CuO NPs. The authors also investigated the effects of hard water and the presence of natural organic matter (NOM). From the treated organisms, the authors investigated several biochemical processes including glutathione activity, glutathione oxidation, thiol compounds, and metallothionein. The glutathione synthase enzyme was inactivated by the CuO NPs; the metallothionine was increased as well as the oxidation of the glutathione. The data suggests that the CuO NPs induced oxidative stress in the organisms. The authors also showed some dissolution of the CuO NPs in solution and concluded the toxicity observed was due to both the dissolved ion and the CuO NPs (Mwaanga et al. 2014). Similar results were observed by the authors when investigating the toxicity of CuO NPs on *Thamnocephalus platyurus* using CuO NPs with an average diameter of 30 nm (Mortimer et al. 2010). Mortimer et al. have investigated the effects of CuO NPs on *Tetrahymena thermophila* (Mortimer et al. 2010). The authors utilized CuO NPs with average diameters of 30 nm. The organisms were treated with Cu^{2+} ions, CuO NPs, and CuO in the bulk form. The concentrations used for the treatments for the CuO NPs ranged from 31.25 to 500 mg/L. The authors showed an EC_{50} of approximately 128 mg/L for the CuO NPs on the organisms. The toxic effects of the CuO NPs were not observed to be time dependent.

20 CuO Toxicity in the Treatment of Bivalves

CuO NP toxicity observed in bivalve species has been shown to be induced by the generation of ROS species inside the organisms as has been observed with other nanoparticles in bivalve species. However, the differentiation between Cu^{2+} ion toxicity and CuO NP toxicity is difficult, which is due to the similar effects. Buffet et al. also investigated the effects of CuO on *Scrobicularia plana* using

200 nm CuO particles (Buffet et al. 2011). In the same study, Buffet et al. studied the effects of CuO NPs on *Hediste diversicolor* treated with approximately 200 nm CuO NPs. The authors were attempting to observe the effects of CuO NPs on behavior and biochemical response to the CuO NPs. The authors treated both organisms with CuO NPs with concentrations that was equivalent to10 ug/L of dissolved Cu^{2+} ions. In addition, the authors also treated both organisms with dissolved Cu^{2+} ions for comparative purposes at a concentration of 10 ug/L. The defense mechanisms for both organisms being exposed to Cu2+ and CuO NPs were elevated compared to controls, which included CAT, GST, SOD, and MTLP. However, the LDH concentrations were only elevated in the *N. diversicolor*. The exposure to the Cu^{2+} ions had a negative effect on the burrowing behavior whereas the CuO NPs did not appear to affect the behavior (Mwaanga et al. 2014). Pradhan et al. investigated the effects of CuO NP with diameters ranging from 30 to 50 nm on *Allogamus ligonifer* (Pradhan et al. 2012). The insects were treated at the highest level of 75 mg/L using either contaminated food or through contaminated water. The treatment of the organisms showed a higher copper accumulation in the larvae bodies than the nontreated organisms. It was determined that the LC_{50} for the CuO NPs was 569 mg/L after 96 h of treatment. The insects treated with the CuO NPs showed decreased appetite and a decrease in the growth rate by approximately 46%. The author noted that the soluble fraction of Cu^{2+} was approximately 10% by body mass (Pradhan et al. 2012). The accumulation of copper oxide NPs in the digestive glands of *Mytilus galloprovincialis* was investigated by Gomes et al. (Gomes et al. 2012). The authors used CuO NPs with an average size distribution of 30 ± 10 nm. The bivalves were treated with 10 ug/L of dissolved Cu^{2+} ion or CuO NPs for a 15-day period. The authors noted that both forms of Cu were effective at causing oxidative stress in the digestive glands; the GPX activities from the exposure to both forms of copper were similar. However, an induction of metallothionein was observed only in the organisms exposed to Cu2+. The authors concluded that the digestive gland is susceptible to CuO NPs and the related oxidative stress and was the main tissue for their accumulation (Gomes et al. 2012). Along the same line, Hu et al. investigated the toxicity of CuO to *Mytilus edulis* (Hu et al. 2014). The authors used 100 nm CuO NPs and administered doses from 400 to 1000 ppb to the mussels. It was noted that copper was located in the gill tissue and small amounts in the digestive gland. The data from the study suggested decreased protein thiol and increased carbonylation (protein oxidation) (Hu et al. 2014). The toxic effects of CuO NPs ranging in size from 40 to 100 nm on *Exaiptasia pallida* was investigated by Siddiqui et al. (Siddiqui et al. 2015). The authors exposed the *E. pallida* to both Cu^{2+} and CuO NPs at concentrations ranges of 10–50 ppm and 50–100 mg/L, respectively. In addition, the treatment was performed over 21 days. The Cu^{2+}-exposed organisms showed higher concentrations of copper accumulation and concentration increased with both the dosage and the time of the treatment. However, overall the organisms treated with the CuO NPs showed higher levels of oxidative stress. The oxidative stress was examined

using standard enzyme tests for the determination of catalase, glutathione peroxidase, glutathione reductase, and carbonic anhydrase. The authors concluded that the behavior of the CuO NPs and Cu^{2+} had differences with respect to the magnitude of the effects on the organism (Siddiqui et al. 2015).

21   CuO Toxicity in the Treatment of Fish and Fish Cell Lines

CuO nanoparticles have shown mixed results in the treatment of different fish species. Like the toxicity observed in different species with ZnO NPs, CuO NPs have the added complexity of the Cu ion toxicity. Oxidative stress appears to be common among different fish species treated with CuO NPs. There appears to be organ accumulation of copper in the liver and kidney was observed after treatment with CuO NPs; the form of the copper is generally not specified in the organs. *Diano rerio* (zebrafish) embryos were investigated by Ganesan et al. for the toxic effects of 51 nm CuO NPs (Ganesan et al. 2016). The authors treated the embryos with concentrations of CuO NPs from 4 to 80 ppm for a total time of 48 h. The authors noticed a 2% death in the embryos at 5 ppm and the LC_{50} for the CuO NPs was determined to be 64 ppm; there was almost a linear increase in mortality with increasing concentration of the NPs. Similarly, the rate of hatching was observed to decrease with nearing CuO NP concentration. Furthermore, the concentration of the CuO NP increased, the heart rate was decreased in the embryos. As the concentration of the CuO NPs increased, the malformation in the embryos increased. The total protein in the fish was observed to increase as the concentration of CuO NPs increased; however, the AcHe and ATPase concentrations were observed to decrease with increasing NP concentrations. Overall indicators of oxidative stress were increased with increasing CuO NP treatment, which included ROS, lipid peroxidation, PCC, and NO. The antioxidant responses to oxidative stress were observed to decrease with increasing CuO NP concentrations (Ganesan et al. 2016). *Carassius auratus* or goldfish were subjected to CuO NPs with an average diameter of 40 nm by Ates et al. to investigate the toxic effects through waterborne and dietary exposure (Ates et al. 2015). The fish were exposed to concentrations of 1 and 10 ppm in either the diet or in the water. The intestine showed the highest levels of copper from the NP treatment from dietary and waterborne exposure. The organ with the next highest levels were the gills followed by the liver tissue. However, the concentration of copper in the heart, brain, and muscle was almost the same as the control samples. The high doses for treatment showed increased malondialdehyde concentrations indicating oxidative stress (Ates et al. 2015). The authors concluded that the waterborne exposure exhibited much higher toxicity than observed through dietary exposure. Villarreal et al. studied the toxic effects of 100 nm CuO NPs on *Oreochromis mossambicus,* a species of Tilapia (Villarreal et al. 2014). The authors used flame synthesized CuO NPs and exposed the fish to concentrations from 0.5 to

5 mg/L. The fish were studied under two different environments: a constant supply of fresh water and an increasing salinity. The fish exposed to the CuO NPs in the saltwater increased in the opercular ventilation. However, the fish exposed to fresh water showed a much smaller response. In addition, the authors noted that the CuO NPs showed effects on the reduced and oxidized glutathione ratios, changes in metal response genes. Furthermore, the organs with increased Cu included the gills and liver, indicating osmotic stress was induced by the CuO NPs (Ates et al. 2015). Isani et al. investigated the effects of 51 nm CuO NPs on *Oncorhynchus mykiss,* commonly known as rainbow trout (Isani et al. 2013). The authors conducted a comparative study between Cu^{2+} ions and CuO NPs. The authors performed in vitro and in vivo studies. The in vitro experiment involved the treatment of red blood cells and treated at multiple CuO concentrations. The treatment concentrations used in the in vivo study were 1 ug/g body weight injected into the fish. Both the ionic form and NP form of the copper increased the hemolysis rate in a concentration-dependent manner for the in vitro studies. The hemolysis was more marked in the cells treated with the Cu^{2+} ions when compared to the CuO NP treatments. The in vivo studies showed high concentrations of the Cu in the gills, kidney, and liver. The data also showed the concentration found in the organs was significantly higher in the ion-treated fish when compared to the NP-treated fish. The toxic effects of 30-40 nm CuO NPs on *Oreochromis niloticus,* a species of tilapia, were investigated by Abdel-Khalek et al. (Abdel-Khalek et al. 2016). The authors exposed the fish to CuO NPs does at 1/10 and 1/20 the LC_{50} determined at 96 h of treatment. The fish were exposed to both CuO NPs and bulk. The Cu tissue accumulation of the following treatment was observed in decreasing order liver, kidney, gills, skin, muscle, which caused organ damage. The authors concluded that the CuO NPs were more toxic than the bulk phase. The NPs had a higher ability to be internalized into the cells compared to the bulk phase. As well the fish treated with the NPs showed intracellular osmotic disorders and increased in a determined blood parameter. Cells from Chinook salmon have been investigated for the toxic effects of CuO NPs with an average diameter of 50 nm, by Srikanth et al. (Srikanth et al. 2016). The authors exposed the CHES-214 cells from the Chinook salmon to CuO NPs at concentrations from 5 ppm to 15 ppm. The cells at all concentrations showed increased oxidative stress after treatment compared to the control cells. In addition, the protein carbonyl concentrations were increased, while the lipid peroxidation were increased. The reduced form of glutathione initially was observed to increase at the 5-ppm treatment and then decrease at 10 and 15 ppm, but all levels of glutathione were at increased levels compared to the control. Furthermore, the concentration of the oxidized glutathione was observed to be lower than the control samples in the study. Both glutathione peroxidase and glutathione sulfo-transferase were increased at all treatments compared to the control samples. As well, all the responses to oxidative stress were at increased levels compared to the control samples. The authors also showed cellular abnormalities with respect to morphological changes. The study concluded that CuO NPs were cytotoxic in a concertation-dependent manner.

22   CuO Toxicity in the Treatment of Mammalian Cells

In cell line studies, CuO NPs have been shown to reduce the cellular viability and induce oxidative stress. Karlsson et al. investigated the effects of CuO on human lung epithelial cell line A549 (Jeng and Swanson 2006). The authors studied the effects of 42 nm CuO nanoparticles at an 80-ppm concentration in solution for 18 h. The study showed a reduce viability of the cells of 96%. In addition, the CuO NPs showed high DNA damage within the cells. In addition, there was a significant increase in the ROS generation within the cells comparted to control treatments. The authors showed that the CuO NPs had an eightfold increase in the dose-response curve when compared to the treatment of the cells with the ionic form of Cu^{2+} from $CuCl_2$.

23   Conclusion

The toxicity of metal oxide NPs can be generally summed up as an oxidative stress process in different species. However, it is not as simple as one mechanism being responsible for the toxicity. In some species such as diatoms, the molting process becomes inhibited causing imminent death. In addition, the behavior of different nanoparticles in solution also has an effect on the toxicity. For example, TiO_2 is insoluble in water, but ZnO and CuO NPs are somewhat water soluble which generate the free metal ion in solution. In discussing the toxicity of ZnO and CuO NPs, the formation of the free ions in solution adds complicity to the question of NP toxicity and the causes. In addition, there does appear to be a link between the phases of the NPs that organisms are exposed to, which has been discussed in the case of TiO_2 NP toxicity. For example, the toxicity of the anatase phase of TiO_2 appears to be higher and has more effects than the rutile phase of TiO_2. The difference in the effect of anatase and rutile may be due to the relatively higher structural stability of rutile when compared to anatase. It is noteworthy that many of the TiO_2 NP toxicity studies have used the Degussa mixture of TiO_2 (anatase and rutile phases in a ratio of about 3:1), which is a common photocatalytic commercial specimen of TiO_2. In addition, the toxicity of NPs does appear to be aggravated in the presence of light. On perusal of literature, it appears that multiple levels of nanomaterial interactions contribute to specific toxicity phenotypes in an organism. In addition, toxic effects also arise from the stabilizing agents utilized in the synthesis and dispersion of NPs in solution. This review thus suggests using multiple levels of interaction studies to develop a full grasp of nanotoxicity.

Acknowledgments   J.G. Parsons is grateful for the generous support provided by a Departmental Grant from the Robert A. Welch Foundation (Grant No. BX-0048). M. Alcoutlabi would like to acknowledge the support from NSF PREM /award/ under grant No. DMR-1523577: UTRGV-UMN Partnership for Fostering Innovation by Bridging Excellence in Research and Student Success.

References

Abdel-Khalek, A. A., Badran, S. R., & Marie, M. (2016). Toxicity evaluation of copper oxide bulk and nanoparticles in Nile tilapia, *Oreochromis niloticus*, using hematological, bioaccumulation and histological biomarkers. *Fish Physiology and Biochemistry, 42*, 1225–1236.

Abdel-Mohsen, A. M., Hrdina, R., Burgert, L., Abdel-Rahman, R. M., Hašová, M., Šmejkalová, D., Kolář, M., Pekar, M., & Aly, A. S. (2013). Antibacterial activity and cell viability of hyaluronan fiber with silver nanoparticles. *Carbohydrate Polymers, 92*, 1177–1187.

Absar, A., Satyajyoti, S., Islam, K. M., Sastry, K. R., & Murali, S. (2005). Extra-/intracellular biosynthesis of gold nanoparticles by an Alkalotolerant fungus, *Trichothecium sp. Journal of Biomedical Nanotechnology, 1*, 47–53.

Adeleye, A. S., Oranu, E. A., Tao, M., & Keller, A. A. (2016). Release and detection of nanosized copper from a commercial antifouling paint. *Water Research, 102*, 374–382.

Akhtar, M. J., Ahamed, M., Kumar, S., Majeed Khan, M. A., Ahmad, J., & Alrokayan, S. A. (2012). Zinc oxide nanoparticles selectively induce apoptosis in human cancer cells through reactive oxygen species. *International Journal of Nanomedicine, 7*, 845–857.

Ali, D., Alarifi, S., Kumar, S., Ahamed, M., & Siddiqui, M. A. (2012). Oxidative stress and genotoxic effect of zinc oxide nanoparticles in freshwater snail *Lymnaea luteola L. Aquatic Toxicology, 124–125*, 83–90.

Arora, S., Sharma, P., Kumar, S., Nayan, R., Khanna, P. K., & Zaidi, M. G. H. (2012). Gold-nanoparticle induced enhancement in growth and seed yield of *Brassica juncea. Plant Growth Regulation, 66*, 303.

Aruoja, V., Dubourguier, H., Kasemets, K., & Kahru, A. (2009). Toxicity of nanoparticles of CuO, ZnO and TiO$_2$ to microalgae *Pseudokirchneriella subcapitata. Science Total Environment, 407*, 1461–1468.

Asharani, P. V., Wu, Y. L., Gong, Z., & Valiyaveettil, S. (2008). Toxicity of silver nanoparticles in zebrafish models. *Nanotechnology, 19*, 255102.

Ates, M., Arslan, Z., Demir, V., Daniels, J., & Farah, I. O. (2015). Accumulation and toxicity of CuO and ZnO nanoparticles through waterborne and dietary exposure of goldfish (*Carassius auratus*). *Environmental Toxicology, 30*, 119–128.

Bhuvaneshwari, M., Iswarya, V., Archanaa, S., Madhu, G. M., Kumar, G. K. S., Nagaraj, R., Chandrasekaran, N., & Mukherjee, A. (2015). Cytotoxicity of ZnO NPs towards fresh water algae *Scenedesmus obliquus* at low exposure concentrations in UV-C, visible and dark conditions. *Aquatic Toxicology, 162*, 29–38.

Borghi, E. B., Morando, P. J., & Blesa, A. (1991). Dissolution of magnetite by Mercaptocarboxylic acids. *Langmuir, 7*, 1652–1659.

Bozich, J. S., Lohse, S. E., Torelli, M. D., Murphy, C. J., Hamers, R. J., & Klaper, R. D. (2014). Surface chemistry, charge and ligand type impact the toxicity of gold nanoparticles to *Daphnia magna. Environmental Science. Nano, 1*, 260–270.

Buerki-Thurnherr, T., Xiao, L., Diener, L., Arslan, O., Hirsch, C., Maeder-Althaus, X., Grieder, K., Wampfler, B., Mathur, S., Wick, P., & Krug, H. F. (2013). In-vitro mechanistic study towards a better understanding of ZnO nanoparticle toxicity. *Nanotoxicology, 7*, 402–416.

Buffet, P., Tankoua OF, Pan, J., Berhanu, D., Herrenknecht, C., Poirier, L., Amiard-Triquet, C., Amiard, J., Berard, J., Risso, C., Guibbolini, M., Romeo, M., Reip, P., Valsami-Jones, E., & Mouneyrac, C. (2011). Behavioural and biochemical responses of two marine invertebrates *Scrobicularia plana* and *Hediste diversicolor* to copper oxide nanoparticles. *Chemosphere, 84*, 166–174.

Buffet, P.-E., Amiard-Triquet, C., Dybowska, A., Risso-de Faverney, C., Guibbolini, M., Valsami-Jones, E., & Mouneyrac, C. (2012). Fate of isotopically labeled zinc oxide nanoparticles in sediment and effects on two endobenthic species, the clam *Scrobicularia plana* and the ragworm *Hediste diversicolor. Ecotoxicology and Environmental Safety, 84*, 191–198.

Calzolai, L., Gilliland, D., & Rossi, F. (2012). Measuring nanoparticles size distribution in food and consumer products: A review. *Journal of Food Additives Contaminants A., 29*, 1183–1193.

Casaletto, M. P., Longo, A., Martorana, A., Prestianni, A., & Venezia, A. M. (2006). XPS study of supported gold catalysts: The role of Au0 and Au+δ species as active sites. *Surface and Iterface Analysis, 38*, 215–218.

Chakraborty, C., Sharma, A. R., Sharma, G., & Lee, S.-S. (2016). Zebrafish: A complete animal model to enumerate the nanoparticle toxicity. *Journal of Nanbiotechnology, 14.* https://doi.org/10.1186/s12951-016-0217-6.

Cheloni, G., Marti, E., & Slaveykova, V. I. (2016). Interactive effects of copper oxide nanoparticles and light to green alga *Chlamydomonas reinhardtii. Aquatic Toxicology, 170*, 120–128.

Chen, J., Dong, X., Zhao, J., & Tang, G. (2009). *In vivo* acute toxicity of titanium dioxide nanoparticles to mice after intraperitioneal injection. *Journal of Applied Toxicology, 29*, 330–337.

Chen, L., Zhou, L., Liu, Y., Deng, S., Wu, H., & Wang, G. (2012). Toxicological effects of nanometer titanium dioxide (nano-TiO2) on *Chlamydomonas reinhardtii. Ecotoxicology and Environmental Safety, 84*, 155–162.

Chen, T.-H., Lin, C.-Y., & Tseng, M.-C. (2011). Behavioral effects of titanium dioxide nanoparticles on larval zebrafish (*Danio rerio*). *Marine Pollution Bulletin, 63*, 303–308.

Chwalibog, A., Sawosz, E., Hotowy, A., Szeliga, J., Mitura, S., Mitura, K., Grodzik, M., Orlowski, P., & Sokolowska, A. (2010). Visualization of interaction between inorganic nanoparticles and bacteria or fungi. *International Journal of Nanomedicine, 5*, 1085–1094.

Clément, L., Hurel, C., & Marmier, N. (2013). Toxicity of TiO$_2$ nanoparticles to cladocerans, algae, rotifers and plants – Effects of size and crystalline structure. *Chemosphere, 90*, 1083–1090.

Currie, H. A., & Silica in Plants, P. C. C. (2007). Biological, biochemical and chemical studies. *Annals of Botany, 100*, 1383–1389.

Dabrunz, A., Duester, L., Prasse, C., Seitz, F., Rosenfeldt, R., Schilde, C., Schaumann, G. E., & Schulz, R. (2011). Biological surface coating and molting inhibition as mechanisms of TiO$_2$ nanoparticle toxicity in *Daphnia magna. PLoS One, 6*, e20112. https://doi.org/10.1371/journal.pone.0020112.

Dasari, T. P., Pathakoti, K., & Hwang, H.-M. (2013). Determination of the mechanism of photoinduced toxicity of selected metal oxide nanoparticles (ZnO, CuO, Co$_3$O$_4$ and TiO$_2$) to *E. coli* bacteria. *Journal of Environmental Sciences, 25*, 882–888.

Deng, Q., Tang, H., Liu, G., Song, X., Kang, S., Wang, H., Ng, D. H. L., & Wang, G. (2015). Photocatalytic degradation of 2,4,40-trichlorobiphenyl into long-chain alkanes using Ag nanoparticle decorated flower-like ZnO microspheres. *New Journal of Chemistry, 39*, 7781.

Durán, N., Durán, M., Jesus, M. B., Seabra, A. B., Fávaro, W. J., & Nakazato, G. (2016). Silver nanoparticles: A new view on mechanistic aspects on antimicrobial activity. *Nanomedicine: Nanotechnology, Biology and Medicine, 12*, 789–799.

Durán, N., Marcato, P. D., De Cont, R., Alves, O. L., Costa, F. T. M., & Brocchi, M. (2010). Potential use of silver nanoparticles on pathogenic bacteria, their toxicity and possible mechanisms of action. *Journal of the Brazilian Chemical Society, 21*, 949–959.

Esmaeillou, M., Moharamnejad, M., Hsankhani, R., Tehrani, A. A., & Maadi, H. (2013). Toxicity of ZnO nanoparticles in healthy adult mice. *Environmental Toxicology and Pharmacology, 35*, 67–71.

Fabrega, J., Fawcett, S. R., Renshaw, J. C., & Lead, J. R. (2009). Silver nanoparticle impact on bacterial growth: effect of pH, concentration, and organic matter. *Environmental Science & Technology, 43*, 7285–7290.

Fafarman, A. T., Koh, W.-K., Diroll, B. T., Kim, D. K., Ko, D.-K., Oh, S. J., Ye, X., Doan-Nguyen, V., Crump, M. R., Reifsnyder, D. C., Murray, C. B., & Kagan, C. R. (2011). Thiocyanate-capped nanocrystal colloids: Vibrational reporter of surface chemistry and solution-based route to enhanced coupling in nanocrystal solids. *Journal of the American Chemical Society, 133*, 15753–15761.

Fröhlich, E., & Roblegg, E. (2012). Models for oral uptake of nanoparticles in consumer products. *Toxicology, 291*, 10–17.

Fu, L., Hamzeh, M., Dodard, S., Zhao, Y. H., & Sunahara, G. I. (2015). Irradiation effects of TiO$_2$ nanoparticles on ROS production and growth inhibition using freshwater green algae pre-exposed to UV irradiation. *Environmental Toxicology and Pharmacology, 39*, 1074–1080.

Ganesan, S., Thirumurthi, N. A., Raghunath, A., Vijayakumar, S., & Perumal, E. (2016). Acute and sub-lethal exposure to copper oxide nanoparticles causes oxidative stress and teratogenicity in zebrafish embryos. *Journal of Applied Toxicology, 36*, 554–567.

Gardea-Torresdey, J. L., Parsons, J. G., Gomez, E., Peralta-Videa, J., Troiani, H. E., Santiago, T. P., & Yacaman, M. J. (2002). Formation and growth of au nanoparticles inside live alfalfa plants. *Nano Letters, 2*, 397–401.

Gerloff, K., Fenoglio, I., Carella, E., Kolling, J., Albrecht, C., Boots, A. W., Förster, I., & Schins, R. P. F. (2012). Distinctive toxicity of TiO₂ rutile/Anatase mixed phase nanoparticles on Caco-2 cells. *Chemical Research in Toxicology, 25*, 646–655.

Gomes, T., Pereira, C. G., Cardoso, C., Pinheiro, J. P., Cancio, I., & Bebianno, M. J. (2012). Accumulation and toxicity of copper oxide nanoparticles in the digestive gland of *Mytilus galloprovincialis*. *Aquatic Toxicology, 118-119*, 72–79.

Govindaraju, K., Khaleel, S., Vijayakumar, B., Kumar, G., & Singaravelu, G. (2008). Silver, gold and bimetallic nanoparticles production using single-cell protein (Spirulina platensis) Geitler. *Journal of Materials Science, 43*, 5115–5122.

Gregory, J. (2005). *Particles in water: Properties and processes*. CRC Press.

Greulich, C., Braun, D., Peetsch, A., Diendorf, J., Siebers, B., Epple, M., & Köller, M. (2012). The toxic effect of silver ions and silver nanoparticles towards bacteria and human cells occurs in the same concentration range. *RSC Advances, 2*, 6981–6987.

Griffin, S., Masood, M. I., Nasim, M. J., Sarfraz, M., Ebokaiwe, A. P., Schäfer, K. H., Keck, C. M., & Jacob, C. (2018). Natural nanoparticles: A particular matter inspired by nature. *Antioxidants (Basel), 7*, 3.

Grigore, M. E., Biscu, E. R., Holban, A. M., Gestal, M. C., & Grumezescu, A. M. (2016). Methods of Synthesis, properties and biomedical applications of CuO nanoparticles. *Pharmaceuticals (Basel), 9*, 75.

Grzelczak, M., Pérez-Juste, J., Mulvaney, P., & Liz-Marzán, L. M. (2008). Shape control in gold nanoparticle synthesis. *Chemical Society Reviews, 37*, 1783–1791.

Hajipour, M. J., Fromm, K. M., Ashkarran, A. A., de Aberasturi, D. J., de Larramendi, I. R., Rojo, T., Serpooshan, V., Parak, W. J., & Mahmoudi, M. (2012). Antibacterial properties of nanoparticles. *Trends in Biotechnology, 30*, 499–511.

Haldorai, Y., Kim, B.-K., Jo, Y.-L., & Shim, J.-J. (2014). Ag@graphene oxide nanocomposite as an efficient visible-light plasmonic photocatalyst for the degradation of organic pollutants: A facile green synthetic approach. *Materials Chemistry and Physics, 143*, 1452–1461.

Heike, L. M., Norbert, T., Daniela, J., Vladimir, D., Baranov, I., & Kneipp, J. (2014). Trends in single-cell analysis by use of ICP-MS. *Analytical and Bioanalytical Chemistry, 206*, 6963.

Heinlaan, M., Ivask, A., Blinova, I., Dubourguier, H., & Kahru, A. (2008). Toxicity of nanosized and bulk ZnO, CuO and TiO₂ to bacteria *Vibrio fischeri* and crustaceans *Daphnia magna* and *Thamnocephalus platyurus*. *Chemosphere, 71*, 1308–1316.

Heinlaan, M., Kahru, A., Kasemets, K., Arbeille, B., Prensier, G., & Dubourguier, H. (2011). Changes in the *Daphnia magna* midgut upon ingestion of copper oxide nanoparticles: A transmission electron microscopy study. *Water Research, 45*, 179–190.

Ho, C.-M., Yau, S. K.-W., Lok, C.-N., So, M.-H., & Che, C. M. (2010). Oxidative dissolution of silver nanoparticles by biologically relevant oxidants: A kinetic and mechanistic study. *Chemistry, an Asian Journal, 5*, 285–293.

Hough, R. M., Noble, R. R. P., Hitchen, G. J., Hart, R., Reddy, S. M., Saunders, M., Clode, P., Vaughan, D., Lowe, J., Gray, D. J., An, R. R., Butt, C. R. M., & Verrall, M. (2008). Naturally occurring gold nanoparticles and nanoplates. *Geology, 36*, 571–574.

Hsiao, I.-L., & Huang, Y.-J. (2011). Effects of various physicochemical characteristics on the toxicities of ZnO and TiO₂ nanoparticles toward human lung epithelial cells. *Science of the Total Environment, 409*, 1219–1228.

Hu, W., Culloty, S., Darmody, G., Lynch, S., Davenport, J., Ramirez-Garcia, S., Dawson, K. A., Lynch, I., Blasco, J., & Sheehan, D. (2014). Toxicity of copper oxide nanoparticles in the blue mussel, *Mytilus edulis*: A redox proteomic investigation. *Chemosphere, 108*, 289–299.

Hu, Y., Xie, J., Tong, Y. W., & Wang, C.-H. (2007). Effect of PEG conformation and particle size on the cellular uptake efficiency of nanoparticles with the HepG2 cells. *Journal of Controlled Release, 118*, 7–17.

Hurst, J. K., & Lymar, S. V. (1997). Toxicity of Peroxynitrite and related reactive nitrogen species toward Escherichia coli. *Chemical Research in Toxicology, 10*, 802–810.

Hwang, E. T., Lee, J. H., Chae, Y. J., Kim, Y. S., Kim, B. C., & Sang, B.-I. (2008). Gu MB. Analysis of the toxic mode of action of silver nanoparticles using stress-specific bioluminescent Bacteria. *Small, 4*, 746–750.

Institute of Medicine, Food and Nutrition Board. (1998). *Dietary reference intakes: Thiamin, riboflavin, niacin, vitamin B6, folate, vitamin B12, pantothenic acid, biotin, and choline.* Washington, DC: National Academy Press. Copper related health conditions.

Isani, G., Falcioni, M. L., Barucca, G., Sekar, D., Andreani, G., Carpene, E., & Falcioni, G. (2013). Comparative toxicity of CuO nanoparticles and $CuSO_4$ in rainbow trout. *Ecotoxicology and Environmental Safety, 97*, 40–46.

Jansa, H., & Huo, Q. (2012). Gold nanoparticle-enabled biological and chemical detection and analysis. *Chemical Society Reviews, 41*, 2849–2866.

Jeevanandam, J., Barhoum, A., Chan, Y. S., Dufresne, A., & Danquah, M. K. (2018). Review on nanoparticles and nanostructured materials: History, sources, toxicity and regulations. *Beilstein Journal of Nanotechnology, 9*, 1050–1074.

Jeng, H. A., & Swanson, J. (2006). Toxicity of metal oxide nanoparticles in mammalian cells. *Journal of Environmental Science and Health, Part A, 41*, 2699–2711.

Ji, J., Longa, Z., & Lin, D. (2011). Toxicity of oxide nanoparticles to the green algae *Chlorella sp. Chemical Engineering Journal, 170*, 525–530.

Jovanovic, B., Whitley, E. M., Kimura, K., Crumpton, A., & Pali, D. (2015). Titanium dioxide nanoparticles enhance mortality of fish exposed to bacterial pathogens. *Environmental Pollution, 203*, 153e164.

Karlsson, H. L., Cronholm, P., Gustafsson, J., & Moller, L. (2008). Copper oxide nanoparticles are highly toxic: A comparison between metal oxide nanoparticles and carbon nanotubes. *Chemical Research in Toxicology, 21*, 1726–1732.

Kerkmann, M., Costa, L. T., Richter, C., Rothenfusser, S., Battiany, J., Hornung, V., Johnson, J., Engler, S., Ketterer, T., Heckl, W., Thalhammer, S., Endres, S., & Hartmann, G. (2004). Spontaneous formation of nucleic acid-based nanoparticles is responsible for high interferon-α induction by CpG-A in Plasmacytoid dendritic cells. *The Journal of Biological Chemistry, 280*, 8086–8093.

Khan, R., Inam, M. A., Park, D. R., Khan, S., Akram, M., & Yeom, I. T. (2019). The removal of CuO nanoparticles from water by conventional treatment C/F/S: The effect of pH and natural organic matter. *Molecules, 24*, 914.

Khashan, K. S., Sulaiman, G. M., & Abdulameer, F. A. (2016). Synthesis and antibacterial activity of CuO nanoparticles suspension induced by laser ablation in liquid. *Arabian Journal for Science and Engineering, 41*, 301–310.

Kimling, J., Maier, M., Okenve, B., Kotaidi, V., Ballot, H., & Plech, A. (2006). Turkevich method for gold nanoparticle synthesis revisited. *The Journal of Physical Chemistry. B, 110*, 15700–15707.

Kimura, H., Sawada, T., Oshima, S., Kozawa, K., Ishioka, T., & Kato, M. (2005). Toxicity and roles of reactive oxygen species. *Current Drug Targets. Inflammation and Allergy, 4*, 489–495.

Kitchens, C. L., McLeod, M. C., & Roberts, C. B. (2005). Chloride ion effects on synthesis and directed assembly of copper nanoparticles in liquid and compressed alkane microemulsions. *Langmuir, 21*, 5166–5173.

Kumar, A., Swarup, D., & Maji, K. (2014). Adhikary B. γ-Fe_2O_3 nanoparticles: An easily recoverable effective photo-catalyst for the degradation of rose bengal and methylene blue dyes in the waste-water treatment plant. *Materials Research Bulletin, 49*, 28–34.

Kumar, V., Guleria, P., Vinay, G., Sudesh, K., & Yadav, K. (2013). Gold nanoparticle exposure induces growth and yield enhancement in *Arabidopsis thaliana. Science of Total Environment, 461–462*, 462–468.

Lammel, T., & Sturve, J. (2018). Assessment of titanium dioxide nanoparticle toxicity in the rainbow trout (*Onchorynchus mykiss*) liver and gill cell lines RTL-W1 and RTgill-W1 under particular consideration of nanoparticle stability and interference with fluorometric assays. *NanoImpact, 11*, 1–19.

Le, T. T., Nguyen, K.-H., Jeon, J.-R., Francis, A. J., & Chang, Y.-S. (2015). Nano/bio treatment of polychlorinated biphenyls with evaluation of comparative toxicity. *Journal of Hazardous Materials, 287*, 335–341.

Lee, J., Lilly, G. D., Doty, R. C., Podsiadlo, P., & Kotov, N. A. (2009). *In vitro* toxicity testing of nanoparticles in 3D cell culture. *Small, 5*, 1213–1221.

Lee, W.-M., & An, Y.-J. (2013). Effects of zinc oxide and titanium dioxide nanoparticles on green algae under visible, UVA, and UVB irradiations: No evidence of enhanced algal toxicity under UV pre-irradiation. *Chemosphere, 91*, 536–544.

Li, F., Liang, Z., Zheng, X., Zhao, W., Wu, M., & Wang, Z. (2015). Toxicity of nano-TiO$_2$ on algae and the site of reactive oxygen species production. *Aquatic Toxicology, 158*, 1–13.

Li, X., Liu, Y., Song, L., & Liu, J. (2003). Responses of antioxidant systems in the hepatocytes of common carp (Cyprinus carpio L.) to the toxicity of microcystin-LR. *Toxicon, 42*, 85–89.

Lim, B., Xiong, Y., & Xia, Y. (2007). A water-based synthesis of octahedral, decahedral, and icosa- hedral Pd nanocrystals. *Angewandte Chemie, International Edition, 46*, 9279–9282.

Love, S. A., Maurer-Jones, M. A., Thompson, J. W., Lin, Y.-S., & Haynes, C. L. (2012). Assessing nanoparticle toxicity. *Annual Review of Analytical Chemistry, 5*, 181–205.

Luis, I. P.-S., & Liz-Marzán, M. (2002). Synthesis of silver Nanoprisms in DMF. *Nano Letters 2, 8*, 903–905.

Mansfield, C. M., Alloy, M. M., Hamilton, J., Verbeck, G. F., Newton, K., Klaine, S. J., & Roberts, A. P. (2015). Photo-induced toxicity of titanium dioxide nanoparticles to *Daphnia magna* under natural sunlight. *Chemosphere, 120*, 206–210.

Manzo, S., Miglietta, M. L., Rametta, G., Buono, S., & Francia, G. D. (2013). Toxic effects of ZnO nanoparticles towards marine algae *Dunaliella tertiolecta*. *Science of Total Environment, 445–446*, 371–376.

Marisa, I., Marin, M. G., Caicci, F., Franceschinis, E., Martucci, A., & Matozzo, V. (2015). *In vitro* exposure of haemocytes of the clam *Ruditapes philippinarum* to titanium dioxide (TiO$_2$) nanoparticles: Nanoparticle characterisation, effects on phagocytic activity and internalisation of nanoparticles into haemocytes. *Marine Environment Research, 103*, 11e17.

Marquis, B. J., Love, S. A., Braun, K. L., & Haynes, C. L. (2009). Analytical methods to assess nanoparticle toxicity. *Analyst, 134*, 425–439.

McDonogh, R. M., Fell, C. J. D., & Fane, A. G. (1984). Surface charge and permeability in the ultrafiltration of non-flocculating colloids. *Journal of Membrane Science, 21*, 285–294.

Melegari, S. P., Perreault, F., Ribeiro Costa, R. H., Popovic, R., & Matias, W. G. (2013). Evaluation of toxicity and oxidative stress induced by copper oxide nanoparticles in the green alga *Chlamydomonas reinhardtii*. *Aquatic Toxicology, 142*, 431–440.

Merrifield, R. C., Stephan, C., & Lead, J. R. (2018). Quantification of Au nanoparticle biop- take and distribution to freshwater algae using single cell—ICP-MS. *Environmental Science & Technology, 52*, 2271–2277.

Metzler, D. M., Li, M., Erdem, A., & Huang, C. P. (2011). Responses of algae to photocatalytic nano-TiO$_2$ particles with an emphasis on the effect of particle size. *Chemical Engineering Journal, 170*, 538–546.

Miller, R. J., Lenihan, H. S., Muller, E. B., Tseng, N., Hanna, S. K., & Keller, A. A. (2010). Impacts of metal oxide nanoparticles on marine phytoplankton. *Environmental Science & Technology, 44*, 7329–7334.

Misra, S. K., Dybowska, A., Berhanu, D., Luoma, S. N., & Valsami-Jones, E. (2012). The com- plexity of nanoparticle dissolution and its importance in nanotoxicological studies. *Science of the Total Environment, 438*, 225–232.

Mortimer, M., Kasemets, K., & Kahru, A. (2010). Toxicity of ZnO and CuO nanoparticles to cili- ated protozoa *Tetrahymena thermophila*. *Toxicology, 269*, 182–189.

Mwaanga, P., Carraway, E. R., & van den Hurk, P. (2014). The induction of biochemical changes in *Daphnia magna* by CuO and ZnO nanoparticles. *Aquatic Toxicology, 150*, 201–209.

National Research Council Committee on Copper in Drinking Water. (2000). *Copper in Drinking Water*. Washington, DC: National Academies Press.

Nosaka, Y., & AY, N. (2017). Generation and detection of reactive oxygen species in photocatalysis. *Chemical Reviews, 117*(17), 11302–11336.

Parke, D. V., & Sapota, A. (1996). Chemical toxicity and reactive oxygen species. *International Journal of Occupational Medicine Environmental Health, 9*, 331–340.

Pauksch, L., Hartmann, S., Rohnke, M., Szalay, G., Alt, V., Schnettler, R., & Lips, K. S. (2014). Biocompatibility of silver nanoparticles and silver ions in primary human mesenchymal stem cells and osteoblasts. *Acta Biomaterialia, 10*, 439–449.

Peng, X., Palma, S., Fisher, N. S., & Wong, S. S. (2011). Effect of morphology of ZnO nanostructures on their toxicity to marine algae. *Aquatic Toxicology, 102*, 186–196.

Personick, M. L., Langille, M. R., Zhang, J., Harris, N., Schatz, G. C., & Mirkin, C. A. (2011). Synthesis and isolation of {110}-faceted gold bipyramids and rhombic dodecahedra. *Journal of the American Chemical Society, 133*, 6170–6173.

Postma, D. (1993). The reactivity of iron oxides in sediments: A kinetic approach. *Geochimica et Cosmochimica Acta, 57*, 5027–5034.

Prabhu, S., & Poulose, E. K. (2012). Silver nanoparticles: Mechanism of antimicrobial action, synthesis, medical applications, and toxicity effects. *International Nano Letters, 2*. https://doi.org/10.1186/2228-5326-2-32.

Pradhan, A., Seena, S., Pascoal, C., & Cássio, F. (2012). Copper oxide nanoparticles can induce toxicity to the freshwater shredder *Allogamus ligonifer*. *Chemosphere, 89*, 1142–1150.

Raffi, M., Hussain, F., Bhatti, T. M., Akhter, J. I., Hameed, A., & Hasan, M. M. (2008). Antibacterial characterization of silver nanoparticles against E. Coli ATCC-15224. *Journal of Materials Science and Technology, 24*, 192–196.

Reeves, J. F., Davies, S. J., Dodd, N. J. F., & Jha, A. N. (2008). Hydroxyl radicals (•OH) are associated with titanium dioxide (TiO$_2$) nanoparticle-induced cytotoxicity and oxidative DNA damage in fish cells. *Mutation Research, 640*, 113–122.

Rico, C. M., Lee, S. C., Rubenecia, R., Mukherjee, A., Hong, J., Peralta-Videa, J. R., & Gardea-Torresdey, J. L. (2014). Cerium oxide nanoparticles impact yield and modify nutritional parameters in wheat (*Triticum aestivum L.*). *Journal of Agricultural and Food Chemistry, 62*, 9669–9675.

Rodriguez, A., Bruno, C. A., Marie-José, C., Lecante, C. P., & Bradley, J. S. (1996). Synthesis and isolation of cuboctahedral and icosahedral platinum nanoparticles. Ligand-dependent structures. *Chemistry of Materials, 8*, 1978–1986.

Rogers, M. A. (2016). Naturally occurring nanoparticles in food. *Current Opinion in Food Science, 7*, 14–19.

Roy, R., Kumar, D., Sharma, A., Gupta, P., Chaudhari, B. P., Tripathi, A., Das, M., & Dwivedi, P. D. (2014). ZnO nanoparticles induced adjuvant effect via toll-like receptors and Src signaling in Balb/c mice. *Toxicology Letters, 230*, 421–433.

Sadiq, M., Dalai, S., Chandrasekaran, N., & Mukherjee, A. (2011). Ecotoxicity study of titania (TiO$_2$) NPs on two microalgae species: *Scenedesmus sp.* and *Chlorella sp. Ecotoxicology and Environmental Safety, 74*, 1180–1187.

Saif, H. S., Singh, S., Parikh, R. Y., Dharne, M. S., Patole, M., Prasad, B. L. V., & Shouche, Y. S. (2008). Bacterial synthesis of copper/copper oxide nanoparticles. *Journal of Nanoscience and Nanotechnology, 8*, 3191–3196.

Sankar, R., Prasath, B. B., Nandakumar, R., Santhanam, P., Shivashangari, K. S., & Ravikumar, V. (2014). Growth inhibition of bloom forming cyanobacterium *Microcystis aeruginosa* by green route fabricated copper oxide nanoparticles. *Environmental Science and Pollution Research, 21*, 14232–14240.

Scarabelli, L., Grzelczak, M., & Liz-Marzán, L. M. (2013). Tuning gold nanorod synthesis through prereduction with salicylic acid. *Chemistry of Materials, 25*, 4232–4238.

Schnippering, M., Powell, H. V., Zhang, M., Macpherson, J. V., Unwin, P. R., Mazurenka, M., & Mackenzie, S. R. (2008). Surface assembly and redox dissolution of silver nanoparticles monitored by evanescent wave cavity ring-down spectroscopy. *Journal of Physical Chemistry C, 112*, 15274–15280.

Sharma, V. K., Sayes, C. M., Guo, B., Pillai, S., Parsons, J. G., Wang, C., Yan, B., & Ma, X. (2019). Interactions between silver nanoparticles and other metal nanoparticles under environmentally relevant conditions: A review. *Science of Total Environment, 653*, 1042–1051.

Shi, W., Han, Y., Guo, C., Zhao, X., Liu, S., Su, W., Zha, S., Wang, Y., & Liu, G. (2017). Immunotoxicity of nanoparticle nTiO$_2$ to a commercial marine bivalve species, *Tegillarca granosa. Fish & Shellfish Immunology, 66*, 300–306.

Shibata, S., Miyajima, K., Kimura, Y., & Yano, T. (2004). Heat-induced precipitation and light-induced dissolution of metal (Ag & Au) nanoparticles in hybrid film. *Journal of Sol-Gel Science and Technology, 31*, 123–130.

Siddiqui, S., Goddard, R. H., & Bielmyer-Fraser, G. K. (2015). Comparative effects of dissolved copper and copper oxide nanoparticle exposure to the sea anemone, *Exaiptasia pallida. Aquatic Toxicology, 160*, 205–213.

Song, K. C., Lee, S. M., Park, T. S., & Lee, B. S. (2009). Preparation of colloidal silver nanoparticles by chemical reduction method. *Korean Journal of Chemical Engineering, 26*, 153.

Song, W., Zhang, J., Guo, J., Zhang, J., Ding, F., Li, L., & Sun, Z. (2010). Role of the dissolved zinc ion and reactive oxygen species in cytotoxicity of ZnO nanoparticles. *Toxicology Letters, 199*, 389–397.

Srikanth, K., Pereira, E., Duarte, A. C., & Rao, J. V. (2016). Evaluation of cytotoxicity, morphological alterations and oxidative stress in Chinook salmon cells exposed to copper oxide nanoparticles. *Protoplasma, 253*, 873–884.

Strasser, P., Koh, S., Anniyev, T., Greeley, J., More, K., Yu, C., Liu, Z., Kaya, S., Nordlund, D., Ogasawara, H., Toney, M. F., & Nilsson, A. (2010). Lattice-strain control of the activity in dealloyed core–shell fuel cell catalysts. *Nature Chemistry, 2*, 454–460.

Stumm, W., & Morgan, J. J. (1996). *Aquatic chemistry: Chemical equilibria and rates in natural waters.* New York: Wiley.

Sulzberger, B., & Laubscher, H. (1995). Reactivity of various types of iron(III) (hydr)oxides towards light-induced dissolution. *Marine Chemistry, 50*, 103–115.

Sun, R. W.-Y., Chen, R., Chung, N. P.-Y., Ho, C.-M., Lin, C.-L. S., & Che, C.-M. (2005). Silver nanoparticles fabricated in Hepes buffer exhibit cytoprotective activities toward HIV-1 infected cells. *Chemical Communications, 40*, 5059–5061.

Suppi, S., Kasemets, K., Ivask, A., Kuennis-Beres, K., Sihtmaee, M., Kurvet, I., Aruoja, V., & Kahru, A. (2015). A novel method for comparison of biocidal properties of nanomaterials to bacteria, yeasts and algae. *Journal of Hazardous Materials, 286*, 75–84.

Tetard, L., Passian, A., Venmar, K. T., Lynch, R. M., Voy, B. H., Shekhawat, G., Dravid, V. P., & Thundat, T. (2008). Imaging nanoparticles in cells by nanomechanical holography. *Nature Nanotechnology, 3*, 501–505.

Tulve, N. S., Stefaniak, A. B., Vance, M. E., Rogers, K., Mwilu, S., LeBouf, R. L., Schwegler-Berry, D., Willis, R., Thomas, T. A., & Marr, L. C. (2015). Characterization of silver nanoparticles in selected consumer products and its relevance for predicting children's potential exposures. *International Journal of Hygiene and Environmental Health, 218*, 345–357.

Vance, M. E., Kuiken, T., Vejerano, E. P., McGinnis, S. P., Hochella, M. F., Jr., Rejeski, D., & Hull, M. S. (2015). Nanotechnology in the real world: Redeveloping the nanomaterial consumer products inventory. *Beilstein Journal of Nanotechnology, 6*, 1769–1780.

Villarreal, F. D., Das, G. K., Abid, A., Kennedy, I. M., & Kultz, D. (2014). Sublethal effects of CuO nanoparticles on Mozambique tilapia (*Oreochromis mossambicus*) are modulated by environmental salinity. *PLoS One, 9*, 88723.

Wang, Z., von dem Bussche, A., Kabadi, P. K., Kane, A. B., & Hurt, R. H. (2013). Biological and environmental transformations of copper-based nanomaterials. *ACS Nano, 7*, 8715–8727.

Warheit, D. B., Hoke, R. A., Finlay, C., Donner, E. M., Reed, K. L., & Sayes, C. M. (2007). Development of a base set of toxicity tests using ultrafine TiO2 particles as a component of nanoparticle risk management. *Toxicology Letters, 171*, 99–110.

Warheit, D. B., Webb, T. R., Sayes, C. M., Colvin, V. L., & Reed, K. L. (2006). Pulmonary instillation studies with nanoscale TiO_2 rods and dots in rats: Toxicity is not dependent upon particle size and surface area. *Toxicological Sciences, 91*, 227–236. https://doi.org/10.1093/toxsci/kfj140.

Wu, J., Liu, W., Xue, C., Zhou, S., Lan, F., Bi, L., Xu, H., Yang, X., & Zeng, F.-D. (2009). Toxicity and penetration of TiO_2 nanoparticles in hairless mice and porcine skin after subchronic dermal exposure. *Toxicology Letters, 191*, 1–8.

Wuithschick, M., Birnbaum, A., Witte, S., Sztuck, M., Vainio, U., Pinna, N., Rademann, K., Emmerling, F., Kraehnert, R., & Polte, J. (2015). Turkevich in New Robes: Key questions answered for the most common gold nanoparticle synthesis. *ACS Nano, 9*, 7052–7071.

Yang, S. P., Bar-Ilan, O., Peterson, R. E., Heideman, W., Hamers, R. J., & Pedersen, J. A. (2013). Influence of humic acid on titanium dioxide nanoparticle toxicity to developing zebrafish. *Environmental Science & Technology, 47*, 4718–4725.

Yasun, E., Chunmei Li, C., Barut, I., Janvier, D., Qiu, L., Cuia, C., & Tan, W. (2015). BSA modification to reduce CTAB induced nonspecificity and cytotoxicity of aptamer-conjugated gold nanorods. *Nanoscale, 22*, 10240–10248.

Zhang, S., Zhang, X., Jiang, G., Zhu, H., Guo, S., Su, D., Lu, G., & Sun, S. (2014). Tuning nanoparticle structure and surface strain for catalysis optimization. *Journal of the American Chemical Society, 136*, 7734–7739.

Zhao, X., Wang, S., Wu, Y., You, H., & Lv, L. (2013). Acute ZnO nanoparticles exposure induces developmental toxicity, oxidative stress and DNA damage in embryo-larval zebrafish. *Aquatic Toxicology, 136–137*, 49–59.

Zhong, Z., Patskovskyy, S., Bouvrette, P., Luong, J. H. T., & Gedanken, A. (2004). The surface chemistry of Au colloids and their interactions with functional amino acids. *The Journal of Physical Chemistry. B, 108*, 4046–4052.

Zhu, X., Chang, Y., & Chen, Y. (2010). Toxicity and bioaccumulation of TiO_2 nanoparticle aggregates in *Daphnia magna*. *Chemosphere, 78*, 209–215.

Zhu, X., Zhu, L., Duan, Z., Qi, R., Li, Y., & Lang, Y. (2008). Comparative toxicity of several metal oxide nanoparticle aqueous suspensions to zebrafish (*Danio rerio*) early developmental stage. *Journal of Environmental Science and Health, Part A, 43*, 278–284.

Part II
Nanomaterial Interactions in Plants and Agricultural Systems

Chapter 6
Responses of Terrestrial Plants to Metallic Nanomaterial Exposure: Mechanistic Insights, Emerging Technologies, and New Research Avenues

Keni Cota-Ruiz, Carolina Valdes, Ye Yuqing, Jose A. Hernandez-Viezcas, Jose R. Peralta-Videa, and Jorge L. Gardea-Torresdey

K. Cota-Ruiz
Department of Chemistry and Biochemistry, The University of Texas at El Paso, El Paso, TX, USA

UC Center for Environmental Implications of Nanotechnology (UC CEIN), The University of Texas at El Paso, El Paso, TX, USA

C. Valdes
Department of Chemistry and Biochemistry, The University of Texas at El Paso, El Paso, TX, USA

Y. Yuqing
UC Center for Environmental Implications of Nanotechnology (UC CEIN), The University of Texas at El Paso, El Paso, TX, USA

J. A. Hernandez-Viezcas · J. R. Peralta-Videa
Department of Chemistry and Biochemistry, The University of Texas at El Paso, El Paso, TX, USA

UC Center for Environmental Implications of Nanotechnology (UC CEIN), The University of Texas at El Paso, El Paso, TX, USA

Environmental Science and Engineering Ph.D. program, The University of Texas at El Paso, El Paso, TX, USA

J. L. Gardea-Torresdey (✉)
Department of Chemistry and Biochemistry, The University of Texas at El Paso, El Paso, TX, USA

UC Center for Environmental Implications of Nanotechnology (UC CEIN), The University of Texas at El Paso, El Paso, TX, USA

Environmental Science and Engineering Ph.D. program, The University of Texas at El Paso, El Paso, TX, USA

NSF-ERC Nanotechnology-Enabled Water Treatment Center (NEWT), Houston, TX, USA
e-mail: jgardea@utep.edu

© Springer Nature Switzerland AG 2021   165
N. Sharma, S. Sahi (eds.), *Nanomaterial Biointeractions at the Cellular, Organismal and System Levels*, Nanotechnology in the Life Sciences, https://doi.org/10.1007/978-3-030-65792-5_6

Contents

1   Metallic Nanoparticles: An Introduction

Nanomaterials have structures with at least one dimension at the "nanoscale" (<100 nm). Examples of these tiny materials include nanoparticles (NPs), nanowires, and nano-films, which possess zero (0D), one (1D), and two dimensions (2D) outer of the "nanoscale," respectively (Pokropivny and Skorokhod 2008). Owing to their greater surface/volume ratio, compared to the micro-scale materials, the NPs become more reactive and they can also achieve the functionality of their micro-sized counterparts with lesser mass (Crane and Scott 2012; Masciangioli and Zhang 2003).

Nanomaterials are used in many applications including electronics, medicine, cosmetics, and agriculture, among others (Hong et al. 2013; Peters et al. 2016; Rao et al. 2006). Particularly, the metal oxide NPs have been gaining more attention in agriculture due to their potential to deliver macro and/or micronutrients. Thus, they can be used as nano-fertilizers (Sabir et al. 2014; White and Gardea-Torresdey 2018). Some recent investigations have evaluated their functionality by chemically adding functional groups. The advantages of these modifications open possible biotechnological applications such as gene or chemical delivery inside plant cells (Mody et al. 2010). Meanwhile, risks and concerns have also been expressed by researchers since NPs are accumulating in agricultural soils (Tourinho et al. 2012). Thus, it is necessary to gain knowledge about the impact of NPs on terrestrial plants.

In this chapter, we will first review the recent literature that describes the application of metallic NPs on terrestrial plants, underlining their positive and negative effects on plants and microorganisms. This work includes the use of metallic NPs as nano-fertilizers and potential agents in the control of plant disease, and management of biotic and abiotic stress. We will outline how soil productivity is modulated by the interactions of NPs, microbiota, plants, and abiotic soil components. Further, the mechanistic responses of plant cells to NPs and NPs fate in plants will be examined in light of recent investigations based on spectroscopic and high-throughput technologies.

2 Positive Effects of Metallic NPs on Plant Physiology

The physiological responses of plants exposed to metallic NPs depend on different factors such as the concentration, size, and type of NPs, soil properties, and plant species (Reddy et al. 2016). The documented positive effects on plants include seed germination enhancement, macro and micronutrients uptake, and yield increase (de la Rosa et al. 2017). Details about researches evaluating the application of different metallic nanoparticles and their positive and negative effects on crops and microorganisms are discussed below.

2.1 The Use of Nano-Fertilizers

Nano-fertilizers are nano-enabled chemicals that can deliver nutrients to plants by either encapsulating the nutrient with a subsequent release or delivering macro/microelements from their own nano-structures (DeRosa et al. 2010). Among the different nano-fertilizers, the ZnO NPs have shown more positive effects on crops. For instance, ZnO NPs at 500 mg L^{-1} increased up to 30% the root growth of soybean (*Glycine max*) seedlings (López-Moreno et al. 2010). Similarly, the highest shoot dry weight in chickpea (*Cicer arietinum L* var. HC-1) was obtained when ZnO NPs was foliarly applied to the plants at 1.5 mg L^{-1} (Burman et al. 2013). Also, the foliar application of ZnO NPs increased leaf mineral concentration and improved the fruit physical/chemical properties in pomegranates (*Punica granatum* cv. Ardestani) (Davarpanah et al. 2016). Therefore, ZnO NPs have been suggested as a potential nano-fertilizer for crops (Burman et al. 2013; Peralta-videa et al. 2014). However, more researches evaluating the nano-fertilizers effects on crops yield are still needed.

The usage of other NPs such as Fe NPs has also shown positive effects on plants. When soybean (*Glycine Max* (L.) Merr.) plants were treated with Fe NPs at a concentration equal or greater than 500 mg L^{-1} (500 mg L^{-1}, 1000 mg L^{-1}, 2000 mg L^{-1}), the root elongation was increased (Alidoust and Isoda 2013). Also, the citrate-coated Fe_2O_3 NPs enhanced the photosynthetic parameters and the dry weight yield of the plants in comparison to the control. Similarly, the usage of Au NPs showed positive effects on the germination index and the root elongation of both cucumber and lettuce seedlings (Barrena et al. 2009). Recently, colloidal Au NPs were applied on 6-year-old ginseng plants and it was demonstrated that Au NPs improved the synthesis of ginsenosides, an active compound in red ginseng, which enhanced the anti-inflammatory ability of the treated plants (Kang et al. 2016). Furthermore, the TiO_2 NPs, which are among the most produced NPs, have been investigated to evaluate their effects on plant physiology. In a study conducted in basil (*Ocimum basilicum*), it was demonstrated that the shoot growth was increased when plants were exposed to hydrophobic TiO_2 NPs at 750 mg kg^{-1} in soil (Tan et al. 2018a). Several other studies testing TiO_2 NPs at concentrations below 0.6% have shown positive effects on spinach. For example, TiO_2 NPs promoted the photosynthetic rate reaction (Gao et al. 2006), which correlated with an increment of the Hill reaction activity in chloroplasts (Hong et al. 2005). Also, it was reported that TiO_2 NPs enhanced the expression of Rubisco activase mRNA, which promotes photosynthesis in spinach (Linglan et al. 2008).

2.2 Plant Disease Control

Plant disease is becoming a major worldwide concern for agriculture. The synthetic products to control pathogenic agents that affect plants have been widely used, and thus accumulated in the environment. Unfortunately, they affect the fate of biological systems. An alternative currently being explored is their substitution by silver nanoparticles (Ag NPs), which have demonstrated antimicrobial properties at much lesser concentrations (Prabhu and Poulose 2012). Thus far, several investigations have shown that the antifungal activity of Ag NPs on plant pathogenic fungi depends on the NP concentration used and in the plant species (Gajbhiye et al. 2009; Jo et al. 2009; Kim et al. 2012; Lamsal et al. 2011). According to an in vitro study, it was proposed that the released Ag$^+$ from silver nanoparticles affects the bacteria cell by either attaching to the cysteine-containing proteins of the cell membrane or by penetrating the cell and interacting with sulfur and phosphorus-containing compounds such as DNA (Morones et al. 2005; Ocsoy et al. 2013). Additionally, the Ag NPs have also been proven to suppress other pathogenic diseases like the powdery mildew that affects cucumber and pumpkin (Lamsal et al. 2011). With the aim to fulfill the antimicrobial effect, Ag NPs have been prepared with various sizes, shapes, surface-coatings, or conjugations with other materials. For example, the double-stranded DNA-Ag NPs composites, synthesized on graphene oxide, showed a significant reduction in the viability of bacteria (*Xanthomonas perforans*) at in vitro and in greenhouse experiments, at the 16 ppm and 100 mg L^{-1} doses, respectively, compared to the controls (Ocsoy et al. 2013).

Besides silver, other metallic nanoparticles have also controlled pathogen and fungal diseases in both in vitro and in vivo experiments. For instance, the use of Zn NPs at 500 mg L^{-1} significantly reduced the growth of *Fusarium graminearum* in agar media (Dimkpa et al. 2013). In another experiment, Zn NPs at doses above ~250 mg L^{-1} significantly inhibited the growth of *Botrytis cinerea* and *Penicillium expansum* in plating assays (He et al. 2011). Similarly, Cu-based nanoparticles have shown antimicrobial activity against different fungi with higher inhibition rates compared to commercial fungicides (Kanhed et al. 2014). Meanwhile, magnesium oxide nanoparticles (MgO NPs) have displayed an increased antibacterial activity as the particle size decreases (Huang et al. 2005). Remarkably, in field studies, some metallic nanoparticles have exhibited similar capabilities to control bacterial growth as in in vitro. For example, in a greenhouse experiment using TiO_2 NPs, the disease infection in cucumber caused by *Pseudomonas lachrymans* and *Psilocybe cubensis* was significantly reduced by NPs as they formed a powerful bactericidal film on the leaves surfaces (Cui et al. 2009). Similarly, tomatoes infected with *Xanthomonas perforans* showed significantly decreased bacterial spots than controls after they were foliar sprayed with TiO_2/Zn NPs at 500–800 mg L^{-1} (Paret et al. 2013).

2.3 Enhancement of Abiotic Stress

Abiotic stressors such as cold, salinity, or drought, among others, cause severe impairments to the physiology of plants. It has been documented that one of the major causes of crop losses worldwide is triggered by abiotic stresses (Sunkar et al. 2007; Tuteja 2010). In addition to their fertilizing and disease prevention properties in plants, metallic NPs have also shown favorable effects against abiotic stressors; they have also been investigated as potential "protectant" to help plants to overcome the negative effects caused by abiotic pressures.

Previous research described that flooding decreased the normal growth of *Crocus sativus* but Ag NPs alleviated this negative effect as they allowed plants to overcome the growth deficit (Rezvani et al. 2012). In another study, the water insufficiency that led to a decrease of seed number production in wheat was overcome when the plants were foliarly sprayed with TiO_2 (0.01%, 0.02%, 0.03%) (Jaberzadeh et al. 2013). Moreover, lower concentrations of TiO_2 (<0.02%) increased the biomass compared to the control groups. Therefore, it was recommended by the authors to use TiO_2 NPs as a strategy to get over physiological impairments caused by water-deficient conditions.

Among the different abiotic stressors, salt stress has been suggested as one of the most significant affecting global agriculture production (Zhu 2001). Interestingly, some metallic NPs have relieved the negative impacts caused by salt stress. In a study with lentil (*Lens culinaris*), the germination rate and seedling growth were reduced by salinity while these affectations were reversed by using Si NPs (Sabaghnia and Janmohammasi 2015). A recent research performed in five sunflowers cultivars showed that salinity decreased shoot and root dry weight, leaf area, and plant height; however, after foliar exposure to Zn NPs at a concentration of 2 g L^{-1}, the leaf area and shoot dry weight were increased compared to the bulk ZnO and to the control treatments (Torabian et al. 2016).

3   Negative Effects of Metallic NPs to Plant Physiology

The application of metallic nanoparticles in agriculture, in the form of nano-fertilizers or nano-agents to control plant disorders, will also increase the exposure of microorganisms, plants, and other higher organisms including humans to NPs. Thus, while exploring the favorable applications of metallic nanoparticles, it is also necessary to consider their negative effects. Herein, studies showing direct and indirect adverse effects of metallic nanoparticles on fungi, plants, and even animal models are described.

Some NPs can penetrate the plasma membrane in a, thus far, undescribed mechanism. Once inside the cell, they can be attached to different organelles and trigger several biochemical responses. Mainly, they cause toxicity by their chemical composition, their size, or by their high surface reactivity (Navarro et al. 2008). For example, the toxicity of TiO_2 NPs showed size-dependent effects on the green algae *Desmodesmus subspicatus*; the greater the particle size, the lesser the toxicity (Hund-Rinke and Simon 2006). Meanwhile, in a study conducted to evaluate the effect of the ZnO NPs suspension and their soluble fraction (supernatant), it was found that the NP-free supernatant did not show any toxicity. Contrarily, ZnO NPs suspension caused physiological affectations to radish, rape, and ryegrass (Lin and Xing 2007). These latter results suggest that the toxicity was produced by NPs rather than the released Zn^{2+}. The NPs can also affect the nutrients absorption by releasing into the cells "similar" ions with different biological roles or by interfering with macro/micronutrient cell absorption. In addition, they may elicit the enhancement of the reactive oxygen species (ROS) production. For instance, the Ag^+ released from Ag NPs triggered the generation of ROS in yeast, *Escherichia coli*, and *Staphylococcus aureus* (Kim et al. 2007). Other photocatalytic NPs such as TiO_2 have shown the capacity to increase the ROS production in bacteria after UV exposure (Adams et al. 2006).

The toxicity of Ag NPs on *Arabidopsis thaliana* was evaluated in comparison with silver ions at 3 mg L^{-1} of both Ag NPs and Ag^+ ions (Qian et al. 2013). The Ag NPs had higher inhibitory effects in root elongation than silver ions, and silver NPs accumulated in leaves, which could damage the thylakoid membrane structure, as proposed by the authors. Moreover, in the same study, the expression of antioxidant and aquaporin genes in the plants was altered by Ag NPs. Another molecular study was conducted in radish (*Raphanus sativus*), perennial ryegrass (*Lolium perenne*), and annual ryegrass (*Lolium rigidum*) seedlings by evaluating the DNA damage under CuO NPs treatments (Atha et al. 2012). The seeds were germinated in CuO NPs or bulk CuO suspensions at the concentration of 10, 100, 500, and 1000 mg L^{-1}, and in Cu^{2+} solution at 1, 10, and 50 mg L^{-1}. The results showed that all Cu treatments reduced the plant growth and caused DNA lesions; however, the CuO NPs caused more oxidative damage than bulk CuO and Cu^{2+}. In another study, the cilantro (*Coriandrum sativum* L.) plants were used to assess the toxicity of CeO_2 NPs at 0, 62.5, 125, 250, and 500 mg kg^{-1} in a full life cycle (Morales et al. 2013). The study evidenced that nutritional properties of cilantro were reduced and that Ce was accumulated in almost all tissues when the plants were treated with CeO_2 NPs at 500 mg kg^{-1}.

4   The Influence of External Factors in NPs-Plant Interactions

Due to the exponential use of NPs and their unknown toxicity to the plants, many recent types of investigations have been conducted to evaluate their effects on terrestrial plants. As previously mentioned, the NPs have a larger area/volume ratio in comparison to their bulk counterparts, which make them sufficiently reactive to undergo transformations in both biological and environmental systems (Louie et al. 2014). In addition to the NPs, the performance of plants can be affected by multiple factors such as the type of soil, the microbial diversity, and the plant variety/species, among others (Cota-Ruiz et al. 2018a). This section summarizes the physiological effects as enumerated above.

4.1   Soils and Metallic NPs

Soils are becoming a major destination site for NPs (Priester et al. 2017; Reddy et al. 2016), as the application of metal-based nanomaterials dominates the industrial products. Once the metallic NPs end up in soils, they are subjected to chemical, physical, and biological transformations and can also interact with other macromolecules (Fig. 6.1) (Amde et al. 2017). The main component of soil that affects the structure (and thus the "reactivity") of the metallic NPs is the natural organic matter (NOM). The NOM derives from the slow decomposition of living organisms including microbes. Their components can promote the aggregation and/or disaggregation of the metallic NPs (Yu et al. 2018). The aggregation phenomenon can occur as homo-aggregation (between the same NPs) or as hetero-aggregation (between the interaction of NPs with different NPs and/or particles). Often, the aggregation increases the NPs sedimentation rate, reducing their mobility into the ecosystems. On the other hand, when metallic NPs are not aggregated, their solubility in the water column is habitually increased, which promotes their bioavailability (Del Real et al. 2016; Joo and Zhao 2017). Additionally, the aggregation process can be affected by physical and chemical parameters such as temperature, pH, and ionic strength, among others. For instance, a previous report demonstrated that the aggregation of CeO NPs increased as the pH and ionic strength (IS) were augmented, from 6.0 to 9.0 and from ~2 to 40 mM, respectively (Van Hoecke et al. 2011).

The most abundant components of the NOMs are the humic acids (HA) and fulvic acids (FA). These compounds, also referred to as humic substances, are generally

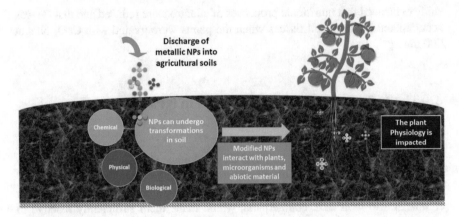

Fig. 6.1 Schematic representation of the chemical, physical, and biological transformations that NPs may suffer once they are incorporated in soils. The modified compounds will interact with terrestrial plants, microorganisms, and the rest of non-living substances influencing the physiology of the plants in a very assorted fashion. (Modified from Amde et al. (2017))

hydrophobic organic acids whose functional groups are phenolic and/or carboxylic substituents (Ritchie and Michael Perdue 2003). These functional groups are the most significant sites that determine their binding properties (Masini et al. 1998). One of the most studied metallic NP, with respect to NPs and humic substances interactions, has been the TiO_2 NP. Zhou et al. (2013) evaluated the adsorption behavior of different TiO_2 NPs to HA substances, as a function of size, crystal structure, and shape of TiO_2 molecules, demonstrating that TiO_2 NPs are poorly adsorbed by HA substances, probably due to their great amount of TiO_2 negative charges at the evaluated pH that was 8.0. Indeed, by lowering the pH at 5.7, in a matrix-sand containing HA, Chen et al. (2012) proved that the absorption of TiO_2 NPs is greatly increased by HA substances, which occur likely by electrostatic interactions. In another study conducted to evaluate the interaction of metallic oxide NPs with HA, it was observed that the TiO_2, Al_2O_3, and ZnO NPs were adsorbed by HA molecules at pH 5.0, and it was also noted that the adsorption was reduced while the pH augmented (Yang et al. 2009). With the use of Fourier transform infrared (FTIR), the previous study demonstrated that the COOH was involved in binding to ZnO NPs; the phenolic OH from HA was the ligand for TiO_2 NPs; and both, the COOH and the phenolic and aliphatic OH from HA were the ligands that interacted with Al_2O_3 NPs. In another study using CeO_2 NPs, it was shown that the pH also significantly affected the NOM-NP adsorption, being higher at pH 6.0 than pH 9.0 (Van Hoecke et al. 2011). These studies point out that acidic soils tend to adsorb metallic NPs to a greater extent in comparison to the neutral or basic soils. Presumably, the protonation/deprotonation of the specific functional groups is affected by changing the pH which leads to the electrostatic interaction between the metal oxide NPs and the ionizable groups from humic substances (Jayalath et al. 2018).

4.2   The Role of Microorganisms

The microorganisms play a pivotal role not only in the dynamics of nutrients in soils but also in the decomposition of waste and in promoting plant health (Lead et al. 2008). Clearly, the incorporation of metallic NPs into the soils will modify the microorganisms' biological activity. The scientific evidence has shown that metallic NPs reduce the activity of microbial enzymes involved in primordial processes such as cellular respiration and biogeochemical cycles (Hegde et al. 2016; Simonin and Richaume 2015). Thus, the measurement of the activity of enzymes such as the β-glucosidase and urease, as well as dehydrogenases, is a good predictor to describe the magnitude of the biological-soil processes perturbations caused by NPs (Chen et al. 2014). However, the studies dealing with the effects of metallic NPs on micro-organisms metabolism are increasing gradually, they are still limited. Moreover, most investigations have been conducted to evaluate the antimicrobial role of metal NPs on human pathogenic microorganisms (Guzman et al. 2012; Ruparelia et al. 2008; Salem et al. 2015). Concerning environmentally relevant microorganisms, many studies have been conducted generally to evaluate the effect of a single NP on a particular microorganism species, which differ from the natural scenario due to the presence of different metallic nanomaterials and the complex abundance of microbial communities on soils.

Some recent studies have explored the effects of metallic NPs on soil microbial communities. The Ag NPs are among the most studied NPs regarding microbial toxicity. They exhibit a marked antimicrobial capacity due to their larger surface area that allows an increased-contact with microorganisms. Indeed, it has been reported that the use of Ag ions does not cause significant impairments on the physi-ology of soil microorganisms in comparison to their Ag NPs counterparts (Colman et al. 2013). The Ag NPs affect microbial growth since they can either be attached to the cell membrane causing malfunctions in its activity, or they can penetrate cells interfering with primordial metabolic cell processes such as the cell division or the respiratory chain (Kim et al. 2007; Rai et al. 2009). Also, Ag NPs inhibit the activity of enzymes of microorganisms involved in key pathways such as nutrient cycles and energy metabolism (Shin et al. 2012).

In addition to silver NPs, other metallic NPs such as CuO, ZnO, and NiO also cause damage to microorganisms (Baek and An 2011; Shen et al. 2015), even at lower doses in comparison to the carbon-based nanoparticles (Simonin and Richaume 2015). For instance, CuO NPs displayed in vitro inhibitory effects on the growth of both gram-positive and gram-negative bacteria, and this impairment increased as particle sizes decreased (Azam et al. 2012). In another study, the reduc-tion of dehydrogenases activity and acid phosphatases enzymes was recorded in soils treated with 1000 mg kg^{-1} of Zn NPs (García-Gómez et al. 2015). Additionally, a recent study reported that nickel oxide nanoparticles (NiO NPs) inhibited the growth of *Klebsiella pneumoniae* (gram-negative) and *Staphylococcus aureus* (gram-positive) in agar medium (Anitha et al. 2018). A study conducted by Xu et al. (2015) reported that applications of CuO NPs and TiO$_2$ NPs affected the enzyme activity and biomass of microorganisms inhabiting paddy fields soils. Moreover, in a study using high-throughput sequencing, it was found that the abundance of essential representative groups of bacteria such as Actinobacteria, Cyanobacteria, Acidobacteria, and Nitrospirae was reduced with increasing concentrations of Ag NPs (from 10 to 100 mg kg^{-1} soil) (Wang et al. 2017). Interestingly, the authors also found that other groups such as Proteobacteria and Planctomycetes were increased.

The microorganisms also have the capability to synthesize NPs using metabolic pathways that involve redox reactions. For example, in some bacteria, the oxidation of Fe (II) leads to the production of FeO NPs (Nowack and Bucheli 2007). The synthesis of NPs can occur in an extra- or intracellular manner; in both cases, the enzymes are the main factors catalyzing this process (Singh et al. 2016). Interestingly, different species of bacteria such as *Pseudomonas* and *Lactobacillus* have the capability to produce Ag NPs, and can also accumulate them within the periplasmic space, as it occurs with *Bacillus* sp. (Narayanan and Sakthivel 2010). The latter scenario suggests that the antimicrobial properties that several NPs exhibit on bacteria is dependent on the type and on the concentration of NPs, as well as on the target microorganism.

Previous reports have shown that NPs can form complexes with heavy metals (Tang et al. 2014). Thus, the incorporation of NPs to heavy metal contaminated soils will modify in an unknown manner the biochemical responses of the microorganisms inhabiting those soils. Among several possible scenarios, a positive one would be if hetero-aggregates are large enough to cross cell membranes, resulting in no uptake of such substances by microorganisms (Louie et al. 2014). However, on the other hand, the chemistry of hetero-aggregates could modify the soil environment conditions reducing the availability of nutrients and consequently lowering the survival possibilities of key soil microorganisms. Also, the incorporation of NPs into soils affects root nodules in leguminous plants which could severally affect their growth (Holden et al. 2018). So far, to the best of the authors knowledge, no studies have been performed to evaluate the interaction of humic substances, heavy metals, soil microorganisms, and metallic NPs in the soil. However, an appreciable volume of literature has been reviewed recently (Adeleye et al. 2016; Hua et al. 2012; Tang et al. 2014), where the interaction of heavy metals, humic substances, and/or NPs in water columns was analyzed and summarized. Additionally, the use of humic adsorbent nanofibers can be suitable to remove heavy metals or dyes (Ayyildiz et al. 2017). Interestingly, the use of nanocomposites (Fe$_3$O$_4$/montmorillonite/humic acid) has been proved to be efficient in the removal of Cr(vi) in water environments (Lu et al. 2018). Moreover, modified TiO$_2$ semiconductors have shown potential to adsorb organic pollutants and to modify the oxidation state of heavy metals (Yan et al. 2017).

4.3   The Effect of NPs on Plants: A Mechanistic Approach

The mechanistic responses of the plant cell under NPs stress are mainly driven by biomolecules such as proteins, lipids, and nucleic acids, among others. Hitherto, the researches exploring the molecular effects of NPs on cells have been more focused on studying the response in mammalian cells than in bacteria or plants. Owing to the importance of terrestrial crops as a main reservoir of the food chain and due to the accumulation of NPs in soils, recent investigators have turned their attention to assessing the molecular responses of plant cells triggered by NP exposures. To understand the cell performance, the gene expression studies have become a very useful approach (Van Aken 2015). Thus, in this section, we explore the responses of plant cells exposed to NPs at the gene-level regulation.

Most of the studies regarding transcript quantification in plants treated with NPs have been performed in *A. thaliana*. A recent transcriptomic study exposed *A. thaliana* plants to Al_2O_3 NPs (10 mg L^{-1}) for 10 days. The results evidenced that the genes involved in the antioxidant defense mechanisms were not changed by NPs, but were differentially affected by ionic exposure (Jin et al. 2017). Also, the RNA-Seq analysis showed that ionic treatments greatly affected the expression of more transcription factors in comparison to Al_2O_3 NPs. Interestingly, the authors found that the increased root growth under Al_2O_3 NPs treatment was associated with an upregulation of genes involved in nutrient uptake and in root development. In another study, García-Sánchez et al. (2015) used microarrays to investigate the effect of 0.2 mg L^{-1} Ag NPs and 20 mg L^{-1} TiO_2 NPs in *A. thaliana* hydroponically cultivated. The results showed that the transcriptional changes (i.e., early defense signaling genes) were comparable to the responses observed when the plants were cultivated with a pathogen or under salt stress conditions. In a similar microarray study, *A. thaliana* plants were exposed for 7 days to 4 mg L^{-1} of ionic, bulk, and nano Zn compounds (Landa et al. 2015). The results indicated that all Zn forms upregulated genes involved in salt stress, water deprivation, and osmotic stress; however, the ionic form had higher effects on these gene responses. In another gene expression study, it was shown that 4-week-old *A. thaliana* plants treated with ZnSe QDs (100 and 250 µM) had no apparent toxicity and overexpressed the genes involved in antioxidant responses (Kolackova et al. 2019). Interestingly, the applied NPs doses inhibited the growth of *Agrobacterium tumefaciens,* the species known to cause the crown gall disease in plants. In agreement with these results, young *A. thaliana* plants incubated with CuO NPs (0–20 mg L^{-1}) exhibited an upregulation of the genes involved in the oxidative stress responses and in the glutathione biosynthesis (Nair and Chung 2014). Likewise, the gene that codifies for the Heat Shock Protein 70 (HSP70) in *A. thaliana* was upregulated when the plants were treated with 500 mg L^{-1} of CeO, CuO, and/or La_2O_3 (Pagano et al. 2016).

In other species, the upregulation of antioxidant or heavy metal-responsive genes has also been registered in response to NPs treatments. For instance, the exposure of cucumber plants to 25 mg of the nano-fungicide $Cu(OH)_2$ resulted in a significant upregulation of the superoxide dismutase (SOD), glutathione peroxidases, monode-hydroascorbate reductase, and peroxidase (Zhao et al. 2017). The induction of anti-oxidant genes by CuO NPs have similarly been reported in soybean (*Glycine max* L.) (Nair and Chung 2014a), and in green pea (*Pisum sativum* L.) (Nair and Chung 2015), both treated at 100–400 mg L^{-1} of the NPs, and also in *Oryza sativa* exposed to 25–1000 mg L^{-1} (Da Costa and Sharma 2016). Additionally, Ag NPs augmented the mRNA content of Cu/Zn SOD in roots of mung bean (*Vigna radiata* L.) sprouts exposed to 10–20 mg L^{-1} of Ag NPs (Nair and Chung 2015a). Moreover, Cu(OH)$_2$ nanowires upregulated the expression of SOD in alfalfa seedlings at 25 mg kg^{-1}, and also the ionic compounds derived from CuSO$_4$ and Cu(NO$_3$)$_2$ (at 75 mg L^{-1}) increased the content of metallothionein transcripts (Cota-Ruiz et al. 2018b), an enzyme involved in metal detoxification.

The exposure of plants to NPs and their released ions cause overproduction of free radicals such as hydroxyl radicals ($^{\bullet}$OH) and non-radical molecules as hydrogen peroxide (H$_2$O$_2$) (Zuverza-Mena et al. 2017). These molecules, also known as ROS, may cause disruptions to biomolecules such as proteins or DNA. From the examples shown above, it has been proposed that plant cells counteract the production of ROS by activating a signaling cascade that promotes the activation of antioxidant genes (Zhao et al. 2017) (Fig. 6.2). The antioxidant enzymes, as the final products of these stimulated genes, also increase their activities under metallic NPs stress to control the excess of ROS production (Rico et al. 2013; Sharma et al. 2012). Indeed, the enzymes represent the earliest line of defense to fight ROS. Thus, the transcriptional and enzyme activations exemplify a coordinated set of defense mechanisms against ROS (Fig. 6.2). Additionally, the evidence suggests that ionic forms at "higher" concentrations (more than 50 mg kg^{-1}/mg L^{-1}) cause more damages to plant cells as they can affect their physiology, resulting in a reduction of growth and biomass production (Bandyopadhyay et al. 2015; Bradfield et al. 2017). On the other hand, a substantial number of consensus in literature suggests that lower doses of NPs may promote beneficial effects in cells since they can activate the genes involved in the production of growth factors, nutrient uptake, and glutathione biosynthesis, among others (Jasim et al. 2017; Reddy et al. 2016) (Fig. 6.2).

5 Spectroscopic Techniques to Explore Plant-NP Interactions

To understand the interaction of nanoparticles with crops, the closely related microscopy and spectroscopy techniques have been very useful to elucidate physiological and biochemical plant responses under metallic NPs stress (Fig. 6.3). These analytical techniques give both quantitative and qualitative results, which depend on several factors such as the instrumental sensitivity, the use of accurate standards, and the sample preparation, among others. This section briefly describes the conventional techniques used to characterize NPs and their derivatives, and those

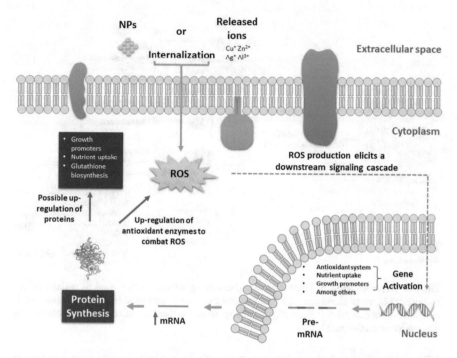

Fig. 6.2 The diagram shows some possible molecular perturbations in plant cells in response to NP exposure or the released ions. The NPs or the ionic compounds elicit the production of ROS that are immediately counteracted by the action of the antioxidant enzyme system. Additionally, a downstream signaling cascade is activated resulting in the upregulation of genes involved in defense machinery against ROS. The cells when exposed to low amounts of NPs or the released ions, the genes involved in nutrient uptake or growth regulators and other beneficial processes are stimulated. (Adapted and modified from Van Aken (2015) and Zhao et al. (2017))

approaches that are gaining momentum; as well as presents some examples of investigations conducted to gain deeper knowledge about metallic NP-Plant interactions.

5.1 Microscopy Sample Characterization

Microscopy techniques are those used to obtain images of plant tissues and/or NPs (Fig. 6.3). Among microscopy instruments, electron microscopes such as scanning electron microscopes (SEM) and transmission electron microscopes (TEM) are the most commonly used. These microscopes are useful in the characterization of plant tissue morphology. While SEM is suitable for surface analysis, TEM allows higher magnification and sample penetration depth. On the other hand, the energy dispersive spectroscopy (EDS), which consists of an attachment to electron microscopes, deciphers sample surface elemental composition. Under proper imaging conditions,

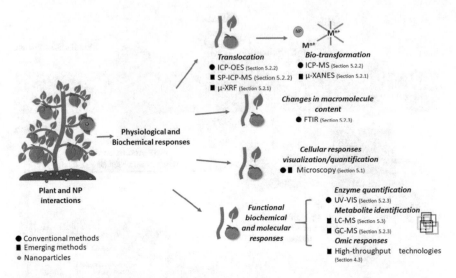

Fig. 6.3 Representation of conventional and emerging methodologies used to understand the interactions between plants and NPs. The physiological and biochemical responses of plants are in bold and italics and the corresponding techniques to assess them are in regular font, followed by the consistent section number

NPs can be observed clustering in certain cellular regions, as has been previously reported by Dai et al. (2018) who identified CuO in cells by using EDS. The EDS is appropriate for heavy metals; however, lower atomic weight elements, down to boron, can be determined with the use of higher sensitive detectors. One of the limitations in the use of EDS is that it does not distinguish the particulate state of the analyzed elements or compounds. Under certain circumstances, the presence of NPs within the sample can be confirmed by using specific dyes/fluorophores, which attach to NPs and emit a signal when excited at the specific wavelength (Zhao et al. 2012).

The confocal microscopy techniques have proven suitable to elucidate the effects of NPs in plant tissue. For instance, NPs can affect the production of nitric oxide (NO), which is a small signaling molecule involved in root growth and root hair development. The effect of NP on NO accumulation can be appropriately examined by the use of dye 4-amino-5-methylamino-20 and 70-difluorofluorescein diacetate (DAF-FM DA) to tag NO before observation under a confocal microscope. In a previous study when alfalfa (*Medicago sativa*) seedlings were given treatments of different nano, bulk, and ionic Cu forms, the use of DAF-FM DA dye coupled with confocal microscopy was suitable as it demonstrated that NO accumulation in seedlings was reduced at the treatment of 75 ppm of all Cu forms. This finding was consistent with reductions in root growth (Cota-Ruiz et al. 2018b) as observed. The fluorescein isothiocyanate (FITC) dye has been previously used in fluorescent confocal microscopy experiments to monitor compounds uptake and translocation in plant tissues. For example, when evaluating *Zea mays* crops incubated with

FITC-treated CeO_2 NPs, it was determined that Ce was accumulated in the cell walls, which suggested an apoplastic pathway for CeO_2 uptake (Zhao et al. 2012). In the latter experiment, the use of micro X-ray fluorescence (μ-XRF) and micro X-ray absorption near edge structure (μ-XANES) was essential to confirm that Ce accumulation inside cells was in the same oxidation state as the NPs. Although the confocal microscopy is very useful to determine target compounds inside cells, it has to be coupled with other techniques to confirm the chemical species. In a recent work with sweet potato, two-photon microscopy demonstrated that both the nano and the bulk CuO can be detected (Bonilla-Bird et al. 2018). The advantage of using two-photon over confocal microscopy is that two-photon microscopy has a lower energy emission source and does not require samples to be fluorescently tagged, which minimizes sample damage and preparation.

5.2 Spectroscopy Sample Characterization

5.2.1 Synchrotron Techniques

Microscopy alone is not sufficient to fully understand NP and plant interactions. Therefore, the use of additional spectroscopy techniques is convenient to complement microscopy studies. Plants may be subjected to NP exposure either through direct interaction in the soil or via foliar contact. Thus, localizing NPs is crucial to determine NP transport pathways in plants. The localization and speciation of NPs in plants can be evaluated using synchrotron techniques. These techniques are characterized by the use of powerful X-rays as their radiation source, which is generated from the high-energy electrons that circulate within the synchrotron. Various synchrotron techniques involve the use of X-ray diffraction, X-ray absorption spectroscopy, and photoelectron spectroscopy, among others. Particularly, for NP localization and speciation, μ-XRF and μ-XANES can be used in conjunction (Fig. 6.3). While μ-XRF is useful for NP localization, μ-XANES is used to determine chemical species of NP in plants (Castillo-Michel et al. 2017; Hernandez-Viezcas et al. 2013; Servin et al. 2012; Zhao et al. 2012). The performance of these studies, however, is limited to synchrotron facilities. When mapping and evaluating the speciation of CeO_2 and ZnO NPs in soybean grown in soil amended with these NPs, it was determined under μ-XRF that low amounts of Ce were translocated to the pod, while Zn was found in the outer pod. Additionally, μ-XANES confirmed that Ce was stored in the pods as CeO_2 NPs while Zn signals resembled Zn-citrate spectra, suggesting Zn-O ligation and biotransformation of this NP (Hernandez-Viezcas et al. 2013). In a separate study conducted with TiO_2 NPs in cucumber, it was concluded under μ-XRF that Ti was translocated from roots to leaves, while μ-XANES verified the presence of Ti as TiO_2 NPs, indicating that biotransformation of the NPs did not occur (Servin et al. 2012).

5.2.2 Inductively Coupled Plasma Techniques

Inductively coupled plasma optical emission spectroscopy (ICP-OES) and inductively coupled plasma mass spectrometry (ICP-MS) can be used to detect and quantify several elements and to localize NPs, respectively (Fig. 6.3). Although both techniques are sensitive, the benefits of ICP-MS over the ICP-OES is the lower detection limit (ppt-ppb) that the former can achieve in comparison to the ICP-OES (ppb-ppm).

The ICP techniques have been broadly used to analyze elements in plant tissues such as root, leaves, shoot, and fruit, among others. For instance, in a study conducted with kidney beans (*Phaseolus vulgaris*), the ICP-OES technique was conclusive to determine the acquisition of Zn when the plants were exposed to different Zn compounds (Medina-Velo et al. 2017). In this experiment, the plants were grown for 45 days in soils containing commercially coated ZnO (Zn-COTE and Zn-COTE HP1), amphiphilic uncoated ZnO NP, hydrophobic ZnO coated with triethoxy-caprylylsilane, bulk ZnO, and ionic $ZnCl_2$ compounds. When the plants were exposed to 125 ppm of Z-COTE, Zn increased by 203%, 139%, and 76% in nodules, stems, and leaves, respectively. At the same concentration, Zn-COTE HP1 caused an increase of Zn by 89%, 97%, and 103% in roots, stems, and leaves, respectively. The aforementioned investigation was followed by a transgenerational evaluation of the effect of ZnO nanomaterials, where it was determined that Z-COTE at 500 ppm and Z-COTE HP1 at 125 and 500 ppm reduced Ni in second-generation seeds by 60%, 41%, and 74%, respectively (Medina-Velo et al. 2018).

An emerging technique for metal-based NP sample analysis is the ICP-MS with single particle (SP) capability (Lee et al. 2014), which has the capacity of quantifying and sizing NPs (Fig. 6.3). Certain NPs are considered to occur environmentally at very low concentrations ranging from parts per trillion (ng L^{-1}) to parts per billion (µg L^{-1}). Thus, method development and sample preparation are critical for proper nanoparticle evaluation. A normal ICP-MS can analyze these concentrations. More than the detection limit, the ICP-MS with SP capability can differentiate the ionic species from a nanoparticle, and define the size of the nanoparticle. For this latter determination, it is important to note that a reference nanoparticle standard of known size is necessary to perform accurate determinations. Although nanoparticle size may be visualized via microscopy techniques, ICP-MS with SP capability facilitates the direct work with liquid samples and is useful to analyze complex biological matrices such as plant tissues (Dan et al. 2016). However, when analyzing plant tissues, the extraction of NPs without altering their properties is necessary. In a previous study in hydroponically grown plants (cucumber, tomato, soybean, and pumpkin) exposed to CeO_2 NPs, the shoots were evaluated to explore CeO_2 biotransformation using SP-ICP-MS (Dan et al. 2016). The authors detected a signal from both particulate and dissolved Ce, inferring that CeO_2 was bio-transformed to other Ce species; however, the composition of the particulate was not identified.

5.2.3 UV-VIS/FTIR Spectroscopy

Despite emerging tools such as "omics" are being tested in the determination of NP-plants interactions, certain conventional techniques are still commonly employed. Among them, the UV-vis spectrophotometric techniques are being used for the quantification of a huge number of metabolites and biomolecules in plants (Fig. 6.3). For instance, by first extracting the chlorophyll from leaves, the chlorophyll a and b content can be evaluated by measuring their absorbance from 645 to 664 nm (Apodaca et al. 2017; Tamez et al. 2018; Tan et al. 2018b). Since NPs may elicit the production of ROS within cells, the enzymes involved in the oxygen homeostasis are commonly evaluated using UV-vis techniques. Enzymes evaluated as stress indicators include catalase, polyphenol oxidase, ascorbic peroxidase, and superoxide dismutase, among others (Adisa et al. 2018; Apodaca et al. 2017; Tamez et al. 2018).

Analysis FTIR spectroscopy though not a new technique has been reliably appropriated for the elucidation of structural properties of macromolecules (Fig. 6.3). Based on the distinct IR absorption or transmittance of plants exposed to NPs, conformation changes in macromolecules can be identified. Since every functional group has its unique absorption/transmittance wavenumber, the specific spectral regions are evaluated. For instance, the 3000–2800 cm^{-1}, 1700–1500 cm^{-1}, and 1300–1180 cm^{-1} regions have been extensively used for lipids, proteins, and lignin, respectively. To appropriately compare the relative intensities of all spectra, data must be previously normalized (Zuverza-Mena et al. 2016).

5.3 Sample Separation and Mass Spectrometry for Sample Characterization

Plant metabolism constitutes all chemical processes occurring within each plant cell to maintain them alive. The metabolites are the products of all of these chemical processes. Plant exposure to different environments such as those imparted by NPs may elicit different plant responses resulting in the generation of diverse metabolites. Separation techniques including liquid chromatography (LC) and gas chromatography (GC), especially when accompanied by an MS detector, can be very useful to distinguish the production of different metabolites (Fig. 6.3). Separation is crucial for the evaluation of substances because of plants posse hundreds of them. In a study with the foliar irrigation of Ag compounds on cucumber (*Cucumis sativus*), the plant oxidative stress responses were measured by the method of GC-MS. Overall, 268 metabolites were identified in plant leaves under this investigation. It was determined that there was an upregulation of phenolic and phytol compounds in response to Ag NP exposure, the former enhancing the antioxidant defense while the later downregulating photosynthesis, respectively (Zhang et al. 2018). Furthermore, the metabolic effects of foliar Cu nano-pesticide application were evaluated in lettuce

(*Lactuca sativa*), where it was determined that nCu(OH)$_2$ disturbed the tricarboxylic (TCA) cycle and reduced the antioxidant levels of cis-caffeic acid, chlorogenic acid, and dehydroascorbic acid (Zhao et al. 2016). The authors discussed that the decrease of antioxidants compared to controls was attributed to oxidative stress damage caused by the NPs. GC-MS techniques can also be used for the evaluation of essential oils in plants. In a study of basil (*Ocimum sanctum*) with the foliar spray of various Cu compounds, the Cu(OH)$_2$ nanowire treatment reduced the essential eugenol and 2-methylundecanal oils by 57% and 71%, respectively, in comparison to their respective controls (Tan et al. 2018b).

In summary, instrumental analysis is restricted by sample preparation. When investigating NP-plant interactions, sample matrices become very complex and thus proper sample preparation needs to be attended prior to instrumental analysis. Additionally, it is useful to analyze the substances of interest with several instrumental techniques to fully comprehend NP and plant interactions.

6   Conclusions and Remarks

The metal oxide NPs have been extensively used in recent years, and a great proportion of which incorporates into the environment, mainly in soils. Many recent investigations have explored the impact of metallic NPs on terrestrial plants, and diverse positive and negative outcomes have been reported. These responses depend on multiple factors; however, they are greatly determined by the type of NP, the nature of the soil, and the plant species. Interestingly, the field of nanotechnology has created tangible options to increase agricultural productivity using nano-fertilizers and different nanomaterials to control plant disease. Furthermore, other promising options to guarantee food production are emerging by the use of agrochemical smart release NPs or by increasing stress tolerance in crops via NP-genetic material delivery. With the use of high-throughput technologies, it has been now suitable to assess the molecular, metabolic, and physiological responses of plant and microorganisms under NP exposure. Meanwhile, the use of spectroscopic techniques including the novel synchrotron approaches is very useful to monitor the uptake, fate, transport, and persistence of the NPs in plant systems. The upcoming research should take advantage of the new technologies to assess comprehensively the broad and positive effects of NPs on agricultural crops.

Acknowledgments   K- Cota-Ruiz is supported by a ConTex postdoctoral fellowship from the UT System and Conacyt, grant #1000001931. C-Valdes acknowledges the Consejo Nacional de Ciencia y Tecnología (CONACyT), México. The authors acknowledge the National Science Foundation and the Environmental Protection Agency under Cooperative Agreement Number DBI-1266377. Any opinions, findings, and conclusions or recommendations expressed in this material are those of the author(s) and do not necessarily reflect the views of the National Science Foundation or the Environmental Protection Agency. This work has not been subjected to EPA review and no official endorsement should be inferred. The authors also acknowledge the USDA grant 2016-67021-24985 and the NSF Grants ERC-1449500, CHE-0840525 and DBI-1429708.

Partial funding was provided by the NSF ERC on Nanotechnology-Enabled Water Treatment (ERC-1449500). This work was also supported by Grant 2G12MD007592 from the National Institutes on Minority Health and Health Disparities (NIMHD), a component of the National Institutes of Health (NIH). J. L. Gardea-Torresdey acknowledges the Dudley family for the Endowed Research Professorship and the Academy of Applied Science/US Army Research Office, Research and Engineering Apprenticeship Program (REAP) at UTEP, grant #W11NF-10-2-0076, sub-grant 13-7. J. L. Gardea-Torresdey acknowledges to the University of Texas System's STARs Retention Award.

References

Adams, L. K., Lyon, D. Y., & Alvarez, P. J. J. (2006). Comparative eco-toxicity of nanoscale TiO$_2$, SiO$_2$, and ZnO water suspensions. *Water Research, 40*, 3527–3532. https://doi.org/10.1016/j.watres.2006.08.004.

Adeleye, A. S., Conway, J. R., Garner, K., Huang, Y., Su, Y., & Keller, A. A. (2016). Engineered nanomaterials for water treatment and remediation: Costs, benefits, and applicability. *Chemical Engineering Journal, 286*, 640–662. https://doi.org/10.1016/j.cej.2015.10.105.

Adisa, I. O., Pullagurala, V. L., Rawat, S., Hernandez-viezcas, J. A., Elmer, W. H., White, J. C., Peralta-videa, J. R., & Gardea-torresdey, J. L. (2018). Role of cerium compounds in fusarium wilt suppression and growth enhancement in tomato (*Solanum lycopersicum*). *Journal of Agricultural and Food Chemistry, 66*, 5959–5970. https://doi.org/10.1021/acs.jafc.8b01345.

Alidoust, D., & Isoda, A. (2013). Effect of γFe$_2$O$_3$ nanoparticles on photosynthetic characteristic of soybean (*Glycine max* (L.) Merr.): Foliar spray versus soil amendment. *Acta Physiologiae Plantarum, 35*, 3365–3375. https://doi.org/10.1007/s11738-013-1369-8.

Amde, M., Liu, J. f., Tan, Z. Q., & Bekana, D. (2017). Transformation and bioavailability of metal oxide nanoparticles in aquatic and terrestrial environments. A review. *Environmental Pollution, 230*, 250–267. https://doi.org/10.1016/j.envpol.2017.06.064.

Anitha, S., Suganya, M., Prabha, D., Srivind, J., Balamurugan, S., & Balu, A. R. (2018). Synthesis and characterization of NiO-CdO composite materials towards photoconductive and antibacterial applications. *Materials Chemistry and Physics, 211*, 88–96. https://doi.org/10.1016/j.matchemphys.2018.01.048.

Apodaca, S. A., Tan, W., Dominguez, O. E., Hernandez-Viezcas, J. A., Peralta-Videa, J. R., & Gardea-Torresdey, J. L. (2017). Physiological and biochemical effects of nanoparticulate copper, bulk copper, copper chloride, and kinetin in kidney bean (*Phaseolus vulgaris*) plants. *Science Total Environment, 599–600*, 2085–2094. https://doi.org/10.1016/j.scitotenv.2017.05.095.

Atha, D. H., Wang, H., Petersen, E. J., Cleveland, D., Holbrook, R. D., Jaruga, P., Dizdaroglu, M., Xing, B., & Nelson, B. C. (2012). Copper oxide nanoparticle mediated DNA damage in terrestrial plant models. *Environmental Science & Technology, 46*, 1819–1827. https://doi.org/10.1021/es202660k.

Ayyildiz, H. F., Ozcan, F., Ertul, S., Kara, H., Aygun, A. (2017). Use of humic acid-based nanofibers for dye removal and transport. Proc. 2017 IEEE 7th Int. Conf. Nanomater. Appl. Prop. 3–5. https://doi.org/10.1109/NAP.2017.8190224.

Azam, A., Ahmed, A. S., Oves, M., Khan, M. S., & Memic, A. (2012). Size-dependent antimicrobial properties of CuO nanoparticles against gram-positive and -negative bacterial strains. *International Journal of Nanomedicine, 7*, 3527–3535. https://doi.org/10.2147/IJN.S29020.

Baek, Y. W., & An, Y. J. (2011). Microbial toxicity of metal oxide nanoparticles (CuO, NiO, ZnO, and Sb$_2$O$_3$) to *Escherichia coli, Bacillus subtilis,* and *Streptococcus aureus. Science of the Total Environment, 409*, 1603–1608. https://doi.org/10.1016/j.scitotenv.2011.01.014.

Bandyopadhyay, S., Plascencia-Villa, G., Mukherjee, A., Rico, C. M., José-Yacamán, M., Peralta-Videa, J. R., & Gardea-Torresdey, J. L. (2015). Comparative phytotoxicity of ZnO NPs, bulk

ZnO, and ionic zinc onto the alfalfa plants symbiotically associated with *Sinorhizobium meliloti* in soil. *Science of the Total Environment, 515–516*, 60–69. https://doi.org/10.1016/j. scitotenv.2015.02.014.

Barrena, R., Casals, E., Colón, J., Font, X., Sánchez, A., & Puntes, V. (2009). Evaluation of the ecotoxicity of model nanoparticles. *Chemosphere, 75*, 850–857. https://doi.org/10.1016/j. chemosphere.2009.01.078.

Bonilla-Bird, N. J., Paez, A., Reyes, A., Hernandez-Viezcas, J. A., Li, C., Peralta-Videa, J. R., & Gardea-Torresdey, J. L. (2018). Two-photon microscopy and spectroscopy studies to determine the mechanism of copper oxide nanoparticle uptake by Sweetpotato roots during postharvest treatment. *Environmental Science & Technology, 52*, 9954–9963. https://doi.org/10.1021/acs. est.8b02794.

Bradfield, S. J., Kumar, P., White, J. C., & Ebbs, S. D. (2017). Zinc, copper, or cerium accumulation from metal oxide nanoparticles or ions in sweet potato: Yield effects and projected dietary intake from consumption. *Plant Physiology and Biochemistry, 110*, 128–137. https:// doi.org/10.1016/j.plaphy.2016.04.008.

Burman, U., Saini, M., & Kumar, P. (2013). Effect of zinc oxide nanoparticles on growth and antioxidant system of chickpea seedlings. *Toxicological and Environmental Chemistry, 95*, 605–612. https://doi.org/10.1080/02772248.2013.803796.

Castillo-Michel, H. A., Larue, C., Pradas del Real, A. E., Cotte, M., & Sarret, G. (2017). Practical review on the use of synchrotron based micro- and nano- X-ray fluorescence mapping and X-ray absorption spectroscopy to investigate the interactions between plants and engineered nanomaterials. *Plant Physiology and Biochemistry, 110*, 13–32. https://doi.org/10.1016/j. plaphy.2016.07.018.

Chen, G., Liu, X., & Su, C. (2012). Distinct effects of humic acid on transport and retention of TiO_2 rutile nanoparticles in saturated sand columns. *Environmental Science & Technology, 46*, 7142–7150. https://doi.org/10.1021/es204010g.

Chen, H., Zhuang, R., Yao, J., Wang, F., Qian, Y., Masakorala, K., Cai, M., & Liu, H. (2014). Short-term effect of aniline on soil microbial activity: A combined study by isothermal microcalorimetry, glucose analysis, and enzyme assay techniques. *Environmental Science and Pollution Research, 21*, 674–683. https://doi.org/10.1007/s11356-013-1955-8.

Colman, B. P., Arnaout, C. L., Anciaux, S., Gunsch, C. K., Hochella, M. F., Kim, B., Lowry, G. V., McGill, B. M., Reinsch, B. C., Richardson, C. J., Unrine, J. M., Wright, J. P., Yin, L., & Bernhardt, E. S. (2013). Low concentrations of silver nanoparticles in biosolids cause adverse ecosystem responses under realistic field scenario. *PLoS One, 8*, e57189. https://doi. org/10.1371/journal.pone.0057189.

Cota-Ruiz, K., Delgado-Rios, M., Martínez-Martínez, A., Núñez-Gastelum, J. A., Peralta-Videa, J. R., & Gardea-Torresdey, J. L. (2018a). Current findings on terrestrial plants - engineered nanomaterial interactions: Are plants capable of Phytoremediating nanomaterials from soil? *Current Opinion in Environmental Science and Health, 6*, 9–15. https://doi.org/10.1016/j. coesh.2018.06.005.

Cota-Ruiz, K., Hernández-Viezcas, J. A., Varela-Ramírez, A., Valdés, C., Núñez-Gastélum, J. A., Martínez-Martínez, A., Delgado-Rios, M., Peralta-Videa, J. R., & Gardea-Torresdey, J. L. (2018b). Toxicity of copper hydroxide nanoparticles, bulk copper hydroxide, and ionic copper to alfalfa plants: A spectroscopic and gene expression study. *Environmental Pollution, 243*, 703–712. https://doi.org/10.1016/J.ENVPOL.2018.09.028.

Crane, R. A., & Scott, T. B. (2012). Nanoscale zero-valent iron: Future prospects for an emerging water treatment technology. *Journal of Hazardous Materials, 211*, 112–125. https://doi. org/10.1016/j.jhazmat.2011.11.073.

Cui, H., Zhang, P., Gu, W., Jiang, J. (2009). Application of Anatase TiO_2 Sol Derived from Peroxotitannic Acid in Crop Diseases Control and Growth Regulation. NSTI-Nanotech 2, 286–26+6.

Da Costa, M. V. J., & Sharma, P. K. (2016). Effect of copper oxide nanoparticles on growth, morphology, photosynthesis, and antioxidant response in *Oryza sativa*. *Photosynthetica, 54,* 110–119. https://doi.org/10.1007/s11099-015-0167-5.

Dai, Y., Wang, Z., Zhao, J., Xu, L., Xu, L., Yu, X., Wei, Y., & Xing, B. (2018). Interaction of CuO nanoparticles with plant cells: Internalization, oxidative stress, electron transport chain disruption, and toxicogenomic responses. *Environmental Science. Nano, 5,* 2269–2281. https://doi.org/10.1039/c8en00222c.

Dan, Y., Ma, X., Zhang, W., Liu, K., Stephan, C., & Shi, H. (2016). Single particle ICP-MS method development for the determination of plant uptake and accumulation of CeO_2 nanoparticles. *Analytical and Bioanalytical Chemistry, 408,* 5157–5167. https://doi.org/10.1007/s00216-016-9565-1.

Davarpanah, S., Tehranifar, A., Davarynejad, G., Abadía, J., & Khorasani, R. (2016). Effects of foliar applications of zinc and boron nano-fertilizers on pomegranate (*Punica granatum* cv. Ardestani) fruit yield and quality. *Science of Horticulture (Amsterdam), 210,* 57–64. https://doi.org/10.1016/j.scienta.2016.07.003.

de la Rosa, G., García-Castañeda, C., Vázquez-Núñez, E., Alonso-Castro, Á. J., Basurto-Islas, G., Mendoza, Á., Cruz-Jiménez, G., & Molina, C. (2017). Physiological and biochemical response of plants to engineered NMs: Implications on future design. *Plant Physiology and Biochemistry, 110,* 226–235. https://doi.org/10.1016/j.plaphy.2016.06.014.

Del Real, A. E. P., Castillo-Michel, H., Kaegi, R., Sinnet, B., Magnin, V., Findling, N., Villanova, J., Carriére, M., Santaella, C., Fernández-Martínez, A., Levard, C., & Sarret, G. (2016). Fate of Ag-NPs in sewage sludge after application on agricultural soils. *Environmental Science & Technology, 50,* 1759–1768. https://doi.org/10.1021/acs.est.5b04550.

DeRosa, M. C., Monreal, C., Schnitzer, M., Walsh, R., & Sultan, Y. (2010). Nanotechnology in fertilizers. *Nature Nanotechnology, 5,* 91. https://doi.org/10.1038/nnano.2010.2.

Dimkpa, C. O., McLean, J. E., Britt, D. W., & Anderson, A. J. (2013). Antifungal activity of ZnO nanoparticles and their interactive effect with a biocontrol bacterium on growth antagonism of the plant pathogen *Fusarium graminearum*. *Biometals, 26,* 913–924. https://doi.org/10.1007/s10534-013-9667-6.

Gajbhiye, M., Kesharwani, J., Ingle, A., Gade, A., & Rai, M. (2009). Fungus-mediated synthesis of silver nanoparticles and their activity against pathogenic fungi in combination with fluconazole. *Nanomedicine Nanotechnology, Biology Medicine, 5,* 382–386. https://doi.org/10.1016/j.nano.2009.06.005.

Gao, F., Hong, F., Liu, C., Zheng, L., Su, M., Wu, X., Yang, F., Wu, C., & Yang, P. (2006). Mechanism of nano-anatase TiO_2 on promoting photosynthetic carbon reaction of spinach: Inducing complex of rubisco-rubisco activase. *Biological Trace Element Research, 111,* 239–253. https://doi.org/10.1385/BTER:111:1:239.

García-Gómez, C., Babin, M., Obrador, A., Álvarez, J. M., & Fernández, M. D. (2015). Integrating ecotoxicity and chemical approaches to compare the effects of ZnO nanoparticles, ZnO bulk, and $ZnCl_2$ on plants and microorganisms in a natural soil. *Environmental Science and Pollution Research, 22,* 16803–16813. https://doi.org/10.1007/s11356-015-4867-y.

García-Sánchez, S., Bernales, I., & Cristobal, S. (2015). Early response to nanoparticles in the *Arabidopsis* transcriptome compromises plant defence and root-hair development through salicylic acid signalling. *BMC Genomics, 16.* https://doi.org/10.1186/s12864-015-1530-4.

Guzman, M., Dille, J., & Godet, S. (2012). Synthesis and antibacterial activity of silver nanoparticles against gram-positive and gram-negative bacteria. *Nanomedicine Nanotechnology, Biology Medicine, 8,* 37–45. https://doi.org/10.1016/j.nano.2011.05.007.

He, L., Liu, Y., Mustapha, A., & Lin, M. (2011). Antifungal activity of zinc oxide nanoparticles against *Botrytis cinerea* and *Penicillium expansum*. *Microbiological Research, 166,* 207–215. https://doi.org/10.1016/j.micres.2010.03.003.

Hegde, K., Brar, S. K., Verma, M., & Surampalli, R. Y. (2016). Current understandings of toxicity, risks and regulations of engineered nanoparticles with respect to environmental microorganisms. *Nanotechnology for Environmental Engineering, 1*, 5. https://doi.org/10.1007/s41204-016-0005-4.

Hernandez-Viezcas, J. A., Castillo-Michel, H., Andrews, J. C., Cotte, M., Rico, C., Peralta-Videa, J. R., Ge, Y., Priester, J. H., Holden, P. A., & Gardea-Torresdey, J. L. (2013). *In situ* synchrotron X-ray fluorescence mapping and speciation of CeO_2 and ZnO nanoparticles in soil cultivated soybean (*Glycine max*). *ACS Nano, 7*, 1415–1423. https://doi.org/10.1021/nn305196q.

Holden, P. A., Mortimer, M., & Wang, Y. (2018). Engineered nanomaterials and symbiotic dinitrogen fixation in legumes. *Current Opinion in Environmental Science and Health, 6*, 54–59. https://doi.org/10.1016/J.COESH.2018.07.012.

Hong, F., Zhou, J., Liu, C., Yang, F., Wu, C., Zheng, L., & Yang, P. (2005). Effect of Nano-TiO_2 on photochemical reaction of chloroplasts of spinach. *Biological Trace Element Research, 105*, 269–279. https://doi.org/10.1385/BTER:105:1-3:269.

Hong, J., Peralta-Videa, J. R., & Gardea-Torresdey, J. L. (2013). Nanomaterials in agricultural production: Benefits and possible threats? *ACS Symposium Series, 1124*, 73–90. https://doi.org/10.1021/bk-2013-1124.ch005.

Hua, M., Zhang, S., Pan, B., Zhang, W., Lv, L., & Zhang, Q. (2012). Heavy metal removal from water/wastewater by nanosized metal oxides: A review. *Journal of Hazardous Materials, 211–212*, 317–331. https://doi.org/10.1016/j.jhazmat.2011.10.016.

Huang, L., Li, D. Q., Lin, Y. J., Wei, M., Evans, D. G., & Duan, X. (2005). Controllable preparation of Nano-MgO and investigation of its bactericidal properties. *Journal of Inorganic Biochemistry, 99*, 986–993. https://doi.org/10.1016/j.jinorgbio.2004.12.022.

Hund-Rinke, K., & Simon, M. (2006). Ecotoxic effect of photocatalytic active nanoparticles (TiO_2) on algae and Daphnids. *Environmental Science and Pollution Research*, 1–8. https://doi.org/10.1065/espr2006.06.311.

Jaberzadeh, A., Moaveni, P., Tohidi Moghadam, H. R., & Zahedi, H. (2013). Influence of bulk and nanoparticles titanium foliar application on some agronomic traits, seed gluten and starch contents of wheat subjected to water deficit stress. *Not. Bot. Horti Agrobot. Cluj-Napoca, 41*, 201–207. https://doi.org/10.15835/NBHA4119093.

Jasim, B., Thomas, R., Mathew, J., & Radhakrishnan, E. K. (2017). Plant growth and diosgenin enhancement effect of silver nanoparticles in fenugreek (*Trigonella foenum-graecum* L.). Saudi. *The Pharmaceutical Journal, 25*, 443–447. https://doi.org/10.1016/j.jsps.2016.09.012.

Jayalath, S., Wu, H., Larsen, S. C., & Grassian, V. H. (2018). Surface adsorption of Suwannee River humic acid on TiO_2 nanoparticles: A study of pH and particle size. *Langmuir, 34*, 3136–3145. https://doi.org/10.1021/acs.langmuir.8b00300.

Jin, Y., Fan, X., Li, X., Zhang, Z., Sun, L., Fu, Z., Lavoie, M., Pan, X., & Qian, H. (2017). Distinct physiological and molecular responses in *Arabidopsis thaliana* exposed to aluminum oxide nanoparticles and ionic aluminum. *Environmental Pollution, 228*, 517–527. https://doi.org/10.1016/j.envpol.2017.04.073.

Jo, Y.-K., Kim, B. H., & Jung, G. (2009). Antifungal activity of silver ions and nanoparticles on Phytopathogenic fungi. *Plant Disease, 93*, 1037–1043. https://doi.org/10.1094/PDIS-93-10-1037.

Joo, S. H., & Zhao, D. (2017). Environmental dynamics of metal oxide nanoparticles in heterogeneous systems: A review. *Journal of Hazardous Materials, 322*, 29–47. https://doi.org/10.1016/j.jhazmat.2016.02.068.

Kang, H., Hwang, Y. G., Lee, T. G., Jin, C. R., Cho, C. H., Jeong, H. Y., & Kim, D. O. (2016). Use of gold nanoparticle fertilizer enhances the ginsenoside contents and anti-inflammatory effects of red ginseng. *Journal of Microbiology and Biotechnology, 26*, 1668–1674. https://doi.org/10.4014/jmb.1604.04034.

Kanhed, P., Birla, S., Gaikwad, S., Gade, A., Seabra, A. B., Rubilar, O., Duran, N., & Rai, M. (2014). *In vitro* antifungal efficacy of copper nanoparticles against selected crop pathogenic fungi. *Materials Letters, 115*, 13–17. https://doi.org/10.1016/j.matlet.2013.10.011.

Kim, J. S., Kuk, E., Yu, K. N., Kim, J. H., Park, S. J., Lee, H. J., Kim, S. H., Park, Y. K., Park, Y. H., Hwang, C. Y., Kim, Y. K., Lee, Y. S., Jeong, D. H., & Cho, M. H. (2007). Antimicrobial effects of silver nanoparticles. *Nanomedicine Nanotechnology, Biology Medicine, 3*, 95–101. https://doi.org/10.1016/j.nano.2006.12.001.

Kim, S. W., Jung, J. H., Lamsal, K., Kim, Y. S., Min, J. S., & Lee, Y. S. (2012). Antifungal effects of silver nanoparticles (AgNPs) against various plant pathogenic fungi. *Mycobiology, 40*, 53–58. https://doi.org/10.5941/MYCO.2012.40.1.053.

Kolackova, M., Moulick, A., Kopel, P., Dvorak, M., Adam, V., Klejdus, B., & Huska, D. (2019). Antioxidant, gene expression and metabolomics fingerprint analysis of *Arabidopsis thaliana* treated by foliar spraying of ZnSe quantum dots and their growth inhibition of *Agrobacterium tumefaciens*. *Journal of Hazardous Materials, 365*, 932–941. https://doi.org/10.1016/j.jhazmat.2018.11.065.

Lamsal, K., Kim, S. W., Jung, J. H., Kim, Y. S., Kim, K. S., & Lee, Y. S. (2011). Application of silver nanoparticles for the control of Colletotrichum species *In Vitro* and pepper anthracnose disease in field., *39*, 194–199. https://doi.org/10.5941/MYCO.2011.39.3.194.

Landa, P., Prerostova, S., Petrova, S., Knirsch, V., Vankova, R., & Vanek, T. (2015). The transcriptomic response of *Arabidopsis thaliana* to zinc oxide: A comparison of the impact of nanoparticle, bulk, and ionic zinc. *Environmental Science & Technology, 49*, 14537–14545. https://doi.org/10.1021/acs.est.5b03330.

Lead, J. R., Batley, G. E., Alvarez, P. J. J., Croteau, M. N., Handy, R. D., McLaughlin, M. J., Judy, J. D., & Schirmer, K. (2008). Nanomaterials in the environment: Behavior, fate, bioavailability, and effects—An updated review. *Environmental Toxicology and Chemistry, 37*, 2029–2063. https://doi.org/10.1002/etc.4147.

Lee, S., Bi, X., Reed, R. B., Ranville, J. F., Herckes, P., & Westerhoff, P. (2014). Nanoparticle size detection limits by single particle ICP-MS for 40 elements. *Environmental Science & Technology, 48*, 10291–10300. https://doi.org/10.1021/es502422v.

Lin, D., & Xing, B. (2007). Phytotoxicity of nanoparticles: Inhibition of seed germination and root growth. *Environmental Pollution, 150*, 243–250. https://doi.org/10.1016/j.envpol.2007.01.016.

Linglan, M., Chao, L., Chunxiang, Q., Sitao, Y., Jie, L., Fengqing, G., & Fashui, H. (2008). Rubisco activase mRNA expression in spinach: Modulation by nanoanatase treatment. *Biological Trace Element Research, 122*, 168–178. https://doi.org/10.1007/s12011-007-8069-4.

López-Moreno, M. L., de la Rosa, G., Hernández-Viezcas, J. á., Castillo-Michel, H., Botez, C. E., Peralta-Videa, J. R., & Gardea-Torresdey, J. L. (2010). Evidence of the differential biotransformation and genotoxicity of ZnO and CeO_2 nanoparticles on soybean (*Glycine max*) plants. *Environmental Science and Technology, 44*, 7315–7320. https://doi.org/10.1021/es903891g.

Louie, S. M., Ma, R., & Lowry, G. V. (2014). Transformations of nanomaterials in the environment. *Frontier of Nanoscience, 7*, 55–87. https://doi.org/10.1016/B978-0-08-099408-6.00002-5.

Lu, H., Wang, J., Li, F., Huang, X., Tian, B., & Hao, H. (2018). Highly efficient and reusable montmorillonite/Fe_3O_4/humic acid nanocomposites for simultaneous removal of Cr (VI) and aniline. *Nanomaterials, 8*, 537. https://doi.org/10.3390/nano8070537.

Masciangioli, T., & Zhang, W. (2003). Environmental Technologies at the Nanoscale. *Environmental Science & Technology*, 102A–108A.

Masini, J. C., Abate, G., Lima, E. C., Hahn, L. C., Nakamura, M. S., Lichtig, J., & Nagatomy, H. R. (1998). Comparison of methodologies for determination of carboxylic and phenolic groups in humic acids. *Analytica Chimica Acta, 364*, 223–233. https://doi.org/10.1016/S0003-2670(98)00045-2.

Medina-Velo, I. A., Barrios, A. C., Zuverza-Mena, N., Hernandez-Viezcas, J. A., Chang, C. H., Ji, Z., Zink, J. I., Peralta-Videa, J. R., & Gardea-Torresdey, J. L. (2017). Comparison of the effects of commercial coated and uncoated ZnO nanomaterials and Zn compounds in kidney bean (*Phaseolus vulgaris*) plants. *Journal of Hazardous Materials, 332*, 214–222. https://doi.org/10.1016/j.jhazmat.2017.03.008.

Medina-Velo, I. A., Zuverza-Mena, N., Tamez, C., Ye, Y., Hernandez-Viezcas, J. A., White, J. C., Peralta-Videa, J. R., & Gardea-Torresdey, J. L. (2018). Minimal transgenerational effect of ZnO nanomaterials on the physiology and nutrient profile of *Phaseolus vulgaris*. *ACS Sustainable Chemistry & Engineering, 6*, 7924–7930. https://doi.org/10.1021/acssuschemeng.8b01188.

Mody, V., Siwale, R., Singh, A., & Mody, H. (2010). Introduction to metallic nanoparticles. *Journal of Pharmacy & Bioallied Sciences, 2*, 282–289. https://doi.org/10.4103/0975-7406.72127.

Morales, M. I., Rico, C. M., Hernandez-Viezcas, J. A., Nunez, J. E., Barrios, A. C., Tafoya, A., Flores-Marges, J. P., Peralta-Videa, J. R., & Gardea-Torresdey, J. L. (2013). Toxicity assessment of cerium oxide nanoparticles in cilantro (*Coriandrum sativum* L.) plants grown in organic soil. *Journal of Agricultural and Food Chemistry, 61*, 6224–6230. https://doi.org/10.1021/jf401628v.

Morones, J. R., Elechiguerra, J. L., Camacho, A., Holt, K., Kouri, J. B., Ramírez, J. T., & Yacaman, M. J. (2005). The bactericidal effect of silver nanoparticles. *Nanotechnology, 16*, 2346–2353. https://doi.org/10.1088/0957-4484/16/10/059.

Nair, P. M. G., & Chung, I. M. (2014). Impact of copper oxide nanoparticles exposure on *Arabidopsis thaliana* growth, root system development, root lignificaion, and molecular level changes. *Environmental Science and Pollution Research, 21*, 12709–12722. https://doi.org/10.1007/s11356-014-3210-3.

Nair, P. M. G., & Chung, I. M. (2014a). A mechanistic study on the toxic effect of copper oxide nanoparticles in soybean (*Glycine max* L.) root development and lignification of root cells. *Biological Trace Element Research, 162*, 342–352. https://doi.org/10.1007/s12011-014-0106-5.

Nair, P. M. G., & Chung, I. M. (2015). The responses of germinating seedlings of green peas to copper oxide nanoparticles. *Biologia Plantarum, 59*, 591–595. https://doi.org/10.1007/s10535-015-0494-1.

Nair, P. M. G., & Chung, I. M. (2015a). Physiological and molecular level studies on the toxicity of silver nanoparticles in germinating seedlings of mung bean (*Vigna radiata* L.). *Acta Physiologiae Plantarum, 37*, 1–11. https://doi.org/10.1007/s11738-014-1719-1.

Narayanan, K. B., & Sakthivel, N. (2010). Biological synthesis of metal nanoparticles by microbes. *Advances in Colloid and Interface Science, 156*, 1–13. https://doi.org/10.1016/j.cis.2010.02.001.

Navarro, E., Baun, A., Behra, R., Hartmann, N. B., Filser, J., Miao, A. J., Quigg, A., Santschi, P. H., & Sigg, L. (2008). Environmental behavior and ecotoxicity of engineered nanoparticles to algae, plants, and fungi. *Ecotoxicology, 17*, 372–386. https://doi.org/10.1007/s10646-008-0214-0.

Nowack, B., & Bucheli, T. D. (2007). Occurrence, behavior and effects of nanoparticles in the environment. *Environmental Pollution, 150*, 5–22. https://doi.org/10.1016/j.envpol.2007.06.006.

Ocsoy, I., Paret, M. L., Ocsoy, M. A., Kunwar, S., Chen, T., You, M., & Tan, W. (2013). Nanotechnology in plant disease management: DNA-directed silver nanoparticles on graphene oxide as an antibacterial against *Xanthomonas perforans*. *ACS Nano, 7*, 8972–8980. https://doi.org/10.1021/nn4034794.

Pagano, L., Servin, A. D., De La Torre-Roche, R., Mukherjee, A., Majumdar, S., Hawthorne, J., Marmiroli, M., Maestri, E., Marra, R. E., Isch, S. M., Dhankher, O. P., White, J. C., & Marmiroli, N. (2016). Molecular response of crop plants to engineered nanomaterials. *Environmental Science & Technology, 50*, 7198–7207. https://doi.org/10.1021/acs.est.6b01816.

Paret, M.L., Vallad, G.E., Averett, D.R., Jones, J.B., Olson, S.M. (2013). Photocatalysis: Effect of light-activated nanoscale formulations of TiO$_2$ on *Xanthomonas perforans* and control of bacterial spot of Tomato. *Phytopathology, 103*(3), 228–236.

Peralta-videa, J. R., Hernandez-viezcas, J. A., Zhao, L., Corral, B., Ge, Y., Priester, J. H., Ann, P., & Gardea-torresdey, J. L. (2014). Cerium dioxide and zinc oxide nanoparticles alter the nutritional value of soil cultivated soybean plants. *Plant Physiology and Biochemistry, 80*, 128–135. https://doi.org/10.1016/j.plaphy.2014.03.028.

Peters, R. J. B., Bouwmeester, H., Gottardo, S., Amenta, V., Arena, M., Brandhoff, P., Marvin, H. J. P., Mech, A., Moniz, F. B., Pesudo, L. Q., Rauscher, H., Schoonjans, R., Undas, A. K., Vettori, M. V., Weigel, S., & Aschberger, K. (2016). Nanomaterials for products and application in agriculture, feed and food. *Trends in Food Science and Technology, 54*, 155–164. https://doi.org/10.1016/j.tifs.2016.06.008.

Pokropivny, V. V., & Skorokhod, V. V. (2008). New dimensionality classifications of nanostruc-tures. *Physica E Low-Dimensional Systems and Nanostructures, 40*, 2521–2525. https://doi.org/10.1016/j.physe.2007.11.023.

Prabhu, S., & Poulose, E. K. (2012). Silver nanoparticles: Mechanism of antimicrobial action, synthesis, medical applications, and toxicity effects. *International Nano Letters, 2*, 32. https://doi.org/10.1186/2228-5326-2-32.

Priester, J. H., Moritz, S. C., Espinosa, K., Ge, Y., Wang, Y., Nisbet, R. M., Schimel, J. P., Susana Goggi, A., Gardea-Torresdey, J. L., & Holden, P. A. (2017). Damage assessment for soybean cultivated in soil with either CeO_2 or ZnO manufactured nanomaterials. *Science of the Total Environment, 579*, 1756–1768. https://doi.org/10.1016/j.scitotenv.2016.11.149.

Qian, H., Peng, X., Han, X., Ren, J., Sun, L., & Fu, Z. (2013). Comparison of the toxicity of silver nanoparticles and silver ions on the growth of terrestrial plant model *Arabidopsis thaliana*. *Journal of Environmental Sciences (China), 25*, 1947–1955. https://doi.org/10.1016/S1001-0742(12)60301-5.

Rai, M., Yadav, A., & Gade, A. (2009). Silver nanoparticles as a new generation of antimicrobials. *Biotechnology Advances, 27*, 76–83. https://doi.org/10.1016/j.biotechadv.2008.09.002.

Rao, C. N. R., Müller, A., & Cheetham, A. K. (2006). *The chemistry of nanomaterials: Synthesis, properties and applications*. John Wiley & Sons.

Reddy, P. V. L., Hernandez-Viezcas, J. A., Peralta-Videa, J. R., & Gardea-Torresdey, J. L. (2016). Lessons learned: Are engineered nanomaterials toxic to terrestrial plants? *Science of the Total Environment, 568*, 470–479. https://doi.org/10.1016/j.scitotenv.2016.06.042.

Rezvani, N., Sorooshzadeh, A., & Farhadi, N. (2012). Effect of Nano-silver on growth of saffron in flooding stress. *World Academy of Science, Engineering and Technology, 6*, 11–16.

Rico, C. M., Morales, M. I., McCreary, R., Castillo-Michel, H., Barrios, A. C., Hong, J., Tafoya, A., Lee, W. Y., Varela-Ramirez, A., Peralta-Videa, J. R., & Gardea-Torresdey, J. L. (2013). Cerium oxide nanoparticles modify the antioxidative stress enzyme activities and macromol-ecule composition in rice seedlings. *Environmental Science & Technology, 47*, 14110–14118. https://doi.org/10.1021/es4033887.

Ritchie, J. D., & Michael Perdue, E. (2003). Proton-binding study of standard and reference fulvic acids, humic acids, and natural organic matter. *Geochimica et Cosmochimica Acta, 67*, 85–93. https://doi.org/10.1016/S0016-7037(02)01044-X.

Ruparelia, J. P., Chatterjee, A. K., Duttagupta, S. P., & Mukherji, S. (2008). Strain specificity in antimicrobial activity of silver and copper nanoparticles. *Acta Biomaterialia, 4*, 707–716. https://doi.org/10.1016/j.actbio.2007.11.006.

Sabaghnia, N., & Janmohammasi, M. (2015). Effect of nano-silicon particles application on salin-ity tolerance in early growth of some lentil genotypes. *Annales UMCS, Biology, 69*, 39–55.

Sabir, S., Arshad, M., & Chaudhari, S. K. (2014). Zinc oxide nanoparticles for revolutioniz-ing agriculture: Synthesis and applications. *Scientific World Journal, 2014*, 1–8. https://doi.org/10.1155/2014/925494.

Salem, W., Leitner, D. R., Zingl, F. G., Schratter, G., Prassl, R., Goessler, W., Reidl, J., & Schild, S. (2015). Antibacterial activity of silver and zinc nanoparticles against *Vibrio cholerae* and enterotoxic *Escherichia coli*. *International Journal of Medical Microbiology, 305*, 85–95. https://doi.org/10.1016/j.ijmm.2014.11.005.

Servin, A. D., Castillo-Michel, H., Hernandez-Viezcas, J. A., Diaz, B. C., Peralta-Videa, J. R., & Gardea-Torresdey, J. L. (2012). Synchrotron micro-XRF and micro-XANES confirmation of the uptake and translocation of TiO_2 nanoparticles in cucumber (*Cucumis sativus*) plants. *Environmental Science & Technology, 46*, 7637–7643. https://doi.org/10.1021/es300955b.

Sharma, P., Bhatt, D., Zaidi, M. G. H., Saradhi, P. P., Khanna, P. K., & Arora, S. (2012). Silver nanoparticle-mediated enhancement in growth and antioxidant status of *Brassica jun-cea*. *Applied Biochemistry and Biotechnology, 167*, 2225–2233. https://doi.org/10.1007/s12010-012-9759-8.

Shen, Z., Chen, Z., Hou, Z., Li, T., & Lu, X. (2015). Ecotoxicological effect of zinc oxide nanoparticles on soil microorganisms. *Frontiers of Environmental Science & Engineering, 9,* 912–918. https://doi.org/10.1007/s11783-015-0789-7.

Shin, Y. J., Kwak, J. I., & An, Y. J. (2012). Evidence for the inhibitory effects of silver nanoparticles on the activities of soil exoenzymes. *Chemosphere, 88,* 524–529. https://doi.org/10.1016/j.chemosphere.2012.03.010.

Simonin, M., & Richaume, A. (2015). Impact of engineered nanoparticles on the activity, abundance, and diversity of soil microbial communities: A review. *Environmental Science and Pollution Research, 22,* 13710–13723. https://doi.org/10.1007/s11356-015-4171-x.

Singh, P., Kim, Y. J., Zhang, D., & Yang, D. C. (2016). Biological synthesis of nanoparticles from plants and microorganisms. *Trends in Biotechnology, 34,* 588–599. https://doi.org/10.1016/j.tibtech.2016.02.006.

Sunkar, R., Chinnusamy, V., Zhu, J., & Zhu, J. K. (2007). Small RNAs as big players in plant abiotic stress responses and nutrient deprivation. *Trends in Plant Science, 12,* 301–309. https://doi.org/10.1016/j.tplants.2007.05.001.

Tamez, C., Morelius, E. W., Hernandez-Viezcas, J. A., Peralta-Videa, J. R., & Gardea-Torresdey, J. (2018). Biochemical and physiological effects of copper compounds/nanoparticles on sugarcane (*Saccharum officinarum*). *Science of the Total Environment.* https://doi.org/10.1016/J.SCITOTENV.2018.08.337.

Tan, W., Du, W., Darrouzet-nardi, A. J., Hernandez-viezcas, J. A., Ye, Y., Peralta-videa, J. R., & Gardea-torresdey, J. L. (2018a). Effects of the exposure of TiO₂ nanoparticles on basil (*Ocimum basilicum*) for two generations. *Science of Total Environment, 636,* 240–248. https://doi.org/10.1016/j.scitotenv.2018.04.263.

Tan, W., Gao, Q., Deng, C., Wang, Y., Lee, W.-Y., Hernandez-Viezcas, J. A., Peralta-Videa, J. R., & Gardea-Torresdey, J. L. (2018b). Foliar exposure of cu(OH)₂ nanopesticide to basil (*Ocimum basilicum*): Variety dependent copper translocation and biochemical responses. *Journal of Agricultural and Food Chemistry, 66,* 3358–3366. https://doi.org/10.1021/acs.jafc.8b00339.

Tang, W. W., Zeng, G. M., Gong, J. L., Liang, J., Xu, P., Zhang, C., & Huang, B. B. (2014). Impact of humic/fulvic acid on the removal of heavy metals from aqueous solutions using nanomaterials: A review. *Science of the Total Environment, 468–469,* 1014–1027. https://doi.org/10.1016/j.scitotenv.2013.09.044.

Torabian, S., Zahedi, M., & Khoshgoftar, A. H. (2016). Effects of foliar spray of two kinds of zinc oxide on the growth and ion concentration of sunflower cultivars under salt stress. *Journal of Plant Nutrition, 39,* 172–180. https://doi.org/10.1080/01904167.2015.1009107.

Tourinho, P. S., van Gestel, C. A. M., Lofts, S., Svendsen, C., Soares, A. M. V. M., & Loureiro, S. (2012). Metal-based nanoparticles in soil: Fate, behavior, and effects on soil invertebrates. *Environmental Toxicology and Chemistry, 31,* 1679–1692. https://doi.org/10.1002/etc.1880.

Tuteja, N. (2010). Cold, salinity, and drought stress. Plant stress biol. From genomics to. *Systematic Biology, 444,* 137–159. https://doi.org/10.1002/9783527628964.ch7.

Van Aken, B. (2015). Gene expression changes in plants and microorganisms exposed to nanomaterials. *Current Opinion in Biotechnology, 33,* 206–219. https://doi.org/10.1016/j.copbio.2015.03.005.

Van Hoecke, K., De Schamphelaere, K. A. C., Van Der Meeren, P., Smagghe, G., & Janssen, C. R. (2011). Aggregation and ecotoxicity of CeO₂ nanoparticles in synthetic and natural waters with variable pH, organic matter concentration and ionic strength. *Environmental Pollution, 159,* 970–976. https://doi.org/10.1016/j.envpol.2010.12.010.

Wang, J., Shu, K., Zhang, L., & Si, Y. (2017). Effects of silver nanoparticles on soil microbial communities and bacterial nitrification in suburban vegetable soils. *Pedosphere, 27,* 482–490. https://doi.org/10.1016/S1002-0160(17)60344-8.

White, J. C., & Gardea-Torresdey, J. (2018). Achieving food security through the very small. *Nature Nanotechnology, 13,* 627–629. https://doi.org/10.1038/s41565-018-0223-y.

Xu, C., Peng, C., Sun, L., Zhang, S., Huang, H., Chen, Y., & Shi, J. (2015). Distinctive effects of TiO$_2$ and CuO nanoparticles on soil microbes and their community structures in flooded paddy soil. *Soil Biology and Biochemistry, 86*, 24–33. https://doi.org/10.1016/j.soilbio.2015.03.011.

Yan, M., Zeng, G., Li, X., Zhao, C., Yang, G., Gong, J., Chen, G., Tang, L., & Huang, D. (2017). Titanium dioxide nanotube arrays with silane coupling agent modification for heavy metal reduction and persistent organic pollutant degradation. *New Journal of Chemistry, 41*, 4377–4389. https://doi.org/10.1039/c6nj03196j.

Yang, K., Lin, D., & Xing, B. (2009). Interactions of humic acid with Nanosized inorganic oxides. *Langmuir, 25*, 3571–3576. https://doi.org/10.1016/j.envpol.2008.11.007.

Yu, S., Liu, J., Yin, Y., & Shen, M. (2018). Interactions between engineered nanoparticles and dissolved organic matter: A review on mechanisms and environmental effects. *Journal of Environmental Sciences, 63*, 198–217. https://doi.org/10.1016/j.jes.2017.06.021.

Zhang, H., Du, W., Peralta-videa, J. R., Gardea-torresdey, J. L., White, J. C., Keller, A. A., Guo, H., Ji, R., & Zhao, L. (2018). Metabolomics reveals how cucumber (*Cucumis sativus*) reprograms metabolites to cope with silver ions and silver nanoparticle-induced oxidative stress. *Environmental Science & Technology, 52*, 8016–8026. https://doi.org/10.1021/acs.est.8b02440.

Zhao, L., Hu, Q., Huang, Y., Fulton, A. N., Hannah-Bick, C., Adeleye, A. S., & Keller, A. A. (2017). Activation of antioxidant and detoxification gene expression in cucumber plants exposed to a cu(OH)$_2$ nanopesticide. *Environmental Science. Nano, 4*, 1750–1760. https://doi.org/10.1039/C7EN00358G.

Zhao, L., Ortiz, C., Adeleye, A. S., Hu, Q., Zhou, H., Huang, Y., & Keller, A. A. (2016). Metabolomics to detect response of lettuce (*Lactuca sativa*) to cu(OH)$_2$ Nanopesticides: Oxidative stress response and detoxification mechanisms. *Environmental Science & Technology, 50*, 9697–9707. https://doi.org/10.1021/acs.est.6b02763.

Zhao, L., Peralta-Videa, J. R., Varela-Ramirez, A., Castillo-Michel, H., Li, C., Zhang, J., Aguilera, R. J., Keller, A. A., & Gardea-Torresdey, J. L. (2012). Effect of surface coating and organic matter on the uptake of CeO$_2$ NPs by corn plants grown in soil: Insight into the uptake mechanism. *Journal of Hazardous Materials, 225–226*, 131–138. https://doi.org/10.1016/j.jhazmat.2012.05.008.

Zhou, D., Ji, Z., Jiang, X., Dunphy, D. R., Brinker, J., & Keller, A. A. (2013). Influence of material properties on TiO$_2$ nanoparticle agglomeration. *PLoS One, 8*, 1–7. https://doi.org/10.1371/journal.pone.0081239.

Zhu, J. K. (2001). Plant salt tolerance. *Trends in Plant Science, 6*, 66–71. https://doi.org/10.1016/S1360-1385(00)01838-0.

Zuverza-Mena, N., Armendariz, R., Peralta-Videa, J. R., & Gardea-Torresdey, J. L. (2016). Effects of silver nanoparticles on radish sprouts: Root growth reduction and modifications in the nutritional value. *Frontiers in Plant Science, 7*, 1–11. https://doi.org/10.3389/fpls.2016.00090.

Zuverza-Mena, N., Martínez-Fernández, D., Du, W., Hernandez-Viezcas, J. A., Bonilla-Bird, N., López-Moreno, M. L., Komárek, M., Peralta-Videa, J. R., & Gardea-Torresdey, J. L. (2017). Exposure of engineered nanomaterials to plants: Insights into the physiological and biochemical responses-a review. *Plant Physiology and Biochemistry, 110*, 236–264. https://doi.org/10.1016/j.plaphy.2016.05.037.

Chapter 7
Cerium and Titanium Oxide Nanoparticles Increase Algal Growth and Nutrient Accumulation in Rice Paddy Environments

Elliott Duncan and Gary Owens

Contents

1 Introduction

Despite the widely accepted notion that engineered nanoparticles (ENPs) can exhibit a range of deleterious ecotoxicological effects if released into the environment or accumulated by humans (Matranga and Corsi 2012; Maurer-Jones et al. 2013), their application in agriculture is rapidly increasing (Huang et al. 2015; Liu

E. Duncan · G. Owens (✉)
Environmental Contaminants Group, Future Industries Institute, University of South Australia, Mawson Lakes, South Australia, Australia
e-mail: Elliott.duncan@unisa.edu.au; gary.owens@unisa.edu.au

© Springer Nature Switzerland AG 2021
N. Sharma, S. Sahi (eds.), *Nanomaterial Biointeractions at the Cellular, Organismal and System Levels*, Nanotechnology in the Life Sciences, https://doi.org/10.1007/978-3-030-65792-5_7

and Lal 2015; Shalaby et al. 2016; Wang et al. 2016). Thus, although many studies have outlined the potentially deleterious effects of ENPs on biota (including humans), in recent years many more studies are increasingly reporting the potential agricultural benefits (i.e. productivity, nutritional quality) of ENP-based amendments (i.e. fertilisers, pesticides) (Liu and Lal 2015; Wang et al. 2016).

However, while research on the use of ENP-based agricultural amendments is increasing, far fewer studies have assessed the indirect or secondary effects of these materials in plant-soil systems. Recent studies have suggested that ENPs such as cerium oxide ($nCeO_2$) and titanium dioxide ($nTiO_2$) have the ability to significantly alter nutrient (N, P, Zn) and contaminant (As, Pb) phytoavailability in agricultural soils (Duncan and Owens 2019; Nikoo Jamal 2018). In addition, it has also been observed that a range of ENPs can alter the abundance and activity of various soil microbial communities (Ge et al. 2011). This has the potential to influence nutrient phytoavailability and pathogen loads in the soil, both of which can indirectly effect agricultural productivity. Finally, there is also evidence to suggest that certain ENPs (e.g. $nCeO_2/nTiO_2$) can stimulate photosynthesis (Owolade et al. 2008; Rico et al. 2015; Zheng et al. 2005) and/or reduce oxidative stress (Xia et al. 2008) in plants and thus influence productivity.

Since research on ENP-based agricultural amendments is in its infancy, it is currently uncertain whether these amendments can specifically influence the growth of 'target' crop species (e.g. wheat, rice, corn) or instead indiscreetly target all plant species in the environment (i.e. weeds, algae). In previous studies, rice (*Oryza sativa*) has responded favourably when exposed to $nCeO_2$ and $nTiO_2$ in terms of both overall productivity and the nutritional quality of the grain produced (Nikoo Jamal 2018). However, there was also visual evidence in these studies that $nCeO_2$ and $nTiO_2$ increased the growth of the unicellular algae that coexist with rice in paddy environments. The conditions present in rice paddies, e.g. standing water, high temperatures, high nutrient loads and elongated day-length, are in fact 'tailor-made' to facilitate algal growth (Roger and Kulasooriya 1980). Unicellular algae are thus ubiquitous in rice paddy environments (Choudhury and Kennedy 2004) and thus have the potential to both benefit and hinder rice production.

There are a number of symbiotic algae-rice processes that can assist the rice plant in resisting pathogen attack (Kim 2006) and also since many cyanobacteria species fix atmospheric N which can be utilised by plants, increase nutrient phytoavailability (Mandal et al. 1999). When paddies are drained to initiate the harvest of rice, the algal biomass is not removed and is largely returned intact to the soil (Gaydon et al. 2012) which means that nutrients accumulated by the algal biomass are potentially then available to the crop that follows rice in the sequence. The capture of fertiliser nutrients by algae during rice growth can also reduce nutrient losses to the environment and increase the overall use efficiency of fertiliser applications which is a major issue for rice cultivation worldwide (Cassman et al. 1996).

Overgrowths of algae in rice paddies can, however, have a deleterious effect on rice production as they can compete with rice for nutrients, light and oxygen (Roger and Kulasooriya 1980). This can result in decreased grain yield production as a result of nutrient deficiencies (Baligar et al. 2001) and can also decrease the establishment rate (number of established plants/number of seeds sown) of rice plants as algal overgrowths can form a physical barrier at the soil-water interface which cannot be penetrated by the rice seedling (Grant et al. 2006; Roger and Kulasooriya 1980). Rice growth, nutrient phytoavailability and algal growth are normally in equilibrium in most established farming systems; however, significant changes in the environment (e.g. high nutrient loads, high temperatures, low water level), which can also be induced by the accumulation of $nCeO_2$ and $nTiO_2$, have the potential to alter this equilibrium.

Thus, the current study investigates the potential impact of elevated concentrations of either $nCeO_2$ or $nTiO_2$ on rice and unicellular algae communities in rice paddy microcosms by monitoring changes in growth and nutrient utilisation. The hypothesis is that the presence of $nCeO_2$ and $nTiO_2$ directly alters the equilibrium between algae and rice in soil microcosms, which facilitates the growth and accumulation of nutrients by algae, thus indirectly reducing rice establishment, biomass production and the nutritional quality.

2 Materials and Methods

2.1 Rice Paddy Microcosm Experiments

A red chromosol collected from Halbury, South Australia, Australia (Table 7.1) was used in the paddy microcosms. Approximately 30 kg of the soil was air dried and sieved to <2 mm prior to fertilising with N and spiking with ENPs.

The dried soil stock was initially divided into 5 kg aliquots ($n = 6$). Three of these aliquots were fertilised with urea at a rate of 300 mg N kg^{-1} soil. Two of these N-fertilised aliquots were then spiked with either $nCeO_2$ (15–30 nm) or $nTiO_2$ (anatase 15–30 nm) (Nanostructured and Amorphous Materials Inc., USA) at a concentration of 50 mg ENP kg^{-1} soil. Both ENPs were applied to the soils as a fine mist after suspension in Milli-Q water. The third N-fertilised aliquot was retained as a no-ENP control. The remaining three 5 kg aliquots, having no added N input, were spiked with $nCeO_2$ or $nTiO_2$ at a concentration of 50 mg ENP kg^{-1} soil as described above, with the final aliquot being retained as a no-ENP, no-N control soil.

N and ENP spiked soils (500 g dry mass) were then transferred to clear plastic containers which had a diameter of 11.5 cm and a height of 9.5 cm for use as microcosms. Ten ($n = 10$) microcosms were prepared per treatment ($n = 6$), with

Table 7.1 Selected physical and chemical characteristics of the Halbury soil used in microcosms

Type	Texture (sand/silt/clay %)	pH (1:5 H$_2$O)	C%	N%	Mineral N (mg kg^{-1})	Total Ce (mg kg^{-1})	Total Ti (mg kg^{-1})
Chromosol	81/10/9 (sandy loam)	7.0	1.3	0.1	45	16	1470

N.B. mineral N and total Ce/Ti concentrations were determined before the application of urea and ENPs to soils deionised water (300 mL) added to each microcosm to simulate paddy environments. Five pre-germinated seeds of the rice (*Oryza sativa*) cultivar Sherpa were planted in each pot. After the addition of water, all microcosms were weighed and water was added as required to maintain a constant water level over the six-week experiment (Fig. 7.1).

All microcosms were incubated in a temperature-controlled glasshouse at the University of South Australia, Mawson Lakes, South Australia (SA), Australia. The environmental conditions within the glasshouse included a temperature regime of $27/17 \pm 2$ (°C) (day/night) and had a natural photoperiod (maximum day length approximately 13.5 h). After 6 weeks growth algal communities were harvested by hand and by centrifuging the overlying rice paddy water at 4500 g for 10 min (Heraeus Multifuge 3 S-R, Thermo Fisher, Australia). At this time, all established rice plants were also removed from the microcosm by hand.

2.2   Determination of N, Ce and Ti in Algal and Rice Tissue Samples

All harvested algal and rice tissues were oven dried at 70 °C for 48 h and ground to a fine powder using an IKEA A11 micromill (MEDOS, Australia). As insufficient algal and rice biomass was produced in individual microcosms, all elemental analyses were performed on three pooled samples ($n = 3$) from each treatment.

A microwave assisted 2:1 (v/v) H_2SO_4:HNO_3 was used to determine total Ce and Ti concentrations in vegetative tissues (Nikoo Jamal 2018). Dried and ground vegetative tissue (0.5 g) was weighed directly into 55 mL polytetrafluoroacetate (PTFE) digestion vessels (CEM, USA) and a 2:1 (v/v) H_2SO_4:HNO_3 acid mixture (5 mL) added. The mixture was then heated in a microwave oven (MARS, CEM, USA) using a two-step time and temperature programme which consisted of ramping to 200 °C over a 20 min period followed by a 45 min holding period at 200 °C. After digestion, samples were allowed to cool to room temperature and then diluted to 10 mL with deionised water. Digests were stored in polyethylene vials in a cool room (~4 °C) until further analysis.

Fig. 7.1 Aerial photograph of the microcosms used in this study

Total Ce and Ti concentrations were measured using an Inductively Coupled Plasma Mass Spectrometer (ICP-MS) using an Agilent 8800 Triple Quadrupole ICP-MS. Experimental conditions for ICP-MS analysis were: plasma flow 1.07 L min^{-1}, sampling depth 8 mm, power 1550 W, pump rate 0.1 rps, and stabilisation delay 25 s. Sulphur oxide ($S^{32}O^{16}$) interferences on the major Ti isotope ($m/z = 48$) were removed, using an ammonia (NH_3) reaction gas during ICP-MS analysis for Ti with Ti-NH_3 clusters measured at $m/z = 114$ while Ce ($m/z = 140$) concentration were measured in the presence of Oxygen (O_2) gas with Ce-O_2 clusters at m/z = 156.

Quantification was performed using external standards prepared by diluting appropriate volumes of a 1 mg mL^{-1} stock solutions of Ti^{4+} and Ce^{4+} high purity ICP-MS standards to create a seven-point calibration curve over a range of concentrations from 0.1 µg L^{-1} to 1 mg L^{-1}, which was a wide enough range to include the estimated values of each sample. A 2% (v/v) HNO$_3$ solution was used as an instrument blank, and standard solutions having a concentration of 10 µg L^{-1} Ti (IV) and Ce (IV) were used for quality control checks (Continuous Check of Variation (CCV)) during ICP-MS analyses. For quality control a commercially available Certified Reference Materials (CRMs) – Bush Branches (NCS DC 73349) and one in-house wheat grain sample spiked with approximately 100 mg nTiO$_2$ kg^{-1} grain (66.6 mg Ti kg^{-1} grain) were also used (Table 7.2).

Total N concentrations in plant tissues were measured using a TruMac Carbon/Nitrogen Determinator (Trumac C/N, LECO Corporation, model 630–300-400). Oven dried soil or plant material (0.5 g) was weighed into a clean ceramic vial for combustion analysis. Ethylenediamine tetraacetic acid (EDTA) was included as a standard reference material together with blanks at a rate of 5% per batch. The measured N content (%) of the EDTA SRM was 9.56 ± 0.03% which was in good agreement with the certified concentration (9.52–9.60%).

Table 7.2 Recoveries of Ce and Ti from the CRMs used in this study

CRM	Measured Ce (mg kg^{-1})	Certified Ce (mg kg^{-1})	% recovery	Measured Ti (mg kg^{-1})	Certified Ti (mg kg^{-1})	% recovery
Bush Branches (NCS DC 73349) ($n = 14$)	1.7 ± 0.1	2.0 ± 0.1	85 ± 5	60 ± 12	95 ± 20	63 ± 13
nTiO$_2$ spiked wheat grain ($n = 15$)	0.03 ± 0.04	N.D.	N/A	55 ± 19	66 ± 6	83 ± 29

N.D. not determined, N/A not applicable

2.3   Statistical Analyses

Factorial ANOVA procedures were used to determine the effect of ENP type, N fertiliser rate and the interaction between ENP type and N fertiliser rate on plant biomass (rice and algae), rice establishment rate and tissue N, Ce and Ti concentrations. Differences between treatments were statistically significant when $P < 0.05$. All data was log-transformed to ensure data was normally distributed.

3   Results

3.1   Algal Biomass Production

Mean algal biomass (dry mass) ranged from <0.01 to 1.10 g microcosm^{-1} (Fig. 7.2). The interaction between ENP type and fertiliser N exposure significantly influenced algal biomass production (DF = 5; F = 22.22; P < 0.001). Algal biomass production was highest in microcosms treated with both ENPs and fertiliser N ($\approx$ 1.00–1.10 g microcosm^{-1}), which was significantly higher than biomass production in microcosms that only received fertiliser N ($\approx$ 0.67 g microcosm^{-1}), which was in turn higher than biomass production in microcosms that did not receive fertiliser N (<0.13 g microcosm^{-1}) (Fig. 7.2).

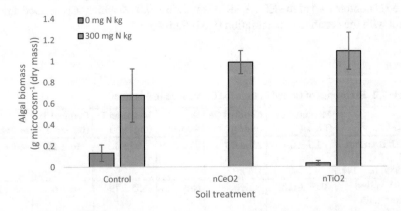

Fig. 7.2 Mean algal biomass (g microcosm^{-1} (dry mass)) produced in microcosms treated with different combinations of N fertiliser and ENPs. Values are means ± SD (n = 10)

3.2   Accumulation of Nitrogen, Cerium and Titanium by Algae

Nitrogen: Algal tissue N concentrations ranged from 583 to 1198 mg N kg^{-1} tissue (dry mass) (Fig. 7.3a). Differences across treatments were statistically significant (DF = 5; F = 75.4; P < 0.001) with concentrations in all microcosms that received fertiliser N ($\approx$ 1000–1200 mg N kg^{-1}) higher than those in microcosms that did not receive fertiliser N ($\approx$ 580–700 mg N kg^{-1}) (Fig. 7.3a). The presence of nCeO$_2$ and nTiO$_2$ in microcosms resulted in a near twofold and threefold increase, respectively, in the total N removed from microcosms in N-fertilised microcosms (Fig. 7.3b). In non N-fertilised microcosms, however, there was little to no effect of ENPs on overall N removal (Fig. 7.3b).

Cerium: Algal Ce concentrations ranged from 33 to 138 mg Ce kg^{-1} tissue (dry mass) (Fig. 7.4a). Cerium concentrations in algae harvested from nCeO$_2$ amended microcosms were significantly higher (DF = 3; F = 113.6; P < 0.001) than those in non-nCeO$_2$ amended microcosms.

Titanium: Algal Ti concentrations ranged from 1643 to 1992 mg Ti kg^{-1} tissue (dry mass) (Fig. 7.4b). There was no significant differences in Ti concentrations present (DF = 3; F = 0.41; P = 0.83) across treatments.

3.3   Rice Establishment and Biomass Production

Establishment rates: Rice establishment rates significantly differed across treatments as a result of both ENP (DF = 2; F = 28.50; P < 0.001) and fertiliser N treatments (DF = 1; F = 32.00; P < 0.001) (Fig. 7.5a). Establishment rates in control and nTiO$_2$ treated plants (60–65%) were significantly higher than those in nCeO$_2$ treated plants (25%), while rates in microcosms that received fertiliser N (37%) were lower than rates in microcosms that did not receive fertiliser N (66%).

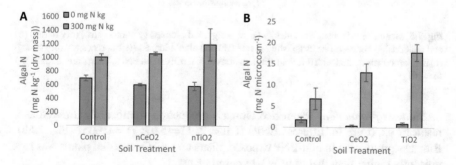

Fig. 7.3 Algal N concentrations (mg N kg^{-1} (dry mass)) (**a**) and total algal N removal (mg N microcosm^{-1}) (**b**) in microcosms treated with different combinations of N fertiliser and ENPs. Values are means ± SD (n = 3)

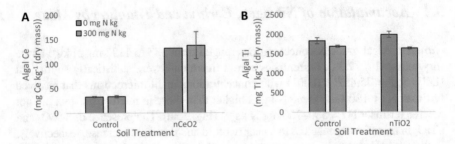

Fig. 7.4 Algal Ce concentrations (mg Ce kg^{-1} (dry mass)) (**a**) and Algal Ti concentrations (mg Ti kg^{-1} (dry mass)) (**b**) in microcosms treated with different combinations of N fertiliser and ENPs. Values are means ± SD ($n = 3$)

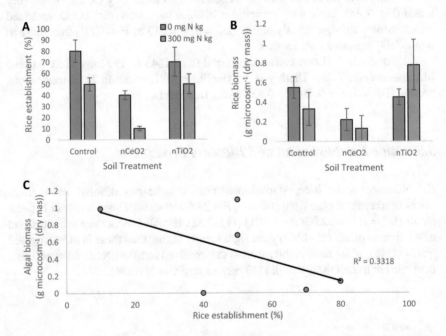

Fig. 7.5 Mean rice establishment rate (%) (**a**) above ground biomass (g microcosm^{-1} (dry mass)) (**b**) and relationship between rice establishment rate (%) and algal biomass ((g microcosm^{-1} (dry mass)) (**c**) in microcosms treated with different combinations of N fertiliser and ENPs. Values are means ± SD ($n = 10$)

Biomass production: Mean rice biomass production significantly differed as a result of exposure to different ENPs (DF = 2; $F = 5.97$; $P = 0.005$) (Fig. 7.5b). Biomass production in non-ENP exposed plants and nTiO$_2$ exposed plants was significantly higher than that of nCeO$_2$ exposed plants.

3.4 Accumulation of Nitrogen, Cerium and Titanium by Rice

Nitrogen: Total N concentrations in rice were significantly influenced by interactions between ENP type and fertiliser N treatment (DF = 2; F = 33.19; P < 0.001) (Fig. 7.6a). All plants grown in the presence of fertiliser N contained significantly higher tissue N concentrations than those grown without fertiliser N. In N-supplied microcosms tissue N concentrations in non-ENP treated plants > nTiO$_2$ treated plants > nCeO$_2$ treated plants, while in non N-supplied microcosms tissue N concentrations in non-ENP treated plants > nTiO$_2$ treated plants (Fig. 7.6a).

Total N removal by rice was, however, higher in non-ENP supplied microcosms when fertiliser N was applied (Fig. 7.6b). In non N-fertilised microcosms there was no difference in total N removal across treatments (Fig. 7.6b).

Cerium: Ce concentrations ranged from 0.3 to 0.8 mg Ce kg^{-1} tissue (dry mass) (Fig. 7.7a) with concentrations in nCeO$_2$ exposed plants significantly higher (DF = 2; F = 10.00; P = 0.003) than those in non-nCeO$_2$ exposed plants.

Titanium: Ti concentrations ranged from 10 to 23 mg Ti kg^{-1} tissue (dry mass) (Fig. 7.7b) with no significant differences in Ti concentrations across treatments (DF = 2; F = 2.45; P = 0.10).

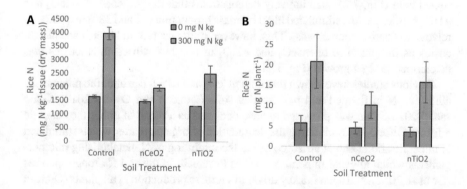

Fig. 7.6 Rice N concentrations (mg N kg^{-1} (dry mass)) (**a**) and total rice N removal (mg N microcosm^{-1}) (**b**) in microcosms treated with different combinations of N fertiliser and ENPs. Values are means ± SD (n = 3)

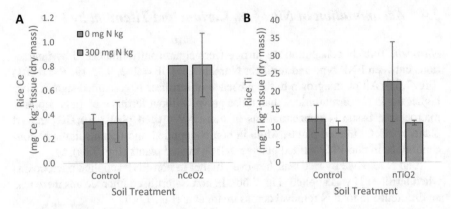

Fig. 7.7 Rice Ce concentrations (mg Ce kg^{-1} (dry mass)) (**a**) and rice Ti concentrations (mg Ti kg^{-1} (dry mass)) (**b**) in microcosms treated with different combinations of N fertiliser and ENPs. Values are means ± SD ($n = 3$)

4 Discussion

4.1 Effects of ENPs on Algal Biomass Production and Nutrient Accumulation

Prior to this study there was visual non-quantitative evidence that nCeO$_2$ and nTiO$_2$ promoted the growth of unicellular algae in rice paddy environments. The data presented here (Fig. 7.2) quantitatively demonstrates that the presence of nCeO$_2$ and nTiO$_2$ in microcosms stimulated algal biomass production (31 and 38% respectively relative to control microcosms). This, however, only occurred in N-fertilised microcosms as the presence of nCeO$_2$ and nTiO$_2$ in non-N fertilised microcosms was deleterious to algal growth (Fig. 7.2).

Previous studies have shown that both nCeO$_2$ and nTiO$_2$ can alter the phytoavailability of N in flooded and non-flooded soils (Duncan and Owens 2019; Nikoo Jamal 2018). This was proposed to have occurred as a result of either the catalytic effects of nCeO$_2$/TiO$_2$ initiating the degradation of organic matter which stimulated N mineralisation (Zhu et al. 2005) or via the inhibition of N metabolising microorganisms which allowed mineral N species to remain in the soil for longer periods (Ge et al. 2011). Nitrogen is a key driver of biomass production in all plants (Duncan et al. 2018) and N-rich waters can promote the proliferation of algae and eutrophication of aquatic environments (Daniel 2018) and thus an increase in mineral N in microcosms is a logical explanation for the observed increased algal growth.

When N was supplied to microcosms as fertiliser there was no difference in total N concentrations between ENP- and non-ENP exposed algae (Fig. 7.3A). The overall algal N removal from microcosms was, however, 92% and 162% higher in nCeO$_2$/TiO$_2$ treated microcosms than in non-ENP treated microcosms (Fig. 7.3B). This implied that while the ENPs themselves did not increase the capacity for algal cells to accumulate N on a per cell or per gram basis, the presence of ENPs in the soil did dramatically increase the total N available for algae to accumulate over a microcosm scale. Many unicellular algae species that inhabit rice paddy environments are known to fix N from the atmosphere (Choudhury and Kennedy 2004; Mandal et al. 1999; Tripathi et al. 2008). It is therefore plausible that ENPs, N and algae exhibited a complex relationship within microcosms whereby ENPs increased phytoavailable N which increased algal biomass which in turn increased the total N fixed by algae from the atmosphere.

Although N phytoavailability was not specifically measured in this experiment, comparing the algal biomass and algal N accumulation in fertiliser N-supplied and non-fertiliser N-supplied microcosms, provided a clue as to the possible cause of increased algal growth. In fertiliser N-supplied microcosms there was a clear increase in algal N accumulation in the presence of either ENP, which suggested that additional N resources were present in the microcosm which were either liberated from the soil or fixed from the atmosphere. In non-fertiliser N-supplied microcosms, however, there were no significant differences in algal biomass production (Fig. 7.2) or overall N accumulation (Fig. 7.3B) across treatments. There are examples of ENPs such as $nTiO_2$ inhibiting the ability of cyanobacteria to fix atmospheric N (Cherchi and Gu 2010) while there are a number of examples of $nCeO_2/TiO_2$ inhibiting the growth of unicellular algae and other microorganisms (Bour et al. 2016; Ge et al. 2011; Hartmann et al. 2010; Lee and An 2013; Pelletier et al. 2010; Xu et al. 2015). This implied that in the absence of fertiliser N, ENPs do not significantly increase N availability, which suggests that ENPs are allowing mineral N species from fertilisers to remain in the soil for longer periods, i.e. inhibiting denitrification rather than increasing the mineralisation of organic N pools.

From this data, the question that therefore arises is whether the stimulation of algal growth resulted in decreased rice production or whether the presence of $nCeO_2/TiO_2$ in microcosms increased nutrient phytoavailability to the point where both plant species (algae and rice) could prosper.

4.2 Effects of ENPs on Rice Establishment, Biomass Production and Nutrient Accumulation

The presence of $nCeO_2$ in microcosms reduced the above ground biomass of rice by approximately 60% (Fig. 7.5b) regardless of whether fertiliser N was supplied. The presence of $nTiO_2$ in microcosms also reduced rice biomass by approximately 18% in non-fertiliser N-supplied microcosms; however, in fertiliser N-supplied microcosms rice biomass was doubled when exposed to $nTiO_2$ (Fig. 7.5b).

The results for $nCeO_2$ implied that algae and rice were in direct competition for resources as increases in algal biomass occurred as a consequence of decreased rice growth (Figs. 7.2 and 7.5). Similar trends were also apparent in a nutritional sense, whereby increased N accumulation by algae corresponded with decreased N accumulation by rice (Figs. 7.3 and 7.6). Finally, the presence of nCeO$_2$ also decreased the establishment rate (%) of rice (Fig. 7.5a).

The presence of nTiO$_2$ in microcosms, particularly those also supplied with fertiliser N, stimulated and sustained the growth of both plant species and thus behaved very differently from those amended with nCeO$_2$ (Figs. 7.2 and 7.5). This implied that sufficient N resources were available to support the growth of both species; however, as N phytoavailability was not measured specifically it could not be assumed that nTiO$_2$ was more successful than nCeO$_2$ in retaining fertiliser N in the soil. In fact, previous studies using exactly the same soil had shown no significant differences in N phytoavailability in nCeO$_2$ and nTiO$_2$ treated flooded and non-flooded soils (Duncan and Owens 2019; Nikoo Jamal 2018).

Alternatively, nTiO$_2$ itself potentially stimulated the growth of both plant species. This is not inconceivable since numerous reports have previously proposed that the photocatalytic properties of TiO$_2$ based materials can result in the stimulation of photosynthesis in a range of plant species (Carvajal and Alcaraz 1998; Frutos et al. 1996; Lyu et al. 2017), while evidence also exists that these materials can stimulate N-fixation in some plants (Zheng et al. 2005).

In this study there were no significant differences in tissue Ti concentrations in algae (Fig. 7.4) or rice (Fig. 7.7) when grown in the presence or absence of nTiO$_2$, or when grown in the presence and absence of fertiliser N (Figs. 7.4 and 7.7). Superficially, this observation makes it seem unlikely that the accumulation of nTiO$_2$ by plants facilitated biomass production. However, since the soil used here had a high natural background Ti concentration (1470 mg Ti/kg soil) (Table 7.1) it seems likely that a significant proportion of the plant-accumulated Ti observed was that naturally available in the soil and was not indicative of the nTiO$_2$ specifically amended to the microcosm. It is therefore possible that small concentrations of nTiO$_2$ were accumulated by both species, which was sufficient to stimulate photosynthesis relative to that of plants not exposed to nTiO$_2$. However, when fertiliser N was not supplied nTiO$_2$ had no effect on biomass production in rice (Fig. 7.5b) and algae (Fig. 7.2) which again suggested that nTiO$_2$-induced changes to N phytoavailability facilitated the observed growth increases.

4.3   Implications for Rice Cultivation, Environmental Health and Opportunities for Other Industries

This work clearly demonstrated that algal biomass in rice paddy environments was strongly stimulated by co-exposure to ENPs (nCeO$_2$/TiO$_2$) and fertiliser N (Fig. 7.2). This increase in algal growth resulted in the accumulation of significant N concentrations by the algal biomass (Fig. 7.3), which in some instances simultaneously reduced rice biomass production (Fig. 7.5b), establishment rates (Fig. 7.5a) and N accumulation (Fig. 7.6), particularly when nCeO$_2$ was involved. In this instance ENPs were thus deleterious to rice production which is a significant outcome given that ENPs will almost certainly come into direct contact with agricultural plants with increasing regularity in the future (Huang et al. 2015). In addition, the growth of rice (and algae) was reduced when both ENPs were present in microcosms in the absence of fertiliser N (Figs. 7.2 and 7.5). This observation is also significant as it implies that the use of fertiliser N can mitigate the deleterious ecotoxicological effects of ENPs. Thus, in low nutrient or organic farming systems there is a real risk of ENPs having a deleterious effect on crop growth.

However, ENPs were not deleterious to rice production in all of the microcosms studied, and as a result of the complex interactions between algae and rice in paddy environments there were also some potential positive results attributable to increased algal growth. For example, even though $nTiO_2$ stimulated algal growth (Fig. 7.2), rice growth was not compromised and actually increased relative to non-exposed plants (Fig. 7.5). In this instance the presence of $nTiO_2$ could actually benefit the farming system as a whole as rice production increased while the algae within the paddy acted as a 'N-sink' which almost certainly minimised N-losses to the atmosphere and also potentially returned N (and other nutrients) to the soil which could be utilised by the next crop in the sequence.

Unicellular algae also facilitate the eutrophication of aquatic systems, which usually occurs in periods of high temperatures, light and nutrient availability (Daniel 2018). Algal 'blooms' are potentially detrimental to both flora and fauna that inhabit aquatic systems as they can limit light penetration to benthic zones, can remove all oxygen from the water-column and the algae themselves can also release a range of toxins that can inhibit the growth of a range of biota (Anderson et al. 2002). Aquatic ecosystems share many similarities with rice paddy environments, and thus if ENPs accumulate in these systems and stimulate algal growth as observed here, the effects could be equally deleterious in riverine environments. In many regions rice is produced in close proximity to major river systems due the high water requirement of rice. The use of ENPs such as $nCeO_2$ and $nTiO_2$ as a component of agricultural products and fertilisers thus increases the possibility of these materials accumulating in aquatic systems and thus careful risk management needs to be implemented to ensure these materials if used do not escape from farmer's fields.

However, increased algal proliferation can also have a range of environmental benefits. The nitrogen use efficiency (NUE) of rice is notoriously poor with most estimates suggesting that up to 70% of N applied as fertiliser is lost to the environment, mainly through denitrification (Cassman et al. 1996; Peng et al. 2006). These losses are significant in an environmental sense as they can contribute to the effects of climate change given that some volatile N species are potent greenhouse gases (Dalal et al. 2003). Increased algal growth in paddies, however, has the potential to capture some of this excess N and return it to the soil once paddies are drained prior to harvest (Gaydon et al. 2012). Thus, ENPs, particularly $nTiO_2$, could be used to reduce the potentially deleterious environmental effects of broadacre rice cultivation; however, as detailed above significant field trials are required to ascertain whether these effects are also likely to occur in situ.

Increased algal growth and nutrient accumulation in the presence of $nCeO_2$ and $nTiO_2$ could be beneficial to industries that cultivate algae for industrial purposes, e.g. biofuel and aquaculture. If unicellular algae can be cultivated in the presence of ENPs in a way in which the end product (i.e. oil, food-source) does not become contaminated, there is the potential for economic benefits for the industry itself and potentially for consumers. For example, if $nCeO_2/nTiO_2$ increase the NUE of algae, industrial quantities of algal biomass could be produced using far less fertiliser N, which provides an economic benefit to the specific algal cultivation industry.

5 Conclusions

In summary, while $nCeO_2$ and $nTiO_2$ were both shown to significantly increase algal biomass production in rice paddy microcosms, this only occurred when fertiliser N was applied in conjunction with ENPs. Increased algal growth predictably lead to increased N accumulation by the algal biomass, which for $nCeO_2$ significantly reduced the biomass production of rice. However, for reasons that are currently uncertain, in the presence of $nTiO_2$ rice biomass production increased despite increased algal growth. Future research should thus focus on determining how $nTiO_2$ is able to stimulate growth in both algae and rice species. In addition, extensive field trials are also required to determine any potential effects on rice growth, nutrient phytoavailability and the impact of algal overgrowths on the productivity of this important cropping system.

References

Anderson, D. M., Glibert, P. M., & Burkholder, J. M. (2002). Harmful algal blooms and eutrophication: Nutrient sources, composition, and consequences. *Estuaries, 25,* 704–726. https://doi.org/10.1007/bf02804901.

Baligar, V. C., Fageria, N. K., & He, Z. L. (2001). Nutrient use efficiency in plants. *Communications in Soil Science and Plant Analysis, 32,* 921–950. https://doi.org/10.1081/CSS-100104098.

Bour, A., Mouchet, F., Cadarsi, S., Silvestre, J., Verneuil, L., Baqué, D., Chauvet, E., Bonzom, J.-M., Pagnout, C., & Clivot, H. (2016). Toxicity of CeO₂ nanoparticles on a freshwater experimental trophic chain: A study in environmentally relevant conditions through the use of mesocosms. *Nanotoxicology, 10,* 245–255.

Carvajal, M., & Alcaraz, C. (1998). Why titanium is a beneficial element for plants. *Journal of Plant Nutrition, 21,* 655–664.

Cassman, K., Gines, G., Dizon, M., Samson, M., & Alcantara, J. (1996). Nitrogen-use efficiency in tropical lowland rice systems: Contributions from indigenous and applied nitrogen. *Field Crops Research, 47,* 1–12.

Cherchi, C., & Gu, A. Z. (2010). Impact of titanium dioxide nanomaterials on nitrogen fixation rate and intracellular nitrogen storage in Anabaena variabilis. *Environmental Science & Technology, 44,* 8302–8307.

Choudhury, A., & Kennedy, I. (2004). Prospects and potentials for systems of biological nitrogen fixation in sustainable rice production. *Biology and Fertility of Soils, 39,* 219–227.

Dalal, R. C., Wang, W., Robertson, G. P., & Parton, W. J. (2003). Nitrous oxide emission from Australian agricultural lands and mitigation options: A review. *Soil Research, 41,* 165–195.

Ding, S., Chen, M., Gong, M., Fan, X., Qin, B., Xu, H., Gao, S-S., Jin, Z., Tsang, D. C. W., & Zhang, C. (2018). Internal phosphorus loading from sediments causes seasonal nitrogen limitation for harmful algal blooms. *Science of the Total Environment, 625,* 872–884. https://doi.org/10.1016/j.scitotenv.2017.12.348.

Duncan, E., Owens, G. (2019). Metal oxide nanomaterials used to remediate heavy metal contaminated soils have strong effects on nutrient and trace element phytoavailability. *Science of the Total Environment, 678,* 430–437. https://doi.org/10.1016/j.scitotenv.2019.04.442.

Duncan, E. G., O'Sullivan, C. A., Roper, M. M., Biggs, J. S., & Peoples, M. B. (2018). Influence of co-application of nitrogen with phosphorus, potassium and Sulphur on the apparent efficiency of nitrogen fertiliser use, grain yield and protein content of wheat: Review. *Field Crops Research, 226,* 56–65. https://doi.org/10.1016/j.fcr.2018.07.010.

Frutos, M., Pastor, J., Martínez-Sánchez, F., & Alcaraz, C. (1996). Improvement of the nitrogen uptake induced by titanium (iv) leaf supply in nitrogen-stressed pepper seedlings. *Journal of Plant Nutrition, 19,* 771–783.

Gaydon, D., Probert, M., Buresh, R., Meinke, H., & Timsina, J. (2012). Modelling the role of algae in rice crop nutrition and soil organic carbon maintenance. *European Journal of Agronomy, 39,* 35–43.

Ge, Y., Schimel, J. P., & Holden, P. A. (2011). Evidence for negative effects of TiO2 and ZnO nanoparticles on soil bacterial communities. *Environmental Science & Technology, 45,* 1659–1664. https://doi.org/10.1021/es103040t.

Grant, A. J., Pavlova, M., Wilkinson-White, L., Haythornthwaite, A., Grant, I., Ko, D., Sutton, B., & Hinde, R. (2006). In RIRaD Corporation (Ed.), *Ecology and biology of nuisance algae in rice fields.* Canberra: Rural Industries Research and Development Corporation.

Hartmann, N., Von der Kammer, F., Hofmann, T., Baalousha, M., Ottofuelling, S., & Baun, A. (2010). Algal testing of titanium dioxide nanoparticles—Testing considerations, inhibitory effects and modification of cadmium bioavailability. *Toxicology, 269,* 190–197.

Huang, S., Wang, L., Liu, L., Hou, Y., & Li, L. (2015). Nanotechnology in agriculture, livestock, and aquaculture in China. A review. *Agronomy for Sustainable Development, 35,* 369–400.

Kim, J.-D. (2006). Screening of cyanobacteria (blue-green algae) from rice paddy soil for antifungal activity against plant pathogenic fungi. *Mycobiology, 34,* 138–142. https://doi.org/10.4489/MYCO.2006.34.3.138.

Lee, W.-M., & An, Y.-J. (2013). Effects of zinc oxide and titanium dioxide nanoparticles on green algae under visible, UVA, and UVB irradiations: No evidence of enhanced algal toxicity under UV pre-irradiation. *Chemosphere, 91,* 536–544.

Liu, R., & Lal, R. (2015). Potentials of engineered nanoparticles as fertilizers for increasing agronomic productions. *Science of the Total Environment, 514,* 131–139.

Lyu, S., Wei, X., Chen, J., Wang, C., Wang, X., & Pan, D. (2017). Titanium as a beneficial element for crop production. *Frontiers in Plant Science, 8,* 597.

Mandal, B., Vlek, P., & Mandal, L. (1999). Beneficial effects of blue-green algae and Azolla, excluding supplying nitrogen, on wetland rice fields: A review. *Biology and Fertility of Soils, 28,* 329–342.

Matranga, V., & Corsi, I. (2012). Toxic effects of engineered nanoparticles in the marine environment: Model organisms and molecular approaches. *Marine Environmental Research, 76,* 32–40.

Maurer-Jones, M. A., Gunsolus, I. L., Murphy, C. J., & Haynes, C. L. (2013). Toxicity of engineered nanoparticles in the environment. *Analytical Chemistry, 85,* 3036–3049.

Nikoo Jamal, N. (2018) Interactions of engineered nanoparticles with economically important food crops. Future Industry Institute (FII), Division of Information Technology, Engineering, and the Environment. University of South Australia, Adelaide, Australia.

Owolade, O., Ogunleti, D., & Adenekan, M. (2008). Titanium dioxide affects disease development and yield of edible cowpea. *EJEAF Chemistry, 7,* 2942–2947.

Pelletier, D. A., Suresh, A. K., Holton, G. A., McKeown, C. K., Wang, W., Gu, B., Mortensen, N. P., Allison, D. P., Joy, D. C., & Allison, M. R. (2010). Effects of engineered cerium oxide nanoparticles on bacterial growth and viability. *Applied and Environmental Microbiology, 76,* 7981–7989.

Peng, S., Buresh, R. J., Huang, J., Yang, J., Zou, Y., Zhong, X., Wang, G., & Zhang, F. (2006). Strategies for overcoming low agronomic nitrogen use efficiency in irrigated rice systems in China. *Field Crops Research, 96,* 37–47.

Rico, C. M., Barrios, A. C., Tan, W., Rubenecia, R., Lee, S. C., Varela-Ramirez, A., Peralta-Videa, J. R., & Gardea-Torresdey, J. L. (2015). Physiological and biochemical response of soil-grown barley (Hordeum vulgare L.) to cerium oxide nanoparticles. *Environmental Science and Pollution Research, 22,* 10551–10558.

Roger, P.-A., Kulasooriya, S. (1980) Blue-green algae and rice. International Rice Research. Institute.

Shalaby, T. A., Bayoumi, Y., Abdalla, N., Taha, H., Alshaal, T., Shehata, S., Amer, M., Domokos-Szabolcsy, É., & El-Ramady, H. (2016). *Nanoparticles, soils, plants and sustainable agriculture. Nanoscience in food and agriculture 1.* Springer.

Tripathi, R., Dwivedi, S., Shukla, M., Mishra, S., Srivastava, S., Singh, R., Rai, U., & Gupta, D. (2008). Role of blue green algae biofertilizer in ameliorating the nitrogen demand and fly-ash stress to the growth and yield of rice (Oryza sativa L.) plants. *Chemosphere, 70,* 1919–1929.

Wang, P., Lombi, E., Zhao, F.-J., & Kopittke, P. M. (2016). Nanotechnology: A new opportunity in plant sciences. *Trends in Plant Science, 21,* 699–712. https://doi.org/10.1016/j.tplants.2016.04.005.

Xia, T., Kovochich, M., Liong, M., Mädler, L., Gilbert, B., Shi, H., Yeh, J. I., Zink, J. I., & Nel, A. E. (2008). Comparison of the mechanism of toxicity of zinc oxide and cerium oxide nanoparticles based on dissolution and oxidative stress properties. *ACS Nano, 2,* 2121–2134.

Xu, C., Peng, C., Sun, L., Zhang, S., Huang, H., Chen, Y., & Shi, J. (2015). Distinctive effects of TiO2 and CuO nanoparticles on soil microbes and their community structures in flooded paddy soil. *Soil Biology and Biochemistry, 86,* 24–33.

Zheng, L., Hong, F., Lu, S., & Liu, C. (2005). Effect of nano-TiO2 on strength of naturally aged seeds and growth of spinach. *Biological Trace Element Research, 104,* 83–91.

Zhu, X., Castleberry, S. R., Nanny, M. A., & Butler, E. C. (2005). Effects of pH and catalyst concentration on photocatalytic oxidation of aqueous ammonia and nitrite in titanium dioxide suspensions. *Environmental Science & Technology, 39,* 3784–3791.

Chapter 8
Effects of Engineered Nanoparticles at Various Growth Stages of Crop Plants

Swati Rawat, Jesus Cantu, Suzanne A. Apodaca, Yi Wang, Chaoyi Deng, Martha L. Lopez-Moreno, Jose R. Peralta-Videa, and Jorge L. Gardea-Torresdey

Contents

S. Rawat · S. A. Apodaca · C. Deng
Environmental Science and Engineering PhD Program, The University of Texas at El Paso, El Paso, TX, USA

University of California Centre for Environmental Implications of Nanotechnology, The University of Texas at El Paso, El Paso, TX, USA

J. Cantu · Y. Wang
University of California Centre for Environmental Implications of Nanotechnology, The University of Texas at El Paso, El Paso, TX, USA

Chemistry and Biochemistry Department, The University of Texas at El Paso, El Paso, TX, USA

M. L. Lopez-Moreno
Department of Chemistry, University of Puerto Rico at Mayaguez, Mayaguez, PR, Puerto Rico

J. R. Peralta-Videa · J. L. Gardea-Torresdey (✉)
Environmental Science and Engineering PhD Program, The University of Texas at El Paso, El Paso, TX, USA

University of California Centre for Environmental Implications of Nanotechnology, The University of Texas at El Paso, El Paso, TX, USA

Chemistry and Biochemistry Department, The University of Texas at El Paso, El Paso, TX, USA
e-mail: jgardea@utep.edu

© Springer Nature Switzerland AG 2021
N. Sharma, S. Sahi (eds.), *Nanomaterial Biointeractions at the Cellular, Organismal and System Levels*, Nanotechnology in the Life Sciences,
https://doi.org/10.1007/978-3-030-65792-5_8

1   Introduction

The interaction between ENPs and plants can vary greatly with nanoparticle (NP) type, plant species, and growth stage. As a result, it is difficult to draw concrete conclusions about their effects (Rastogi et al. 2017). To this purpose, factors that are independent from NP material such as properties (size, shape, concentration, stability, coating, zeta potential, and surface characteristics), exposure (concentration, method, and time), and plant type can be isolated, and inferences can be made (Rawat et al. 2017). Due to their properties, engineered nanoparticles can act as promotors and inhibitors in agricultural plants (Gardea-Torresdey et al. 2014; Reddy et al. 2016). Therefore, investigations of engineered nanoparticles (ENPs) in agriculture aim towards determination of possible effects of ENPs on crops and to improve harvest yield by reduction of micro- and macronutrient deficiency along with disease suppression (Singh et al. 2018; Adisa et al. 2018, 2020). A wide array of ENPs have been studied for their potential effects on different terrestrial plants such as Cu, CuO, ZnO, MnOx, FeOx, Fe_3O_4, TiO_2, CeO_2, Al_2O_3, ZrO_2, SiO_2, Ag nanoparticles (NPs), and carbon-based ENPs like single-walled and multiwalled carbon nanotubes, fullerol, graphene, to name a few (Parveen and Rao 2015; Xiang et al. 2015; Yang et al. 2015; Liu et al. 2016; Karunakaran et al. 2016; Du et al. 2017a; Rawat et al. 2018a, 2019; Deng et al. 2020; Wang et al. 2020). Although there has been extensive research of ENPs in plants, there is a gap that has not been bridged, because the toxicity of ENPs cannot be generalized for all plant species. It is in fact species and ENP dependent (Zuverza-Mena et al. 2017).

By means of this chapter, the authors attempt to probe into various studies of the effects of ENPs on plants at different growth stages, namely germination, seedling, half life cycle, full life cycle studies, and finally transgenerational effects (Scheme 8.1).

2   Effects of ENPs on Germination

Germination studies are vital to assess the overall nano-particle and plant interaction. They are indicative of the beneficial or detrimental effects the nano-particles could have at the rudimentary stage of seed germination. These effects could be prolonged and have an influence on the plant growth and nutrition. It is equally possible that the plant's internal defense mechanism counteracts the detrimental effects and outgrows them. Two plant examples of corn and rice have been covered here.

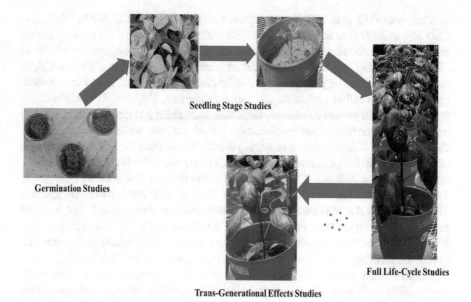

Scheme 8.1 A representation of plant growth stages and nanomaterial interaction studies

2.1   Germination of Corn

Corn is a cereal crop ranked third in world production, after rice and wheat (Food and Agricultural Organization of the Unites Nations 2020). From the total production worldwide in 2018–19, United States was the leading producer with 366.3 million metric tons followed by China with 257.3 million metric tons (Shahbandeh 2020). It is used as a food source for people and livestock, and therefore has an important role in food security. With the extensive use of nanoparticles in commercial products, determination of fate and transport of ENPs has become a concern. Therefore, germination of corn has been extensively studied in the presence of different ENPs and different media to determine the toxicity and potential benefits. For instance, the effects of ZnO on the germination of corn (*Zea mays* L.) depend on the dose of the ENPs. Germination was studied in a petri dish format, with the seeds being exposed to an aqueous suspension or solution on the germination paper. As a result, concentrations below 1000 ppm have shown to have no effect whereas a dose of 1500 ppm increases the germination from approximately 40% to 80% (Subbaiah et al. 2016). On the other hand, according to Yang et al. (2015), 2000 ppm ZnO had no effect on germination. This test was done in a petri dish format again. The presence of Ag NPs ranging from 50 to 2500 ppm had no effect on the germination, except for 500 ppm which decreased it from 96% to 86% (Almutairi and Alharbi 2015). Li et al. (2016) showed that an increase in concentration of γ-Fe_2O_3 from 0 to 50 ppm had little to no effect on the germination rate of corn seeds while 100 ppm γ-Fe_2O_3 decreased the rate by 17%. A similar trend was observed with the germination energy and germination index for the same study (Li et al. 2016).

According to Yang et al. (2015), the presence of TiO_2, SiO_2, CeO_2, Fe_3O_4, Al_2O_3, and CuO NPs at 2000 ppm in direct contact with maize (petri dish study) had no effect on the germination rate of the seeds. Karunakaran et al. (2016) investigated the effects of ZrO_2, SiO_2, Al_2O_3, and TiO_2 under different media. It was found that SiO_2 and ZrO_2 either had little effect or promoted the germination of maize in all three media tested: petri dish, cotton ball as a substrate, and soil. Furthermore, Al_2O_3 and TiO_2 inhibited the germination in all growth conditions, but it was determined that direct contact with the nanoparticles (petri dish study) inhibited the growth the most while the soil media was the least severe. Karunakaran et al. (2016) further investigated the metal oxide uptake of the seeds from the petri dish method via XRF and showed a high uptake for μSiO_2 and $nSiO_2$ with 4.79% and 2.36% compared to the control. The uptake of the Al_2O_3, ZrO_2, and TiO_2 for both microparticles (MPs) and NPs were as follows: $TiO_2 > Al_2O_3 > ZrO_2$. Pariona et al. (2017) investigated the effects of 0, 1, 2, 4, 6 g/L of hematite and ferrihydrite nanoparticles on the germination of maize seeds. In the study, it was determined that the presence of hematite had no negative effects on the germination of the seedlings while ferrihydrite increased the germination slightly.

Carbon-based nanoparticles have attracted a lot of attention due to their wide applicability such as: medicine, electronics, optics, energy storage, and agriculture (O'Connell 2006; De Volder et al. 2013). The use of carbon nanoparticles will eventually get into the environment and interact with living organisms such as plants. Therefore, carbon-based nanoparticles have also been extensively investigated for their effects on corn in the germination stages and throughout the entire life cycle (Lahiani et al. 2013; Lahiani et al. 2015). Lahiani et al. (2013) investigated the effects of carbon nanotubes on seeds from barley, corn, and soybean in contact with nanotubes via two different methods, contact on agar media and coated via airspray. From the study, it was determined that the germination of corn significantly increased when in contact with the multi-walled carbon nanotubes (MWCNTs) at 50, 100, and 200 ppm (Fig. 8.1, Panel 1). They also demonstrated the aggregation of CNTs inside the seed tissue by means of Raman spectroscopy (Fig. 8.1, Panel 2). In another study, Lahiani et al. (2015) investigated the effects of single-walled carbon nano-horns (SWCNHs) (0, 25, 50, and 100 ppm) on the germination of barley, corn hybrid N79Z 300 GT, rice, soybean, switch grass, and tomato. A similar trend was observed, in which the germination of corn increased in the presence of the carbon nano-horns. Furthermore, at higher concentrations, the germination occurred within an earlier time period than the control (Lahiani et al. 2015).

2.2   Germination of Rice

Rice (*Oryza sativa* L.) is a vital grain crop and is normally grown and over 715 million tons is produced annually (Muthayya et al. 2014). Due to its importance in food security, studies have focused on investigating possible benefits and toxicity of engineered nanoparticles on the growth of rice plants. Materials that have been investigated include various carbon-based and metal-based nanoparticles (Thuesombat et al. 2014; Wang

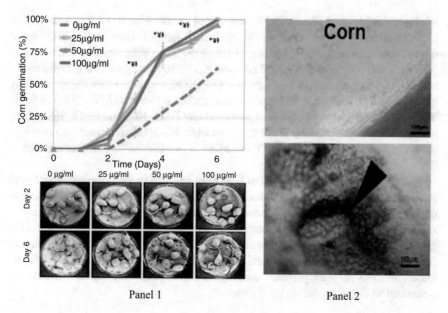

Panel 1   Panel 2

Fig. 8.1 Panel 1. Quantitative and qualitative representation of corn seed germination in MWCNT treatment by airspray technique with relation to time. Results are shown as an average measurement of 24 seeds per treatment. Aqueous solutions of MWCNTs at 50, 100, and 200 µg/mL concentration were used for airspray. *: $p < 0.05$, CNTs (25 µg/mL) compared to control. ¥: $p < 0.05$, MWCNT (50 µg/mL) compared to control. ‡: $p < 0.05$, MWCNT (100 µg/mL) compared to control. Phenotypic comparison of the corn seeds with control or CNT treatment on day 2 and day 6 of the airspray treatment. Panel 2. Optical Raman spectroscopic images of corn seeds from control vs. the MWCNT treatment to visualize the aggregation of CNT (black arrow) inside the seed. (Reprinted (adapted) with permission from Lahiani, M. H., Dervishi, E., Chen, J., Nima, Z., Gaume, A., Biris, A. S., Khodakovskaya, M. V., 2013. Impact of carbon nanotube exposure to seeds of valuable crops. *ACS Applied Materials Interfaces, 5*(16), 7965–7973. Copyrights 2013 American Chemical Society)

et al. 2016a; Gupta et al. 2018). For instance, Liu et al. (2018) investigated the interaction of nano-CuO with rice seeds to alleviate the antagonistic effects of arsenic ions. In the study, it was found that As alone decreases the germination of the rice seeds, but in the presence of CuO, the germination increased with an increase in CuO concentration with the exception of the As + CuO 1.0 ppm treatment which was similar to the treatment containing As alone (Liu et al. 2018). The effects of carbon nanotubes (CNTs) on the germination of rice were studied by Nair et al. (2010) and Jiang et al. (2014), and it was determined that at low concentrations (0–100 ppm) the seed germination was promoted. Additionally, Adhikari et al. (2013) determined that 0–100 ppm of SiO_2 and 0–600 ppm Mo NPs had no adverse effect on the germination of the rice seeds. Shaw and Hossain (2013) investigated the effects of nano-CuO at 0.5, 1.0, and 1.50 mM (39.77, 79.55, and 119.32 ppm) on the germination of rice seeds using cotton ball as the media. From the study, it was determined that the germination decreased as the concentration of the nano-CuO was increased. For instance, the controls exhibited 91.6%

while the samples had 78.6%, 75.6%, and 71.6% germination at 0.5, 1.0, and 1.5 mM CuO, respectively (Shaw and Hossain 2013). In a study, Lahiani et al. (2015) determined that the germination of rice seeds increased with an increase in the concentration (0, 25, 50, and 100 ppm) of SWCNTs. Hao et al. (2016) investigated the germination of rice seeds in the presence of Fe_2O_3 nano-cubes (NCs), short nanorods (SRs), and long nanorods (LRs), TiO_2 NPs, and multi-walled carbon nanotubes (MWCNTs) at 5, 10, 30, 50, 100, and 150 mg/L. It was determined that Fe_2O_3 NCs and Fe_2O_3 SRs had no adverse effect on the germination of the seeds while Fe_2O_3 LRs inhibited their development. Similarly, the presence of TiO_2 and MWCNTs inhibited the germination ratio of the rice seeds (Hao et al. 2016).

To summarize, the germination of both corn and rice in the presence of different engineered nanoparticles proved that the germination is dosage and material dependent. Higher concentrations of the metal-based nanoparticles tend to inhibit the germination of the seeds while low concentrations either had no effect or promoted the growth. Furthermore, the media used plays an important role in the germination process. For instance, the use of soil and cotton as a substrate helps alleviate the toxic effects of the nanoparticles on the germ, thus producing higher germination compared to direct contact in petri dishes.

3 Effects of ENPs on Seedling Growth

At the advent of nanoparticle and plant interaction studies, the researchers first started out by conducting germination and seedling stage studies. As the field evolved with time, the researchers gradually moved towards long-term exposure studies or full life cycle studies. The initial kinds of analyses still hold their value since they are indicative of early stage endpoints in the toxicity studies. The following section elaborates on the response of seedlings to metal-based nanoparticles, predominantly silver nanoparticles.

3.1 Response of Seedlings to Ag NPs

The global market for Ag NPs is estimated to reach 400–800 tons by 2025, with main uses as a conductive material (electronic devices and paints) or antimicrobial agent (textiles, packaging, and commercial products) (Pulit-Prociak and Banach 2016). The unique antimicrobial activity of Ag NPs also endorses its implementation into the food and agricultural industry against pathogens. Given that Ag NPs can enter the environment through direct or indirect means, it is essential to understand their influence on plants since it can impact industry and/or public health (Wilson 2018). Initial growth of the germinated seed represents an early developmental stage in the plant's life cycle. Once plants begin to photosynthesize, they are no longer dependent on the seed's energy reserves and must rely on external sustenance (Gascon et al. 2015). Seedlings typically have high mortality rates due to their small size, which leaves them susceptible to resource competition and environmental variability (Hanley et al. 2004).

Beneficial results following exposure to Ag NPs have been conveyed in a variety of terrestrial plants including, but not limited to, radish, fenugreek, wheat, rice, and Indian mustard (Pandey et al. 2014; Hojjat and Hojjat 2015; Bahri et al. 2016; Sabir et al. 2018; Gupta et al. 2018). Low concentrations can stimulate plant growth and development. Maximum root and shoot growth were observed in radish seedlings pretreated with 10 mgL^{-1} of Ag NPs grown under in vitro culture conditions (Bahri et al. 2016). In a similar experimental setup, 10 µg/mL of Ag NPs enhanced root length, fresh weight, and dry weight in fenugreek up to 46.6% from controls (Hojjat and Hojjat 2015). Similarly, concentrations ranging from 25 to 100 mgL^{-1} improved seedling vigor index (12.2–116.3%), root fresh weight (19–37.7%), and root biomass (21.7–79.5%) in wheat (Sabir et al. 2018). The interaction between Ag NPs and plants is facilitated by the relationship between plant physiology and biochemistry. Low levels of reactive oxygen species (ROS) in combination with raised antioxidative enzyme activities and gene expression levels could explain the phyto-stimulatory effect observed by 10 to 40 mgL^{-1} of Ag NPs in rice seedlings (Gupta et al. 2018). Treatment with Ag NPs could have induced a more efficient redox balance, supplementing growth parameters. This notion is substantiated by Pandey et al. (2014), where a slight elongation in root and shoot length of Indian mustard corresponded with elevated chlorophyll and protein content up to 1000 mgL^{-1} in leaves and 100 mgL^{-1} in roots, respectively.

Several studies have found Ag NPs to be toxic towards plants, typically under high-dose exposure (Parveen and Rao 2015; Kim et al. 2018; Al-Huqail et al. 2018; Qian et al. 2013; Zuverza-Mena et al. 2016). Growth of wheat and mung bean seedlings was negatively affected by 10 to 80 mg/L of Ag NPs in a dose-dependent manner (Kim et al. 2018). Under 30 mg/L, changes were minimal and nonsignificant; from 60 to 80 mg/L, root and shoot length were inhibited. According to Al-Huqail et al. (2018), white lupin was also found to be less tolerant to high concentrations of Ag NPs. Seedling fresh weight, length, and dry matter were reduced 18.2–62.4% at 300–500 mg/kg, in addition to vigor indices (23.6–48.9%). These results could be due to the application method (transplantation), where in both instances the roots came into direct contact with the exposure medium. The small size of NPs allows them to be adsorbed onto roots, crossing through the epidermis and root cortex via the apoplastic pathway and the endodermis by protoplasts, where they reach the central portion of the roots and are translocated to aerial parts of the plant (Parveen and Rao 2015). This process has been visually explained by means of the diagram as shown in Fig. 8.2 (Pullagurala et al. 2018). Transmission electron microscopy (TEM) and metal content analysis indicated that Ag NPs accumulated in the leaves of thale cress and were able to disrupt thylakoid membrane structure, oxidant-antioxidant system balance, and homeostasis of water and other small molecules (Qian et al. 2013). Total chlorophyll content was suppressed by 32–36.5% and gene expression of antioxidant enzymes and aquaporins (water channels) was upregulated. Remarkably, these effects were exclusive to Ag NP treatments, even when compared to ionic Ag⁺ treatments. Moreover, Zuverza-Mena et al. (2016) demonstrated that Ag NPs can alter nutrient content and macromolecule conformation. Radish seedlings exposed to 500 mg/L of Ag NPs had significantly less Ca, Mg, B, Cu, Mn, and Zn (10–52.5%) and infrared (IR) band of lipids, proteins, and structural plant cell components (lignin, pectin, and cellulose) were modified. In addition, Wu et al. (2020) conducted a study on lettuce seedlings, to compare the toxicity due to Ag NPs and ionic Ag under a 15-day-long root and foliar exposure. Results indicated that root exposure caused significantly higher toxicity than foliar exposure; nano form was more actively uptaken and translocated compared the ionic form. Also, toxicity in the ionic treatments was due to oxidative stress.

Mixed observations have also been described throughout literature (Almutairi and Alharbi 2015; Wang et al. 2018). Exposure of Ag NPs to corn, watermelon, and zucchini had significant positive and negative impacts on plant growth (Almutairi and Alharbi 2015). While root length in corn seedlings was decreased (17.2–36.5%) by 0.05–2.5 mg/L of Ag NPs, fresh weight was increased (3.3–34.6%) by 0.5–2 mg/L. Although root length and fresh weight in watermelon was stimulated

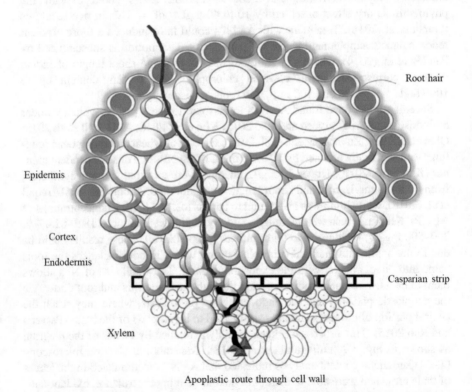

Fig. 8.2 Apoplastic route of the uptake of Ag nanoparticles from the soil solution to the plant tissue, through the epidermis, cortex, endodermis, and translocated to the aerial parts of the plant (Reprinted (adapted) with permission from Pullagurala et al. 2018. Copyrights @ 2018 Elsevier Ltd.)

45.1–161.1% and 12.9–25.1% at 1 to 2.5 mgL^{-1} of Ag NPs, the same concentrations also reduced dry weight 4.3–27.1% from controls. Improvements in zucchini root length, fresh weight, and dry weight ranged from 4.3–27.1%, 48.3–95.8%, and 62.1–85.8%, respectively, within the Ag NP concentration gradient of 0.05 to 2 mg/L. Feasibility of Ag NPs in agricultural applications (crop improvement and food production) is dependent on their biocompatibility with different plant species. Even at the nanoscale, size is a critical physicochemical determinant in plant response. Large particle sized Ag NPs (70 nm) exerted stronger toxicity than small particle sized Ag NPs (30 nm) on leaf length and width in lettuce (Wang et al. 2018). This could be due to the time scale of the study (10 days). Larger NPs generally have a lower tendency to aggregate and stronger stability; thus, they can accumulate within soil and hence plants more readily than smaller NPs (Rawat et al. 2018b).

3.2   Response of Seedlings to Other ENPs

Lin and Xing (2007) tested the effects of nano-sized Zn, ZnO, multiwalled carbon nanotubes (MWCNT), aluminum, and alumina on ryegrass, radish, grape, cucumber, lettuce, and corn for a seedling stage study. They saw clear inhibitions of root development in all the seedlings after 5 days of exposure, especially at 2000 mg/L nZn and nZnO. The seedling growth was hampered by nZn in ryegrass and by nZnO in corn at 2000 mgL^{-1} (Lin and Xing 2007). The broad bean (*Vicia faba* L.) seedling were grown hydroponically under carboxylated multiwalled carbon nanotube (MWCNT-COOH) exposure (2.5, 5, and 10 mg/L) along with heavy metals like lead (Pb, 20 μM) and cadmium (Cd, 5 μM). During the 20-day exposure period, MWCNT-COOH were observed to cause oxidative stress in leaves. The combined exposure of MWCNT-COOH and heavy metals increased biochemical and subcellular damage to the leaves compared to heavy metal alone (Wang et al. 2014). In another study conducted to assess the physiological responses of *Ulmus elongata* to nano-anatase TiO$_2$ by foliar application, it was observed that the net photosynthetic rate was deteriorated by 0.1%, 0.2%, and 0.4% (w/v) of the foliar spray compared to the control (Gao et al. 2013). Similarly, cabbage, tomato, red spinach, and lettuce were grown under graphene exposure at 500–2000 mg/L. After 20 days of exposure, significant adverse effects on plant morphology, growth, and biomass were observed on all the seedlings except for lettuce. These effects included reduction in leaf number and sizes, dose-dependent rise in oxidative stress, cell death, and visible necrotic lesions in the plant tissue (Begum et al. 2011). Additionally, Nair et al. (2014) exposed mung bean (*Vigna radiata* L.) to various concentrations of CuO NPs in a hydroponic environment for 21 days. Among various other effects, they observed hampering of seedling root growth and biomass at all the concentrations. They observed a disruption of shoot growth and biomass at the two highest concentrations, 200 and 500 mg/L. The positive effects of metal-based nanoparticles have also been investigated in some plant studies. In one such study, the nano-sized silicon was found to counter the biotic and abiotic stress caused by UV-B including photosynthetic and enzyme activities in wheat (*Triticum aestivum*) seedlings to a much higher degree as compared to conventional silicon (Tripathi et al. 2017). Pariona et al. (2017) conducted germination as well as seedling stage study to elucidate the uptake mechanisms and effects of iron-based NPs on *Zea mays*. To observe the uptake of NPs and their aggregates on the maize seedlings, they conducted confocal laser-scanning microscopy on stem samples from the treatments and the controls (Fig. 8.3). The minute red-colored signals represent the hematite and ferrihydrite aggregates uptaken into the stem of the seedlings. According to the authors' analysis of the microscopy images, the NPs were uptaken both by the apoplastic route to the epidermis and the symplastic route to the vascular system, i.e., the xylem and phloem of the plant (Pariona et al. 2017).

To summarize, differences in the effects of NPs to terrestrial plants depend on their properties, plant species and ages, and exposure media and concentrations. Though both advantageous and adverse outcomes were reported, it can be concluded that the most significant factors are concentration and plant type. Understanding the interactions of NPs with plants is pivotal for scalability and applicability in agriculture, as well as determining environmental consequences from industrial and commercial use.

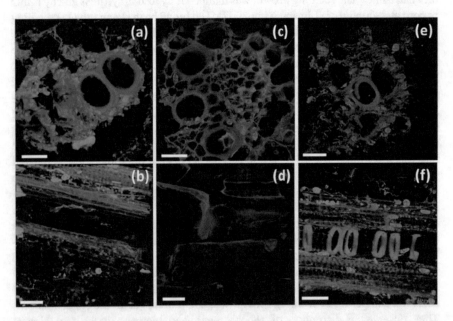

Fig. 8.3 Confocal microscopy images of transversal (top) and longitudinal (bottom) sections of maize seedlings that were grown in control (**a**, **b**), hematite NP (**c**, **d**), and ferrihydrite NP (**e**, **f**) treatments for 12 days past the germination stage. (Reprinted with permission from Pariona et al. 2017. Copyrights @ 2017 Elsevier Ltd.)

4   Effects of ENPs on Plants: Full Life Cycle Growth

The scientists took to more long-term exposure studies in the last decade, in the course of studying the nano-plant interactions. These studies are more useful to the field of application of ENPs for the use of agricultural amendments like pesticides, fertilizers, herbicides, or site-specific delivery systems for the same. Some of these long-term exposure studies and the effects of nanoparticles on physiology, agronomy, gene expression, and biochemistry have been summarized for cucumber, tomato, beans and peas, and food crops. The plant matrix, type of plant, physiochemical properties of the nanoparticles, and duration of exposure are important factors that influence the effects.

4.1   Effects of ENPs on Cucumber

The effects of Ag, CeO_2, CuO, ZnO, and TiO_2 ENPs on the growth of cucumber to full maturity has been extensively investigated. For instance, foliar exposure of Ag ENPs at 500, 1000, 1500, 2000, 2500, and 3000 ppm caused a significant increase in the amount of fruit production as well as the weight of each fruit per plant with increasing concentrations of the ENPs (Shams et al. 2013). The presence of 3000 ppm Ag increased the number of fruit production, fruit diameter, and fruit weight by 2.00 cm, 1.3 cm, and 20 g, respectively, when compared to the control. Additionally, a similar trend was observed in which the plant height of the cucumbers treated with 3000 ppm Ag increased by 19 cm. On the other hand, when exposed with ZnO or TiO_2 ENPs via soil application; CeO_2 or CuO via a foliar application, no significant alteration of the plant biomass or yield was found (Zhao et al. 2013; Servin et al. 2013; Zhao et al. 2014; Hong et al. 2016). However, soil exposure of CeO_2 ENPs at 800 mg/kg reduced the fruit yield by 31.6% when compared to the control ($p \leq 0.07$) (Zhao et al. 2013, 2014). Treatment of CuO or CeO_2 ENPs via foliar exposure caused a reduction in the firmness of the cucumber fruit (Hong et al. 2016).

The elemental content within the fruit was examined, and it was found that the Mo content was reduced by up to 51% with the following treatments: soil exposure of 400 and 800 mg/kg for both CeO_2 and ZnO and foliar exposure of 50, 100, and 200 mgL^{-1} for both CeO_2 and CuO ENPs (Zhao et al. 2013, 2014; Hong et al. 2016). Additionally, the P content in the fruit increased with foliar exposure of CeO_2 and CuO as well with soil treatment of TiO_2 ENPs (Servin et al. 2013; Hong et al. 2016).

It has been demonstrated that Ag, CeO_2, and ZnO ENPs after treatment via either foliar or soil exposure can be uptaken and translocated to the other parts of the cucumber plant. Shams et al. (2013) detected the presence of Ag ENPs in cucumber fruit, fruit skin, and roots after foliar application of 500, 1000, 1500, 2000, 2500, and 3000 ppm. Similarly, it was determined that CeO_2 ENPs can be translocated from the root to leaf to fruit and vice versa when treated via either foliar or soil application (Zhao et al. 2014; Hong et al. 2016). A similar trend was observed when cucumber plants were exposed to ZnO ENPs via soil application (Zhao et al. 2013). On the other hand, CuO ENPs accumulated in the leaves of the plant without any translocation through foliar exposure (Hong et al. 2016).

4.2 Effects of ENPs on Growth of Tomato

The effects of CeO_2, ZnO, and TiO_2 ENPs on the growth of tomato plants have shown to enhance fruit yield. For instance, both foliar and soil exposure of CeO_2 ENPs have resulted in an increase in tomato fruit yield (Barrios et al. 2016; Adisa et al. 2018). In another study, exposure of the tomato plants to CeO_2 ENPs via soil application caused a lower number of fruits but increased their size and weight, leading to an enhancement in yield (Wang et al. 2012). The presence of citric acid coating on CeO_2 reduced the tomato fruit dry weight when treated via soil application (Barrios et al. 2016). Furthermore, soil and foliar exposure of ZnO and TiO_2 increased the number of total fruits and size compared to the controls (Raliya et al. 2015).

When exposed to ZnO and TiO_2, the lycopene content in the fruit was enhanced when treated via both foliar and soil applications (Raliya et al. 2015). Soil exposure of CeO_2 ENPs increased the reducing sugars but decreased the boron content in the tomato fruits (Barrios et al. 2016). On the other hand, in the same study, CeO_2-CA ENPs reduced the total sugars, reducing sugars, and starch but increased the boron content (Barrios et al. 2016). In various studies, CeO_2 altered tomato plant nutrient element contents by increasing B, Ca, K, Mg, Na, P, S, and Fe in root; Ca in stem; while reducing B, Fe, Mn, Al, and Ca in fruit (Antisari et al. 2015; Barrios et al. 2016; Adisa et al. 2018).

After soil exposure, Ag and CeO_2 ENPs were translocated to the tomato fruit (Wang et al. 2012; Antisari et al. 2015). After exposure of ZnO and TiO_2 via either soil or foliar application, it was determined that the ENPs can be translocated from the roots to the leaves and vice versa, respectively (Raliya et al. 2015). Other ENPs such as Co and Ni accumulated in the roots and shoots, while SnO_2 mainly accumulated in the roots after soil exposure (Antisari et al. 2015).

4.3 Effect of ENPs on Bean and Pea

The effects and interactions of ENPs on bean and pea plants have been examined with CeO_2, CuO, and ZnO ENPs as described. Soil exposure of CeO_2 ENPs at 62.5–500 mg/kg increased the kidney bean plant yield and seed moisture (Peralta-Videa et al. 2014; Priester et al. 2017). On the other hand, soil treatment of CuO ENPs at 50 and 100 mg/kg reduced green pea pod biomass (Ochoa et al. 2017).

The presence of ZnO and CeO$_2$ had different effects on the fruit elemental content of the soybeans. For instance, ZnO ENPs at 100, 500, and 1000 mg/kg increased the Zn, Mn, and Cu content in the pods, while CeO$_2$ ENPs at the same concentrations decreased the Ca but increased both P and Cu content (Peralta-Videa et al. 2014; Priester et al. 2017). In a different study via soil exposure, kidney bean pods exhibited an increase in S and P content while the accumulation of Cu, Fe, Ni, Mo, and Na were negatively affected (Majumdar et al. 2015). However, opposite results were found when green pea plant was treated with CuO ENPs at 50 and 100 mg/kg in which the Fe and Ni concentrations were increased (Ochoa et al. 2017).

Both bean and pea plants have been reported to uptake ENPs such as CeO$_2$, CuO, and ZnO via soil exposure (Peralta-Videa et al. 2014; Majumdar et al. 2015; Ochoa et al. 2017; Priester et al. 2017; Medina-Velo et al. 2017a; b). Among these ENPs, CeO$_2$ and ZnO were translocated upward from the root to the pods (Peralta-Videa et al. 2014; Majumdar et al. 2015; Priester et al. 2017; Medina-Velo et al. 2017a; b). However, no evidence has been found for the translocation of CuO ENPs (Ochoa et al. 2017).

4.4 Effect of ENPs on Food Crops

Exposure of TiO$_2$, CeO$_2$, ZnO, and Mn$_2$O$_3$ ENPs on food crops have been investigated to determine the possible effects and interactions. Treatment of barley with CeO$_2$ ENPs at 500 mg/kg inhibited the formation of grains while exposure to TiO$_2$ ENPs at 500 and 1000 mg/kg enhanced the grain yield (Du et al. 2011; Pošćić et al. 2016). On the other hand, exposure of CeO$_2$ ENPs to wheat increased the shoot biomass, grain yield, number of spikelets per spike, and amount of grains per spike (Rico et al. 2014). The effects of ZnO with corn varied depending on the route of exposure (foliar or soil). Foliar application increased the corn yield while soil exposure decreased the yield (Zhao et al. 2015; Subbaiah et al. 2016).

The food crops mentioned above could uptake ZnO, TiO$_2$, and CeO$_2$ ENPs via soil exposure while Mn$_2$O$_3$ and ZnO via foliar applications (Du et al. 2011; Rico et al. 2013, 2014; Du et al. 2015; Zhao et al. 2015; Subbaiah et al. 2016; Du et al. 2017b; Zahra et al. 2017; Dimkpa et al. 2018). Furthermore, CeO$_2$ ENPs were translocated from the roots to stem in corn and rice from soil exposure (Rico et al. 2013; Zhao et al. 2015). The same trend was observed in which ZnO and TiO$_2$ were also able to be translocated from the roots to leaves in corn and rice plants, respectively (Zhao et al. 2015; Du et al. 2017b). In addition, translocation of Mn$_2$O$_3$ ENPs from leaves to grain occurred in wheat when treated via a foliar application (Dimkpa et al. 2018).

5   Transgenerational Effects of ENPs

A handful of studies have looked at the varied transgenerational effects of nano Cu treatments on plants. Wang et al. (2016b) exposed three different eco-types (Col-0, Bay-0, Ws-2) of *Arabidopsis thaliana* to nCuO (20 and 50 mg/L), bulk CuO (50 mg/L) and ionic Cu (0.15 mg/L). The carry over effects of the Cu treatments onto the second generation of plants were investigated by testing for pollen and seeds viability. Pollen grain from the 50 mg/L nCuO treated plants had the germination inhibited statistically significantly by 10%, 10%, and 18% for Col-0, Bay-0, and Ws-2 ecotypes, respectively. In addition, nCuO at both 20 and 50 mg/L significantly reduced the germination ratios of the harvested seeds. The root elongation in the seedlings was also significantly reduced at all the Cu treatments.

In another such study, the effect of nano-ceria ($nCeO_2$) exposure on the growth parameters of second-generation tomato seedlings was investigated by Wang et al. (2013). Seeds from tomato plants grown under 10 mg/L $nCeO_2$ treatment were retained for the second-generation seedling stage study with and without the 10 mg/L nanoceria treatment. Therein, the authors observed that the second-generation treated seedlings from treated parent gave a significantly lower biomass, lower transpiration, and a slightly higher reactive oxygen species activity. Morphological differences were observed between untreated and treated seedlings in terms of root elongation and density, them being distinctly higher for the treated seedlings. Additionally, the second-generation treated seedlings exhibited higher Ce bioaccumulation than the untreated ones, though not statistically significant.

Similarly, Tan et al. (2018) exposed basil (*Ocimum basilicum*) in pristine, hydrophilically coated, and hydrophobically coated nano-TiO_2 treatments through soil amendments. A second-generation study was undertaken with the seeds from the full life cycle study. Similar exposure pattern was adopted in the second-generation study and yielded various results. Plants that were treated with coated (hydrophobic and hydrophilic) TiO_2 had reduced chlorophyll levels but increased sugar levels as compared to those that were unexposed in both the generations. Sequential exposure to nano-TiO_2 yielded stomatal conductance higher by 214% ($p \leq 0.10$) in basil plants compared to those that were never exposed. Similarly, sequential exposure to coated TiO_2 resulted in negative influence on plant photosynthesis but improved plant growth and proliferation when compared to the pristine treatment.

Geisler-Lee et al. (2014) examined the possible reproductive toxicity, among other things, induced by silver nanoparticles in soil grown *Arabidopsis thaliana* plants. Significant reductions in seed germination rate over three generations were observed. The initial generation was untreated plants. Thereafter, the plants from first-generation seeds were irrigated with 75 or 300 µg/L Ag NP treatments along with the corresponding ionic treatments. The seeds from 300 µg/L treatment gave the lowest germination rate among all the treatments that were significantly low compared to control (seeds from the untreated plants). In the following generation, seeds from 75 and 300 µg/L gave a significantly low germination rate compared to the corresponding control. In the third generation, seeds from 75 µg/L had germination rate as low as 70% of the initial generation. This generational effect testifies the strong toxicity the Ag NPs could induce in plants.

Additionally, Medina-Velo et al. (2018) conducted a generational study with ZnO NPs (coated and uncoated) and their effect on second-generation kidney bean (*Phaseolus vulgaris*) seeds. The soil grown plants were treated with coated and uncoated nano ZnO, bulk ZnO, and $ZnCl_2$ at 0–500 mg/kg in the first generation. Seeds from the first-generation harvest were grown in clean soil without any treatment to observe any residual effect of the first-generation treatments. No effects on seed nutrient quality, Zn accumulation, time to maturity, or sugar, starch, and protein content in the seeds were observed. None of the treatments affected the ascorbate peroxidase, catalase, and superoxide dismutase activities in the second-generation seeds except for 500 mg/Kg $ZnCl_2$ that increased SOD in the seeds by 28%. Overall, a low residual transgenerational effect was observed in kidney beans seeds grown under nano-ZnO treatments.

6 Summary

A range of positive and negative effects of metal-based and carbon-based ENPs have been observed in different plant species. The analysis was organized by the growth stage at which the plant was investigated. The plants' growth characteristics, biochemistry, elemental uptake and bioaccumulation, nutrition characteristics, gene construct could be improved or deteriorated depending on the dosage of the chemical, plant species, growth matrix, and other factors. The current need requires moving towards application of nanotechnology to benefit agriculture by making agricultural processes nano-enabled. It could be the usage as nano-fertilizer, nano-pesticides, nano-fungicide, or nano-enabled target specific delivery systems to curtail wastage in the field. Another important application area is nano-enabled disease suppression to improve crop yields. Our efforts now and in the near future need to be geared towards the abovementioned avenues. Scientific advancement in the areas of nanotechnology and crop science can be judiciously tied together to feed the burgeoning human populations on the earth.

Acknowledgments The authors acknowledge the USDA grant 2016-67021-24985 and the NSF Grant CHE-0840525. Partial funding was also provided by the NSF ERC on Nanotechnology-Enabled Water Treatment (ERC-1449500). This work was also supported by the National Science Foundation and the Environmental Protection Agency under Cooperative Agreement Number DBI-1266377. This work has not been subjected to EPA review, and no official endorsement should be inferred. The Grant 2G12MD007592 from the National Institutes on Minority Health and Health Disparities (NIMHD), a component of the National Institutes of Health (NIH), also supported this work. J. L. Gardea-Torresdey (J-GT) acknowledges the Dudley family for the Endowed Research Professorship. Also, J-GT acknowledges the LERR and STARs Retention Award (2018) of the University of Texas System.

References

Adhikari, T., Kundu, S., & Rao, A. S. (2013). Impact of SiO_2 and mo nano particles on seed germination of rice (*Oryza sativa* L.). *International Journal of Agriculture Food Science and Technology, 4*(8), 809–816.

Adisa, I. O., Rawat, S., Pullagurala, V. L. R., Dimkpa, C. O., Elmer, W. H., White, J. C., Hernandez-Viezcas, J. A., Peralta-Videa, J. R., & Gardea-Torresdey, J. L. (2020). Nutritional status of tomato (*Solanum lycopersicum*) fruit grown in fusarium-infested soil: Impact of cerium oxide nanoparticles. *Journal of Agricultural and Food Chemistry, 68*(7), 1986–1997.

Adisa, I. O., Reddy Pullagurala, V. L., Rawat, S., Hernandez-Viezcas, J. A., Dimkpa, C. O., Elmer, W. H., White, J. C., Peralta-Videa, J. R., & Gardea-Torresdey, J. L. (2018). Role of cerium compounds in fusarium wilt suppression and growth enhancement in tomato (*Solanum lycopersicum*). *Journal of Agricultural and Food Chemistry, 66*(24), 5959–5970.

Al-Huqail, A. A., Hatata, M. M., Al-Huqail, A. A., & Ibrahim, M. M. (2018). Preparation, characterization of silver phyto nanoparticles and their impact on growth potential of *Lupinus termis* L. seedlings. *Saudi Journal of Biological Sciences, 25*(2), 313–319.

Almutairi, Z. M., & Alharbi, A. (2015). Effect of silver nanoparticles on seed germination of crop plants. *International Journal of Nuclear and Quantum Engineering, 9*, 6.

Antisari, L. V., Carbone, S., Gatti, A., Vianello, G., & Nannipieri, P. (2015). Uptake and translocation of metals and nutrients in tomato grown in soil polluted with metal oxide (CeO_2, Fe_3O_4, SnO_2, TiO_2) or metallic (Ag, co, Ni) engineered nanoparticles. *Environmental Science and Pollution Research, 22*(3), 1841–1853.

Bahri, S., Bhatia, S., Moitra, S., Sharma, N., & Bhatt, R. (2016). Influence of silver nanoparticles on seedlings of *Vigna radiata* (L.) R. wilczek. *DU Journal of Undergraduate Research and Innovation, 2*, 142–148.

Barrios, A. C., Rico, C. M., Trujillo-Reyes, J., Medina-Velo, I. A., Peralta-Videa, J. R., & Gardea-Torresdey, J. L. (2016). Effects of uncoated and citric acid coated cerium oxide nanoparticles, bulk cerium oxide, cerium acetate, and citric acid on tomato plants. *Science of the Total Environment, 563*, 956–964.

Begum, P., Ikhtiari, R., & Fugetsu, B. (2011). Graphene phytotoxicity in the seedling stage of cabbage, tomato, red spinach, and lettuce. *Carbon, 49*(12), 3907–3919.

De Volder, M. F., Tawfick, S. H., Baughman, R. H., & Hart, A. J. (2013). Carbon nanotubes: Present and future commercial applications. *Science (New York, N.Y.), 339*(6119), 535–539. https://doi.org/10.1126/science.1222453.

Deng, C., Wang, Y., Cota-Ruiz, K., Reyes, A., Sun, Y., Peralta-Videa, J. R., Hernandez-Viezcas, J. A., Turley, R. S., Niu, G., Li, C., & Gardea-Torresdey, J. L. (2020). Bok choy (*Brassica rapa*) grown in copper oxide nanoparticles-amended soils exhibits toxicity in a phenotype-dependent manner: Translocation, biodistribution and nutritional disturbance. *Journal of Hazardous Materials, 398*, 122978.

Dimkpa, C. O., Singh, U., Adisa, I. O., Bindraban, P. S., Elmer, W. H., Gardea-Torresdey, J. L., & White, J. C. (2018). Effects of manganese nanoparticle exposure on nutrient acquisition in wheat (*Triticum aestivum* L.). *Agronomy, 8*(9), 158.

Du, W., Gardea-Torresdey, J. L., Ji, R., Yin, Y., Zhu, J., Peralta-Videa, J. R., & Guo, H. (2015). Physiological and biochemical changes imposed by CeO_2 nanoparticles on wheat: A life cycle field study. *Environmental Science Technology, 49*(19), 11884–11893.

Du, W., Gardea-Torresdey, J. L., Xie, Y., Yin, Y., Zhu, J., Zhang, X., Ji, R., Gu, K., Peralta-Videa, J. R., & Guo, H. (2017b). Elevated CO_2 levels modify TiO_2 nanoparticle effects on rice and soil microbial communities. *Science of the Total Environment, 578*, 408–416.

Du, W., Sun, Y., Ji, R., Zhu, J., Wu, J., & Guo, H. (2011). TiO_2 and ZnO nanoparticles negatively affect wheat growth and soil enzyme activities in agricultural soil. *Journal of Environmental Monitoring, 13*(4), 822–828.

Du, W., Tan, W., Peralta-Videa, J. R., Gardea-Torresdey, J. L., Ji, R., Yin, Y., & Guo, H. (2017a). Interaction of metal oxide nanoparticles with higher terrestrial plants: Physiological and biochemical aspects. *Plant Physiology and Biochemistry, 110*, 210–225.

Food and Agricultural Organization of the Unites Nations, 2020. Crop Production-Year 2018. http://www.fao.org/faostat/en/#rankings/commodities_by_country

Gao, J., Xu, G., Qian, H., Liu, P., Zhao, P., & Hu, Y. (2013). Effects of nano-TiO_2 on photosynthetic characteristics of *Ulmus elongata* seedlings. *Environmental Pollution, 176*, 63–70.

Gardea-Torresdey, J. L., Rico, C. M., & White, J. C. (2014). Trophic transfer, transformation, and impact of engineered nanomaterials in terrestrial environments. *Environmental Science Technology, 48*(5), 2526–2540.

Gascon, C., Brooks, T. M., Contreras-MacBeath, T., Heard, N., Konstant, W., Lamoreux, J., Launay, F., Maunder, M., Mittermeier, R. A., & Molur, S. (2015). The importance and benefits of species. *Current Biology, 25*(10), R431–R438.

Geisler-Lee, J., Brooks, M., Gerfen, J., Wang, Q., Fotis, C., Sparer, A., Ma, X., Berg, R., & Geisler, M. (2014). Reproductive toxicity and life history study of silver nanoparticle effect, uptake and transport in *Arabidopsis thaliana*. *Nanomaterials, 4*(2), 301–318.

Gupta, S. D., Agarwal, A., & Pradhan, S. (2018). Phytostimulatory effect of silver nanoparticles (AgNPs) on rice seedling growth: An insight from antioxidative enzyme activities and gene expression patterns. *Ecotoxicology and Environmental Safety, 161*, 624–633.

Hanley, M., Fenner, M., Whibley, H., & Darvill, B. (2004). Early plant growth: Identifying the end point of the seedling phase. *New Phytologist, 163*(1), 61–66.

Hao, Y., Zhang, Z., Rui, Y., Ren, J., Hou, T., Wu, S., Rui, M., Jiang, F., Liu, L., 2016. Effect of different nanoparticles on seed germination and seedling growth in rice. Paper presented at the *2nd Annual International Conference on Advanced Material Engineering (AME 2016)*.

Hojjat, S. S., & Hojjat, H. (2015). Effect of nano silver on seed germination and seedling growth in fenugreek seed. *International Journal of Food Engineering, 1*(2), 106–110.

Hong, J., Wang, L., Sun, Y., Zhao, L., Niu, G., Tan, W., Rico, C. M., Peralta-Videa, J. R., & Gardea-Torresdey, J. L. (2016). Foliar applied nanoscale and microscale CeO_2 and CuO alter cucumber (*Cucumis sativus*) fruit quality. *Science of the Total Environment, 563*, 904–911.

Jiang, Y., Hua, Z., Zhao, Y., Liu, Q., Wang, F., Zhang, Q. (2014) The effect of carbon nanotubes on rice seed germination and root growth. Paper presented at the *Proceedings of the 2012 International Conference on Applied Biotechnology (ICAB 2012)*, 1207–1212.

Karunakaran, G., Suriyaprabha, R., Rajendran, V., & Kannan, N. (2016). Influence of ZrO_2, SiO_2, Al_2O_3 and TiO_2 nanoparticles on maize seed germination under different growth conditions. *IET Nanobiotechnology, 10*(4), 171–177.

Kim, D., Saratale, R. G., Shinde, S., Syed, A., Ameen, F., & Ghodake, G. (2018). Green synthesis of silver nanoparticles using laminaria japonica extract: Characterization and seedling growth assessment. *Journal of Cleaner Production, 172*, 2910–2918.

Lahiani, M. H., Chen, J., Irin, F., Puretzky, A. A., Green, M. J., & Khodakovskaya, M. V. (2015). Interaction of carbon nanohorns with plants: Uptake and biological effects. *Carbon, 81*, 607–619.

Lahiani, M. H., Dervishi, E., Chen, J., Nima, Z., Gaume, A., Biris, A. S., & Khodakovskaya, M. V. (2013). Impact of carbon nanotube exposure to seeds of valuable crops. *ACS Applied Materials Interfaces, 5*(16), 7965–7973.

Li, J., Hu, J., Ma, C., Wang, Y., Wu, C., Huang, J., & Xing, B. (2016). Uptake, translocation and physiological effects of magnetic iron oxide (γ-Fe2O3) nanoparticles in corn (*Zea mays* L.). *Chemosphere, 159*, 326–334.

Lin, D., & Xing, B. (2007). Phytotoxicity of nanoparticles: Inhibition of seed germination and root growth. *Environmental Pollution, 150*(2), 243–250.

Liu, J., Dhungana, B., & Cobb, G. P. (2018). Copper oxide nanoparticles and arsenic interact to alter seedling growth of rice (*Oryza sativa* japonica). *Chemosphere, 206*, 330–337.

Liu, R., Zhang, H., & Lal, R. (2016). Effects of stabilized nanoparticles of copper, zinc, manganese, and iron oxides in low concentrations on lettuce (*Lactuca sativa*) seed germination: Nanotoxicants or nanonutrients? *Water, Air Soil Pollution, 227*(1), 42.

Majumdar, S., Almeida, I. C., Arigi, E. A., Choi, H., VerBerkmoes, N. C., Trujillo-Reyes, J., Flores-Margez, J. P., White, J. C., Peralta-Videa, J. R., & Gardea-Torresdey, J. L. (2015). Environmental effects of nanoceria on seed production of common bean (*Phaseolus vulgaris*): A proteomic analysis. *Environmental Science Technology, 49*(22), 13283–13293.

Medina-Velo, I. A., Barrios, A. C., Zuverza-Mena, N., Hernandez-Viezcas, J. A., Chang, C. H., Ji, Z., Zink, J. I., Peralta-Videa, J. R., & Gardea-Torresdey, J. L. (2017a). Comparison of the effects of commercial coated and uncoated ZnO nanomaterials and Zn compounds in kidney bean (*Phaseolus vulgaris*) plants. *Journal of Hazardous Materials, 332*, 214–222.

Medina-Velo, I. A., Dominguez, O. E., Ochoa, L., Barrios, A. C., Hernández-Viezcas, J. A., White, J. C., Peralta-Videa, J. R., & Gardea-Torresdey, J. L. (2017b). Nutritional quality of bean seeds harvested from plants grown in different soils amended with coated and uncoated zinc oxide nanomaterials. *Environmental Science: Nano, 4*(12), 2336–2347.

Medina-Velo, I. A., Zuverza-Mena, N., Tamez, C., Ye, Y., Hernandez-Viezcas, J. A., White, J. C., Peralta-Videa, J. R., & Gardea-Torresdey, J. L. (2018). Minimal transgenerational effect of ZnO nanomaterials on the physiology and nutrient profile of *Phaseolus vulgaris*. *ACS Sustainable Chemistry Engineering, 6*(6), 7924–7930.

Muthayya, S., Sugimoto, J. D., Montgomery, S., & Maberly, G. F. (2014). An overview of global rice production, supply, trade, and consumption. *Annals of the New York Academy of Sciences, 1324*(1), 7–14.

Nair, P. M. G., Kim, S., & Chung, I. M. (2014). Copper oxide nanoparticle toxicity in mung bean (*Vigna radiata* L.) seedlings: Physiological and molecular level responses of in vitro grown plants. *Acta Physiologiae Plantarum, 36*(11), 2947–2958.

Nair, R., Varghese, S. H., Nair, B. G., Maekawa, T., Yoshida, Y., & Kumar, D. S. (2010). Nanoparticulate material delivery to plants. *Plant Science, 179*(3), 154–163.

O'Connell, M. J. (2006). *Carbon nanotubes: Properties and applications.* CRC press.

Ochoa, L., Medina-Velo, I. A., Barrios, A. C., Bonilla-Bird, N. J., Hernandez-Viezcas, J. A., Peralta-Videa, J. R., & Gardea-Torresdey, J. L. (2017). Modulation of CuO nanoparticles toxicity to green pea (*Pisum sativum* fabaceae) by the phytohormone indole-3-acetic acid. *Science of the Total Environment, 598*, 513–524.

Pandey, C., Khan, E., Mishra, A., Sardar, M., & Gupta, M. (2014). Silver nanoparticles and its effect on seed germination and physiology in *Brassica juncea* L. (Indian mustard) plant. *Advanced Science Letters, 20*(7–8), 1673–1676.

Pariona, N., Martinez, A. I., Hdz-García, H., Cruz, L. A., & Hernandez-Valdes, A. (2017). Effects of hematite and ferrihydrite nanoparticles on germination and growth of maize seedlings. *Saudi Journal of Biological Sciences, 24*(7), 1547–1554.

Parveen, A., & Rao, S. (2015). Effect of nanosilver on seed germination and seedling growth in *Pennisetum glaucum*. *Journal of Cluster Science, 26*(3), 693–701.

Peralta-Videa, J. R., Hernandez-Viezcas, J. A., Zhao, L., Diaz, B. C., Ge, Y., Priester, J. H., Holden, P. A., & Gardea-Torresdey, J. L. (2014). Cerium dioxide and zinc oxide nanoparticles alter the nutritional value of soil cultivated soybean plants. *Plant Physiology and Biochemistry, 80*, 128–135.

Pošćić, F., Mattiello, A., Fellet, G., Miceli, F., & Marchiol, L. (2016). Effects of cerium and titanium oxide nanoparticles in soil on the nutrient composition of barley (*Hordeum vulgare* L.) kernels. *International Journal of Environmental Research and Public Health, 13*(6), 577.

Priester, J. H., Moritz, S. C., Espinosa, K., Ge, Y., Wang, Y., Nisbet, R. M., Schimel, J. P., Goggi, A. S., Gardea-Torresdey, J. L., & Holden, P. A. (2017). Damage assessment for soybean cultivated in soil with either CeO_2 or ZnO manufactured nanomaterials. *Science of the Total Environment, 579*, 1756–1768.

Pulit-Prociak, J., & Banach, M. (2016). Silver nanoparticles–a material of the future...? *Open Chemistry, 14*(1), 76–91.

Pullagurala, V. L. R., Rawat, S., Adisa, I. O., Hernandez-Viezcas, J. A., Peralta-Videa, J. R., & Gardea-Torresdey, J. L. (2018). Plant uptake and translocation of contaminants of emerging concern in soil. *Science of the Total Environment, 636*, 1585–1596.

Qian, H., Peng, X., Han, X., Ren, J., Sun, L., & Fu, Z. (2013). Comparison of the toxicity of silver nanoparticles and silver ions on the growth of terrestrial plant model *Arabidopsis thaliana*. *Journal of Environmental Sciences, 25*(9), 1947–1956.

Raliya, R., Nair, R., Chavalmane, S., Wang, W., & Biswas, P. (2015). Mechanistic evaluation of translocation and physiological impact of titanium dioxide and zinc oxide nanoparticles on the tomato (*Solanum lycopersicum* L.) plant. *Metallomics, 7*(12), 1584–1594.

Rastogi, A., Zivcak, M., Sytar, O., Kalaji, H. M., He, X., Mbarki, S., & Brestic, M. (2017). Impact of metal and metal oxide nanoparticles on plant: A critical review. *Frontiers in Chemistry, 5*, 78.

Rawat, S., Adisa, I. O., Wang, Y., Sun, Y., Fadil, A. S., Niu, G., Sharma, N., Hernandez-Viezcas, J. A., Peralta-Videa, J. R., & Gardea-Torresdey, J. L. (2019). Differential physiological and biochemical impacts of nano vs micron cu at two phenological growth stages in bell pepper (*Capsicum annuum*) plant. *NanoImpact, 14*, 100161.

Rawat, S., Apodaca, S. A., Tan, W., Peralta-Videa, J. R., & Gardea-Torresdey, J. L. (2017). Terrestrial nanotoxicology: Evaluating the nano-biointeractions in vascular plants. In *Bioactivity of engineered nanoparticles* (21). Springer.

Rawat, S., Pullagurala, V. L. R., Hernandez-Molina, M., Sun, Y., Niu, G., Hernandez-Viezcas, J. A., Peralta-Videa, J. R., & Gardea-Torresdey, J. L. (2018a). Impacts of copper oxide nanoparticles on bell pepper (*Capsicum annum* L.) plants: A full life cycle study. *Environ Sci Nano, 5*(1), 83–95.

Rawat, S., Pullagurala, V. L., Adisa, I. O., Wang, Y., Peralta-Videa, J. R., & Gardea-Torresdey, J. L. (2018b). Factors affecting fate and transport of engineered nanomaterials in a terrestrial environment. *Current Opinion in Environmental Science Health, 6*, 47–53.

Reddy, P. V. L., Hernandez-Viezcas, J. A., Peralta-Videa, J. R., & Gardea-Torresdey, J. L. (2016). Lessons learned: Are engineered nanomaterials toxic to terrestrial plants? *Science of the Total Environment, 568*, 470–479.

Rico, C. M., Lee, S. C., Rubenecia, R., Mukherjee, A., Hong, J., Peralta-Videa, J. R., & Gardea-Torresdey, J. L. (2014). Cerium oxide nanoparticles impact yield and modify nutritional parameters in wheat (*Triticum aestivum* L.). *Journal of Agricultural and Food Chemistry, 62*(40), 9669–9675.

Rico, C. M., Morales, M. I., Barrios, A. C., McCreary, R., Hong, J., Lee, W., Nunez, J., Peralta-Videa, J. R., & Gardea-Torresdey, J. L. (2013). Effect of cerium oxide nanoparticles on the quality of rice (*Oryza sativa* L.) grains. *Journal of Agricultural and Food Chemistry, 61*(47), 11278–11285.

Sabir, S., Arshad, M., Satti, S. H., & I.J.A.A.R. (2018). Effect of green synthesized silver nanoparticles on seed germination and seedling growth in wheat. *International Journal of Agronomy Research, 12*(4), 1–7.

Servin, A. D., Morales, M. I., Castillo-Michel, H., Hernandez-Viezcas, J. A., Munoz, B., Zhao, L., Nunez, J. E., Peralta-Videa, J. R., & Gardea-Torresdey, J. L. (2013). Synchrotron verification of TiO$_2$ accumulation in cucumber fruit: A possible pathway of TiO$_2$ nanoparticle transfer from soil into the food chain. *Environmental Science & Technology, 47*(20), 11592–11598.

Shahbandeh (2020). Global corn production in 2019/2020, by country. https://www.statista.com/statistics/254292/global-corn-production-by-country/

Shams, G., Ranjbar, M., & Amiri, A. (2013). Effect of silver nanoparticles on concentration of silver heavy element and growth indexes in cucumber (*Cucumis sativus* L. negeen). *Journal of Nanoparticle Research, 15*(5), 1630.

Shaw, A. K., & Hossain, Z. (2013). Impact of nano-CuO stress on rice (*Oryza sativa* L.) seedlings. *Chemosphere, 93*(6), 906–915.

Singh, A., Singh, N., Afzal, S., Singh, T., & Hussain, I. (2018). Zinc oxide nanoparticles: A review of their biological synthesis, antimicrobial activity, uptake, translocation and biotransformation in plants. *Journal of Materials Science, 53*(1), 185–201.

Subbaiah, L. V., Prasad, T. N. V. K. V., Krishna, T. G., Sudhakar, P., Reddy, B. R., & Pradeep, T. (2016). Novel effects of nanoparticulate delivery of zinc on growth, productivity, and zinc biofortification in maize (*Zea mays* L.). *Journal of Agricultural and Food Chemistry, 64*(19), 3778–3788.

Tan, W., Du, W., Darrouzet-Nardi, A. J., Hernandez-Viezcas, J. A., Ye, Y., Peralta-Videa, J. R., & Gardea-Torresdey, J. L. (2018). Effects of the exposure of TiO₂ nanoparticles on basil (*Ocimum basilicum*) for two generations. *Science of the Total Environment, 636*, 240–248.

Thuesombat, P., Hannongbua, S., Akasit, S., & Chadchawan, S. (2014). Effect of silver nanoparticles on rice (*Oryza sativa* L. cv. KDML 105) seed germination and seedling growth. *Ecotoxicology and Environmental Safety, 104*, 302–309.

Tripathi, D. K., Singh, S., Singh, V. P., Prasad, S. M., Dubey, N. K., & Chauhan, D. K. (2017). Silicon nanoparticles more effectively alleviated UV-B stress than silicon in wheat (*Triticum aestivum*) seedlings. *Plant Physiology and Biochemistry, 110*, 70–81.

Wang, C., Jiang, K., Wu, B., Zhou, J., & Lv, Y. (2018). Silver nanoparticles with different particle sizes enhance the allelopathic effects of Canada goldenrod on the seed germination and seedling development of lettuce. *Ecotoxicology, 27*(8), 1116–1125.

Wang, C., Liu, H., Chen, J., Tian, Y., Shi, J., Li, D., Guo, C., & Ma, Q. (2014). Carboxylated multi-walled carbon nanotubes aggravated biochemical and subcellular damages in leaves of broad bean (*Vicia faba* L.) seedlings under combined stress of lead and cadmium. *Journal of Hazardous Materials, 274*, 404–412.

Wang, J., Fang, Z., Cheng, W., Yan, X., Tsang, P. E., & Zhao, D. (2016a). Higher concentrations of nanoscale zero-valent iron (nZVI) in soil induced rice chlorosis due to inhibited active iron transportation. *Environmental Pollution, 210*, 338–345.

Wang, Q., Ebbs, S. D., Chen, Y., & Ma, X. (2013). Trans-generational impact of cerium oxide nanoparticles on tomato plants. *Metallomics, 5*(6), 753–759.

Wang, Q., Ma, X., Zhang, W., Pei, H., & Chen, Y. (2012). The impact of cerium oxide nanoparticles on tomato (*Solanum lycopersicum* L.) and its implications for food safety. *Metallomics, 4*(10), 1105–1112.

Wang, Y., Deng, C., Cota-Ruiz, K., Peralta-Videa, J. R., Sun, Y., Rawat, S., Tan, W., Reyes, A., Hernandez-Viezcas, J. A., Niu, G., Li, C., & Gardea-Torresdey, J. L. (2020). Improvement of nutrient elements and allicin content in green onion (*Allium fistulosum*) plants exposed to CuO nanoparticles. *Science of the Total Environment, 725*, 138387.

Wang, Z., Xu, L., Zhao, J., Wang, X., White, J. C., & Xing, B. (2016b). CuO nanoparticle interaction with *Arabidopsis thaliana*: Toxicity, parent-progeny transfer, and gene expression. *Environmental Science Technology, 50*(11), 6008–6016.

Wilson, N. (2018). Nanoparticles: Environmental problems or problem solvers? *Bioscience, 68*(4), 241–246.

Wu, J., Wang, G., Vijver, M. G., Bosker, T., & Peijnenburg, W. J. G. M. (2020). Foliar versus root exposure of AgNPs to lettuce: Phytotoxicity, antioxidant responses and internal translocation. *Environmental Pollution, 261*, 114117.

Xiang, L., Zhao, H., Li, Y., Huang, X., Wu, X., Zhai, T., Yuan, Y., Cai, Q., & Mo, C. (2015). Effects of the size and morphology of zinc oxide nanoparticles on the germination of chinese cabbage seeds. *Environmental Science and Pollution Research, 22*(14), 10452–10462.

Yang, Z., Chen, J., Dou, R., Gao, X., Mao, C., & Wang, L. (2015). Assessment of the phytotoxicity of metal oxide nanoparticles on two crop plants, maize (*Zea mays* L.) and rice (*Oryza sativa* L.). *International Journal of Environmental Research and Public Health, 12*(12), 15100–15109.

Zahra, Z., Waseem, N., Zahra, R., Lee, H., Badshah, M. A., Mehmood, A., Choi, H., & Arshad, M. (2017). Growth and metabolic responses of rice (*Oryza sativa* L.) cultivated in phosphorus-deficient soil amended with TiO₂ nanoparticles. *Journal of Agricultural and Food Chemistry, 65*(28), 5598–5606.

Zhao, L., Peralta-Videa, J. R., Rico, C. M., Hernandez-Viezcas, J. A., Sun, Y., Niu, G., Servin, A., Nunez, J. E., Duarte-Gardea, M., & Gardea-Torresdey, J. L. (2014). CeO_2 and ZnO nanoparticles change the nutritional qualities of cucumber (*Cucumis sativus*). *Journal of Agricultural and Food Chemistry, 62*(13), 2752–2759.

Zhao, L., Sun, Y., Hernandez-Viezcas, J. A., Hong, J., Majumdar, S., Niu, G., Duarte-Gardea, M., Peralta-Videa, J. R., & Gardea-Torresdey, J. L. (2015). Monitoring the environmental effects of CeO_2 and ZnO nanoparticles through the life cycle of corn (*Zea mays*) plants and in situ μ-XRF mapping of nutrients in kernels. *Environmental Science Technology, 49*(5), 2921–2928.

Zhao, L., Sun, Y., Hernandez-Viezcas, J. A., Servin, A. D., Hong, J., Niu, G., Peralta-Videa, J. R., Duarte-Gardea, M., & Gardea-Torresdey, J. L. (2013). Influence of CeO_2 and ZnO nanoparticles on cucumber physiological markers and bioaccumulation of Ce and Zn: A life cycle study. *Journal of Agricultural and Food Chemistry, 61*(49), 11945–11951.

Zuverza-Mena, N., Armendariz, R., Peralta-Videa, J. R., & Gardea-Torresdey, J. L. (2016). Effects of silver nanoparticles on radish sprouts: Root growth reduction and modifications in the nutritional value. *Frontiers in Plant Science, 7*, 90.

Zuverza-Mena, N., Martínez-Fernández, D., Du, W., Hernandez-Viezcas, J. A., Bonilla-Bird, N., López-Moreno, M. L., Komarek, M., Peralta-Videa, J. R., & Gardea-Torresdey, J. L. (2017). Exposure of engineered nanomaterials to plants: Insights into the physiological and biochemical responses-A review. *Plant Physiology and Biochemistry, 110*, 236–264.

Chapter 9
Application of Metal Oxide Nanomaterials in Agriculture: Benefit or Bane?

Nazanin Nikoo Jamal, Elliott Duncan, and Gary Owens

Contents

1   Background

Food security, the provision of an ever-increasing world population with sufficient food that is affordable, nutritious, and safe, is a major global challenge. By the end of the twenty-first century, in excess of 9.5 billion people will inhabit the earth, which will result in an increased demand for food, and a reduction in the area and resources available for food production. Currently, more than 1 billion people are at risk of or already suffer from food shortages and an additional 1 billion people are at risk of malnutrition (Rengel et al. 1999). By 2050, the number of people with

N. N. Jamal
Environmental Contaminates Group, Future industries Institute, The University of South Australia, Mawson Lakes, South Australia, Australia

School of Natural and Built Environments, The University of South Australia, Mawson Lakes, South Australia, Australia
e-mail: nazanin.nikoo_jamal@mymail.unisa.edu.au

E. Duncan · G. Owens (✉)
Environmental Contaminates Group, Future industries Institute, The University of South Australia, Mawson Lakes, South Australia, Australia
e-mail: elliott.duncan@unisa.edu.au; gary.owens@unisa.edu.au

© Springer Nature Switzerland AG 2021
N. Sharma, S. Sahi (eds.), *Nanomaterial Biointeractions at the Cellular, Organismal and System Levels*, Nanotechnology in the Life Sciences, https://doi.org/10.1007/978-3-030-65792-5_9

insecure food supplies and/or food with poor nutritional value will only increase and thus world food production will need to increase by 40% in the next 20 years and 70% by 2050. To combat this impending crisis the production of plants used in food production will essentially need to double relative to current rates of production (Rosegrant and Cline 2003). These increases must, however, be achieved sustainably and not compromise the nutritional quality of food or increase the environmental degradation resulting from broadacre agriculture.

While many studies have advocated engineered nanomaterials (ENMs) as a game changing, novel means of improving agricultural productivity (Liu and Lal 2015), many other studies have equally advocated ENMs as agents for environmental disaster due to significant physical and chemical interactions with both inorganic and organic species in aquatic and soil environments (Engates and Shipley 2011; Shipley et al. 2011). This chapter examines whether the presence of ENMs, specifically metal oxide nanomaterials, can significantly affect agricultural crop productivity and environmental health in terrestrial agricultural systems.

2 The Rise of Nanotechnology

The field of nanotechnology is developing rapidly and the extensive use of engineered nanomaterials (ENMs) in industrial (i.e. electronics, medicine, and agriculture) and domestic products has raised worldwide concern about their inevitable release and accumulation in the environment (Nowack and Bucheli 2007).

Whether dispersed in gaseous, liquid, or solid media, ENMs by definition must have at least one dimension <100 nm (Arruda et al. 2015). Currently, ENMs are categorised into four groups (Klaine et al. 2008), which include: (a) carbon-based materials including fullerenes, single-walled carbon nanotubes (SWCNT), and multi-walled carbon nanotubes (MWCNT); (b) metal oxides such as nTiO$_2$, nZnO and nCeO$_2$; (c) dendrimers which are synthetic polymers with tree-like structure for performing specific chemical function such as drug delivery (Esfand and Tomalia 2001), and (d) composites which are a combination of ENMs with much larger materials, such as concrete and ceramics (Lin and Xing 2007).

In comparison to their bulk equivalents, ENMs have unique properties. For example, they have a large specific surface area, high surface area to volume ratio, and variable surface charge (Ma et al. 2010). Due to these unique properties, ENMs have been proposed to have very different environmental fates and behaviour compared to common contaminants (Ma et al. 2010).

Most ENMs are released into the environment inadvertently as a result of industrial processes or the use of domestic products; however, activities such as water purification or soil remediation can intentionally release ENMs into the environment (Farré et al. 2009; Colvin 2003). Once released to the environment, soils are likely to be a major 'sink' for ENMs and thus it is likely that terrestrial plants will be exposed to increasing concentrations of ENMs over time. Exposure of ENMs to plants may have adverse effects such as reducing photosynthesis and affecting nutrient uptake translocation. In addition, the accumulation of ENMs in plants may result in 'trophic transfer' of ENMs within food webs which could be detrimental to human and ecological health.

Currently, both positive and negative effects of ENMs on plants have been reported. It has been postulated that ENMs may have significant applications in agriculture and horticulture as they may inhibit the growth of weed species, protect against soil pathogens, and also assist in the bio-fortification of crops with essential trace elements (Tang 2013). Conversely, some ENMs (e.g. alumina and nZnO) are likely to be phytotoxic, and as a result, they have been observed to inhibit seed germination and root elongation in some plant species (Yang and Watts 2005; Lin and Xing 2008). Much less information is, however, known about how ENMs are accumulated and translocated by plants over their entire life cycle. In addition, there is still much to learn regarding 'typical' ENM concentrations found in edible plant tissues and whether ENMs are accumulated and sequestered by plants in a 'pristine' form or whether they are biotransformed in- or ex-situ. Therefore, the efflux of ENMs to soil, air, and water as a result of agricultural activities has raised inevitable concerns regarding their interaction with terrestrial plants and therefore their entrance into agricultural food webs.

As an emerging field of research, understanding how ENMs influence plant physiology and whether ENMs accumulate in plant tissues is of considerable importance agriculturally and to the wider community. This chapter focuses on whether metal-based ENMs (especially nCeO$_2$ and nTiO$_2$) are accumulated in commercially important agricultural crops. Given their global importance as a food source particular interest is also given to whether ENMs accumulate in cereal grains such as rice and wheat. In addition, this chapter also focuses on whether nCeO$_2$ and nTiO$_2$ have the potential to alter plant physiology, specifically, grain yields due to the economic importance of grain production to growers and industry.

3 Phytotoxicity of ENMs Via Seed Germination and Root Elongation Tests

The exposure of plant species to ENMs has received considerable attention in recent years (Stampoulis et al. 2009; Zhu et al. 2008) where ENMs can interact with plants not only through uptake and accumulation but also through root adsorption and absorption (Ma et al. 2010). Consequently, the biological impact of ENMs on plants is commonly investigated via simple short-term seed germination and root elongation assays (Tang 2013). Although these studies lack the completeness of longer-term hydroponic or soil growth studies, they are often a useful screening tool for potential ENM toxicity. According to Lin and Xing (2007) (Mushtaq 2011), the toxicity of ENMs may be as a result of three actions: (1) generation of reactive oxygen species (ROS), (2) cellular degradation from the penetration of ENMs through cell membranes, and (3) the release of toxic metal ions from ENM dissolution (Mushtaq 2011).

The phytotoxicity of two ENMs, nCeO$_2$ and nano-lanthanum oxide (nLa$_2$O$_3$), of similar particle size (25 nm) on cucumber seedlings showed that nCeO$_2$ had no effect on root or shoot elongation and biomass at all concentrations tested (0–2000 mg L^{-1}), while nLa$_2$O$_3$ reduced root at concentrations ≥2 mg L^{-1}, shoot elongation at 2000 mg L^{-1}, and biomass at 20 mg L^{-1} (Table 9.1) (Ma et al. 2015). Nanoscaled lanthanum oxide (nLa$_2$O$_3$) also increased the H$_2$O$_2$ content which resulted in increased cell death. It was hypothesised that the higher dissolution of nLa$_2$O$_3$ may have facilitated a phytotoxic response (Ma et al. 2015).

In another study, nTiO$_2$ had no effect on root elongation of wheat when exposed to concentrations between 1 and 500 mg L^{-1} (Feizi et al. 2012). Furthermore, shoot length when exposed to 2 and 10 mg L^{-1} nTiO$_2$ was higher than an untreated control and larger bulk counterparts (Table 9.1). It was hypothesised that at specific concentrations nTiO$_2$ can improve seed germination and seedling growth of wheat (Feizi et al. 2012). The result was in agreement with Zheng et al., 2005 (Zheng et al. 2005) who showed enhanced spinach germination upon exposure to nTiO$_2$.

When rice seeds were soaked in nTiO$_2$ suspensions at three concentrations (50, 100, 5000 mg L^{-1}) for 3 days, nTiO$_2$ had no effect on seed germination and caused only a slight decrease in root length (Boonyanitipong et al. 2011) which is consistent with the work of Lin and Xing (Lin and Xing 2007). Overall, root elongation was more sensitive to ENM exposure than seed germination because the grain surface restricts the passage of ENMs into the grain (Boonyanitipong et al. 2011).

The effect of five metal oxide ENMs (nCo$_3$O$_4$, nCuO, nFe$_2$O$_3$, nNiO, nTiO$_2$) at different concentrations (up to 5000 mg L^{-1}) on seed germination and root elongation of three common vegetables (cucumber, lettuce, and radish) was investigated by Tang et al. (Krug and Wick 2011). In this study, even at lower concentrations (< 1000 mg L^{-1}), both nCuO and nNiO were more toxic than all of the other ENMs tested. It was postulated that smaller ENMs would have higher surface energy and would, therefore, be more toxic to cells (Krug and Wick 2011). In this study, a large concentration of all tested ENMs was adsorbed on to the surface of the roots and shoots in all experiments. This was attributed to coagulation in an aqueous solution which minimises the total surface energy and therefore minimises ENM toxicity (Tang 2013). Conversely, exposure to nCo$_3$O$_4$ improved root elongation of radish even at high concentrations (Table 9.1). The authors hypothesised that the positive effect of ENMs on plants was due to increased water uptake by seeds in the presence of high ENM concentrations (Tang 2013).

The hydrodynamic diameter and ENM aggregation behaviour are important variables to consider in phytotoxicity studies. In a study by Song et al. (Song et al. 2013), nTiO$_2$ showed no phytotoxic effects on either germination or root elongation when tomato seeds were soaked in six nTiO$_2$ suspensions (0, 50, 100, 500, 1000, 2500, 5000 mg L^{-1}), while nAg solutions inhibited tomato root elongation even at relatively low concentrations (< 100 mg L^{-1}) (Song et al. 2013). In this study, the hydrodynamic diameter and agglomeration behaviour of ENMs was proposed to have regulated phytotoxicity. Since nAg was accumulated at much higher concentrations than nTiO$_2$, it was hypothesised that the decreased aggregation of nAg allowed it to be accumulated, whereas larger nTiO$_2$ aggregates were not absorbed (Song et al. 2013).

Table 9.1 Phytotoxicity of common metal oxide ENMs to different plant species

Studied ENMs	Size (nm)	Concentration (mg L^{-1})	Plant type	Exposure time	Plant response	References
nTiO$_2$ nFe$_2$O$_3$	30–50	0–5000	Cucumber	6 days	– 40% reduction in seed germination at 5000 mg L^{-1} nTiO$_2$ – 65% reduction in root length when exposed to 5000 mg L^{-1} nTiO$_2$ – [a]GI reduction with increase nTiO$_2$ concentration	Mushtaq (2011)
nTiO$_2$	21	1–500	Wheat	8 days	– No significant effect on root length – Shoot length and seedling length at 2 and 10 mg L^{-1} nTiO$_2$ were 8% and 7.3% respectively higher than control and 10.2 and 7% respectively higher than larger bulk counterparts in comparison to control	Feizi et al. (2012)
nTiO$_2$ nZnO	<100	50–5000	Rice	7 days	– 100% seed germination by all treatments. – Root length reduction when exposed to increasing nZnO concentrations	Boonyanitipong et al. (2011)
nTiO$_2$ nCo$_3$O$_4$ nCuO nFe$_2$O$_3$ nNiO	10–50	0–5000	Cucumber Lettuce Radish	3 days	– nCuO and nNiO were far more toxic (−100% GI) – Lettuce was more sensitive to all ENM treatments – Up to 50% increase in root elongation of radish seedlings when exposed to Co$_3$O$_4$	Tang (2013)
nTiO$_2$	10–27	0-5000	Tomato	12 days	– No significant difference in germination rate	Song et al. (2013)

(continued)

Table 9.1 (continued)

Studied ENMs	Size (nm)	Concentration (mg L^{-1})	Plant type	Exposure time	Plant response	References
nCeO$_2$ nLa$_2$O$_3$	25	0–2000	Cucumber	5 and 14 days	– nCeO$_2$ had no phytotoxicity to cucumber – nLa$_2$O$_3$ reduced root, shoot biomass, and shoot elongation by 65.8% and 42.8%, 18.3% respectively, at 2000 mg L^{-1} – nLa$_2$O$_3$ reduced root elongation by 65.8% at 2 mg L^{-1}	Ma et al. (2015)
nCeO$_2$ nTiO$_2$	14–34	250–1000	Cabbage, carrot, corn, cucumber, lettuce, oats, onions, ryegrass, soybean, and tomato	Time is not stated.	– nCeO$_2$ did not influence germination – 30% and 20% increase in germination rate of cabbage and oats respectively when exposed to nTiO$_2$ – 10% and 5% reduction in root length of cucumber and onion respectively when exposed to nTiO$_2$	Andersen et al. (2016)

[a]Mode of exposure = petri dish with moist filter paper. *GI*: Germination Index; which combines seed germination and root growth

Among ten common agronomic plant species (cabbage, carrot, corn, cucumber, lettuce, oats, onions, ryegrass, soybean, and tomatoes) recently studied by Andersen et al. (Andersen et al. 2016), carrot and ryegrass did not display any signs of toxicity upon nTiO$_2$ exposure (Table 9.1). While exposure to nCeO$_2$ decreased the root length of lettuce and tomato, it increased root length in ryegrass and onion. The authors concluded that the average root length was more sensitive to ENM exposure than seed germination and that nTiO$_2$ and nCeO$_2$ have different effects on plant growth at early stages of the life cycle, while effects at later growth stages were still unknown (Andersen et al. 2016).

Clearly, the interaction of ENMs with agronomically important plants is a complex process and depends on plant species, ENMs properties, exposure time, and concentration, and while ENMs may have a short-term positive, negative, or neutral effect on establishment, germination, and root elongation, their overall long-term effects on plant physiology are poorly understood.

4   Uptake, Accumulation, and Translocation of Nanoparticles

Recent studies have shown that ENMs may be translocated to different plant parts via root to shoot pathways (Du et al. 2016). The vast majority of root to shoot translocation studies have been carried out either in hydroponic or soil media (Du et al. 2016). Many such studies have indicated that ENMs are absorbed through the root's endodermis via apoplastic and symplastic pathways and are thereafter transferred to the vascular cylinder or xylem (Servin et al. 2013; Zhao et al. 2012). However other reports suggest that ENMs can also penetrate through the stomata in the cuticle of the leaves, into the stems and finally into the roots through phloem vessels (Larue et al. 2014). To some extent the media in which such experiments were conducted (i.e. hydroponic or soil) may influence the active accumulation pathways.

4.1   Hydroponic Studies

Hydroponics is a plant culturing technique whereby plants are grown in an aquatic nutrient-rich matrix rather than in soil. The benefit of this technique is that concentrations of nutrients, contaminants, and, in this case, ENMs can be maintained at desired levels with more certainty as immobilisation and mineralisation processes that occur in soil are eliminated via the use of liquid culture. Thus, hydroponic experiments are a useful starting point for the examination of ENM-plant interactions. Different ENM factors such as hydrodynamic size, crystalline phase, and method of application (e.g. foliar application) have been investigated in a number of hydroponic studies (Krug and Wick 2011; Du et al. 2016; Servin et al. 2013; Zhao et al. 2012; Larue et al. 2014, 2011a, 2012a; Mccutcheon and Schnoor 2004).

   The hydrodynamic diameter of ENMs almost certainly influences the accumulation of ENMs by root systems and the subsequent re-distribution of ENMs in above-ground and reproductive tissues. Hydroponic studies in which wheat, rapeseed, and Thale cress (Arabidopsis thaliana) were exposed to nTiO$_2$ demonstrated that smaller nTiO$_2$ aggregates were taken up by all three plant species and partitioned in the parenchymal of the roots and also in the vascular cylinder (Table 9.2). However, the accumulation of nTiO$_2$ aggregates had no effect on germination at concentrations between 50 and 1000 mg L^{-1} (Larue et al. 2011a). This study also investigated the behaviour of wheat and rapeseed when exposed to 14 and 25 nm nTiO$_2$ (100 mg L^{-1}) either through leaves (foliar application) or roots under hydroponic conditions. Smaller ENMs with higher surface activity were hypothesised to enlarge root pores and thus elevate nutrient uptake and plant biomass production; therefore nTiO$_2$ (mostly 14 nm) was internalised in roots and translocated to leaves (Table 9.2) with the translocation of Ti believed to be linked to water-flow within leaves (Larue et al. 2012a). Across plant species, rapeseed clearly accumulated more nTiO$_2$ than wheat which was hypothesised to occur as a result of rapeseeds ability to hyper-accumulate certain elements in their tissues (Mccutcheon and Schnoor 2004). The effect of

Table 9.2 Uptake and translocation of common metal oxide ENMs in different hydroponically grown plant species

ENMs	Primary size (nm)	Concentration (mgL^{-1})	Plant type	Exposure time (days)	Plant response	References
nTiO$_2$ (anatase and rutile)	12 and 25	0–1000	Wheat	7	– nTiO$_2$ did not significantly alter germination and root elongation – nTiO$_2$ (12 nm) was accumulated in the parenchyma and vascular cylinder of wheat roots	Larue et al. (2011a)
nTiO$_2$ (anatase and rutile)	14–655	100	Wheat	7	– nTiO$_2$ above 140 nm were not accumulated in wheat root – nTiO$_2$ above 36 nm were accumulated in roots but not translocated – nTiO$_2$ did not affect germination – nTiO$_2$ increased root elongation by 50%	Larue et al. (2012b)
nTiO$_2$ (anatase and rutile)	14 and 25	100	Wheat Rapeseed	7	– Parenchymal region of wheat root accumulated more nTiO$_2$ (14 nm) – Vascular cylinder of wheat root accumulate more nTiO$_2$ (25 nm) – 14 nm nTiO$_2$ (14 nm) was translocated to leaves more efficiently than 25 nm particles – Exposure to nTiO$_2$ 14 nm increased root length in both species (68% in wheat, and 31% in rapeseed) – Rapeseed accumulated greater nTiO$_2$concentrations than wheat	Larue et al. (2012a)

(continued)

Table 9.2 (continued)

ENMs	Primary size (nm)	Concentration (mgL⁻¹)	Plant type	Exposure time (days)	Plant response	References
nCeO$_2$ (powder and suspensión)	8 ± 1	0.98 and 2.94 g m^{-3} (as powder) 20–320 mg L^{-1} (as suspensions)	Cucumber	15	– Increased tissue Ce (up to 50%) content with increased nCeO$_2$ exposure – Ce from nCeO$_2$ translocated to cucumber stem and roots when exposed either as a powder or suspension	Hong et al. (2014)
nCeO$_2$, and Ce^{3+}	10–30	10	Radish	21	– nCeO$_2$ had no effect on radish growth – Ce^{3+} had negative effect on radish growth – nCeO$_2$ and Ce^{3+} accumulated in edible root and shoot	Zhang et al. (2015)
nCuO	43 ± 9	100	Rice	14	– nCuO was accumulated, translocated and speciation by rice – The Cu content in mature leaves, stem and young leaves were 4.3, 2.3, and 1.9 times respectively greater than non-Cu exposed plants	Peng et al. (2015)

nTiO$_2$ crystalline phase (14 to 655 nm rutile and anatase) on wheat showed no impact on seed germination, vegetative development, and photosynthesis (Table 9.2). In another study (Larue et al. 2012b), nTiO$_2$ particles with a dimension of 36 nm were able to be translocated from roots–shoots/leaves. The authors of this study postulated that nTiO$_2$ accumulation in wheat roots and leaves could be as a result of hypo-osmotic stress. Exposure of wheat plantlets to nTiO$_2$ caused hypo-osmotic stress which increased the root pore size to ≈ 40 nm which allowed 36 nm-ENMs to enter the cell and also resulted in increased water intake by roots and turgor (Larue et al. 2012b).

Other studies have focused on the uptake of ENMs from exposure through leaves and translocation to other tissues. Hong et al. (Hong et al. 2014) found that hydroponically grown cucumbers accumulated Ce from nCeO$_2$ and translocated it to different plant parts when it was applied as a foliar powder or suspension (Table 9.2). In this study it was hypothesised that the size of the $nCeO_2$ particles (8 ± 1 nm) was smaller than the stomatal opening (~21 μm length, ~13 μm width, pore length of ~12 μm, and pore aperture of ~1.23 μm.) (Hong et al. 2014). This result is in agreement with the results of foliar application of $nTiO_2$ on wheat and rapeseed by Larue et al. (Larue et al. 2012a).

In some studies the effect of ENMs and their corresponding bulk counterparts have been investigated. For example, the uptake and accumulation of three different forms of Ce (Bulk CeO_2, $nCeO_2$ and Ce^{3+}) at a concentration of 10 mg L^{-1} in radish demonstrated that while exposure to all forms of Ce increased tissue Ce concentrations in the edible roots and leaves, exposure to $nCeO_2$ did not affect radish growth; however, exposure to ionic Ce (Ce^{3+}) decreased plant growth (Zhang et al. 2015).

In a recent study by Peng et al. (Peng et al. 2015) the translocation pathway of $nCuO$ and Cu speciation in rice plants grown hydroponically was investigated (Peng et al. 2015). The Cu content in rice roots exposed to 100 mg L^{-1} $nCuO$ was nearly 23 times higher than in non-Cu exposed plants. In this study $nCuO$ was adsorbed on the root surface; however, it was not absorbed into the vascular cylinder. It was hypothesised that the casparian strips of endodermis acted as an apoplastic pathway between the root surface and vascular tissue which minimised $nCuO$ translocation. Cu concentrations in mature leaves and young leaves were between 1.9 and 4.3 times greater than in non-Cu exposed plants.

Thus, a critical review of the literature has indicated that, minimal research has investigated the effect of ENM exposure on grain yield production in cereals. Based on the findings of this review, variables such as size and type of ENMs, environmental conditions, plant species, and mode of application (foliar or traditional ENMs application) may dictate whether ENMs are accumulated by plants and whether any negative or positive physiological responses occur.

4.2 Soil Studies

Compared to hydroponic studies long-term growth studies using soil as a growth media are far more representative of agricultural ecosystems. Such studies are important because, as a result of the extensive use of ENMs in consumer products and industry, the accumulation of ENMs in agricultural soils is inevitable. Since plants, along with soil and water, are the major components of agricultural ecosystems, they will certainly play an important role in the transfer of ENMs within food webs, where the phytoavailability and phytotoxicity of ENMs largely depend on how strongly they associate with the solid phases of the soil matrix (Hund-Rinke et al. 2012). Thus, while ENMs may have significant potential for application in both agriculture and horticulture, and may both enhance the growth of target species and inhibit that of pest-species (i.e. weeds) (Tang 2013), it is essential to understand the potential eco-toxicological effects of the accumulation of ENMs in soils (Du et al. 2011).

Nanoscale TiO_2 inhibited soil enzymatic processes (e.g. protease, catalase, and peroxidase) and limited plant growth in wheat grown in a soil-based pot trial (Du et al. 2011). It was hypothesised that the presence of $nTiO_2$ in wheat cells and accumulation on cell walls generated ROS which damaged cell membranes (Du et al. 2011). The Ti content in wheat tissues was not significantly different from non-Ti exposed plants as $nTiO_2$ did not dissolve and instead adhered to root cell wall (Skrabal and Terry 2002). In addition, background Ti concentrations in soils are often in the % range and thus tend to mask any contributions from the much smaller concentrations on $nTiO_2$ added. TEM imagery also demonstrated that some $nTiO_2$ particles (50 ± 10 nm) penetrated into the root cell wall and aggregated in situ to a size several times bigger than their initial or 'pristine' size (Du et al. 2011).

One of the first studies on the influence of $nCeO_2$ exposure on the nutritional quality of rice demonstrated that the effects of $nCeO_2$ were generally negative. In this study, exposure to $nCeO_2$ increased grain Ca and K concentrations. However, exposure also reduced plant Fe and S concentrations and the concentrations of prolamin, glutelin, lauric, and valeric acids and starch in rice grains (Table 9.3). In regions in which diets are largely based on rice, the reduction in Fe concentrations in the presence of $nCeO_2$ could increase the prevalence of Fe deficiencies, while the reduction in grain S could affect grain protein content and the antioxidant capacity of grains (Rico et al. 2013b).

In another study the translocation of $nCeO_2$ from roots to above-ground tissues was limited when wheat was grown in $nCeO_2$-amended soil (Rico et al. 2014). An application of 500 mg $nCeO_2$ kg^{-1} improved plant height, biomass, and grain yield (Table 9.3). In addition, $nCeO_2$ applications at concentrations of 125–500 mg kg^{-1} reduced S content in grains, while Mn storage in grains was reduced only at 250 mg kg^{-1}. Even though the authors failed to determine the cause of S and Mn reduction, it will ultimately have a negative effect on food quality (Rico et al. 2014), which is in agreement with another previous study on rice (Rico et al. 2013b).

Few studies have examined the phytotoxicity of ENMs in comparison to the phytotoxicity of the ionic forms of each metal. Studies of carrots grown in sand treated with either $nZnO$, $nCuO$, and $nCeO_2$ or ionic Zn^{2+}, Cu^{2+}, and Ce^{4+} showed that the total biomass and uptake in roots decreased with increasing concentrations of Zn^{2+} and $nZnO$ (Ebbs et al. 2016). Conversely, Cu^{2+} and Ce^{4+} had stronger negative effects on shoot biomass when compared with $nCuO$ and $nCeO_2$ (Ebbs et al. 2016). In this study, since metal accumulation and penetration in the carrot taproot and translocation to shoots was consistently higher in ionic treatments relative to ENM treatments, it was concluded that ENMs were less toxic than their ionic counterparts. One limitation of this study was that the authors did not consider the dissolution of ENMs; therefore, no conclusion can be made whether ENMs were transported within the taproot or translocated to shoots. Consequently, a good understanding of the ionic transport in plants does not necessarily predict ENM transport (Ebbs et al. 2016). However, in direct contrast, lettuce, radish, and cucumber exposed to metal oxide ENMs (nCo_3O_4, $nCuO$, nFe_2O_3, $nNiO$ and $nTiO_2$) exhibited higher phytotoxicity than when exposed to metal ions of the same concentration (Tang 2013).

Table 9.3 Uptake and translocation of common metal oxide ENMs by plant species grown in soil

ENMs	Initial Size(nm)	Concentration (mg ENM kg^{-1} soil)	Plant type	Soil type	Exposure time	Plant response	References
nTiO$_2$	20–100	90.9	Wheat	Loamy clay (pH 7.3)	7 months	– Wheat biomass production decreased by 15% when plants were exposed to nTiO$_2$ – Biomass of the wheat shoot in soil containing nTiO$_2$ was (277 g) which was not significantly different from plants grown in un-spiked soil (320 g) – Soil protease, catalase, and peroxide activities were reduced 90%, 60%, and 33% respectively. But urease activity was increased by 15%	Du et al. (2011)
nCeO$_2$	8 ± 1	500	Rice	Potting mix (Earthgro)	6 months	– Ce accumulation in rice grains treated with nCeO$_2$ is 805% higher than untreated grains – K and Ca concentration were 8.8% and 25.5% higher in treated grains than in non-Ce controls – S and Fe were 5.9 and 30.4% lower in the grains of nCeO$_2$ treated plants than in non-Ce exposed controls – Reduction in prolamin, glutelin, lauric and valeric acids, and starch by 5.9%, 17%, 22%, 3%, and 7.8% respectively	Rico et al. (2013a)
nCeO$_2$	8 ± 1	0–500	Wheat	Potting mix (Miraclegro)	Not stated (estimated 5–6 months)	– Ce accumulation in root increased with increased nCeO$_2$ exposure (from 1974 to 111,121 µg mg^{-1}) – Improved plant growth, shoot biomass, grain yield by 9%, 12.7%, and 36.6% respectively. – nCeO$_2$ exposure of 250 mg kg^{-1} reduced Mn accumulation in grain up to 15% – S accumulation in wheat grain reduced by up to 17% at all nCeO$_2$ concentration	Rico et al. (2014)
nZnO, nCuO, nCeO$_2$, ionic Zn^{2+}, Cu^{2+}, and Ce^{4+}	30–40 25–55 30–50	0.5–500	Carrot	Coarse sand	4 months	– No sign of toxicity in any treatments – Ionic treatment of Zn^{2+} and Cu^{2+} decreased the biomass – Lower accumulation of Zn, Cu, and Ce from ENMs compared with ionic counterparts	Ebbs et al. (2016)

5 Biotransformation of ENMs during Plant Uptake and Translocation

Biotransformation is a critical factor that may modify the environmental fate, behaviour, and toxicity of ENMs. Few studies have been published on the biotransformation of ENMs within vegetative tissues and, in addition, most studies were not conducted over the complete life cycle of plants from germination to maturity. To elucidate the changes in ENMs speciation, a variety of techniques have been proposed, including μ-X-ray fluorescence, X-ray absorption near edge structure (XANES), and extended X-ray absorption fine structure (López-Moreno et al. 2010; Servin et al. 2012a; Wang et al. 2012; Zhang et al. 2012). For example, Larue et al. (Larue et al. 2011b) investigated anatase and rutile nTiO$_2$ in the vascular system of wheat roots using both μ-XRF mapping and XANES. This study was performed to demonstrate that nTiO$_2$ can be transferred intact through plants by their roots without inducing dissolution or localised changes in crystalline structure.

One of the first studies of biotransformation of metal oxide ENMs on terrestrial plants from germination to maturity was undertaken by Priester et al. (2012). This study investigated soybean grown in soil amended with three different concentrations of nCeO$_2$ and nZnO. This study showed that both ENMs were accumulated in plant tissues and in addition nCeO$_2$ decreased plant growth and reduced the ability of plants to fix atmospheric N (Priester et al. 2012).

Servin et al. (Servin et al. 2012b) also examined the biotransformation of ENMs within vegetative tissue via exposing cucumber (*Cucumis sativus*) to nTiO$_2$ (0–4000 mg L^{-1}) in a hydroponic media. This study also used synchrotron μ-XRF and μ-XANES to characterise the presence and chemical speciation of Ti within the plant tissues. Overall, this study showed that root elongation and root size were increased at all nTiO$_2$ concentrations due to enhanced nitrogen accumulation. The synchrotron μ-XRF results showed that nTiO$_2$ was absorbed by roots and transported to the above-ground plant parts. μ-XANES spectra showed that nTiO$_2$ was not biotransformed as the absorbed Ti was also present as TiO$_2$ in cucumber tissue. This study also found that the type of polymorph present was important since the rutile phase were preferentially translocated to the leaves while the anatase phase remained in the roots (Servin et al. 2012b). The study of Zhang et al. (2012) was one of the first to use transmission electron microscopy (TEM) techniques to study the uptake and distribution of ENMs in plant tissues, specifically the biotransformation of nCeO$_2$ in cucumber. TEM images showed clusters of nCeO$_2$ on cucumber roots after treatment with 2000 mg L^{-1} nCeO$_2$ for 21 days which were subsequently verified to be CePO$_4$ by employing a soft X-ray scanning transmission microscopy (STXM) technique. Near edge X-ray absorption fine structure (XANES) spectra indicated that the Ce in the roots was as 66% CeO$_2$ and 34% CePO$_4$ while Ce in the shoots/leaves was composed of CeO$_2$ and cerium carboxylates (Shoots—86.4% CeO$_2$; 13.6% cerium carboxylates, leaves—78.5% CeO$_2$ and 21.5% cerium carboxylates) (Zhang et al. 2012) .

In another study, a combination of microscopic and spectroscopic techniques was used to investigate the uptake of nZnO and Zn^{2+} ions by maize (Lv et al. 2015). Here X-ray absorption spectroscopy (XAS) showed that Zn^{2+} ions released from nZnO were taken up by roots and accumulated in maize tissues mainly as $ZnPO_4$. Detailed μ-XRF maps indicated that for both treatments the distribution and Zn content in leaves and stems were similar but that greater accumulation of Zn occurred in the root cortex when plants were exposed to nZnO rather than Zn^{2+}. Biotransformation of nZnO to $ZnPO_4$ inside plants inhibited translocation of nZnO into the shoots (Lv et al. 2015).

In a very recent study, a desert plant, Mesquite (*Prosopis juliflora velutina*) was exposed to nCeO$_2$. Ce uptake by roots increased when exposed to concentrations from 500 to 1000 mg L^{-1}; however, when exposed to concentrations between 1000 and 4000 mg L^{-1} root toxicity was observed even though Ce uptake was similar. The translocation of Ce from root to shoot was low even at the higher concentrations considered indicating that root accumulation was potentially restricting transport. Micro X-ray fluorescence (μ-XRF) showed high Ce absorption in the roots through the cortex by the apoplastic pathway. X-ray absorption near edge structure (XANES) indicated that nCeO$_2$ was the main Ce chemical species in mesquite (Hernandez-Viezcas et al. 2016) and that the two peaks at 5729 and 5737 eV attributable to Ce were mostly Ce(IV). Linear combination fitting (LCF) was performed using model compounds of nCeO$_2$, Ce(OH)$_3$, Ce(SO$_4$)$_2$, and Ce (III) acetate and indicated that 81% of the Ce in the analysed spots remained unaltered nCeO$_2$ (Hernandez-Viezcas et al. 2016).

6   Conclusion and Future Directions

To sit idly by while the future world population experiences dramatic decreases in food security is not an option and governments simply must embrace new technologies for increased crop production. While genetically modified food and increases in the water use efficiency of existing plants has a place in this process (Agata Tyczewska et al. 2019) the adoption of new materials to enhance agricultural production must also be seriously considered because the costs of simply doing nothing are too catastrophic to ignore. While current research in the practical applications of ENMs for agricultural purposes is progressing, this is at a much slower pace than is required to meet the forecast global food demand.

The plethora of simple laboratory-scale phytotoxicity studies conducted to date (i.e. root elongation and seed germination) often routinely present conflicting results on the effects of exposure to metal oxide ENMs. Similarly, as the experimental systems become more complex (i.e. moving from hydroponic to pot trials involving plant growth experiments in natural soils), these discrepancies tend to increase. What is clear is that the interactions between plants and metal oxide ENMs are extremely complex and may involve many different individual factors which interactively control the fate and transport of ENMs in agricultural systems. Variables such as size and type of ENMs, environmental conditions, plant species, and mode of application (foliar or traditional ENM application) may dictate whether ENMs are accumulated by plants in the short term and whether any long-term negative or positive physiological responses occur.

6.1   Future Directions

Basic challenges still exist in the provision of robust analytical tools to identify ENMs in plant compartments that are distinguishable from bulk elements commonly present in the environment. This is especially true of elements such as Ti and Fe which form important metal oxide ENMs but are also ubiquitous at high levels in soils. Such methods are also challenged by complex biotransformations that can occur external to the plant as well as subsequently internally following accumulation of ENMs within a plant. Thus the proper examination of the fate and transport of ENMs requires the ability to examine a myriad of potential transformation products often at trace levels.

Although a large number of studies have investigated the effects of ENMs on agriculturally important crops, field-based research investigating the long-term effect of ENMs on agricultural crops as well as soil microcosms are rare, largely due to the limitation of the exposure of ENMs to the field without a complete understanding of their subsequent effect on the ecosystem. Indeed almost all of the experiments conducted to date have been performed under laboratory controlled greenhouse conditions and thus lack the ability to properly assess the role of a wider array of processes that occur under field conditions such as diversity of bacterial community, natural existence of ENMs in soil, and weathering conditions (i.e. rain, temperature, and wind erosion). Thus, there is a need for further investigation on ENM interactions with plants under conditions that are much more representative of those that occur in the field. For example, where field trials are legislatively prohibitive due the concerns of ENM release, experiments using lysimeters, could be conducted to provide more realistic conditions for the study of the environmental behaviour of ENMs and to more accurately predict their transport to crop tissues and eventual effect on yield nutritional quality as well as soil microcosms. In addition, currently available data on the effects of ENMs on grain yield and quality is not adequate and future detailed studies are needed including the effects of specific ENMs on the amino acid, carbohydrate, protein, and sugar content of grains as these are all essential nutrients in the human diet and none have been commonly evaluated to date.

The effects of ENMs on microbial communities in agricultural soils also require further investigation. It now seems increasingly likely that microbial communities have some significant effect on N and P cycling, which potentially influences crop growth. Moreover, the accumulation of ENMs in crop roots means that in the longer-term root decomposition will result in hotspots for ENMs release into the soil which would almost certainly affect both fungal and microbial soil communities.

Thus, future critical research efforts are needed to understand the interactions between metal oxide ENMs and diverse soil microbial communities to quantify the direct potential effects of ENMs on soil microbial communities and the consequential indirect effects on crop production.

Overall, in the agricultural sector, where continuous innovation is required to ensure economic sustainability and long-term food security, nanotechnology is widely recognised as a 'Key Enabling Technology' (Parisi et al. 2015). However, at the moment, the application of ENMs in agriculture as a mature research field remains in its infancy and considerably more research needs to be urgently conducted to ensure the safe practical adoption of this technology in the future.

References

Agata Tyczewska, E. W., Gracz, J., Kuczynski, J., & Twardowski, T. (2019). Towards food security: Current state and future prospects of Agrobiotechnology. *Trends in Biotechnology, 36*(12), 11.

Andersen, C. P., et al. (2016). Germination and early plant development of ten plant species exposed to TiO_2 and CeO_2 nanoparticles. *Environmental Toxicology and Chemistry, 9999*, 1–7.

Arruda, S. C. C., et al. (2015). Nanoparticles applied to plant science: A review. *Talanta, 131*, 693–705.

Boonyanitipong, P., et al. (2011). Toxicity of ZnO and TiO_2 nanoparticles on germinating rice seed Oryza sativa L. *International Journal of Bioscience. Biochemistry and Bioinformatics, 1*(4), 282.

Colvin, V. L. (2003). The potential environmental impact of engineered nanomaterials. *Nature Biotechnology, 21*(10), 1166–1170.

Du, W., et al. (2011). TiO_2 and ZnO nanoparticles negatively affect wheat growth and soil enzyme activities in agricultural soil. *Journal of Environmental Monitoring, 13*(4), 822–828.

Du, W., et al. (2016). Interaction of metal oxide nanoparticles with higher terrestrial plants: Physiological and biochemical aspects. *Plant Physiology and Biochemistry, 110*, 210.

Ebbs, S. D., et al. (2016). Accumulation of zinc, copper, or cerium in carrot (Daucus carota) exposed to metal oxide nanoparticles and metal ions. *Environmental Science: Nano, 3*, 114.

Engates, K. E., & Shipley, H. J. (2011). Adsorption of Pb, Cd, Cu, Zn, and Ni to titanium dioxide nanoparticles: Effect of particle size, solid concentration, and exhaustion. *Environmental Science and Pollution Research, 18*(3), 386–395.

Esfand, R., & Tomalia, D. A. (2001). Poly (amidoamine)(PAMAM) dendrimers: From biomimicry to drug delivery and biomedical applications. *Drug Discovery Today, 6*(8), 427–436.

Farré, M., et al. (2009). Ecotoxicity and analysis of nanomaterials in the aquatic environment. *Analytical and Bioanalytical Chemistry, 393*(1), 81–95.

Feizi, H., et al. (2012). Impact of bulk and nanosized titanium dioxide (TiO_2) on wheat seed germination and seedling growth. *Biological Trace Element Research, 146*(1), 101–106.

Hernandez-Viezcas, J. A., et al. (2016). Interactions between CeO_2 nanoparticles and the desert plant mesquite: A spectroscopy approach. *ACS Sustainable Chemistry & Engineering, 4*, 1187.

Hong, J., et al. (2014). Evidence of translocation and physiological impacts of foliar applied CeO2 nanoparticles on cucumber (Cucumis sativus) plants. *Environmental Science & Technology, 48*(8), 4376–4385.

Hund-Rinke, K., Schlich, K., & Klawonn, T. (2012). Influence of application techniques on the ecotoxicological effects of nanomaterials in soil. *Environmental Sciences Europe, 24*(12), 1–30.

Klaine, S., et al. (2008). Nanomaterials in the environment: Behaviour, fate, bioavialability , and effects. *Environmental Toxicology and Chemistry, 27*(9), 26.

Krug, H. F., & Wick, P. (2011). Nanotoxicology: An interdisciplinary challenge. *Angewandte Chemie International Edition, 50*(6), 1260–1278.

Larue, C., et al. (2011a). Investigation of titanium dioxide nanoparticles toxicity and uptake by plants. in *Journal of Physics: Conference Series*. IOP Publishing.

Larue, C., et al. (2011b). Investigation of titanium dioxide nanoparticles toxicity and uptake by plants. *Journal of Physics: Conference Series, 304*, 012057/1-012057/7.

Larue, C., et al. (2012a). Comparative uptake and impact of TiO_2 nanoparticles in wheat and rapeseed. *Journal of Toxicology and Environmental Health, Part A, 75*(13–15), 722–734.

Larue, C., et al. (2012b). Accumulation, translocation and impact of TiO_2 nanoparticles in wheat (Triticum aestivum spp.): Influence of diameter and crystal phase. *Science of the Total Environment, 431*, 197–208.

Larue, C., et al. (2014). Fate of pristine TiO_2 nanoparticles and aged paint-containing TiO_2 nanoparticles in lettuce crop after foliar exposure. *Journal of Hazardous Materials, 273*, 17–26.

Lin, D., & Xing, B. (2007). Phytotoxicity of nanoparticles: Inhibition of seed germination and root growth. *Environmental Pollution, 150*(2), 243–250.

Lin, D., & Xing, B. (2008). Root uptake and Phytotoxicity of ZnO nanoparticles. *Environmental Science & Technology, 42*(15), 5580–5585.

Liu, R., & Lal, R. (2015). Potentials of engineered nanoparticles as fertilizers for increasing agronomic productions. *Science of the Total Environment, 514*, 131–139.

López-Moreno, M. L., et al. (2010). Evidence of the differential biotransformation and genotoxicity of ZnO and CeO_2 nanoparticles on soybean (Glycine max) plants. *Environmental Science & Technology, 44*(19), 7315–7320.

Lv, J., et al. (2015). Accumulation, speciation and uptake pathway of ZnO nanoparticles in maize. *Environmental Science: Nano, 2*(1), 68–77.

Ma, X., et al. (2010). Interactions between engineered nanoparticles (ENPs) and plants: Phytotoxicity, uptake and accumulation. *Science of the Total Environment, 408*(16), 3053–3061.

Ma, Y., et al. (2015). Origin of the different phytotoxicity and biotransformation of cerium and lanthanum oxide nanoparticles in cucumber. *Nanotoxicology, 9*(2), 262–270.

Mccutcheon, S. C., & Schnoor, J. L. (2004). *Phytoremediation: Transformation and control of contaminants* (Vol. 121). John Wiley & Sons.

Mushtaq, Y. K. (2011). Effect of nanoscale Fe3O4, TiO2 and carbon particles on cucumber seed germination. *Journal of Environmental Science and Health, Part A, 46*(14), 1732–1735.

Nowack, B., & Bucheli, T. D. (2007). Occurrence, behavior and effects of nanoparticles in the environment. *Environmental Pollution, 150*(1), 5–22.

Parisi, C., Vigani, M., & Rodríguez-Cerezo, E. (2015). Agricultural nanotechnologies: What are the current possibilities? *Nano Today, 10*(2), 124–127.

Peng, C., et al. (2015). Translocation and biotransformation of CuO nanoparticles in rice (Oryza sativa L.) plants. *Environmental Pollution, 197*, 99–107.

Priester, J. H., et al. (2012). Soybean susceptibility to manufactured nanomaterials with evidence for food quality and soil fertility interruption. *Proceedings of the National Academy of Sciences, 109*(37), E2451–E2456.

Rengel, Z., Batten, G. D., & Crowley, D. E. (1999). Agronomic approaches for improving the micronutrient density in edible portions of field crops. *Field Crops Research, 60*(1–2), 27–40.

Rico, C. M., et al. (2013a). Effect of cerium oxide nanoparticles on rice: A study involving the antioxidant defense system and in vivo fluorescence imaging. *Environmental Science & Technology, 47*(11), 5635–5642.

Rico, C. M., et al. (2013b). Effect of cerium oxide nanoparticles on the quality of rice (Oryza sativa L.) grains. *Journal of Agricultural and Food Chemistry, 61*(47), 11278–11285.

Rico, C. M., et al. (2014). Cerium oxide nanoparticles impact yield and modify nutritional parameters in wheat (Triticum aestivum L.). *Journal of Agricultural and Food Chemistry, 62*(40), 9669–9675.

Rosegrant, M. W., & Cline, S. A. (2003). Global food security: Challenges and policies. *Science, 302*(5652), 1917–1919.

Servin, A. D., et al. (2012a). Synchrotron micro-XRF and micro-XANES confirmation of the uptake and translocation of TiO₂ nanoparticles in cucumber (Cucumis sativus) plants. *Environmental Science & Technology, 46*(14), 7637–7643.

Servin, A. D., et al. (2012b). Synchrotron micro-XRF and micro-XANES confirmation of the uptake and translocation of TiO₂ nanoparticles in cucumber (Cucumis sativus) plants. *Environmental Science & Technology, 46*(14), 7637–7643.

Servin, A. D., et al. (2013). Synchrotron verification of TiO₂ accumulation in cucumber fruit: A possible pathway of TiO₂ nanoparticle transfer from soil into the food chain. *Environmental Science & Technology, 47*(20), 11592–11598.

Shipley, H. J., Engates, K. E., & Guettner, A. M. (2011). Study of iron oxide nanoparticles in soil for remediation of arsenic. *Journal of Nanoparticle Research, 13*(6), 2387–2397.

Skrabal, S. A., & Terry, C. M. (2002). Distributions of dissolved titanium in porewaters of estuarine and coastal marine sediments. *Marine Chemistry, 77*(2–3), 109–122.

Song, U., et al. (2013). Functional analyses of nanoparticle toxicity: A comparative study of the effects of TiO₂ and Ag on tomatoes *(Lycopersicon esculentum). Ecotoxicology and Environmental Safety, 93*, 60–67.

Stampoulis, D., Sinha, S. K., & White, J. C. (2009). Assay-dependent phytotoxicity of nanoparticles to plants. *Environmental Science & Technology, 43*(24), 9473–9479.

Tang, Y. J. (2013). et al. Phytotoxicity of metal oxide nanoparticles is related to both dissolved metals ions and adsorption of particles on seed surfaces. Journal of Petroleum & Environmental Biotechnology, 2012.

Wang, Z., et al. (2012). Xylem- and phloem-based transport of CuO nanoparticles in maize (Zea mays L.). *Environmental Science & Technology, 46*(8), 4434–4441.

Yang, L., & Watts, D. J. (2005). Particle surface characteristics may play an important role in phytotoxicity of alumina nanoparticles. *Toxicology Letters, 158*(2), 122–132.

Zhang, P., et al. (2012). Biotransformation of ceria nanoparticles in cucumber plants. *ACS Nano, 6*(11), 9943–9950.

Zhang, W., et al. (2015). Uptake and accumulation of bulk and nanosized cerium oxide particles and ionic cerium by radish (Raphanus sativus L.). *Journal of Agricultural and Food Chemistry, 63*(2), 382–390.

Zhao, L., et al. (2012). Transport of Zn in a sandy loam soil treated with ZnO NPs and uptake by corn plants: Electron microprobe and confocal microscopy studies. *Chemical Engineering Journal, 184*, 1–8.

Zheng, L., et al. (2005). Effect of nano-TiO₂ on strength of naturally aged seeds and growth of spinach. *Biological Trace Element Research, 104*(1), 83–91.

Zhu, H., et al. (2008). Uptake, translocation, and accumulation of manufactured iron oxide nanoparticles by pumpkin plants. *Journal of Environmental Monitoring, 10*(6), 713–717.

Chapter 10
The Role of Zinc Oxide Nanoparticles in Plants: A Critical Appraisal

Amit Kumar, Indrakant K. Singh, Rashmi Mishra, Akanksha Singh, Naleeni Ramawat, and Archana Singh

Contents

1   Introduction

Zinc (Zn) is a crucial micronutrient that is required for the appropriate growth and development of plants. Zn not only participates in protein synthesis and metabolism of biomolecules, but it also encourages protection against environmental stresses. Zn acts as cofactor of above 300 proteins, among which majority are DNA and RNA polymerases and zinc finger proteins (Coleman 1998; Lopez Millan et al. 2005).

A. Kumar · A. Singh (✉)
Department of Botany, Hansraj College, University of Delhi, Delhi, India
e-mail: archanasingh@hrc.du.ac.in

I. K. Singh (✉)
Molecular Biology Research Lab, Department of Zoology, Deshbandhu, College, University of Delhi, New Delhi, India
e-mail: iksingh@db.du.ac.in

R. Mishra
Department of Biotechnology, Noida Institute of Engineering and Technology, Greater Noida, India

A. Singh · N. Ramawat
Amity Institute of Organic Agriculture, Amity University Uttar Pradesh, Noida, India

© Springer Nature Switzerland AG 2021
N. Sharma, S. Sahi (eds.), *Nanomaterial Biointeractions at the Cellular, Organismal and System Levels*, Nanotechnology in the Life Sciences, https://doi.org/10.1007/978-3-030-65792-5_10

It is the only metal present which regulates the activity of all six enzyme classes (oxidoreductase, transferase, hydrolases, lyases, isomerases and ligases) (Lacerda et al. 2018). Zn also contributes to cell division, synthesis of tryptophan/ auxin, maintenance of membrane structure and chloroplast function (Lacerda et al. 2018; Marschner 2011). Moreover, Zn also causes regulation of alteration in gene expression during biotic and abiotic stresses. The deficiency of Zn causes reddish-brown patches on the lamina of subterminal leaf, inward curling of leaf, interveinal chlorosis, reduced leaf size, stunted growth, necrosis in root apex, decrease in crop yield and quality (Schutzendubel and Polle 2002; Hafeez et al. 2013). Thus, a sufficient quantity of Zn is necessary for plant growth, development and crop yield.

Application of nanoparticles has gained the momentum as a promising approach for sustainable agriculture to maintain the adequate amount of available form of micronutrients. Use of nanoparticles in the field of agriculture is one of the greatest boons of the twenty-first century. This technology has impacted in every arena of science and technology including agriculture. The application of nanoparticles (NPs) reveals potential advantages for crop improvement, crop protection against abiotic and biotic stresses, and post-harvest management (Usman et al. 2020). Nanoparticles exhibit advantages in terms of their unique properties such as extreme ultra-fine size (1–100 nm) and physiochemical characteristics, in comparison of bulk forms of the same elements (Faizan et al. 2020). Nanoparticles have the potential to overcome the harmful effects of insecticides, pesticides and fertilizers and introduce plant's natural source of nutrition. Nanoparticles help in maintenance of plant's nutrition, disease management and soil remediation. Among the metallic nanoparticles, zinc oxide (ZnO) has received focused attention since they are stable and possess good conductivity and catalytic properties. Additionally, they harbour antimicrobial properties. These features of ZnO NPs are good enough to select it as an entity useful for scientific and industrial applications. Development of engineered nanoparticles based on metals: silver (Ag), iron (Fe), zinc (Zn), copper (Cu) or metal oxides such as titanium dioxide (TiO_2) and zinc oxide (ZnO) has created a pathway for their potential application as nanofertilizers (Liu and Lal 2015).

ZnO plays effective role in physiological and anatomical responses in plants and utilization of zinc oxide particles (ZnO NPs) in agriculture has been accelerated and gained importance (Agarwal et al. 2017). Zinc in the form of capsules of size 1–100 nm can be utilized as a Zinc nanoparticle (Zn NPs). ZnO NPs are safer and relatively less toxic than its other forms such as $ZnSO_4$ NPs (Du et al. 2019). ZnO NPs are distinctive due to their greatly ionic nature, elevated surface areas and extraordinary crystal structures (Klabunde et al. 1996). It is a durable moiety with greater selectivity and heat resistance. Moreover, Zn is also important for human health and ZnO NPs are compatible to human cells (Padmavathy and Vijayaraghavan 2008). Further, ZnO NPs have been shown to enhance nutrient uptake, seed germination, growth of plants and crop yield (Prasad et al. 2012; Laware and Raskar 2014; Upadhyaya et al. 2015, Upadhyaya et al. 2016). Additionally, Zn NPs have been associated with abiotic stress tolerance in plants such as temperature stress, drought, heavy metal stress and salinity by increasing the concentration of antioxidant enzymes and metabolites (Baybordi 2005; Taran et al. 2017; Venkatachalam et al. 2017a, b). Furthermore, Zn nanomaterials strengthen plant defence against pests and pathogens and can be applied as bacteriostatic agents for managing the infection caused due to pathogens (Raskar and Laware 2014). Their roles in improving germination in seed, seedling vigour, radicle and plumule length were recently reported (Singh et al. 2016). These NPs are especially effective in cases where pathogenic microbes have gained multidrug resistance. Many investigators have verified their antibacterial and antifungal properties (Lamsal et al. 2011; Soria-Castro et al. 2019). Zn compounds/Zinc nanopowders are well used as fungicides and pesticides (Raikova et al. 2006). Therefore, in this chapter, we have scrutinized various roles of ZnO NPs as growth boosters and stress relieving agent for plants. We have also discussed the importance of ZnO NPs in agriculture due to its ameliorative effects on biotic and abiotic stresses.

1.1   Uptake and Transport of Zn NPs in Plants

Plants uptake available micronutrients by roots from the soil. Zn, one of the essential micronutrients, is uptaken in the form of Zn^{2+} or in complex with chelates. It is translocated to the shoot through xylem via transporters such as members of ZIP (ZRT IRT-like proteins), NRAMP (Natural resistance-associated macrophage protein), YSL (Yellow stripe like), PCR (Plant cadmium resistance), MTP (Metal tolerance protein) and HMP (Heavy Metal ATPases) families (Blaudez et al. 2003; Krämer 2005; Yoneyama et al. 2015). Further, Zn^{2+} reaches to assimilatory organs through symplastic movement via phloem (Lin and Xing 2008). Plants can also take Zn^{2+} from leaf surface, if Zn compounds / ZnO NPs are sprayed on it (Du et al. 2019). However the mechanism of Zn transport through leaf is not completely understood. ZnO NPs uptake and transport is well studied in tomato, barley, ryegrass, rice and common bean (Milner et al. 2013; Tiong et al. 2014; Lin and Xing 2008; da Cruz et al. 2019).

During soil-to-root uptake, ZnO NPs aggregate in the rhizosphere, enter into the root cells either apoplastically or symplastically and reaches to vascular bundle from where it is transported to its sink (Fig. 10.1). Zn molecules in soil are present in soluble form and roots also exudate organic acid, i.e. mucilage on the root surface, which helps in dissolution of Zn. Mucilage is a hydrated polysaccharide, pectic compound around roots which enhances Zn aggregation on the root surface. This increases the Zn concentration in the nearby root and Zn ions start moving towards the concentration gradient through ion pores in roots. After adsorption, ZnO NPs increase the permeability of cell wall by making "holes" in the wall and move through plasmodesmata showing symplastic movement. Plasmodesmata facilitate the transport of ZnO NPs. Alternatively, ZnO NPs can also enter through apoplastic pathway through epidermis and cortex. NPs enter into the vascular tissue after crossing the protoplast of endodermal cells. In rye grass's cells ZnO NPs (NPs: diameter ~ 19 nm) enter into the cell plasmodesmata with ~40 nm diameter and then move towards the stele (Lin and Xing 2008).

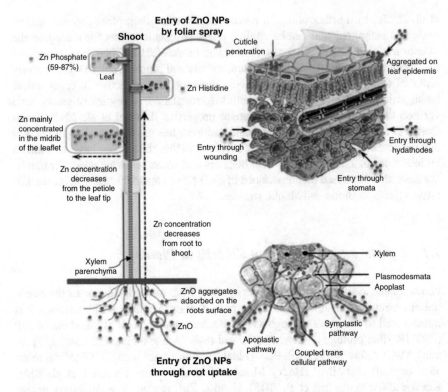

Fig. 10.1 A Schematic model showing the uptake and transport of ZnO NPs from root to shoot

Root-to-shoot translocation occurs because of decreasing concentration gradient of Zn concentration. Most likely, there are two mechanisms of Zn transport from root to shoot: radial Zn movement and axial xylem transport. Axial transport system carries the ions from the lower region of the stem to upper region, via xylem (Fig. 10.1). The Zn content and velocity decreases while getting transported away from root. Simultaneous radial movement delivers the Zn from xylem towards cortical cells. Cortex tissue stores transported Zn for nutrition pathway and for defence purpose strategy, when Zn is not able to reach leaves via axial mode of transport (da Cruz et al. 2019). During transport of Zn into the petiole and leaflet, Zn concentration gradient is created and it decreases from petiole to the tip of leaf. In leaves Zn is mainly present in two forms: Zn phosphate and Zn-histidine complex and Zn-malate.

2   Positive Impact of Zn NPs in Plants

Zn NPs supplementations promote seed germination, plant growth, development and improve crop yield and quality (Fig. 10.2) (Kolenčík et al. 2019). Besides this, it is a source of micro-nutrition to the plants. When seeds of wheat were primed with

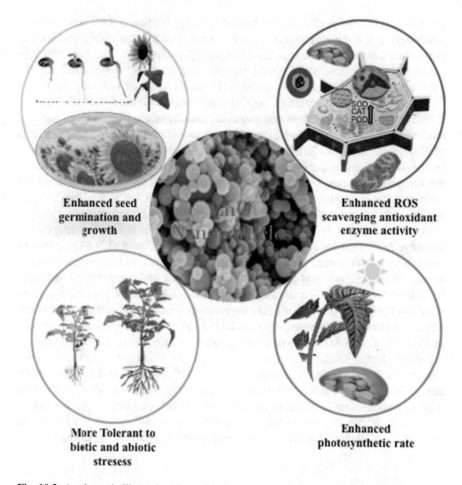

Fig. 10.2 A schematic illustration displaying the effects of ZnO NPs on plant growth, development and tolerance against abiotic and biotic stresses

ZnO NPs, the germinated plants showed increased plant growth, photosynthesis and biomass (Munir et al. 2018). Pre-treatment of seedlings with low and mild concentration of ZnO NPs (0–300 mg/L) revamps germination and seedling development (García-López et al. 2018). Recently, the positive effects of ZnO NPs were also observed on plant growth and performance in fabaceous, cucurbits and solanaceous plants (Dimkpa et al. 2017; Faizan et al. 2018; Sharifan et al. 2019). Spraying of ZnO NPs upgraded fruit harvest in pomegranate (Davarpanah et al. 2016) and mango trees (Zakzouk and Ibrahim 2017). ZnO NPs also improved the quality of cherry tomatoes by suppressing fruit ripening during post-harvest (Guo et al. 2020). ZnO NPs treated plants show early blossoming in onion and produced more number of healthy seeds (Laware and Raskar 2014). In foxtail millet, foliar spray of ZnO NPs improved the grain nutritional quality by slightly enhancing the oil and nitrogen contents (Kolenčík et al. 2019). Additionally, application of ZnO NPs results in manipulation and improvement in protein content and photosynthetic pigments in cluster bean, pearl millet, soybean and in *Cyamopsis tetragonoloba*. ZnO NPs supplementation also increased the total phenolic and flavonoid content and also the photosynthetic pigments in comparison to the treatment given with only ZnO.

In Safflower plant, presence of ZnO NPs increased the level of malondialdehyde. It is vital for activation of many enzymes; the activity of enzymes such as guaiacol oxidase, polyphenol oxide and dehydrogenase increases at different concentration of NPs (Hafizi and Nasr 2018). ZnO NPs as an effective elicitor increased the biosynthesis of tropane alkaloids such as scopolamine and hyoscyamine, upregulating the biosynthetic gene, hyoscyamine-6-β-hydroxylase *(h6h)*. This influenced the concentration of tropane alkaloid, transcript level of *h6h* and antioxidant enzyme activity in *H. reticulatus* L. hairy roots (Asl et al. 2019).

At molecular and biochemical level, ZnO NPs activate antioxidant enzymes to scavenge ROS. Positive effects of ZnO NPs have been identified in *Capsicum chinense* where it improved growth and metabolic markers such as germination of seeds, seedling vigour, and accumulation of biomass and nutraceutical attributes (total flavonoids, condensed tannins, total phenols, and DPPH antioxidant capacity) (García-López et al. 2019). ZnO NPs supplementation upregulated the antioxidant activity in *Portulaca oleracea* (Iziy et al. 2019). ZnO NPs also upgraded the activity of CAT, POX, and SOD enzymes and boosted proline concentration that further enhanced photosynthetic efficiency in tomato plants (Faizan et al. 2018). Soybean also has been demonstrated to enhance SOD, CAT and POX and APX and H_2O_2 scavenging activities under the influence of ZnO NPs. Therefore, ZnO NPs can be used as a novel nanofertilizer in Zn deficient soil for crop improvement (Yusefi-Tanha et al. 2020). Because of these positive effects of ZnO NPs, it is widely used and has many applications from seeds to post-harvest in modern agriculture.

2.1 Impact of ZnO NPs against Abiotic Stress

Abiotic stresses are major constraints for normal growth and development of plants. Drought, salinity, heat, ultraviolet radiation, chilling and metal toxicity are common abiotic stresses which are experienced by plants. ZnO NPs contribute towards alleviation of abiotic stresses (Table 10.1) by stimulating antioxidant activity, increasing osmolytes build up, stress-related unbounded amino acids and nutrients (Fig. 10.2).

Table 10.1 Effect of ZnO NPs during abiotic stress

S. no.	Plant species	Stress	Growth and development	References
1.	*Zea mays*	Drought	Drought tolerance	Sun et al. (2020)
2.	*Sorghum bicolor*	Drought	Fortify edible seeds with N, P and K. enhanced plant growth.	Dimkpa et al. (2019)
3.	*Glycine max*	Drought	Increases germination and; decreases residual and dry weight of seed	Dimkpa et al. (2017)
4.	*Oryza sativa* L.	Water	Stimulates growth and antioxidant responses	Upadhyaya et al. (2020)
5.	*Trigonella foenum-graecum*	Salinity	Depends on ZnO concentration and plant cultivar	Noohpisheh et al. (2020)
6.	*Eleusine coracana* L.	Salinity	Enhanced shoot growth and biomass. Root growth inhibited. High level of antioxidant enzymes, photosynthetic pigments	Haripriya et al. (2018)
7.	*Lupinus termis*	Salinity	Stimulated growth with high level of photosynthetic pigments, organic solutes, phenol, ascorbic acid, Zn and antioxidant enzymes.	Latef et al. (2017)
8.	*Solanum tuberosum* L.	Salinity	Enhanced growth	Mahmoud et al. (2020)
9.	*Gossypium barbadense* L.	Salinity	Enhanced growth parameters and yield. Excess use cause P/Zn imbalance	Hussein and Abou-Baker (2018)
10.	*Mangifera indica* L.	Salinity	Improved plant growth, nutrients uptake and carbon assimilation. High content of proline and antioxidant enzymes. Flower malformation decreases and increase in annual yield.	Elsheery et al. (2020)
11.	*Zea mays* L.	Cadmium	Decreases cd concentration uptake and enhances antioxidant enzymes.	Gowayed (2017)
12.	*Oryza sativa* L.	Arsenic	Resistance to as toxicity; increases germination rate, biomass, Zn and chlorophyll content, decreases MDA content	Wu et al. (2020)
13.	*Triticum aestivum* L.	Cadmium	Enhance the organic amendment effect; reduces cd effect and other toxic trace elements	Bashir et al. (2020); Hussain et al. (2018)
14.	*Triticum aestivum* L.	Cadmium	Minimize the Cd stress along with drought	Khan et al. (2019)
15.	*Spinaciae oleracea*	Cadmium, Lead	Mitigate the uptake of heavy metals; increase of Fe (10%) and Cu (4%) concentration	Sharifan et al. (2020)
16.	*Petroselinum sativum*	Cadmium, Lead	Mitigate the uptake of heavy metals; decrease of Fe (8%) and Cu (1.5%) concentration	Sharifan et al. (2020)

(continued)

Table 10.1 (continued)

S. no.	Plant species	Stress	Growth and development	References
17.	*Coriandrum sativum*	Cadmium, Lead	Mitigate the uptake of heavy metals; increase of Fe (9%) and Cu (8%) concentration	Sharifan et al. (2020)
18.	*Lycopersicon esculentum*	Cadmium	Reduces Cd toxicity; upregulation of antioxidative enzyme, increases biomass, physiological responses	Faizan et al. (2020)
19.	*Oryza sativa* L.	Cadmium, arsenic	Reduces As and Cd in grains	Ma et al. (2020)
20.	*Oryza sativa* L.	Cadmium	Diminished Cd concentration and increase the Zn uptake with biochar; improved biomass and photosynthesis of plants.	Ali et al. (2019)
21.	*Leucaena leucocephala* Lam.	Cadmium, lead	Reduces MDA content and alleviates antioxidative enzyme activity in leaf tissue seedling	Venkatachalam et al. (2017a)
22.	*Lactuca sativa*	Cadmium, lead	Reduces Cd and Pb accumulation; increases Fe uptake	Sharifan et al. (2019)
23.	*Glycine max*	Arsenic	Alleviates As toxicity via modulating antioxidant enzymes involved in ascorbate-glutathione cycle and glyoxalase system	Ahmad et al. (2020)
24.	*Arabidopsis thaliana*	Heat	Genomic instability; transcriptional gene silencing	Wu and Wang (2019)
25.	*Triticum aestivum* L.	Heat	Enhances yield quantity and increases antioxidant enzyme activities	Hassan et al. (2018)

2.1.1 Heavy Metal Stress

Plants are continuously exposed to heavy metals due to soil pollution, which hinders plants growth and development. When high concentrations of heavy metals [such as Arsenic (As), Lead (Pb) and cadmium (Cd)] are present in soil, they cannot be metabolized by plants and cause toxicity. Among all the heavy metal toxicity, Cd and As toxicity are more prevalent. ZnO NPs are suitable to alleviate Cd toxicity and enhancing growth, photosynthesis rate, chlorophyll a and b contents, antioxidant enzymes and protein content in *Lycopersicon esculentum* (Faizan et al. 2020). ZnO NPs (500 mg L^{-1}) protected maize seed germination and seedling growth from Cd toxicity by trespassing with Cd uptake from soil and activating seedling antioxidant mechanism which lowered free radical production (Gowayed 2017). ZnO NPs (10–200 mg L^{-1}) effectively provided resistance against As toxicity by increasing seed germination, biomass, and interfering As uptake in rice (Wu et al. 2020). ZnO NPs are also effective in reducing multi-metal toxicity caused by As and Cd simultaneously in a rice crop (Ma et al. 2020). Similarly, ZnO NPs alleviated Cd and Pb toxicity significantly in *Leucaena leucocephala* and *Lactuca sativa* L. var. *Longifolia* (Venkatachalam et al. 2017a; Sharifan et al. 2019). In the presence of ZnO NPs, upregulation of 24-epibrssinolide brassinosteroids is reported in tomato seedling, which alleviates heavy metal stress (Li et al. 2016). These findings endorse the utilization of ZnO NPs in agricultural fields for the improved crop yield and quality.

2.1.2   Heat Stress

Heat stress is faced by plants when there is a rise in temperature beyond optimum temperature for a particular period of time or prolonged exposure of high intensity light which causes irreversible damage (Wahid et al. 2007). Excess of heat or temperature results in loss of the reactive oxygen species scavenging system (Karuppanapandian et al. 2011). The response to tolerate heat stress due to ZnO NPs supplementation has been studied in few plants. ZnO NPs at a low concentration enhanced the plant genomic instability and triggered epigenetic changes to cope with heat stress (Wu and Wang 2020). ZnO NPs increased the life span of wheat during heat stress increasing the total yield (10 ppm ZnO NPs). The enhancement was associated with increased antioxidant enzyme activities (catalase, superoxide dismutase, Glutathione S transferase, and peroxidase) and decreased lipid peroxidation product: malondialdehyde (Hassan et al. 2018).

Drought Stress

Drought stress is caused due to the lack of rainfall and reduced replenishment of underground water that subjects crop plants to a limited supply of water. The main cause of this stress is anthropogenic activities such as excessive use of underground water, deforestation and many more. It is known that drought stress affects the subcellular organelle structure due to lipid peroxidation (Hu et al. 2018). It affects directly to the ultrastructure of chloroplast by damaging its internal component and results in a decrease in the starch granules accumulation (Xu et al. 2009). Drought stress has been demonstrated to decrease the nutrition and productivity of important crop plants such as wheat and soybean. Supplementation of crops with Zn ions enhances the productivity of the crops (Bagci et al. 2007; Karim et al. 2012; Dimkpa et al. 2017). Application of ZnO NPs on drought stressed maize plants improved photosynthetic rate, stomatal movement and enhances the efficiency of water use. This occurred because of the enhanced activity of enzymes: UDP-Glucose phosphorylase, phosphoglucoisomerase and cytoplasmic invertase. Indirectly, ZnO NPs induce regulation of important enzymes for carbohydrate metabolism to withstand the drought stress (Sun et al. 2020). As described before, during drought stress, modification of subcellular ultrastructure occurs, leading to accumulations of malondialdehyde and osmolytes. In Maize, ZnO NPs (100 mgL^{-1}) promote melatonin biosynthesis and antioxidant activities (enhanced transcript and protein levels of Fe/Mn SOD, Cu/Zn SOD, APX, CAT, TDC, SNAT, COMT, and ASMT), leading to protection from oxidative damage of the cell organelles, mitochondria and chloroplast in maize leaves (Sun et al. 2020). The effect of ZnO NP has been also studied on growth and physiological changes of *Oryza sativa* in relieving PEG induced water stress. In rice water stress can be overcome by ZnO NPs at lower concentrations (Upadhyaya et al. 2020). The role of NPs in mitigating water stress needs further investigation to determine the underlying mechanism.

2.1.3 Salinity Stress

Salinity stress is the topic of major concern in semi-arid and arid regions of the world. Salinity stress causes ionic imbalance and interference in osmosis event (Noohpisheh et al. 2020). The role of ZnO NPs for mitigating salinity stress has been studied in many crops such as millet, cotton and potato (Haripriya et al. 2018; Hussein and Abou-Baker 2018; Mahmoud et al. 2020). Effect of ZnO NPs on salinity stress was studied on *Trigonella foenum-graecum*, which grows mainly in semi-arid regions. Under salinity stress, high concentration of Na^+ inhibits K^+ and Ca^+ in shoot and root due to which lipid peroxidation occurs. ZnO NPs induce proline synthesis which scavenges ROS and free radicals. The effect of ZnO NPs on *T. foenum-graecum* during high salt concentration depends on the optimum levels of ZnO NPs, concentration of NaCl, and the type of cultivar (Noohpisheh et al. 2020). Seed priming with ZnO NPs (60 mg L^{-1}) was also deciphered to be an operative method that can be utilized to alleviate salt stress. Seed priming with ZnO NPs results in enhanced antioxidant enzyme activities, stimulate the growth and development of stressed plants and also led to increase in phenolic compounds, level of photosynthetic pigments, ascorbic acid. ZnO NPs also decreases the contents of salt and MDA in stressed plants (Latef et al. 2017). ZnO NPs supplementation increased the level of photosynthetic pigments and carotenoids in the leaves of finger millet. Moreover, the foliar application of ZnO NPs induced physiological and biochemical changes by activating antioxidative enzymes, which provide tolerance against salt stress in finger millet. The SOD converts superoxide anion (O^{2-}) free radicals into hydrogen peroxide (H_2O_2), which is further degraded by CAT and POX (Haripriya et al. 2018). The enhanced activity of these scavenging enzymes alleviates finger millet plants from oxidative damage (Haripriya et al. 2018). ZnO NPs in combination with Si NPs showed ameliorative effect on mango tree under salinity conditions (Elsheery et al. 2020).

2.2 Impact of ZnO NPs against Biotic Stress

Biotic stress includes pest and pathogen attack on plants. Biotic stresses are caused by virus, bacteria, fungi namatodes and herbivorous insects. Use of nanotechnology to control biotic stress is innovative and novel approach. NPs can also be used for early detection of phytopathogens. ZnO NP has antimicrobial property; therefore, it has been used for enhancing crop nutrition, biomass and overall productivity by inhibiting the growth of pest and pathogens (Anderson et al. 2017).

Table 10.2 Effect of ZnO NPs during biotic stress

S. no.	Plant species	Stress	Growth and development	References
1.	Mung bean broth	*Fusarium graminearum*	Antifungal activity	Dimkpa et al. (2013)
2.	Potato dextrose agar	*Botrytis cinera*, *Penicillium expansum*	Antifungal activity (*P. expansum* more sensitive)	He et al. (2011)
3.	*Coffea* sp.	*Colletotrichum kahawae*	Antifungal	Mosquera-Sánchez et al. (2020)
4.	*Pennisetum glaucum*	*Sclerospora graminicola*	Promote growth	Nandhini et al. (2019)
5.	*Lens culinaris* Medik.	*Alternaria alternate, fusarium oxysporum, Xanthomonas axonopodis, pseudomonas syringae, Meloidogyne incognita*	Enhance growth, pod number/plant, chlorophylls, carotenoid contents and nitrate reductase actvity	Siddiqui et al. (2018)
6.	Rosa 'Noare'	*Xanthomonas* sp.	Reduced bacterial spot	Paret et al. (2013)
7.	*Nicotiana tabacum*	*Peronospora tabacina*	Inhibit spore germination and leaf infectivity	Wagner et al. (2016)
8.	*Pongamia pinnata*	*Callosobruchus maculutus*	Delay developmental process and decreases enzyme activity of insects	Malaikozhundan and Vinodhini (2018)
9.	*Beta vulgaris* L.	*Pectobacterium betavasculorum, pseudomonas syringae Xanthomonas campestris*	Enhanced plant growth and enzyme activity	Siddiqui et al. (2019)
10.	*Citrus paradisi* Macf.	*Xanthomonas citri*	Antimicrobial activity	Graham et al. (2016)

Another advantage of using ZnO NPs lies in the fact that they are less toxic to plants as well as associated soil microbes compared to other metal oxide NPs (Dimkpa et al. 2013). ZnO NPs are also applied in combination with other metal oxide to enhance its activity against microbes (Table 10.2). ZnO NPs have been successfully applied as a nanopesticide for controlling pest attack in plants by regulating stress-related genes. ZnO NPs alleviate fungal, bacterial as well as viral pathogen-related manifestations (Table 10.2). ZnO NPs in combination with anti-fungal bacterial strains could inhibit the growth of *Fusarium graminearum* as a potent fungicide (Dimkpa et al. 2013). The growth of *Penicillium expansum* and *Botrytis cinerea* was significantly inhibited by ZnO NPs (He et al. 2011), *Colletotrichum kahawae* (Mosquera-Sánchez et al. 2020) and *Sclerospora* sp. (Nandhini et al. 2019). ZnO NPs inhibited spore germination and further growth of tobacco pathogen *Peronospora tabacina* (Wagner et al. 2016). The ZnO NPs reduced the levels of Fe-chelating siderophore metabolites of *Pythium hyphae* by increasing the activity of ferric reductase. This indicates toxic effects of the Zn NPs against *Pythium*, especially in the presence of a metal chelator (Zabrieski et al. 2015).

Since ZnO NPs are less toxic and beneficial for controlling plant-pathogens and thus they are considered a better choice among metal NPs for alleviating pathogenic effects. Two derivatives of ZnO NPs, Zinkicide SG4 (plate like) and Zinkicide SG6 (particulate), were studied against *Xanthomonas citri* subsp. *citri* for the control of citrus canker on grapefruit tree. Zinkicides inhibited citrus scab and melanose fungal diseases caused by *Elsinoe fawcettii* and *Diaporthe citri*, respectively (Graham et al. 2016). Biofabricated ZnO NPs were used for seed priming and foliar spray, which suppressed spore germination of downy mildew and upregulated plant defence. Zn NPs elicited systemic resistance in pearl millet against *Sclerospora graminicola* and could serve as suitable method to manage downy mildew (Nandhini et al. 2019). ZnO NPs are efficient inhibitor of *Colletotrichum kahawae*, a fungus on coffee plant. Wherein an inhibition rate of 96% was reported with a dose of 1200 ppm ZnO NPs, when a popular fungicide produced an inhibition of 88% (Mosquera-Sánchez et al. 2020). Zn NPs are used for enhancing the photolytic activity of TiO_2 NPs for controlling *Xanthomonas* sp. which causes leaf spot on rosa 'noare' (Paret et al. 2013).

The response of ZnO NPs have been studied on lentil plant (*Lens culinaris* Medik.) inoculated with various fungal and bacterial pathogens *Meloidogyne incognita, Fusarium oxysporum* f. sp. *lentis, Pseudomonas syringae* pv. *Syringae Xanthomonas axonopodis* pv. *phaseoli, Alternaria alternata* (Siddiqui et al. 2018). Effects of ZnO NPs and titanium dioxide TiO_2 NPs were compared on *Xanthomonas campestris* pv. Beticola, *Pseudomonas syringae* pv. Aptata and *Pectobacterium betavasculorum* on growth, chlorophyll, metabolites contents, antioxidant enzymes and ROS of beetroot. The ZnO NPs exhibited better results with reduced indices of disease and enhanced plant growth than TiO_2 NPs, indicating the fact that ZnO NPs can be a finer option to use for disease management in beetroot (Siddiqui et al. 2019). ZnO–CuSi NPs were also applied in combination to control citrus canker disease at less than half of the harmful metallic rate of the CuO/ZnO antimicrobials (Young et al. 2017).

Moreover, ZnO NPs are equally effective on nematodes and pests. Spray of ZnO NPs results in reduction of leaf spot, wilt, galling, multiplication of nematode, blight disease severity indices and thereby protect plant against multiple pathogens (Siddiqui et al. 2019). Similarly, pesticidal effect of ZnO NPs coated in leaf extract of *Pongamia pinnata* was studied against the Pulse beetle, *Callosobruchus maculatus*. The hatchability and fecundity of *C.Maculatus* reduced on the basis of variable dose. There is significant delay in the larval, pupal and total development period and shows 100% mortality rate at 25 μg mL^{-1}. The result of the doses exhibits that ZnO NPs can be effectively utilized as alternate mode of pest control.

3 Negative Impacts of ZnO NPs

Recently, the use of metal-based NPs in agriculture has amplified significantly, which may replace the conventional toxic agrochemicals: insecticides/pesticides, but they can exacerbate metal-based contaminations side by side. Through contaminations of agricultural land, they can enter the food chain causing biomagnification and health hazard for humans as well as animals. Many ecosystems are already contaminated with heavy metals. Phytotoxicity of ZnO NPs was observed in a number of crop species including lettuce, soybean, cucumber, rice, corn and *Typha latifolia* (López-Moreno et al. 2010; Yin et al. 2017; Ma et al. 2020). Presence of Zn at a concentration more than 400 mg kg^{-1} in plant tissue is considered as Zn toxicity in plants and symptoms of Zn toxicity include stunted growth, interveinal chlorosis and distorted leaves. High concentration of ZnO NPs negatively impact photosynthesis rate by reducing chlorophyll content, hinder water conductance by affecting root activity and reduces transpiration by affecting stomatal movement (Xiao et al. 2019). The ZnO NPs also adversely affected the biomass, chlorophyll and sugar contents of *Arabidopsis*. High concentration of ZnO NPs caused enhanced ROS accumulation and high rate of lipid peroxidation in *Arabidopsis* (Khan et al. 2019). Similarly, high concentration of ZnO NPs supplementation reduced plant growth and photosynthetic efficiency in *Arabidopsis* and *Lycopersicum* (Wang et al. 2016; 2018). Zn toxicity severely affects a plant's metabolic processes. Since Zn is a core constituent of some vital proteins that have roles in RNA and DNA stabilization, excess of Zn may affect the stabilizations of RNA and DNA, finally causing genetic anomalies.

High concentration (500 mg L^{-1}) of ZnO NPs increases the antioxidant enzymatic activity (POD, APX, CAT) and non-enzymatic antioxidant content (phenolics, anthocyanin and flavanoids) activities in potato (Raigond et al. 2017), *Portulaca oleracea* (Iziy et al. 2019), *Capsicum chinense* (García-López et al. 2019) and *Brassica nigra* (Zafar et al. 2016). Moreover, ZnO NPs enhance the concentration of phenolics and anthocyanin content and increase the activity of catalase and peroxidase due to oxidative damage caused by NPs in potato which causes toxicity due to excess of oxidative stress in plants (Raigond et al. 2017). Additionally, accumulation of ZnO NPs in soil adversely affects the rhizospheric environment of soil. NPs alter the secondary metabolite of root-associated microbe of *Pseudomonas chlororaphis,* which has antimicrobial property and involved in plant protection and rhizosphere health (Fang et al. 2013; Goodman et al. 2016). Production of phenazines is also altered by these particles, which protect the plants against pathogens by increasing pyoverdine like siderophore indirectly enhancing Fe availability in the rhizosphere. Effects of ZnO NPs (diameter, ~85 nm) were investigated in the plants *Allium cepa, Nicotiana tabacum,* and *Vicia faba* and the results indicated that it could be cytotoxic and genotoxic and can cause downregulation of antioxidant mechanism and cell-cycle arrest. Long-term effects of ZnO NPs (diameter, ~85 nm) use also included membrane disintegrity, chromosomal aberrations, micronucleus formation, and breakdown of DNA strand in the meristem root of *Allium cepa* cells.

In *Vicia faba* and *Nicotiana tabacum*, high rate of ROS production and lipid peroxidation were also observed in the presence of ZnO NPs (diameter, ~85 nm) (Ghosh et al. 2016). Similarly, ZnO NPs damaged tobacco BY-2 cells due to lipid peroxidation and loss of membrane integrity, dysfunction of endoplasmatic reticulum and mitochondria, leading to programmed cell death (Balážová et al. 2020). High concentrations of ZnO NPs affected seed germination and seedling growth and enhanced the biosynthesis of phenolic compounds, finally showing its phytotoxic nature on *Capsicum annuum* (García-López et al. 2018). High concentration of ZnO NPs also caused genotoxicity in *Sesamum indicum* (Sadasivam et al. 2018).

4   Conclusion and Future Recommendation

Zn is one of the essential micronutrients required for optimal growth and development of plants. If there are low concentrations of bioavailable Zn in soils, which is widespread globally, Zn-fertilizers are augmented to the plot, but this does not serve the purpose as Zn is fixed with insoluble compounds, and plants remain deprived of it ultimately. Therefore, Zn NPs are applied as an alternative approach in order to meet the optimal requirement. Many well-designed studies have demonstrated the role of Zn NPs as an efficient nanofertilizer boosting growth, development and protection against abiotic and biotic stresses. Hence, Zn NPs are used for enhanced Zn Use Efficiency. However, evidence also shows that higher concentrations of Zn NPs, and even lower concentrations in case of some genotypes, are toxic to many plants. In this backdrop, Zn NPs of biological origin (green synthesis) have been tested and found to be less toxic and more effective than chemically synthesized nanomaterials. However, the green synthesis has its own limitations as the product characterization and refinement have not been standardized yet. Another aspect that deserves a serious consideration is about the fate of residual nanomaterials as how can they affect the soil microbiome and other soil-inhabiting invertebrates. Until future research shows the complete understanding of the mechanisms involved in ameliorative effects of Zn NPs, precautions must be exercised in its generous applications, specifically to food crops.

References

Agarwal, H., Kumar, S. V., & Kumar, R. (2017). A review on green synthesis of zinc oxide nanoparticles—An eco-friendly approach. *Resource Efficient Technology, 3*, 406–413.
Ahmad, P., Alyemeni, M., Al-Huqail, A., et al. (2020). Zinc oxide nanoparticles application alleviates arsenic (As) toxicity in soybean plants by restricting the uptake of as and modulating key biochemical attributes, antioxidant enzymes, ascorbate-glutathione cycle and glyoxalase system. *Plants, 9*(7), 825. 1.

Ali, S., Rizwan, M., Noureen, S., et al. (2019). Combined use of biochar and zinc oxide nanoparticle foliar spray improved the plant growth and decreased the cadmium accumulation in rice (*Oryza sativa* L.) plant. *Environmental Science and Pollution Research 26*(11), 11288–99

Anderson, A., McLean, J., Jacobson, A., & Britt, D. (2017). CuO and ZnO nanoparticles modify interkingdom cell signaling processes relevant to crop production. *Journal of Agricultural and Food Chemistry, 66*(26), 6513–6524.

Asl, K., Hosseini, B., Sharafi, A., & Palazon, J. (2019). Influence of nano-zinc oxide on tropane alkaloid production, h6h gene transcription and antioxidant enzyme activity in *Hyoscyamus reticulatus* L. hairy roots. *Engineering in Life Sciences, 19*(1), 73–89.

Bagci, S. A., Ekiz, H., Yilmaz, A., & Cakmak, I. (2007). Effects of zinc deficiency and drought on grain yield of field-grown wheat cultivars in Central Anatolia. *Journal of Agronomy and Crop Science, 193*(3), 198–206.

Balážová, Matej Baláž, Ľ., & Babula, P. (2020). Zinc oxide nanoparticles damage tobacco BY-2 cells by oxidative stress followed by processes of autophagy and programmed cell death. *Nanomaterials, 10*(6), 1066.

Bashir, A., Rizwan, M., Ali, S., et al. (2020). Effect of composted organic amendments and zinc oxide nanoparticles on growth and cadmium accumulation by wheat; a life cycle study. *Environmental Science and Pollution Research*, 1–11.

Baybordi, A. (2005). Effect of zinc, iron, manganese and copper on wheat quality under salt stress conditions. *Journal of Water Soil, 140*, 150–170.

Blaudez, D., Kohler, A., Martin, F., et al. (2003). Poplar metal tolerance protein 1 confers zinc tolerance and is an oligomeric vacuolar zinc transporter with an essential leucine zipper motif. *The Plant Cell, 15*(12), 2911–2928.

Coleman, J. E. (1998). Zinc enzymes. *Current Opinion in Chemical Biology, 2*, 222–234.

da Cruz, T. N. M., Savassa, S. M., Montanha, G. S., et al. (2019). A new glance on root-to-shoot in vivo zinc transport and time-dependent physiological effects of ZnSO4 and ZnO nanoparticles on plants. *Scientific Reports, 9*, 10416. https://doi.org/10.1038/s41598-019-46796-3.

Davarpanah, S., Tehranifar, A., Davarynejad, G., et al. (2016). Effects of foliar applications of zinc and boron nano-fertilizers on pomegranate (Punica granatum cv. Ardestani) fruit yield and quality. *Scientia Horticulturae, 210*, 57–64.

Dimkpa, C. O., Bindraban, P. S., Fugice, J., et al. (2017). Composite micronutrient nanoparticles and salts decrease drought stress in soybean. *Agronomy for Sustainable Development, 37*, 5. https://doi.org/10.1007/s13593-016-0412-8.

Dimkpa, C. O., McLean, J., Britt, D., & Anderson, A. (2013). Antifungal activity of ZnO nanoparticles and their interactive effect with a biocontrol bacterium on growth antagonism of the plant pathogen *Fusarium graminearum*. *Biometals, 26*(6), 913–924.

Dimkpa, C. O., Singh, U., Bindraban, P., et al. (2019). Zinc oxide nanoparticles alleviate drought-induced alterations in sorghum performance, nutrient acquisition, and grain fortification. *Science of the Total Environment, 688*, 926–934.

Du, W., Yang, J., Xiaoping, P., & Mao, H. (2019). Comparison study of zinc nanoparticles and zinc sulphate on wheat growth: From toxicity and zinc biofortification. *Chemosphere, 227*, 109–116.

Elsheery, N. I., Helaly, M., El-Hoseiny, H., & Alam-Eldein, S. (2020). Zinc oxide and silicone nanoparticles to improve the resistance mechanism and annual productivity of salt-stressed mango trees. *Agronomy, 10*(4), 55.

Faizan, M., Faraz, A., & Mir, A. R. (2020). Role of zinc oxide nanoparticles in countering negative effects generated by cadmium in *Lycopersicon esculentum*. *Journal of Plant Growth Regulation*, 1–15.

Faizan, M., Faraz, A., Yusuf, M., et al. (2018). Zinc oxide nanoparticle-mediated changes in photosynthetic efficiency and antioxidant system of tomato plants. *Photosynthetica, 56*(2), 678–686.

Fang, T., Watson, J. L., Goodman, J., Dimkpa, C. O., et al. (2013). Does doping with aluminum alter the effects of ZnO nanoparticles on the metabolism of soil pseudomonads? *Microbiological Research, 168*, 91–98.

García-López, J. I., Niño-Medina, G., Olivares-Sáenz, E., et al. (2019). Foliar application of zinc oxide nanoparticles and zinc sulfate boosts the content of bioactive compounds in habanero peppers. *Plants, 8*(8), 254.

García-López, J. I., Zavala-García, F., Olivares-Sáenz, E., et al. (2018). Zinc oxide nanoparticles boosts phenolic compounds and antioxidant activity of *Capsicum annuum* L. during germination. *Agronomy, 8*(10), 215.

Ghosh, M., Jana, A., Sinha, S., et al. (2016). Effects of ZnO nanoparticles in plants: Cytotoxicity, genotoxicity, deregulation of antioxidant defenses, and cell-cycle arrest. *Mutation Research/ Genetic Toxicology and Environmental Mutagenesis, 807*, 25–32.

Goodman, J., Mclean, J. E., Britt, D. W., & Anderson, A. J. (2016). Sublethal doses of ZnO nanoparticles remodel production of cell signaling metabolites in the root colonizer *Pseudomonas chlororaphis* O6. *Environmental Science: Nano, 3*, 1103–1113.

Gowayed, S. (2017). Impact of zinc oxide nanoparticles on germination and antioxidant system of maize (*Zea mays* L.) seedling under cadmium stress. *Journal of Plant Production Sciences, 6*(1), 1–11.

Graham, J. H., Johnson, E., Myers, M., et al. (2016). Potential of nano-formulated zinc oxide for control of citrus canker on grapefruit trees. *Plant Disease, 100*(12), 2442–2447.

Guo, X., Chen, B., Wu, X., et al. (2020). Utilization of cinnamaldehyde and zinc oxide nanoparticles in a carboxymethylcellulose-based composite coating to improve the postharvest quality of cherry tomatoes. *International Journal of Biological Macromolecules, 160*, 175–182.

Hafeez, B., Khanif, Y., & Saleem, M. (2013). Role of zinc in plant nutrition-a review. *Journal of Experimental Agriculture International, 50*, 374–391.

Hafizi, Z., & Nasr, N. (2018). The effect of zinc oxide nanoparticles on safflower plant growth and physiology. *Engineering, Technology & Applied Science Research, 8*(1), 2508–2513.

Haripriya, P., Stella, P., & Anusuya, S. (2018). Foliar spray of zinc oxide nanopartcles improves salt tolerance in finger millet crops under glasshouse conditon. *SCIOL Biotechnology, 1*, 20–29.

Hassan, N. S., Salah El Din, T., Hendawey, M., et al. (2018). Magnetite and zinc oxide nanoparticles alleviated heat stress in wheat plants. *Current Nanomaterials, 3*(1), 32–43.

He, L., Liu, Y., Mustapha, A., & Lin, M. (2011). Antifungal activity of zinc oxide nanoparticles against Botrytis cinerea and Penicillium expansum. *Microbiological Research, 166*(3), 207–215.

Hu, W., Tian, S., Di, Q., Duan, S., & Dai, K. (2018). Effects of exogenous calcium on mesophyll cell ultrastructure, gas exchange, and photosystem II in tobacco (Nicotiana tabacum Linn.) under drought stress. *Photosynthetica, 56*(4), 1204–1211.

Hussain, A., Ali, S., Rizwan, M., et al. (2018). Zinc oxide nanoparticles alter the wheat physiological response and reduce the cadmium uptake by plants. *Environmental Pollution, 242*, 1518–1526.

Hussein, M. M., & Abou-Baker, N. H. (2018). The contribution of nano-zinc to alleviate salinity stress on cotton plants. *Royal Society Open Science, 5*(8), 171809.

Iziy, E., Majd, A., Reza Vaezi-Kakhkiet, M., et al. (2019). Effects of zinc oxide nanoparticles on enzymatic and nonenzymatic antioxidant content, germination, and biochemical and ultrastructural cell characteristics of Portulaca oleracea L. *Acta Societatis Botanicorum Poloniae, 4*, 88.

Karim, M. R., Zhang, Y.-Q., Zhao, R.-R., et al. (2012). Alleviation of drought stress in winter wheat by late foliar application of zinc, boron, and manganese. *Journal of Plant Nutrition and Soil Science, 175*(1), 142–151.

Karuppanapandian, T., Moon, J., Kim, C., et al. (2011). Reactive oxygen species in plants: Their generation, signal transduction, and scavenging mechanisms. *Australian Journal of Crop Science, 5*, 709–725.

Khan, A. R., Wakeel, A., Muhammad, N., et al. (2019). Involvement of ethylene signaling in zinc oxide nanoparticle-mediated biochemical changes in Arabidopsis thaliana leaves. *Environmental Science: Nano, 6*(1), 341–355.

Klabunde, T., Sträter, N., Fröhlich, R., et al. (1996). Mechanism of Fe (III)–Zn (II) purple acid phosphatase based on crystal structures. *Journal of Molecular Biology, 259*(4), 737–748.

Kolenčík, M., Ernst, D., Komár, M., et al. (2019). Effect of foliar spray application of zinc oxide nanoparticles on quantitative, nutritional, and physiological parameters of foxtail millet (*Setaria italica* l.) under field conditions. *Nanomaterials, 9*(11), 1559.

Krämer, U. (2005). MTP1 mops up excess zinc in Arabidopsis cells. *Trends in Plant Science, 10*(7), 313–315.

Lacerda, J. S., Martinez, H., Pedrosa, A., et al. (2018). Importance of zinc for arabica coffee and its effects on the chemical composition of raw grain and beverage quality. *Crop Science, 58*(3), 1360–1370.

Lamsal, K., Kim, S. W., Jung, J., et al. (2011). Application of silver nanoparticles for the control of Colletotrichum species in vitro and pepper anthracnose disease in field. *Mycobiology, 39*(3), 194–199.

Latef, A., Alhmad, M., & Abdelfattah, K. (2017). The possible roles of priming with ZnO nanoparticles in mitigation of salinity stress in lupine (Lupinus termis) plants. *Journal of Plant Growth Regulation, 36*(1), 60–70.

Laware, S. L., & Raskar, S. (2014). Influence of zinc oxide nanoparticles on growth, flowering and seed productivity in onion. *International Journal of Current Microbiology Science, 3*(7), 874–881.

Li, M., Ahammed, G., Li, C., et al. (2016). Brassinosteroid ameliorates zinc oxide nanoparticles-induced oxidative stress by improving antioxidant potential and redox homeostasis in tomato seedling. *Frontiers in Plant Science, 7*, 615.

Lin, D., & Xing, B. (2008). Root uptake and phytotoxicity of ZnO nanoparticles. *Environmental Science & Technology, 42*(15), 5580–5585.

Liu, R., & Lal, R. (2015). Potentials of engineered nanoparticles as fertilizers for increasing agronomic productions. *Science of the Total Environment, 514*, 131–139.

Lopez Millan, A. F., Ellis, D. R., & Grusak, M. A. (2005). Effect of zinc and manganese supply on the activities of superoxide dismutase and carbonic anhydrase in *Medicago truncatula* wild type and *raz* mutant plants. *Plant Science, 168*, 1015–1022.

López-Moreno, M., de la Rosa, G., Hernández-Viezcas, J., et al. (2010). Evidence of the differential biotransformation and genotoxicity of ZnO and CeO2 nanoparticles on soybean (Glycine max) plants. *Environmental Science & Technology, 44*(19), 7315–7320.

Ma, X., Sharifan, H., Dou, F., & Sun, W. (2020). Simultaneous reduction of arsenic (As) and cadmium (cd) accumulation in rice by zinc oxide nanoparticles. *Chemical Engineering Journal, 384*, 123802.

Mahmoud, A., Abdeldaym, E., Abdelaziz, S., et al. (2020). Synergetic effects of zinc, boron, silicon, and zeolite nanoparticles confer tolerance in potato plants subjected to salinity. *Agronomy, 10*(1), 19.

Malaikozhundan, B., Vinodhini, J. (2018). Nanopesticidal effects of Pongamia pinnata leaf extract coated zinc oxide nanoparticle against the Pulse beetle, *Callosobruchus maculatus*. *Materials Today Communications 14*, 106–15.

Marschner, H. (2011). Marschner's mineral nutrition of higher plants. Academic press: Aug 8.

Milner, M., Seamon, J., Craft, E., & Kochi an, L. (2013). Transport properties of members of the ZIP family in plants and their role in Zn and Mn homeostasis. *Journal of Experimental Botany, 64*(1), 369–381.

Mosquera-Sánchez, L. P., Arciniegas-Grijalba, P., Patiño-Portela, M., et al. (2020). Antifungal effect of zinc oxide nanoparticles (ZnO-NPs) on Colletotrichum sp., causal agent of anthracnose in coffee crops. *Biocatalysis and Agricultural Biotechnology, 25*, 101579.

Munir, T., Rizwan, M., Kashif, M., et al. (2018). Effect of zinc oxide nanoparticles on the growth and zn uptake in wheat (*Triticum aestivum* l.) by seed priming method. *Digest Journal of Nanomaterials & Biostructures, 1*, 13.

Nandhini, M., Rajini, S., Udayashankar, A., et al. (2019). Biofabricated zinc oxide nanoparticles as an eco-friendly alternative for growth promotion and management of downy mildew of pearl millet. *Crop Protection, 121*, 103–112.

Noohpisheh, Z., Amiri, H., Mohammadi, A., & Farhadi, S. (2020). Effect of the foliar application of zinc oxide nanoparticles on some biochemical and physiological parameters of *Trigonella foenum-graecum* under salinity stress. *Plant Biosystems, 1*–14.

Padmavathy, N., & Vijayaraghavan, R. (2008). Enhanced bioactivity of ZnO nanoparticles—An antimicrobial study. *Science and Technology of Advanced Materials, 9*(3), 035004.

Paret, M., Palmateer, A., & Knox, G. (2013). Evaluation of a light-activated nanoparticle formulation of titanium dioxide with zinc for management of bacterial leaf spot on rosa 'Noare'. *Horticulture Science, 48*(2), 189–192.

Prasad, T., Sudhakar, P., Sreenivasulu, Y., et al. (2012). Effect of nanoscale zinc oxide particles on the germination, growth and yield of peanut. *Journal of Plant Nutrition, 35*(6), 905–927.

Raigond, P., Raigond, B., Kaundal, B., et al. (2017). Effect of zinc nanoparticles on antioxidative system of potato plants. *Journal of Environmental Biology, 38*(3), 435.

Raikova, O. P., Panichkin, L. A. and Raikova, N. N. (2006). Studies on the effect of ultrafine metal powders produced by different methods on plant growth and development. Nanotechnologies and information technologies in the 21st century. In *Proceedings of the International Scientific and Practical Conference*, pp. 108–111.

Raskar, S. V., & Laware, S. L. (2014). Effect of zinc oxide nanoparticles on cytology and seed germination in onion. *International Journal of Current Microbiology Applied Science, 3*, 467–473.

Sadasivam, N., Periakaruppan, R., & Sivaraj, R. (2018). *Lantana aculeata* L.-mediated zinc oxide nanoparticle-induced DNA damage in *Sesamum indicum* and their cytotoxic activity against SiHa cell line. In Faisal et al. (Eds.), *Phytotoxicity of nanoparticles* (pp. 347–366). Springer.

Schutzendubel, A., & Polle, A. (2002). Plant responses to abiotic stresses: Heavy metal-induced oxidative stress and protection by mycorrhization. *Journal of Experimental Botany, 53*(372), 1351–1365.

Sharifan, H., Ma, X., Moore, J. M., Habib, M. R., & Evans, C. (2019). Zinc oxide nanoparticles alleviated the bioavailability of cadmium and Lead and changed the uptake of iron in hydroponically grown lettuce (*Lactuca sativa* L. var. Longifolia). *ACS Sustainable Chemistry & Engineering, 7*(19), 16401–16409.

Sharifan, H., Moore, J. M., & Ma, X. (2020). Zinc oxide (ZnO) nanoparticles elevated iron and copper contents and mitigated the bioavailability of lead and cadmium in different leafy greens. *Ecotoxicology and Environmental Safety, 191*, 110177.

Siddiqui, Z., Khan, M., Abd Allah, E., & Parveen, A. (2019). Titanium dioxide and zinc oxide nanoparticles affect some bacterial diseases, and growth and physiological changes of beetroot. *International Journal of Vegetable Science, 25*(5), 409–430.

Siddiqui, Z. A., Khan, A., Khan, M., & Abd-Allah, E. (2018). Effects of zinc oxide nanoparticles (ZnO NPs) and some plant pathogens on the growth and nodulation of lentil (*Lens culinaris* Medik.). *Acta Phytopathologica et Entomologica Hungarica, 53*(2), 195–211.

Singh, A., Singh, N. B., Hussain, I., et al. (2016). Green synthesis of nano zinc oxide and evaluation of its impact on germination and metabolic activity of *Solanum lycopersicum*. *Journal of Biotechnology, 233*, 84–94.

Soria-Castro, M., De la Rosa-García, S. C., Quintana, P., et al. (2019). Broad spectrum antimicrobial activity of ca(Zn(OH)$_3$)$_2$·2H$_2$O and ZnO nanoparticles synthesized by the sol–gel method. *Journal of Sol-Gel Science and Technology, 89*, 284–294. https://doi.org/10.1007/s10971-018-4759-y.

Sun, L., Song, F., Guo, J., et al. (2020). Nano-ZnO-induced drought tolerance is associated with melatonin synthesis and metabolism in maize. *International Journal of Molecular Sciences, 21*(3), 782.

Taran, N., Storozhenko, V., Svietlova, N., et al. (2017). Effect of zinc and copper nanoparticles on drought resistance of wheat seedlings. *Nanoscale Research Letters, 12*(1), 60.

Tiong, J., McDonald, G., Genc, Y., et al. (2014). HvZIP7 mediates zinc accumulation in barley (*Hordeum vulgare*) at moderately high zinc supply. *New Phytologist, 201*(1), 131–143.

Upadhyaya, H., Shome, S., Tewari, S., et al. (2015). *Effect of Zn nano-particles on growth responses of rice, nano technology–novel perspectives and prospects* (Vol. 13, pp. 978–993). New Delhi: McGraw Hill Education India Private Limited. ISBN.

Upadhyaya, H. D., Bajaj, D., Das, S., et al. (2016). Genetic dissection of seed-iron and zinc concentrations in chickpea. *Scientific reports 6*(1), 1–2.

Upadhyaya, H., Shome, S., Tewari, S., et al. (2020). Responses to ZnO nanoparticles during water stress in Oryza sativa L. *Journal of Stress Physiology & Biochemistry, 16*(2), 67.

Usman, M., Farooq, M., Wakeel, A., et al. (2020). Nanotechnology in agriculture: Current status, challenges and future opportunities. *Science of the Total Environment, 721*, 137778.

Venkatachalam, P., Jayaraj, M., Manikandan, R., et al. (2017a). Zinc oxide nanoparticles (ZnONPs) alleviate heavy metal-induced toxicity in *Leucaena leucocephala* seedlings: A physiochemical analysis. *Plant Physiology and Biochemistry, 110*, 59–69.

Venkatachalam, P., Priyanka, N., Manikandan, K., et al. (2017b). Enhanced plant growth promoting role of phycomolecules coated zinc oxide nanoparticles with P supplementation in cotton (*Gossypium hirsutum* L.). *Plant Physiology and Biochemistry, 110*, 118–127.

Wagner, G., Korenkov, V., Judy, J., & Bertsch, P. M. (2016). Nanoparticles composed of Zn and ZnO inhibit *Peronospora tabacina* spore germination *in vitro* and *P. tabacina* infectivity on tobacco leaves. *Nanomaterials, 6*(3), 50.

Wahid, A., Gelani, S., Ashraf, M., et al. (2007). Heat tolerance in plants: An overview. *Environmental and Experimental Botany, 61*(3), 199–223.

Wang, X., Li, Q., Pei, Z., & Wang, S. (2018). Effects of zinc oxide nanoparticles on the growth, photosynthetic traits, and antioxidative enzymes in tomato plants. *Biologia Plantarum, 62*(4), 801–808.

Wang, X., Yang, X., Chen, S., et al. (2016). Zinc oxide nanoparticles affect biomass accumulation and photosynthesis in Arabidopsis. *Frontiers in Plant Science, 6*, 1243.

Wu, F., Fang, Q., Yan, S., et al. (2020). Effects of zinc oxide nanoparticles on arsenic stress in rice (*Oryza sativa* L.): Germination, early growth, and arsenic uptake. *Environmental Science and Pollution Research International, 27*, 26974.

Wu, J., & Wang, T. (2020). Synergistic effect of zinc oxide nanoparticles and heat stress on the alleviation of transcriptional gene silencing in *Arabidopsis thaliana. Bulletin of Environmental Contamination and Toxicology, 104*(1), 49–56.

Xiao, L., Wang, S., Yang, D., et al. (2019). Physiological effects of MgO and ZnO nanoparticles on the Citrus maxima. *Journal of Wuhan University of Technology-Mater. Sci. Ed, 34*(1), 243–253.

Xu, Z. Z., Zhou, G. S., & Shimizu, H. (2009). Effects of soil drought with nocturnal warming on leaf stomatal traits and mesophyll cell ultrastructure of a perennial grass. *Crop Science, 49*, 1843–1851.

Yin, Y., Hu, Z., Du, W., et al. (2017). Elevated CO2 levels increase the toxicity of ZnO nanoparticles to goldfish (Carassius auratus) in a water-sediment ecosystem. *Journal of Hazardous Materials, 327*, 64–70.

Yoneyama, T., Ishikawa, S., & Fujimaki, S. (2015). Route and regulation of zinc, cadmium, and iron transport in rice plants (*Oryza sativa* L.) during vegetative growth and grain filling: Metal transporters, metal speciation, grain Cd reduction and Zn and Fe biofortification. *International Journal of Molecular Sciences, 16*(8), 19111–19129.

Young, M., Ozcan, A., Myers, M., et al. (2017). Multimodal generally recognized as safe ZnO/nanocopper composite: A novel antimicrobial material for the management of citrus phytopathogens. *Journal of Agricultural and Food Chemistry, 66*(26), 6604–6608.

Yusefi-Tanha, E., Fallah, S., Rostamnejadi, A., & Pokhrel, L. (2020). Zinc oxide nanoparticles (ZnONPs) as novel nanofertilizer: Influence on seed yield and antioxidant defense system in soil grown soybean (Glycine max cv. Kowsar). *Science of the Total Environment, 738*, 140240.

Zabrieski, Z., Morrell, E., Hortin, J., et al. (2015). Pesticidal activity of metal oxide nanoparticles on plant pathogenic isolates of Pythium. *Ecotoxicology, 24*(6), 1305–1314.

Zafar, H., Ali, A., Ali, J., et al. (2016). Effect of ZnO nanoparticles on Brassica nigra seedlings and stem explants: Growth dynamics and antioxidative response. *Frontiers in Plant Science, 7*, 535.

Zakzouk, U., & Ibrahim, A. (2017). Improving growth, flowering, fruiting and resistance of malformation of mango trees using nano-zinc. *Middle East Journal of Agriculture Research, 6*(3), 673–681.

Chapter 11
Fate and Effects of Engineered Nanomaterials in Agricultural Systems

Qingqing Li, Chuanxin Ma, Jason C. White, and Baoshan Xing

Contents

Q. Li · B. Xing
Stockbridge School of Agriculture, University of Massachusetts Amherst,
Amherst, MA, USA

J. C. White (✉)
Department of Analytical Chemistry, The Connecticut Agricultural Experiment Station,
New Haven, CT, USA
e-mail: Jason.White@ct.gov

C. Ma
Key Laboratory for City Cluster Environmental Safety and Green Development of the
Ministry of Education, Institute of Environmental and Ecological Engineering,
Guangdong University of Technology, Guangzhou, China

© Springer Nature Switzerland AG 2021
N. Sharma, S. Sahi (eds.), *Nanomaterial Biointeractions at the Cellular,
Organismal and System Levels*, Nanotechnology in the Life Sciences,
https://doi.org/10.1007/978-3-030-65792-5_11

1 Introduction

1.1 General Introduction of Engineered Nanomaterials

According to the US National Nanotechnology Initiative (NNI), nanotechnology is defined as the understanding and control of matter at dimensions between approximately 1 and 100 nanometers National Nanotechnology Initiative. The behavior and properties of particles in the dimension of 1–100 nanometers change dramatically, particularly at the ranges below 10 nanometers because of quantum effects. In addition, NP-properties including melting point, fluorescence, electrical conductivity, magnetic permeability, and chemical reactivity of particles are size-dependent and become more prominent at the smaller scale (Bhushan 2017). Thus, engineered nanomaterials (NMs) have been extensively designed and vastly applied in many fields (DeRosa et al. 2010; Farokhzad and Langer 2009; Mazzola 2003; Serrano et al. 2009). Nanoparticles (NPs) can be classified into different types based on their compositions, shapes, structures, etc (Portehault et al. 2018). In the material-based catalog, there are metal and metal oxide NPs, carbon NPs, inorganic mineral NPs and biopolymeric NPs, etc. Based on their dimensions, there are nano-dots (0 D), nano-wires (1 D), layered nanomaterials (2 D and 3 D), nano-spheres (3 D), nano-cubes, and other irregular or hierarchical 3 D nanomaterials (Makhlouf and Barhoum 2018).

1.2 Applications of Nanomaterials in Food Industry, Pharmaceutical and Self-Care, Environmental Remediation

Due to their high efficiency on inhibition of microorganisms growth, NPs have been used as novel antibacterial or antifouling agents in food industry (Banerjee et al. 2011). The efficiency of NP Ag, TiO_2, Cu, and Fe against drug-resistant bacteria in food packages, containers, and coating materials has not been extensively studied (Anyaogu et al. 2008; Boyer et al. 2010; Kwak et al. 2001; Lee et al. 2007b; Vatanpour et al. 2012). NPs can produce free radicals, mostly known as reactive oxygen species (ROS), to induce oxidative stress, and subsequently cause irreversible damages to bacteria. In addition, both mechanical damages and electrostatic interaction induced by NPs can severely disrupt bacteria membrane integrity (Hajipour et al. 2012). Bio-nanocomposites have been developed to replace conventional non-biodegradable petroleum-based plastic food packaging materials. Nanofillers (e.g., nanostarch, nanocellulose, nanochitosan/nanochitin, nanoproteins, and nanolipids) incorporated nanofilms are edible packaging materials (Jeevahan and Chandrasekaran 2019). Beside, NPs have been used as food additives as well. Synthetic amorphous nano-silica (E551) is the most commonly used one as clarifying agent for beverages, and as a free-flow and anti-caking agent in powdered food items (Peters et al. 2016). Titanium dioxide NPs are another commonly used food additive (E171) especially in dairy product and candy as a pigment to enhance the white color (McClements and Xiao 2017). NPs contain nutritional elements such as iron oxide been added as iron fortification for food as well as food colorant (Chaudhry and Castle 2011). Organic NPs (e.g., lipid, protein, and carbohydrate NPs) have also been designed and developed for food additive applications (McClements and Xiao 2017).

In recent years, nanotherapeutics have drawn great attentions in drug delivery and gene therapy. NPs can be applied as nanomedicine, nanocarrier, diagnostic nanodevices, sensors, etc. It can offer great opportunity to build smart nano systems for precise and targeted treatment for both animals and plants. Sophisticated therapeutic nano system can be constructed around tumor microenvironment in response to varied internal stimuli (pH, temperature, enzyme, redox, and H_2O_2), and external triggers (magnetic, photo and ultrasound) (Lu et al. 2016; Peer et al. 2007; Qiao et al. 2019; Shi et al. 2017). However, the relevant study is still at the early stage and their full use in the clinical practice still needs further investigation due to the lack of validation of the overall safety (Qiao et al. 2019). One extensively studied drug delivery system is Metal-Organic-Frameworks (MOFs)-based biomedicine. Through incorporating therapeutic molecules into nano-MOFs, loaded therapeutic molecules can be released in a controlled manner and free circulation in the blood stream can be ensured (He et al. 2015a; Horcajada et al. 2011; Simon-Yarza et al. 2018). Besides, NPs can be applied as a non-disruptive detoxification approach to achieve attenuated virulence during immune processing (Gou et al. 2014; Hu et al. 2013). Nano-sized hydroxyapatite (nHA) has been commonly used as bone tissues because of its biocompatibility and osteoconductive properties (Venkatesan and Kim 2014).

Owing to the high surface area-to-mass ratio and highly reactive surface, nanomaterials have shown potential in the removal of environmental pollutants and biological contaminants. NPs can serve as highly efficient absorbent for multiple pollutants, and degradation agent in catalytic systems. Carbon-based nanomaterials (CNMs) including graphene family materials (GFMs) (Ali et al. 2019a; Zhao et al. 2011), carbon nanotubes (CNTs) (Gupta et al. 2013; Ren et al. 2011), have been engineered as highly efficient adsorbents. CNMs can be assembled into structures with different dimensions (2D nanowire, 3D hydrogel/ aerosol) to be readily applied on site and to achieve easier retrieve and recovery of both contaminants and materials (He et al. 2018a; Liu et al. 2012). Using as electron transfer interfaces and well dispersive loading vehicles, CNMs have also been incorporated with other reactive and functional molecules to minimize aggregating and stacking induced deficiency of their adsorption, photoactivities, and antimicrobial capacities (Sapsford et al. 2013; Shen et al. 2020; Sun et al. 2012; Wang et al. 2012; Zhu et al. 2020). MOFs (Zhang et al. 2014), nZVI (Sun et al. 2012), MXenes (Zhang et al. 2018), and g-C_3N_4 (Cui et al. 2012; Mamba and Mishra 2016) have been considered novel nanomaterials that can be potentially used as absorbents in environmental remediation (Stefaniuk et al. 2016; Wu et al. 2018).

1.3 Verified and Potential Toxicity of Nanomaterials to Plants

During manufacture, use, transport, and disposal, the release of NPs into soil and aquatic environments will be inevitable (Gottschalk et al. 2013). Associated risk and potential toxicity of NPs to the environmental biota and humans have drawn massive attention. Engineered NPs applied as food packaging, food additive, pharmaceutical, and self-care products are considered the direct access to human body and potentially cause negative impacts on human health, while using as environmental remediation agents could be released into environment unintentionally and expose to microorganisms, plants, animals, and eventually human. The potential damages to organisms caused by NPs could imbalance the ecosystems. Moreover, NP releases from byproducts such as micro/nano plastics into the environment and food chain have raised considerable attentions due to the commonly recognized pollution, toxic effects of plastics as well as the exceptional transport rate and depth of their micro/nano-forms (Besseling et al. 2014; Cole and Galloway 2015; Koelmans et al. 2015; Velzeboer et al. 2014).

In addition to damaging the surface integrity and altering the charge balance, the excess amounts of ROS and the successive DNA damages induced by NPs in organisms have been well demonstrated as the primary toxic mechanisms (Fu et al. 2014; Ma et al. 2018). For metal and metal oxide NPs, the toxicity induced by the ions released from NP dissolution has been considered as a major contributor to the total toxic effect (Ma et al. 2015; 2018). In general, NP induced toxicity is dose-dependent, and high doses of NPs often cause irreversible or lethal damages to the organisms, while low doses or environmentally relevant concentrations of NPs display no impact or biostimulation on boosting plant growth (Dimkpa and Bindraban 2016; Ma et al. 2016). Recently, nano-bio-eco interfacial interactions and NP mediated two-way or three-way interactions have been brought up to fully understand the behavior of NPs in a connected ecosystem and the underlying mechanisms of nanotoxicity. Due to the complexity of multi-interfacial processes, and the accumulation of controversial results from literatures, nanotoxicity is still case-dependent and a general conclusion is unlikely to be revealed (He et al. 2015b; 2018b).

1.4 Current Status of Nanomaterial Application in Agriculture

The projected world population of 9.7 billion in 2050 (United Nations, D. o. E. a. S. A 2017; World Population Prospects 2019) will require a doubled increase in overall agricultural production compared to that of 2005 (Alexandratos and Bruinsma 2012). However, the statistics of Food and Agriculture Organization of the United States (FAO) indicated that 14% food loss in 2019 was mainly because of inadequate storage, processing, packaging, and limited shelf life (Food and Agriculture Organization of the United Nations 2019). Even though the projected agricultural capacity could meet the demand at the time of 2050, it would unlikely be maintained in a sustainable manner (Kah et al. 2018). The environmental pressure of the food system was estimated to increase 50 ~ 92% by 2050 compared to those of 2010 due to the absence of technical improvement and other mitigation measures (Springmann et al. 2018). Global agriculture is regarded as the leading cause of environmental degradation (Clark and Tilman 2017). Chemical fertilizer utilization is still accounted for the most crop production increase but also pose adverse effects on both agricultural environment and biota (Mueller et al. 2012). The main disadvantage of conventional fertilizers is the great loss of nutrients, which subsequently causes low use efficiency by plants.

Nanotechnology has been introduced into agriculture in order to maintain sustainable agri-food systems (Subramanian et al. 2015; Yin et al. 2018). As shown in Fig. 11.1, due to the exceptional properties, nano-scaled agrichemicals have shown potentials being used as (a) *nanofertilizers*, which can be customized with desired chemical components, improve the nutrient use efficiency through slow release, and boost the plant yield, while reduce adverse environmental impacts caused by nutrient leaching (Raliya et al. 2017), (b) *nanopesticides*, which can reduce the indiscriminate use of conventional pesticides, deliver pesticides to only targeted spots, mitigate contamination in food chains, and minimize human exposure, (c) *nanocarriers*, which have shown high loading capacities and can release the key ingredients in a more controlled timeline, (d) *nanosensors*, which can monitor plant health status and resource demands in a more fast and precise way by translating plant chemical signals into readable digital information.

Objectives of the current chapter: As nanomaterials are preferred in various agricultural applications, this chapter aims to examine their interactions at multiple interfaces: nano-foliar/root surface interactions (uptake, transport, deposit, dissolution), and effects of NPs on overall agriculture (effects on crop health and production, effects on agricultural environment–farmland soil, waterbody).

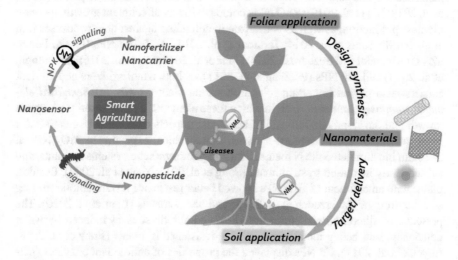

Fig. 11.1 A schematic illustration of the nanomaterial application in agriculture

In the past decades, the transport and fate of NPs in plant-soil systems have been extensively studied (Jassby et al. 2019; Lv et al. 2019; Ma et al. 2018). A common finding was that exposure to high dose and sharp shape of NPs could cause adverse effects on terrestrial plants, while beneficial effects within an appropriate dose and with functionalized modifications have drawn more attention in agriculture (Adisa et al. 2019; Dimkpa and Bindraban 2017; Guo et al. 2018; Kah et al. 2018; Servin and White 2016; Yin et al. 2018). Thus, in the following section, the fate and application of NMs at the nano-bio-eco interfaces have been addressed from the perspectives of NM types, size, and morphologies.

2   Nanomaterials Used for Agricultural Applications

As novel fertilizers, nanomaterials have been further classified as (a) *Macronutrient-incorporated nanomaterials* such as nano-hydroxyapatites (nHA) (Liu and Lal 2014), serve as phosphorus-based nanofertilizer with 18.2% and 41.2% increase in shoot and root biomass, respectively, as compared to its conventional analog $(Ca(H_2PO_4)_2)$. Nano-sized MgO, another essential macronutrient-based fertilizer, can provide enough magnesium and show size-dependent antibacterial activity (Cai et al. 2018; Liao et al. 2019); (b) *Micronutrient-incorporated nanomaterials*. Due to the unique properties, most of the NPs (e.g., metal, polymer, biomacromolecule) can effectively inhibit microbial growth (Servin et al. 2015). Mn, Cu, and Zn NPs have been demonstrated to suppress microorganisms induced diseases by activating the defense enzymes, phenylalanine ammonia lyase (PAL) and polyphenol oxidases (PPO), in plants (Datnoff et al. 2007). Besides, numerous studies have reported that Cu-based NPs, including CuO, Cu $(OH)_2$ nanosheets, and $Cu_3(PO_4)_2 \cdot 3H_2O$ nanosheets, could suppress fungal rot disease induced by *Fusarium* in plants (Elmer et al. 2019; Ma et al. 2019); ZnO NMs could positively affect plant growth, increase biomass production, as well as participate in defending against abiotic stresses (Ali et al. 2019b; Sabir et al. 2014; Tarafdar et al. 2014); Fe-based NPs $(Fe_2O_3, Fe_3O_4,$ nZVI) (Almeelbi and Bezbaruah 2014; Guha et al. 2018; Rui et al. 2016; Tombuloglu et al. 2019) and Mn NPs (Pradhan et al. 2013), as both nanofertilizers and nutrient carriers, also played important roles in boosting plant growth. (c) *Beneficial elements composed nanomaterials*, in which these elements are non-essential for plant growth and development, but exhibit beneficial impacts on helping plants to boost plant growth or defend against stresses. For instance, exposure to TiO_2 NPs at 5–10 nm biostimulated the N metabolism and photosynthetic systems in plants, and subsequently increased fresh biomass (Song et al. 2012; Yang et al. 2007). Besides, plants with amendment of TiO_2 NPs showed better resistance toward abiotic stresses such as drought (Jaberzadeh et al. 2013) and heavy metals (Lian et al. 2020). The presence of silicon NPs significantly alleviated abiotic stresses induced by water deficiency, salt, heavy metals, and lodging-resistance in plants (Frew et al. 2018; Luyckx et al. 2017). Ce NPs displayed the properties of antioxidant enzymes (peroxidase and phosphodiesterase mimetics) by scavenging reactive oxygen species

(ROS) and recovering phosphorus (Janos et al. 2019; Patel et al. 2018). Other NMs, such as hydroxyapatite (Kottegoda et al. 2011; Xiong et al. 2018a) zeolite (Bansiwal et al. 2006; Li et al. 2010; Zwingmann et al. 2011), nanoclay (Sarkar et al. 2015, 2014), and chitosan coated NPs (Abdel-Aziz et al. 2016; Kashyap et al. 2015; Saharan et al. 2016), can also serve as carriers for conventional nutrients (urea, ammonium, calcium phosphate, potassium chloride).

3 Nanomaterial Transport at Nano-Bio-Eco Interfaces

3.1 Nano-Foliar/Root Surface Interactions

It has been generally recognized that plant–NP interactions take place in three steps: (a) NP deposition and transformation on a plant surface (e.g., leaf, root, or stem), (b) NP penetration through the cuticle and epidermis, and (c) NP transport and transformation within the plants (Jassby et al. 2019). The diverse morphological and physiological features of the plant surfaces are results of adaption to environments. Higher plants have epidermis featured similar composition and structure. The above-ground parts of the higher plants are all covered by cuticle, a continuous extracellular membrane, which serves as the first barrier of plants and a multifunctional interface between biosphere and atmosphere. Small pores (<5 nm) can be formed during the deposition of epicuticular waxes and larger apertures, known as stomata (10–100 nm) on leaf surfaces (Eichert and Goldbach 2008; Eichert et al. 2008; Jassby et al. 2019). These apertures allow gas and water exchange and nanoparticles can transport through and penetrate into plant cells (Dietz and Herth 2011). In addition, there are 5–20 nm nanopores on plant cell walls (Fleischer et al. 1999) and 50–60 nm plasmodesma between cells, both of which act as channels for substance transport (Zambryski 2004). These features provide substantial advantages for nanoparticles to enter into plants and transport in plant bodies. Casparian strips on the endodermis can restrict apoplastic water flow into the root and make it difficult for NPs to transport into the intact roots. However, NPs can cause physical damages to plant cells, which could facilitate the NP uptake by plants. In addition, passage cells located near root hairs serve as an entry for water and nutrients transport from root hair to xylem, and thus have been considered as the only possible route for NPs to penetrate the endodermis in intact roots.

Due to the interferences of soil organic matters and minerals with nanoparticles, foliar application has been recognized a more efficient way to deliver nanoparticles into the plants. Upon exposure, nanoparticles deposit on the epicuticular waxes, trichomes, and stomata. The stomatal pathway is the main route for nanoparticles to enter the plants; stomatal loading capacity is considered the rate-limiting step for NPs entry into plants. Beside the stomatal pathway, the cuticular pathway is another important route for NP entry. Several studies were conducted to investigate the cuticular pathway for NPs uptake. For example, pristine TiO_2-NPs were internalized in the leaf cells of lettuce by damaging the cuticles and the cell walls (Larue et al. 2014). Ascribe to its less disruption to the epicuticular waxes, reduced phytotoxicity, and reduced chemical runoff, nanostructured liquid crystalline particles have been applied as an alternative to surfactant-based agrochemical delivery to the epidermis layer underneath the adaxial cuticle (devoid of stomata). For nanoparticles that can be dissolved, they can release ions that can be easily taken up by plants.

It is noteworthy that there are desired uptake and undesired uptake of NPs depending on the nature of whether the NPs are essential to plants. For NPs serve as fertilizer, their uptake and released ions are desired, while for those acting as pesticides, their retention on the plant surfaces are desired rather than their uptake by plants. Both agrochemical NPs are favored to extend their retention time on the plant surfaces for either facilitated uptake or longer reaction time with insects. On the above-ground surfaces of plants, NPs can go through interfacial behavior such as aggregation, dissolution, and transformation mediated by photocatalysis. Besides, due to weather change, NPs can be rinsed off by rainfall. To extend the stay of the NPs on the plant surfaces, modifications to increase the adhesion of NPs are conducted; meanwhile, compositing functional materials with enhanced adhesive property are developed. Chemical adhesive agents are often not favored due to their potential eco-toxicity. Inspired by natural glues such as adhesive protein secreted by mussel, polydopamine self-polymerized functional groups are assembled to the NPs' surface (Lee et al. 2007a). Catechol function groups are bound with carboxylic or hydroxyl groups on the foliage surface through strong hydrogen bond (Liang et al. 2017). Natural protective glues such as silk protein, tannin acid, and pyrogallol are also used to increase the surface retention of NPs (Kundu et al. 2008; Yu et al. 2019).

3.2 Uptake and Transport in Plants

Xylem and phloem are two channels for the long-distance transport of nutrients in vascular plants. Xylem can carry water and minerals (referred as xylem sap) from roots to above-ground parts while phloem is in charge of the translocation of the photosynthates (referred as phloem sap) to the whole plants. The xylem pathway is mainly driven by negative pressure and the water flow is unidirectional, while the phloem pathway is powered by positive hydrostatic pressures and the flow of the organic compounds inside is multi-directional. There are two filtrating structures in these two matter flow channels, perforation plates in xylem and sieve plates in phloem. They serve as the filters in the plants with general pore sizes of 200 nm to 1.5 μm in xylem and 43–340 nm in phloem, respectively (Jassby et al. 2019). Sap composition and flow rate are two important physiological factors that can impact the transport of NPs while they move along with the sap. Organic compounds in the sap can interact with NPs, modify the surface properties, and alter the transporting behavior of the NPs. However, the underlying mechanisms by which sap components affect NP fate and translocation in plants are still largely unknown.

It is noteworthy that the transporting rate of NPs might not be the same as the flow rate of the sap due to the viscosity of the sap, while the existence of NPs can hardly change the flow rate due to their neglectable concentration in the sap.

It is well known that the size, shape, and dimension of NPs can directly determine the NP accumulation in plants regardless of exposure routes. Su et al. reported that to achieve better uptake via root exposure, at least one dimension of the NPs should be less than 50 nm (Jassby et al. 2019). Generally, increased uptake can be observed with decreased NP size. However, 50 nm is unlikely an absolute threshold for each NPs due to the differences among plant species. NPs were observed to move much further outside of the main vasculature in dicot plants than monocot plants, due to the larger airspace volume in dicot leaves (Spielman-Sun et al. 2019). NPs with spherical shape have been extensively applied; however, it may not be the most efficient ones for plant uptake. Currently, little is known regarding the shape impacts of NPs on their uptake by plants while the cellular distribution of NPs with different shapes have been largely reported. Evidences on the shape impacts of NP to their cellular uptake are still controversial. Carnovale et al. systematically studied the effect of size and shape of gold nanoparticles on their uptake and toxicity and concluded that spherical particles were taken up in greater numbers compared to the shapes with broad flat faces (Carnovale et al. 2019). A quantitative study of the uptake of different shaped NPs showed that under drop-cast application, spherical NPs with high aspect ratios were the most efficiently delivered, while conversely, NPs with low aspect ratios could be more effectively taken up by watermelon plants via aerosol application (Raliya et al. 2016). The NP doses can also determine their uptake by plants. The higher the concentrations are, the less the NPs can be absorbed. High concentrations of NPs usually result in homo- and hetero-aggregation, both of which can increase the particle size and subsequently reduce the NP accumulation in plants. In addition, the surface properties of NPs such as charging, coating, and functional groups, could also affect their uptake. For example, positively charged gold NPs exhibited higher adsorption on plant root surfaces and thus lowered their translocation from roots to other plant parts (Zhu et al. 2012). Similar findings were also evident on CeO_2 NPs interaction with plants (Liu et al. 2019; Spielman-Sun et al. 2019).

3.3   Distribution and Fate

NPs can further transform, translocate, and accumulate inside plants. Versatile techniques have been applied to track, quantify the NPs in plants, and determine the ultimate fate of NPs. Stable isotope labeling has been applied to characterize several widely used metal and metal oxide NMs (ZnO, CuO, Ag, and TiO_2) for the purposes of environmental and biological tracing. This strategy can be useful in nanosafety assessment prior to allowing NMs to enter the market and environment, as well as for product authentication and tracking (Zhang et al. 2019b). It is reported that naturally formed Ag NPs have distinctly different isotope fractionations as compared with those of engineered Ag NPs. This finding provides the possibility of distinguishing the Ag NPs from nature or manufacturing source (Zhang et al. 2019b). By using soft X-ray scanning transmission microscopy (STXM) and near edge X-ray absorption fine structure (XANES), it was demonstrated that Ce (III) released from $nCeO_2$ were precipitated with phosphate in intercellular spaces, or formed complexes with carboxyl compounds during translocation to the shoots (Zhang et al. 2012).

The species-dependent uptake of NPs make it more difficult to understand the NP fate and behaviors in plants. It is reported that uptake and translocation of CeO_2 NPs in monocots and dicots were different under phosphate deficiency treatment. The removal of phosphate from the nutrient solutions promoted the upward translocation of Ce from roots to shoots. Transformation of CeO_2 NPs in the plants was also dependent on particle size, the growth medium, and the composition of xylem and root exduate (Zhang et al. 2019a, 2017).

4 Effect of Nanomaterials on Crops

Nanotechnology has shown significant potential in agriculture from the following aspects: (1) increasing the crop productivity through incorporating nutritional elements directly into nano-form or with nano-vectors to achieve better nutrient use efficiencies (NUEs); (2) suppressing plant disease and pests; (3) attenuating environmental side effects via less input of commercial agrichemicals; (4) serving as growth stimuli or enzyme alternative/ mimics; and (5) applications in pathogen and toxin detection in plants and sensors for assessing specific conditions or analytes of interest in plant systems. Generally, nano-enabled agrichemicals offer possibilities to design and modify the nutrients, pesticides, and other agrochemicals in a relatively confined dimension in comparison with their ionic or molecular form. Currently, use of NPs in agriculture has not involved in the structure-mediated function. In this section, effects of NPs on crops and the underlying mechanisms are discussed.

4.1 NUEs of Nanofertilizers Vs. their Conventional Analogs

Plants need external energy such as light, water, and nutrients input to maintain themselves metabolism and growth. Beside carbon, oxygen, and hydrogen, other nutrients need to be acquired through foliage and root transport. Specifically, macronutrients, including nitrogen (N), phosphorus (P) and potassium (K), calcium (Ca), magnesium (Mg), and sulfur (S), are essential for plant growth and consumed in relatively large amount by plants. In addition, plants also require trace amounts of micronutrients, including iron (Fe), boron (B), chlorine (Cl), manganese (Mn), zinc (Zn), copper (Cu), molybdenum (Mo), and nickel (Ni), to establish healthy life cycles.

Unlike wild plants, crop plants seasonally removing nutrients from crop land, and therefore external fertilizers are needed to maintain the soil fertility for a sustainable agriculture. Based on the uptake, transport, and remobilization of the nutrients, nutrient use efficiency (NUE) can be defined as dry biomass produced per unit nutrient uptake/ application (Dobermann 2007; Guo et al. 2018; Hawkesford et al. 2016), to evaluate the quality of fertilizers. Commercial fertilizers have been estimated to contribute 30–50% of the crop yield (Stewart et al. 2005). The inadequate NUEs of conventional fertilizers have always been the primary setback for the development of the modern agriculture (Dobermann 2007; Hawkesford et al. 2016). The loss of the nutrients in the conventional fertilizers not only causes unnecessary economic burden for farmers, but also lowers soil fertility, increases soil pollution and water eutrophication, etc. Statistics from FAO showed that the amount of commercial fertilizers has been increased annually to achieve the food demand for the increasing population (FAO 2019).

Nitrogen and phosphorus are two elements with relatively lower NUEs and therefore the main cause of water eutrophication. The improvements of their use efficiencies are critical to meet the challenges of food security, environmental degradation, and climate changes (Zhang et al. 2015). One typical N-based fertilizer, urea, vanished quickly through volatilization during the agricultural application. It usually requires multiple feeds to preserve sufficient nitrogen for plant uptake in a full life cycle (Zhang et al. 2015). To the best of our knowledge, nitrogen nanofertilizer-related investigation was very limited, one of the most used strategy is to load nanocarrier with conventional N fertilizer to achieve a relatively slow and controlled release (Rop et al. 2018). Kottegoda et al. reported that urea decorated nano-sized hydroxyapatite (nHA) served as a slow release fertilizer for both nitrogen and phosphorus. nHA was recognized as the carrier for urea and the direct nutrient source. More than 50% reduction in urea use could still maintain the yield at notably higher level with N agronomic use efficiency at 48% compared to that of pure urea at 18% (Kottegoda et al. 2017). Another novel nitrogen-based fertilizer was nano-nitrogen chelate (NNC) fertilizers, which showed less nitrate leaching and more sugar production on sugarcane by controlling elemental release and extending the NNC retention (Alimohammadi et al. 2020). Regarding the phosphorus-based nanofertilizer, nHA mentioned above is the most characterized and tested NMs for the purposes of enhancing the P use efficiency in agriculture. Several studies have demonstrated that nHA could efficiently provide P to crops and increase biomass production and capability to defend against abiotic stresses as compared to its conventional analog ($Ca(H_2PO_4)_2$) (Liu and Lal 2014; Xiong et al. 2018a, b; Yoon et al. 2020).

Incorporating nutritional elements into nano-form is the most straightforward approach to preserve nutrients in a protected matrix, while not all the nutrient-incorporated nanomaterials are suitable as fertilizers, especially metal nanomaterial, due to their acute/ long-term toxic effects toward bio-organisms. Ascribe to this "disadvantage," nanomaterial-incorporated pesticides were designed.

4.2 Antimicrobial Effects of Nanopesticides Vs. their Conventional Analogs

It has been estimated that 10–75% loss of applied synthetic pesticides during the delivery to the target, and the lost pesticides caused contamination of surrounding environemnt (Aktar et al. 2009; Pimentel and Burgess 2014). To counteract the main drawback and to assure the crop quality with less application of pesticides, nano-formulated pesticides were developed (Kah et al. 2018). Micronutrients play important roles in alleviating pathogen-induced biotic stresses (Servin et al. 2015). Mn, Cu, and Zn have been well recognized to enhance disease resistance by activating the host defense enzymes phenylalanine ammonia lyase and polyphenol oxidases (Datnoff et al. 2007). Copper ($CuSO_4$) has been applied as fungicide in Europe for more than a century, which caused the accumulation of copper in soils up to 700 ppm, almost ten times higher than the local background concentration (Ruyters et al. 2013). Inefficient and ubiquitous pesticide and herbicide use have induced antimicrobial resistance and potentially reduce biodiversity, disrupt balanced ecosystem functions, and cause environmental contamination as well (Beketov et al. 2013; Goulson 2013; Goulson et al. 2015; Lowry et al. 2019; Stehle and Schulz 2015). Gilbertson et al. investigated effects of the surface reactivity and antimicrobial activity of nano-cupric oxide with different shapes and found that CuO nanosheets have the highest surface reactivity, electrochemical activity, and antimicrobial activity compared to CuO spheres and bulk CuO, but this trend could not be simply explained by surface area alone (Gilbertson et al. 2016). CeO_2 NPs applied to *Fusarium*-infected tomatoes showed significant disease control ability with elevated antioxidant enzyme activities (Adisa et al. 2018). Recently, a novel Cu-based nanopesticide ($Cu_3(PO_4)_2 \cdot 3H_2O$ nanosheets) was used for the suppression of *Fusarium*-infected plants. A possible molecular explanation could be nano-sized Cu induced upregulation of pathogenesis-related genes (Ma et al. 2019). MgO is another macronutrient-incorporated nanomaterials, which not only provides magnesium as an essential macronutrient, but also shows statistically significant antibacterial properties and size-dependent activity ascribe to its nano-form (Cai et al. 2018; Liao et al. 2019).

4.3 Benefits of Using Nanomaterial Carriers

For the past decades, nitrogen, phosphorus, and potassium are usually incorporated into composite fertilizer (NPK fertilizers) to achieve balanced macronutrients uptake. To prolong the release time of the fertilizers, and prevent loss of nutrients through leaching, nanomaterials have been incorporated into NPK composite fertilizer application system as promising high loading capacity carriers or protective coatings (Kashyap et al. 2015; Lowry et al. 2019; Rop et al. 2018). It is reported that the foliar application of an NPK-chitosan composite to wheat notably increased the grain yield by more than 50% as compared to conventional NPK fertilizer; in addition, the crop life cycle was shortened by 23.5%. Chitosan-NPK nanoparticles were observed inside the phloem tissue, especially in sieve tubes, and nutrients were carried in the sugar flow to shoots and roots (Abdel-Aziz et al. 2016). Through computational modeling, plant virus nanoparticles (VNPs) showed superior ability in delivery and controlled release of pesticides compared to the synthetic counterparts. The rod-like tobacco mild green mosaic VNPs can deliver pesticides to the rhizosphere with much greater mobility and highest dye loading capacity; thus, VNPs can be potentially developed as next-generation nanopesticide delivery system toward precision farming (Chariou et al. 2019, 2020). By decorating β-cyclodextrin molecular baskets with quantum dots (QDs), Santana et al. provided insight into loading and delivering broad chemicals to specific target under the biorecognized guidance of peptide motif and the delivery efficiency of these coated QDs can be as high as approximately 75% (Santana et al. 2020).

4.4   NP Sensor Application in Smart Agriculture

Sensors became necessary component to human daily life ascribe to its various industrial and consumer applications. According to National Institute of Food and Agriculture (NIFA), the sensor technology could find their usage in nearly all aspects of production, processing, and management in agricultural and food systems, especially as a necessary tool for precision agriculture (Agriculture, N. I.o. F. a. n.d.). Specifically in the crop growth management, biological recognition and sensing integrated technologies can be used to monitor plant needs in real time, with high stability, fast dynamics, accuracy and reproducibility, for specific analytes of interest (Giraldo et al. 2019; Kwak et al. 2017), and adjusting resource use by interacting with agricultural device of smart nanobiotechnology-based sensors that report plant signaling molecules associated with stress status or resource. Depending on different sensing mechanisms, various nanoparticles have been integrated into optical and electrochemical sensor platforms for application in plants (Srivastava et al. 2018). Promising outcomes have been reported under lab/greenhouse experiments, but so far not much field data available to further verify the efficacy and feasibility of such nanosensors in a real life scenario.

Ascribe to the electronic properties, metallic- and carbon-based nanoparticles have been used as electrode coating materials for **electrochemical nanosensors** to track multiple redox active species in plants, which enable them as a useful tool for plant disease diagnosis and resource (fertilizers or pesticides) reallocation (Kashyap et al. 2019; Lew et al. 2019; Zhu et al. 2015). By wrapping single-walled carbon nanotubes (SWNTs) into compositionally designed polymer, corona phase molecular recognition **(CoPhMoRe) nanosensors** have been invented and successfully detected several plant signal molecules such as dopamine, nitric oxide, H_2O_2, glucose, and proteins (Kwak et al. 2017; Lew et al. 2020). Gold nanoparticles, semiconductor quantum dots (QDs), lanthanide-doped upconversion nanoparticles

(UCNPs), mesoporous silica nanoparticles, and polymer nanoparticles have been used as fluorescence resonance energy transfer (FRET) donors or quenchers in **FRET-based nanosensor** platforms. Signals of metal ions, DNA, sugar, ATP, and phytoestrogens can be monitored with better photostability and emission sensitivity over traditional FRET biosensors (Chen et al. 2012). Possessing strong and tunable plasmon resonance in the NIR region, gold and silver nanoparticles have been extensively functionalized and applied as substrates for in situ surface-enhanced Raman scattering (**SERS**) detection of pesticides on plant surfaces (Pang et al. 2016), phytohormones such as brassinosteroids (BRs) (Chen et al. 2017), VOCs in the tea leaves (Park et al. 2020). Recent years, due to their multidimensional structures and structure-dependent unique electronic, electrocatalytic, and optical properties, transition-metal dichalcogenides (TMDs), especially MoS_2, have attracted newly interests in designing novel biosensing platforms. By assembling MoS_2 into zero-, one-, two-, and three-dimensional nanostructures, promising applications in optical, electrochemical, and electronic biosensors can be expected (Barua et al. 2017; Kwak et al. 2017).

4.5 Potential Side Effects and Possible Ways to Avoid these

Up to now, most nanosafety related studies have been conducted in two distinct epistemic communities. On the nano-bio interface, risk evaluation of nanoparticles interacting with cells, tissues, and organisms in vitro and in vivo is of main interest. Meantime, on the nano-eco interface, risk evaluation of NP interacting with the surrounding abiotic environments such as soil organic matter (SOM), soil minerals, aerosol, and co-existed contaminants covered most of the studies (Lombi et al. 2019). With the development of agri-nanotechnologies, the gap between two communities would be expected to be filled for a more comprehensive understanding of the NP behavior in a connected ecosystem (He et al. 2015b). At the nano-bio-eco interface, unexpected side effects could be generated that neither of these two disciplines can cover or even recognize. For example, Cu (OH)$_2$ has been used as novel nanopesticide ascribe to its high efficiency and nontoxic effects on the treated plants, while it was reported that Cu (OH)$_2$ nanopesticides strongly mitigated the degradation of the problematic insecticide thiacloprid by down-regulating thiacloprid-degradative *nth* gene abundance. Decreased bioavailability of thiacloprid was discovered due to the absorption on the nanopesticides (Zhang et al. 2019c). If such mitigation effects can be applied on other organic toxins in the soil, major concerns should be raised to the potential environmental risks of nanopesticides. Concerns over the joint toxicity of nano-enabled agrochemicals and the co-existing contaminants in environment have also drawn great attentions. The benefit of nano-enabled agrochemicals can be offset by the potential risks of acting as carriers of other pollutants, increase colloid-facilitated transport of pesticides to water bodies and contaminate the human food chains (Deng et al. 2017; He et al. 2015b; Lombi et al. 2019). Other issues include possibility of developing or spreading the antimicrobial resistance, increase of the exposure of off-target wildlife, and change of food characteristics and security with adding of NPs into food industry (Patel et al. 2018). Besides, the dose-dependent effects of NPs have been generally recognized, while during the field application, the control of the dosage can be tricky as the species dependence also takes place along with the dose-dependent response.

5   Conclusion and Perspectives

The age of focusing only the toxicity and negative effects of nanomaterials in the real environment is fading away. It does not mean that the negative effects of uncontrolled/unmonitored release of nanomaterials into biological systems are negligible, but can be compatible with the next level studies on exploring the hiding benefits of nanomaterials in consumer daily life. This chapter has described the interaction between nanomaterials and plants and summarized the current research status of nanomaterial applications in agriculture.

As discussed in the above multiple sections, elaborate and systematic design and tests are warranted to assure safety and efficacy application of nanomaterials in agriculture. Up to now, enormous interests have been focused on the "front end" researches such as design and development of certain nanomaterials for potential agricultural utilizations. However, established concepts or mechanisms from other disciplines or industries still need vast work and specified adaptions in the "back end" to be able to feasibly apply in agriculture. Furthermore, current agricultural requirement for advanced nanotechnology applications is more toward precision, smart, cost-effective and eco-friendly directions, which rise expectations in the outcome products and correspondingly elaboration of the design.

Acknowledgements This work was supported by BARD (IS-4964-16R) and USDA NIFA Hatch Program (CONH00147).

References

Abdel-Aziz, H. M., Hasaneen, M. N., & Omer, A. M. (2016). Nano chitosan-NPK fertilizer enhances the growth and productivity of wheat plants grown in sandy soil. *Spanish Journal of Agricultural Research, 14*(1), 0902.

Adisa, I. O., Pullagurala, V. L. R., Peralta-Videa, J. R., Dimkpa, C. O., Elmer, W. H., Gardea-Torresdey, J., & White, J. (2019). Recent advances in nano-enabled fertilizers and pesticides: A critical review of mechanisms of action. *Environmental Science. Nano, 6*, 2002.

Adisa, I. O., Reddy Pullagurala, V. L., Rawat, S., Hernandez-Viezcas, J. A., Dimkpa, C. O., Elmer, W. H., White, J. C., Peralta-Videa, J. R., & Gardea-Torresdey, J. L. (2018). Role of cerium compounds in Fusarium wilt suppression and growth enhancement in tomato (Solanum lycopersicum). *Journal of Agricultural and Food Chemistry, 66*(24), 5959–5970.

Aktar, W., Sengupta, D., & Chowdhury, A. (2009). Impact of pesticides use in agriculture: Their benefits and hazards. *Interdisciplinary Toxicology, 2*(1), 1–12.

Alexandratos, N., Bruinsma, J. (2012). World agriculture towards 2030/2050: the 2012 revision.

Ali, I., Mbianda, X., Burakov, A., Galunin, E., Burakova, I., Mkrtchyan, E., Tkachev, A., & Grachev, V. (2019a). Graphene based adsorbents for remediation of noxious pollutants from wastewater. *Environment International, 127*, 160–180.

Ali, S., Rizwan, M., Noureen, S., Anwar, S., Ali, B., Naveed, M., Abd Allah, E. F., Alqarawi, A. A., & Ahmad, P. (2019b). Combined use of biochar and zinc oxide nanoparticle foliar spray improved the plant growth and decreased the cadmium accumulation in rice (Oryza sativa L.) plant. *Environmental Science and Pollution Research,* 1–12.

Alimohammadi, M., Panahpour, E., & Naseri, A. (2020). Assessing the effects of urea and nano-nitrogen chelate fertilizers on sugarcane yield and dynamic of nitrate in soil. *Soil Science and Plant Nutrition,* 1–8.

Almeelbi, T., & Bezbaruah, A. (2014). Nanoparticle-sorbed phosphate: Iron and phosphate bio-availability studies with Spinacia oleracea and Selenastrum capricornutum. *ACS Sustainable Chemistry & Engineering, 2*(7), 1625–1632.

Anyaogu, K. C., Fedorov, A. V., & Neckers, D. C. (2008). Synthesis, characterization, and anti-fouling potential of functionalized copper nanoparticles. *Langmuir, 24*(8), 4340–4346.

Banerjee, I., Pangule, R. C., & Kane, R. S. (2011). Antifouling coatings: Recent developments in the design of surfaces that prevent fouling by proteins, bacteria, and marine organisms. *Advanced Materials, 23*(6), 690–718.

Bansiwal, A. K., Rayalu, S. S., Labhasetwar, N. K., Juwarkar, A. A., & Devotta, S. (2006). Surfactant-modified zeolite as a slow release fertilizer for phosphorus. *Journal of Agricultural and Food Chemistry, 54*(13), 4773–4779.

Barua, S., Dutta, H. S., Gogoi, S., Devi, R., & Khan, R. (2017). Nanostructured MoS2-based advanced biosensors: A review. *ACS Applied Nano Materials, 1*(1), 2–25.

Beketov, M. A., Kefford, B. J., Schäfer, R. B., & Liess, M. (2013). Pesticides reduce regional bio-diversity of stream invertebrates. *Proceedings of the National Academy of Sciences, 110*(27), 11039–11043.

Besseling, E., Wang, B., Lürling, M., & Koelmans, A. A. (2014). Nanoplastic affects growth of S. obliquus and reproduction of D. magna. *Environmental Science & Technology, 48*(20), 12336–12343.

Bhushan, B. (2017). *Springer handbook of nanotechnology*. Springer.

Boyer, C., Priyanto, P., Davis, T. P., Pissuwan, D., Bulmus, V., Kavallaris, M., Teoh, W. Y., Amal, R., Carroll, M., & Woodward, R. (2010). Anti-fouling magnetic nanoparticles for siRNA deliv-ery. *Journal of Materials Chemistry, 20*(2), 255–265.

Cai, L., Chen, J., Liu, Z., Wang, H., Yang, H., & Ding, W. (2018). Magnesium oxide nanopar-ticles: Effective agricultural antibacterial agent against Ralstonia solanacearum. *Frontiers in Microbiology, 9*, 790.

Carnovale, C., Bryant, G., Shukla, R., & Bansal, V. (2019). Identifying trends in gold nanoparticle toxicity and uptake: size, shape, capping ligand, and biological Corona. *ACS Omega, 4*(1), 242–256.

Chariou, P. L., Dogan, A. B., Welsh, A. G., Saidel, G. M., Baskaran, H., & Steinmetz, N. F. (2019). Soil mobility of synthetic and virus-based model nanopesticides. *Nature Nanotechnology, 1*.

Chariou, P. L., Ortega-Rivera, O. A., & Steinmetz, N. F. (2020). Nanocarriers for the delivery of medical, veterinary, and agricultural active ingredients. *ACS Nano, 14*(3), 2678–2701.

Chaudhry, Q., & Castle, L. (2011). Food applications of nanotechnologies: An overview of oppor-tunities and challenges for developing countries. *Trends in Food Science & Technology, 22*(11), 595–603.

Chen, M., Zhang, Z., Liu, M., Qiu, C., Yang, H., & Chen, X. (2017). In situ fabrication of label-free optical sensing paper strips for the rapid surface-enhanced Raman scattering (SERS) detection of brassinosteroids in plant tissues. *Talanta, 165*, 313–320.

Chen, N.-T., Cheng, S.-H., Liu, C.-P., Souris, J. S., Chen, C.-T., Mou, C.-Y., & Lo, L.-W. (2012). Recent advances in nanoparticle-based Förster resonance energy transfer for biosensing, molecular imaging and drug release profiling. *International Journal of Molecular Sciences, 13*(12), 16598–16623.

Clark, M., & Tilman, D. (2017). Comparative analysis of environmental impacts of agricultural production systems, agricultural input efficiency, and food choice. *Environmental Research Letters, 12*(6), 064016.

Cole, M., & Galloway, T. S. (2015). Ingestion of nanoplastics and microplastics by Pacific oyster larvae. *Environmental Science & Technology, 49*(24), 14625–14632.

Cui, Y., Ding, Z., Liu, P., Antonietti, M., Fu, X., & Wang, X. (2012). Metal-free activation of H2O2 by g-C3N4 under visible light irradiation for the degradation of organic pollutants. *Physical Chemistry Chemical Physics, 14*(4), 1455–1462.

Datnoff, L. E., Elmer, W. H., & Huber, D. M. (2007). *Mineral nutrition and plant disease.* American Phytopathological Society. APS Press.

Deng, R., Lin, D., Zhu, L., Majumdar, S., White, J. C., Gardea-Torresdey, J. L., & Xing, B. (2017). Nanoparticle interactions with co-existing contaminants: Joint toxicity, bioaccumulation and risk. *Nanotoxicology, 11*(5), 591–612.

DeRosa, M. C., Monreal, C., Schnitzer, M., Walsh, R., & Sultan, Y. (2010). Nanotechnology in fertilizers. *Nature Nanotechnology, 5*(2), 91.

Dietz, K. J., & Herth, S. (2011). Plant nanotoxicology. *Trends in Plant Science, 16*(11), 582–589.

Dimkpa, C. O., & Bindraban, P. S. (2016). Fortification of micronutrients for efficient agronomic production: A review. *Agronomy for Sustainable Development, 36*(1), 7.

Dimkpa, C. O., & Bindraban, P. S. (2017). Nanofertilizers: New products for the industry? *Journal of Agricultural and Food Chemistry, 66*(26), 6462–6473.

Dobermann, A. (2007). Nutrient use efficiency–measurement and management. *Fertilizer Best Management Practices, 1.*

Eichert, T., & Goldbach, H. E. (2008). Equivalent pore radii of hydrophilic foliar uptake routes in stomatous and astomatous leaf surfaces–further evidence for a stomatal pathway. *Physiologia Plantarum, 132*(4), 491–502.

Eichert, T., Kurtz, A., Steiner, U., & Goldbach, H. E. (2008). Size exclusion limits and lateral heterogeneity of the stomatal foliar uptake pathway for aqueous solutes and water-suspended nanoparticles. *Physiologia Plantarum, 134*(1), 151–160.

Elmer, W., Plaza Perez, C., Pagano, L., De La Torre-Roche, R., Zuverza-Mena, N., Ma, C., Borgatta, J., Hamers, R., White, J. (2019). *Nanoparticles of Cu for suppression of Fusarium root diseases*, Phytopathology, 2019; Amer Phytopathological Soc 3340 Pilot Knob Road, St Paul, MN 55121 USA.

FAO. (2019). *World fertilizer trends and outlook to 2022* (p. 40). Rome: FAO.

Farokhzad, O. C., & Langer, R. (2009). Impact of nanotechnology on drug delivery. *ACS Nano, 3*(1), 16–20.

Fleischer, A., O'Neill, M. A., & Ehwald, R. (1999). The pore size of non-graminaceous plant cell walls is rapidly decreased by borate ester cross-linking of the pectic polysaccharide rhamnogalacturonan II. *Plant Physiology, 121*(3), 829–838.

Food and Agriculture Organization of the United Nations. (2019). *The state of food and agriculture: Moving forward on food loss and waste reduction.* Rome: Food and Agriculture Organization of the United Nations.

Frew, A., Weston, L. A., Reynolds, O. L., & Gurr, G. M. (2018). The role of silicon in plant biology: A paradigm shift in research approach. *Annals of Botany, 121*(7), 1265–1273.

Fu, P. P., Xia, Q., Hwang, H.-M., Ray, P. C., & Yu, H. (2014). Mechanisms of nanotoxicity: Generation of reactive oxygen species. *Journal of Food and Drug Anal, 22*(1), 64–75.

Gilbertson, L. M., Albalghiti, E. M., Fishman, Z. S., Perreault, F. O., Corredor, C., Posner, J. D., Elimelech, M., Pfefferle, L. D., & Zimmerman, J. B. (2016). Shape-dependent surface reactivity and antimicrobial activity of nano-cupric oxide. *Environmental Science Technology, 50*(7), 3975–3984.

Giraldo, J. P., Wu, H., Newkirk, G. M., & Kruss, S. (2019). Nanobiotechnology approaches for engineering smart plant sensors. *Nature Nanotechnology, 14*(6), 541–553.

Gottschalk, F., Sun, T., & Nowack, B. (2013). Environmental concentrations of engineered nanomaterials: Review of modeling and analytical studies. *Environmental Pollution, 181*, 287–300.

Gou, M., Qu, X., Zhu, W., Xiang, M., Yang, J., Zhang, K., Wei, Y., & Chen, S. (2014). Bio-inspired detoxification using 3D-printed hydrogel nanocomposites. *Nature Communications, 5*, 3774.

Goulson, D. (2013). An overview of the environmental risks posed by neonicotinoid insecticides. *Journal of Applied Ecology, 50*(4), 977–987.

Goulson, D., Nicholls, E., Botías, C., & Rotheray, E. L. (2015). Bee declines driven by combined stress from parasites, pesticides, and lack of flowers. *Science, 347*(6229), 1255957.

Guha, T., Ravikumar, K., Mukherjee, A., Mukherjee, A., & Kundu, R. (2018). Nanopriming with zero valent iron (nZVI) enhances germination and growth in aromatic rice cultivar (Oryza sativa cv Gobindabhog L.). *Plant Physiology and Biochemistry, 127*, 403–413.

Guo, H., White, J. C., Wang, Z., & Xing, B. (2018). Nano-enabled fertilizers to control the release and use efficiency of nutrients. *Current Opinion in Environmental Science & Health, 6*.

Gupta, V. K., Kumar, R., Nayak, A., Saleh, T. A., & Barakat, M. (2013). Adsorptive removal of dyes from aqueous solution onto carbon nanotubes: A review. *Advances in Colloid and Interface Science, 193*, 24–34.

Hajipour, M. J., Fromm, K. M., Ashkarran, A. A., de Aberasturi, D. J., de Larramendi, I. R., Rojo, T., Serpooshan, V., Parak, W. J., & Mahmoudi, M. (2012). Antibacterial properties of nanoparticles. *Trends in Biotechnology, 30*(10), 499–511.

Hawkesford, M. J., Kopriva, S., & De Kok, L. J. (2016). *Nutrient use efficiency in plants*. Springer.

He, C., Liu, D., & Lin, W. (2015a). Nanomedicine applications of hybrid nanomaterials built from metal–ligand coordination bonds: Nanoscale metal–organic frameworks and nanoscale coordination polymers. *Chemical Reviews, 115*(19), 11079–11108.

He, K., Chen, G., Zeng, G., Chen, A., Huang, Z., Shi, J., Huang, T., Peng, M., & Hu, L. (2018a). Three-dimensional graphene supported catalysts for organic dyes degradation. *Applied Catalysis, B: Environmental, 228*, 19–28.

He, X., Aker, W. G., Fu, P. P., & Hwang, H.-M. (2015b). Toxicity of engineered metal oxide nanomaterials mediated by nano–bio–eco–interactions: A review and perspective. *Environmental Science. Nano, 2*(6), 564–582.

He, X., Fu, P., Aker, W. G., & Hwang, H.-M. (2018b). Toxicity of engineered nanomaterials mediated by nano–bio–eco interactions. *Journal of Environmental Science and Health, Part C, 36*(1), 21–42.

Horcajada, P., Gref, R., Baati, T., Allan, P. K., Maurin, G., Couvreur, P., Ferey, G., Morris, R. E., & Serre, C. (2011). Metal–organic frameworks in biomedicine. *Chemical Reviews, 112*(2), 1232–1268.

Hu, C.-M. J., Fang, R. H., Luk, B. T., & Zhang, L. (2013). Nanoparticle-detained toxins for safe and effective vaccination. *Nature Nanotechnology, 8*(12), 933.

Jaberzadeh, A., Moaveni, P., Moghadam, H. R. T., & Zahedi, H. (2013). Influence of bulk and nanoparticles titanium foliar application on some agronomic traits, seed gluten and starch contents of wheat subjected to water deficit stress. *Notulae Botanicae Horti Agrobotanici Cluj-Napoca, 41*(1), 201–207.

Janos, P., Ederer, J., Dosek, M., Stojdl, J., Henych, J., Tolasz, J., Kormunda, M., & Mazanec, K. (2019). Can cerium oxide serve as a phosphodiesterase-mimetic nanozyme? *Environmental Science. Nano, 6*, 3684.

Jassby, D., Su, Y., Kim, C., Ashworth, V., Adeleye, A. S., Rolshausen, P., Roper, C., & White, J. (2019). Delivery, uptake, fate, and transport of engineered nanoparticles in plants: A critical review and data analysis. *Environmental Science. Nano, 6*, 2311.

Jeevahan, J., & Chandrasekaran, M. (2019). Nanoedible films for food packaging: A review. *Journal of Materials Science, 1*–29.

Kah, M., Kookana, R. S., Gogos, A., & Bucheli, T. D. (2018). A critical evaluation of nanopesticides and nanofertilizers against their conventional analogues. *Nature Nanotechnology, 13*(8), 677.

Kashyap, P. L., Kumar, S., Jasrotia, P., Singh, D., & Singh, G. P. (2019). Nanosensors for plant disease diagnosis: Current understanding and future perspectives. In *Nanoscience for Sustainable Agriculture* (pp. 189–205). Springer.

Kashyap, P. L., Xiang, X., & Heiden, P. (2015). Chitosan nanoparticle based delivery systems for sustainable agriculture. *International Journal of Biological Macromolecules, 77*, 36–51.

Koelmans, A. A., Besseling, E., & Shim, W. J. (2015). Nanoplastics in the aquatic environment. Critical review. In *Marine anthropogenic litter* (pp. 325–340). Cham: Springer.

Kottegoda, N., Munaweera, I., Madusanka, N., & Karunaratne, V. (2011). A green slow-release fertilizer composition based on urea-modified hydroxyapatite nanoparticles encapsulated wood. *Current Science, 101*, 73–78.

Kottegoda, N., Sandaruwan, C., Priyadarshana, G., Siriwardhana, A., Rathnayake, U. A., Berugoda Arachchige, D. M., Kumarasinghe, A. R., Dahanayake, D., Karunaratne, V., & Amaratunga, G. A. (2017). Urea-hydroxyapatite nanohybrids for slow release of nitrogen. *ACS Nano, 11*(2), 1214–1221.

Kundu, S. C., Dash, B. C., Dash, R., & Kaplan, D. L. (2008). Natural protective glue protein, sericin bioengineered by silkworms: Potential for biomedical and biotechnological applications. *Progress in Polymer Science, 33*(10), 998–1012.

Kwak, S.-Y., Kim, S. H., & Kim, S. S. (2001). Hybrid organic/inorganic reverse osmosis (RO) membrane for bactericidal anti-fouling. 1. Preparation and characterization of TiO2 nanoparticle self-assembled aromatic polyamide thin-film-composite (TFC) membrane. *Environmental Science & Technology, 35*(11), 2388–2394.

Kwak, S.-Y., Wong, M. H., Lew, T. T. S., Bisker, G., Lee, M. A., Kaplan, A., Dong, J., Liu, A. T., Koman, V. B., & Sinclair, R. (2017). Nanosensor technology applied to living plant systems. *Annual Review of Analytical Chemistry, 10*, 113–140.

Larue, C., Castillo-Michel, H., Sobanska, S., Trcera, N., Sorieul, S., Cécillon, L., Ouerdane, L., Legros, S., & Sarret, G. (2014). Fate of pristine TiO2 nanoparticles and aged paint-containing TiO2 nanoparticles in lettuce crop after foliar exposure. *Journal of Hazardous Materials, 273*, 17–26.

Lee, H., Dellatore, S. M., Miller, W. M., & Messersmith, P. B. (2007a). Mussel-inspired surface chemistry for multifunctional coatings. *Science, 318*(5849), 426–430.

Lee, S. Y., Kim, H. J., Patel, R., Im, S. J., Kim, J. H., & Min, B. R. (2007b). Silver nanoparticles immobilized on thin film composite polyamide membrane: Characterization, nanofiltration, antifouling properties. *Polymers for Advanced Technologies, 18*(7), 562–568.

Lew, T. T. S., Koman, V. B., Gordiichuk, P., Park, M., & Strano, M. S. (2019). The emergence of plant Nanobionics and living plants as technology. *Advanced Materials Technologies*, 1900657.

Lew, T. T. S., Koman, V. B., Silmore, K. S., Seo, J. S., Gordiichuk, P., Kwak, S.-Y., Park, M., Ang, M. C.-Y., Khong, D. T., & Lee, M. A. (2020). Real-time detection of wound-induced H 2 O 2 signalling waves in plants with optical nanosensors. *Nature Plants, 6*(4), 404–415.

Li, J.-X., Wee, C.-D., & Sohn, B.-K. (2010). Growth response of hot pepper Applicated with ammonium (NH4+) and potassium (K+)-loaded zeolite. *Korean Journal of Soil Science and Fertilizer, 43*(5), 741–747.

Lian, J., Zhao, L., Wu, J., Xiong, H., Bao, Y., Zeb, A., Tang, J., & Liu, W. (2020). Foliar spray of TiO2 nanoparticles prevails over root application in reducing cd accumulation and mitigating cd-induced phytotoxicity in maize (Zea mays L.). *Chemosphere, 239*, 124794.

Liang, J., Yu, M., Guo, L., Cui, B., Zhao, X., Sun, C., Wang, Y., Liu, G., Cui, H., & Zeng, Z. (2017). Bioinspired development of P (St–MAA)–avermectin nanoparticles with high affinity for foliage to enhance folia retention. *Journal of Agricultural and Food Chemistry, 66*(26), 6578–6584.

Liao, Y., Strayer-Scherer, A., White, J., De La Torre-Roche, R., Ritchie, L., Colee, J., Vallad, G., Freeman, J., Jones, J., & Paret, M. (2019). Particle-size dependent bactericidal activity of magnesium oxide against Xanthomonas perforans and bacterial spot of tomato. *Scientific Reports, 9*(1), 1–10.

Liu, F., Chung, S., Oh, G., & Seo, T. S. (2012). Three-dimensional graphene oxide nanostructure for fast and efficient water-soluble dye removal. *ACS Applied Materials & Interfaces, 4*(2), 922–927.

Liu, M., Feng, S., Ma, Y., Xie, C., He, X., Ding, Y., Zhang, J., Luo, W., Zheng, L., & Chen, D. (2019). Influence of surface charge on the phytotoxicity, transformation and translocation of CeO2 nanoparticles in cucumber plants. *ACS Applied Materials & Interfaces, 11*, 16905.

Liu, R., & Lal, R. (2014). Synthetic apatite nanoparticles as a phosphorus fertilizer for soybean (Glycine max). *Scientific Reports, 4*, 5686.

Lombi, E., Donner, E., Dusinska, M., & Wickson, F. (2019). A one health approach to managing the applications and implications of nanotechnologies in agriculture. *Nature Nanotechnology, 14*(6), 523–531.

Lowry, G. V., Avellan, A., & Gilbertson, L. M. (2019). Opportunities and challenges for nanotechnology in the Agri-tech revolution. *Nature Nanotechnology, 14*(6), 517–522.

Lu, Y., Aimetti, A. A., Langer, R., & Gu, Z. (2016). Bioresponsive materials. *Nature Reviews Materials, 2*(1), 1–17.

Luyckx, M., Hausman, J.-F., Lutts, S., & Guerriero, G. (2017). Silicon and plants: Current knowledge and technological perspectives. *Frontiers in Plant Science, 8*, 411.

Lv, J., Christie, P., & Zhang, S. (2019). Uptake, translocation, and transformation of metal-based nanoparticles in plants: Recent advances and methodological challenges. *Environmental Science. Nano, 6*(1), 41–59.

Ma, C., Borgatta, J., De La Torre Roche, R., Zuverza-Mena, N., White, J. C., Hamers, R. J., & Elmer, W. (2019). Time-dependent transcriptional response of tomato (Solanum lycopersicum L.) to cu nanoparticle exposure upon infection with Fusarium oxysporum f. sp. lycopersici. *ACS Sustainable Chemistry & Engineering, 7*(11), 10064–10074.

Ma, C., Liu, H., Guo, H., Musante, C., Coskun, S. H., Nelson, B. C., White, J. C., Xing, B., & Dhankher, O. P. (2016). Defense mechanisms and nutrient displacement in Arabidopsis thaliana upon exposure to CeO 2 and in 2 O 3 nanoparticles. *Environmental Science. Nano, 3*(6), 1369–1379.

Ma, C., White, J. C., Dhankher, O. P., & Xing, B. (2015). Metal-based nanotoxicity and detoxification pathways in higher plants. *Environment Science Technology, 49*(12), 7109–7122.

Ma, C., White, J. C., Zhao, J., Zhao, Q., & Xing, B. (2018). Uptake of engineered nanoparticles by food crops: Characterization, mechanisms, and implications. *Annual Review of Food Science and Technology, 9*, 129–153.

Makhlouf, A. S. H., & Barhoum, A. (2018). *Emerging applications of nanoparticles and architectural nanostructures: Current prospects and future trends*. William Andrew.

Mamba, G., & Mishra, A. (2016). Graphitic carbon nitride (g-C3N4) nanocomposites: A new and exciting generation of visible light driven photocatalysts for environmental pollution remediation. *Applied Catalysis. B, Environmental, 198*, 347–377.

Mazzola, L. (2003). Commercializing nanotechnology. *Nature Biotechnology, 21*(10), 1137–1143.

McClements, D. J., & Xiao, H. (2017). Is nano safe in foods? Establishing the factors impacting the gastrointestinal fate and toxicity of organic and inorganic food-grade nanoparticles. *npj Science of Food, 1*(1), 1–13.

Mueller, N. D., Gerber, J. S., Johnston, M., Ray, D. K., Ramankutty, N., & Foley, J. A. (2012). Closing yield gaps through nutrient and water management. *Nature, 490*(7419), 254.

National Nanotechnology Initiative. What is nanotechnology? https://www.nano.gov/nanotech-101/what/definition.

Pang, S., Yang, T., & He, L. (2016). Review of surface enhanced Raman spectroscopic (SERS) detection of synthetic chemical pesticides. *TrAC Trends in Analytical Chemistry, 85*, 73–82.

Park, J., Thomasson, J. A., Gale, C. C., Sword, G. A., Lee, K.-M., Herrman, T. J., & Suh, C. P.-C. (2020). Adsorbent-SERS technique for determination of plant VOCs from live cotton plants and dried teas. *ACS Omega, 5*(6), 2779–2790.

Patel, V., Singh, M., Mayes, E. L., Martinez, A., Shutthanandan, V., Bansal, V., Singh, S., & Karakoti, A. S. (2018). Ligand-mediated reversal of the oxidation state dependent ROS scavenging and enzyme mimicking activity of ceria nanoparticles. *Chemical Communications,* *54*(99), 13973–13976.

Peer, D., Karp, J. M., Hong, S., Farokhzad, O. C., Margalit, R., & Langer, R. (2007). Nanocarriers as an emerging platform for cancer therapy. *Nature Nanotechnology, 2*(12), 751.

Peters, R. J., Bouwmeester, H., Gottardo, S., Amenta, V., Arena, M., Brandhoff, P., Marvin, H. J., Mech, A., Moniz, F. B., & Pesudo, L. Q. (2016). Nanomaterials for products and application in agriculture, feed and food. *Trends in Food Science & Technology, 54*, 155–164.

Pimentel, D., & Burgess, M. (2014). Environmental and economic costs of the application of pesticides primarily in the United States. In *Integrated pest management* (pp. 47–71). Springer.

Portehault, D., Delacroix, S., Gouget, G., Grosjean, R., & Chan-Chang, T.-H.-C. (2018). Beyond the compositional threshold of nanoparticle-based materials. *Accounts of Chemical Research, 51*(4), 930–939.

Pradhan, S., Patra, P., Das, S., Chandra, S., Mitra, S., Dey, K. K., Akbar, S., Palit, P., & Goswami, A. (2013). Photochemical modulation of biosafe manganese nanoparticles on Vigna radiata: A detailed molecular, biochemical, and biophysical study. *Environmental Science & Technology, 47*(22), 13122–13131.

Qiao, Y., Wan, J., Zhou, L., Ma, W., Yang, Y., Luo, W., Yu, Z., & Wang, H. (2019). Stimuli-responsive nanotherapeutics for precision drug delivery and cancer therapy. *Wiley Interdisciplinary Reviews. Nanomedicine and Nanobiotechnology, 11*(1), e1527.

Raliya, R., Franke, C., Chavalmane, S., Nair, R., Reed, N., & Biswas, P. (2016). Quantitative understanding of nanoparticle uptake in watermelon plants. *Frontiers in Plant Science, 7*, 1288.

Raliya, R., Saharan, V., Dimkpa, C., & Biswas, P. (2017). Nanofertilizer for precision and sustainable agriculture: Current state and future perspectives. *Journal of Agricultural and Food Chemistry, 66*(26), 6487–6503.

Ren, X., Chen, C., Nagatsu, M., & Wang, X. (2011). Carbon nanotubes as adsorbents in environmental pollution management: A review. *Chemical Engineering Journal, 170*(2–3), 395–410.

Rop, K., Karuku, G. N., Mbui, D., Michira, I., & Njomo, N. (2018). Formulation of slow release NPK fertilizer (cellulose-graft-poly (acrylamide)/nano-hydroxyapatite/soluble fertilizer) composite and evaluating its N mineralization potential. *Annals of Agricultural Sciences, 63*(2), 163–172.

Rui, M., Ma, C., Hao, Y., Guo, J., Rui, Y., Tang, X., Zhao, Q., Fan, X., Zhang, Z., & Hou, T. (2016). Iron oxide nanoparticles as a potential iron fertilizer for peanut (Arachis hypogaea). *Frontiers in Plant Science, 7*, 815.

Ruyters, S., Salaets, P., Oorts, K., & Smolders, E. (2013). Copper toxicity in soils under established vineyards in Europe: A survey. *Science of the Total Environment, 443*, 470–477.

Sabir, S., Arshad, M., & Chaudhari, S. K. (2014). Zinc oxide nanoparticles for revolutionizing agriculture: Synthesis and applications. *The Scientific World Journal, 2014*.

Saharan, V., Kumaraswamy, R., Choudhary, R. C., Kumari, S., Pal, A., Raliya, R., & Biswas, P. (2016). Cu-chitosan nanoparticle mediated sustainable approach to enhance seedling growth in maize by mobilizing reserved food. *Journal of Agricultural and Food Chemistry, 64*(31), 6148–6155.

Santana, I., Wu, H., Hu, P., & Giraldo, J. P. (2020). Targeted delivery of nanomaterials with chemical cargoes in plants enabled by a biorecognition motif. *Nature Communications, 11*(1), 1–12.

Sapsford, K. E., Algar, W. R., Berti, L., Gemmill, K. B., Casey, B. J., Oh, E., Stewart, M. H., & Medintz, I. L. (2013). Functionalizing nanoparticles with biological molecules: Developing chemistries that facilitate nanotechnology. *Chemical Reviews, 113*(3), 1904–2074.

Sarkar, S., Datta, S., & Biswas, D. (2015). Effect of fertilizer loaded nanoclay/superabsorbent polymer composites on nitrogen and phosphorus release in soil. *Proceedings of the National Academy of Sciences, India Section B: Biological Sciences, 85*(2), 415–421.

Sarkar, S., Datta, S. C., & Biswas, D. R. (2014). Synthesis and characterization of nanoclay–polymer composites from soil clay with respect to their water-holding capacities and nutrient-release behavior. *Journal of Applied Polymer Science, 131*, 6.

Sensor Applications. https://nifa.usda.gov/sensor-applications

Serrano, E., Rus, G., & Garcia-Martinez, J. (2009). Nanotechnology for sustainable energy. *Renewable and Sustainable Energy Reviews, 13*(9), 2373–2384.

Servin, A., Elmer, W., Mukherjee, A., De la Torre-Roche, R., Hamdi, H., White, J. C., Bindraban, P., & Dimkpa, C. (2015). A review of the use of engineered nanomaterials to suppress plant disease and enhance crop yield. *Journal of Nanoparticle Research, 17*(2), 92.

Servin, A. D., & White, J. C. (2016). Nanotechnology in agriculture: Next steps for understanding engineered nanoparticle exposure and risk. *NanoImpact, 1,* 9–12.

Shen, Y., Zhu, C., & Chen, B. (2020). Immobilizing 1–3 nm ag nanoparticles in reduced graphene oxide aerogel as a high-effective catalyst for reduction of nitroaromatic compounds. *Environmental Pollution, 256,* 113405.

Shi, J., Kantoff, P. W., Wooster, R., & Farokhzad, O. C. (2017). Cancer nanomedicine: Progress, challenges and opportunities. *Nat. Rev. Cancer, 17*(1), 20.

Simon-Yarza, T., Mielcarek, A., Couvreur, P., & Serre, C. (2018). Nanoparticles of metal-organic frameworks: On the road to in vivo efficacy in biomedicine. *Advanced Materials, 30*(37), 1707365.

Song, G., Gao, Y., Wu, H., Hou, W., Zhang, C., & Ma, H. (2012). Physiological effect of anatase TiO_2 nanoparticles on Lemna minor. *Environmental Toxicology and Chemistry, 31*(9), 2147–2152.

Spielman-Sun, E., Avellan, A., Bland, G. D., Tappero, R. V., Acerbo, A. S., Unrine, J. M., Giraldo, J. P., & Lowry, G. V. (2019). Nanoparticle surface charge influences translocation and leaf distribution in vascular plants with contrasting anatomy. *Environmental Science. Nano, 6*(8), 2508–2519.

Springmann, M., Clark, M., Mason-D'Croz, D., Wiebe, K., Bodirsky, B. L., Lassaletta, L., de Vries, W., Vermeulen, S. J., Herrero, M., & Carlson, K. M. (2018). Options for keeping the food system within environmental limits. *Nature, 562*(7728), 519.

Srivastava, A. K., Dev, A., & Karmakar, S. (2018). Nanosensors and nanobiosensors in food and agriculture. *Environmental Chemistry Letters, 16*(1), 161–182.

Stefaniuk, M., Oleszczuk, P., & Ok, Y. S. (2016). Review on nano zerovalent iron (nZVI): From synthesis to environmental applications. *Chemical Engineering Journal, 287,* 618–632.

Stehle, S., & Schulz, R. (2015). Agricultural insecticides threaten surface waters at the global scale. *Proceedings of the National Academy of Sciences, 112*(18), 5750–5755.

Stewart, W., Dibb, D., Johnston, A., & Smyth, T. (2005). The contribution of commercial fertilizer nutrients to food production. *Agronomy Journal, 97*(1), 1–6.

Subramanian, K. S., Manikandan, A., Thirunavukkarasu, M., & Rahale, C. S. (2015). Nanofertilizers for balanced crop nutrition. In *Nanotechnologies in food and Agriculture* (pp. 69–80). Springer.

Sun, H., Liu, S., Zhou, G., Ang, H. M., Tadé, M. O., & Wang, S. (2012). Reduced graphene oxide for catalytic oxidation of aqueous organic pollutants. *ACS Applied Materials & Interfaces, 4*(10), 5466–5471.

Tarafdar, J., Raliya, R., Mahawar, H., & Rathore, I. (2014). Development of zinc nanofertilizer to enhance crop production in pearl millet (Pennisetum americanum). *Agricultural Res., 3*(3), 257–262.

Tombuloglu, H., Slimani, Y., Tombuloglu, G., Almessiere, M., & Baykal, A. (2019). Uptake and translocation of magnetite (Fe_3O_4) nanoparticles and its impact on photosynthetic genes in barley (Hordeum vulgare L.). *Chemosphere, 226,* 110–122.

United Nations, Department of Economic and Social Affairs (2017). Population Division, *World Population Prospects: The 2017 Revision, Key Findings and Advance Tables*. United Nations: New York, US.

Vatanpour, V., Madaeni, S. S., Khataee, A. R., Salehi, E., Zinadini, S., & Monfared, H. A. (2012). TiO_2 embedded mixed matrix PES nanocomposite membranes: Influence of different sizes and types of nanoparticles on antifouling and performance. *Desalination, 292,* 19–29.

Velzeboer, I., Kwadijk, C., & Koelmans, A. (2014). Strong sorption of PCBs to nanoplastics, microplastics, carbon nanotubes, and fullerenes. *Environmental Science & Technology, 48*(9), 4869–4876.

Venkatesan, J., & Kim, S.-K. (2014). Nano-hydroxyapatite composite biomaterials for bone tissue engineering—A review. *Journal of Biomedical Nanotechnology, 10*(10), 3124–3140.

Wang, J., Tsuzuki, T., Tang, B., Hou, X., Sun, L., & Wang, X. (2012). Reduced graphene oxide/ZnO composite: Reusable adsorbent for pollutant management. *ACS Applied Materials & Interfaces, 4*(6), 3084–3090.

World Population Prospects (2019). Highlights. https://population.un.org/wpp/Publications/Files/WPP2019_Highlights.pdf

Wu, Y., Pang, H., Liu, Y., Wang, X., Yu, S., Fu, D., Chen, J., & Wang, X. (2018). Environmental remediation of heavy metal ions by novel-nanomaterials: A review. *Environmental Pollution, 246*, 608.

Xiong, L., Wang, P., Hunter, M. N., & Kopittke, P. M. (2018a). Bioavailability and movement of hydroxyapatite nanoparticles (HA-NPs) applied as a phosphorus fertiliser in soils. *Environmental Science. Nano, 5*(12), 2888–2898.

Xiong, L., Wang, P., & Kopittke, P. M. (2018b). Tailoring hydroxyapatite nanoparticles to increase their efficiency as phosphorus fertilisers in soils. *Geoderma, 323*, 116–125.

Yang, F., Liu, C., Gao, F., Su, M., Wu, X., Zheng, L., Hong, F., & Yang, P. (2007). The improvement of spinach growth by nano-anatase TiO 2 treatment is related to nitrogen photoreduction. *Biological Trace Element Research, 119*(1), 77–88.

Yin, J., Wang, Y., & Gilbertson, L. M. (2018). Opportunities to advance sustainable design of nano-enabled agriculture identified through a literature review. *Environmental Science. Nano, 5*(1), 11–26.

Yoon, H. Y., Lee, J. G., Esposti, L. D., Iafisco, M., Kim, P. J., Shin, S. G., Jeon, J.-R., & Adamiano, A. (2020). Synergistic release of crop nutrients and stimulants from hydroxyapatite nanoparticles functionalized with humic substances: Toward a multifunctional Nanofertilizer. *ACS Omega, 5*, 6598.

Yu, M., Sun, C., Xue, Y., Liu, C., Qiu, D., Cui, B., Zhang, Y., Cui, H., & Zeng, Z. (2019). Tannic acid-based nanopesticides coating with highly improved foliage adhesion to enhance foliar retention. *RSC Advances, 9*(46), 27096–27104.

Zambryski, P. (2004). Cell-to-cell transport of proteins and fluorescent tracers via plasmodesmata during plant development. *The Journal of Cell Biology, 164*(2), 165–168.

Zhang, P., Ma, Y., Xie, C., Guo, Z., He, X., Valsami-Jones, E., Lynch, I., Luo, W., Zheng, L., & Zhang, Z. (2019a). Plant species-dependent transformation and translocation of ceria nanoparticles. *Environmental Science. Nano, 6*(1), 60–67.

Zhang, P., Ma, Y., Zhang, Z., He, X., Zhang, J., Guo, Z., Tai, R., Zhao, Y., & Chai, Z. (2012). Biotransformation of ceria nanoparticles in cucumber plants. *ACS Nano, 6*(11), 9943–9950.

Zhang, P., Misra, S., Guo, Z., Rehkämper, M., & Valsami-Jones, E. (2019b). Stable isotope labeling of metal/metal oxide nanomaterials for environmental and biological tracing. *Nature Protocols, 1*–22.

Zhang, P., Xie, C., Ma, Y., He, X., Zhang, Z., Ding, Y., Zheng, L., & Zhang, J. (2017). Shape-dependent transformation and translocation of ceria nanoparticles in cucumber plants. *Environmental Science & Technology Letters, 4*(9), 380–385.

Zhang, Q., Yu, J., Cai, J., Song, R., Cui, Y., Yang, Y., Chen, B., & Qian, G. (2014). A porous metal–organic framework with–COOH groups for highly efficient pollutant removal. *Chemical Communications, 50*(92), 14455–14458.

Zhang, X., Davidson, E. A., Mauzerall, D. L., Searchinger, T. D., Dumas, P., & Shen, Y. (2015). Managing nitrogen for sustainable development. *Nature, 528*(7580), 51–59.

Zhang, X., Xu, Z., Wu, M., Qian, X., Lin, D., Zhang, H., Tang, J., Zeng, T., Yao, W., & Filser, J. (2019c). Potential environmental risks of nanopesticides: Application of cu (OH) 2 nanopesticides to soil mitigates the degradation of neonicotinoid thiacloprid. *Environment International, 129*, 42–50.

Zhang, Y., Wang, L., Zhang, N., & Zhou, Z. (2018). Adsorptive environmental applications of MXene nanomaterials: A review. *RSC Advances, 8*(36), 19895–19905.

Zhao, G., Jiang, L., He, Y., Li, J., Dong, H., Wang, X., & Hu, W. (2011). Sulfonated graphene for persistent aromatic pollutant management. *Advanced Materials, 23*(34), 3959–3963.

Zhu, C., Xu, J., Song, S., Wang, J., Li, Y., Liu, R., & Shen, Y. (2020). TiO2 quantum dots loaded sulfonated graphene aerogel for effective adsorption-photocatalysis of PFOA. *Science Total Environ, 698*, 134275.

Zhu, C., Yang, G., Li, H., Du, D., & Lin, Y. (2015). Electrochemical sensors and biosensors based on nanomaterials and nanostructures. *Analytical Chemistry, 87*(1), 230–249.

Zhu, Z.-J., Wang, H., Yan, B., Zheng, H., Jiang, Y., Miranda, O. R., Rotello, V. M., Xing, B., & Vachet, R. W. (2012). Effect of surface charge on the uptake and distribution of gold nanoparticles in four plant species. *Environmental Science & Technology, 46*(22), 12391–12398.

Zwingmann, N., Mackinnon, I. D., & Gilkes, R. J. (2011). Use of a zeolite synthesised from alkali treated kaolin as a K fertiliser: Glasshouse experiments on leaching and uptake of K by wheat plants in sandy soil. *Applied Clay Science, 53*(4), 684–690.

Part III
Nanomterial Interactions in Cell Lines and Microorganisms

Chapter 12
Interaction of Food-Grade Nanotitania with Human and Mammalian Cell Lines Derived from GI Tract, Liver, Kidney, Lung, Brain, and Heart

Ananya Sharma and Aniruddha Singh

Contents

1 Introduction

Nanotitania additives (TiO_2 NPs: E171) have gained applications in a wide array of consumer products, including milk, confectionaries, pastries, toothpastes, enhanced food, sunscreen, cosmetics, and medicines under the assumption that they are inert particles that are safe to human health (Fig. 12.1 exhibits food products containing Ti additives). A cursory look at Fig. 12.2 shows content as high as 100 mg Ti per serving for powdered donuts, while other products with the highest Ti contents include sweets or candies, such as chewing gums, chocolate, and products with white icing or powdered sugar toppings. In 2005, the global production of nanoscale TiO_2 was estimated to be 2000 tons (t), worth $70 million, with about 1300 t of which was used in personal care products, such as topical sunscreens and cosmetics.

A. Sharma
Vanderbilt University School of Medicine, Nashville, TN, USA

A. Singh (✉)
Western Kentucky Heart and Lung, Bowling Green, KY, USA

University of Kentucky College of Medicine, Bowling Green, KY, USA

Vanderbilt University School of Medicine, Nashville, TN, USA

© Springer Nature Switzerland AG 2021
N. Sharma, S. Sahi (eds.), *Nanomaterial Biointeractions at the Cellular, Organismal and System Levels*, Nanotechnology in the Life Sciences,
https://doi.org/10.1007/978-3-030-65792-5_12

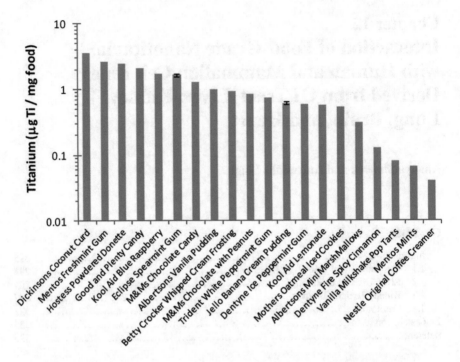

Fig. 12.1 Normalized Ti concentration in food products. For the top 20 products (upper), error bars represent the standard deviation from samples digested in triplicate (Weir et al. 2012)

The production had soared to 5000 tons by the year 2010, and it is expected to accelerate until at least 2025, with greater preference for nanotitania (Weir et al. 2012). Multiple sources of nanoscale TiO_2 can thus result in human exposure through water, air, and soil sediments. Figure 12.2 shows the simulated exposure to TiO_2 for the US population, with an average of 1–2 mg TiO_2/kg bodyweight/day for children under the age of 10 years and approximately 0.2–0.7 mg TiO_2/kg bodyweight/day for the other consumer age groups (Weir et al. 2012). Industrial applications of nano-Ti add enormously to the list of on-the-shelve grocery products.

In October 2010, the National Organic Standards Board recommended discontinuation of the engineered nanomaterials (ENMs) from food products bearing the U.S. Department of Agriculture's Organic label (Kessler 2011). If the USDA adopts this recommendation at some point of time, the consumers will treat ENMs containing foods similar to the controversial category of genetically modified organisms (GMOs) foods. In addition, the exclusivity of organic foods will be questioned as they contain nanotechnology-enabled value additions like flavor- and texture-enhancing ingredients and shelf life-extending packaging nanomaterials. In light of recent research findings on the effect of TiO_2 and the public outcry against its presence in food, France has imposed a ban on the use of this food additive for the year 2020. The USA may lose millions of dollars that come from the export of confectionaries and pastries to France, if the US industry does not remove this additive from their products (USDA-FAS, 2019).

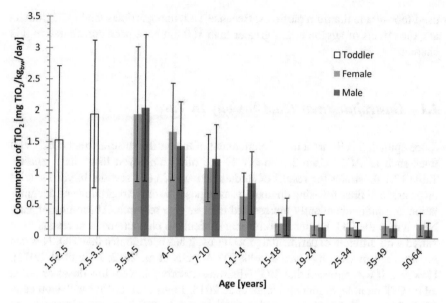

Fig. 12.2 Histogram of the average daily exposure to TiO₂ for the US population (Monte Carlo simulation). Error bars represent the upper and lower boundary scenarios (Weir et al. 2012)

Recent in vitro and in vivo studies based on mammalian cell lines and animal models, respectively, have shown various harmful effects of nano-Ti exposures affecting growth and development, reproduction, metabolism, immune functions, and life span of organisms (Baranowska-Wójcik et al. 2020). Some clinical studies have examined the role of internal (constant) Ti exposures because of dental implants and found their impact on the whole-body health via allergic reactions (Tibau et al. 2019). Many investigations have mapped out various pathways involved in the toxicity of TiO_2 NPs exposures. These include, but are not limited to, cytotoxicity, cellular oxidative stress, inflammatory responses, genotoxicity, and altered expression patterns of genes or proteins (Baranowska-Wójcik et al. 2020). While many studies based on animal models have established the harmful effects of nanotitania, there are few studies directly connecting to the clinical diagnosis of a medical condition or mortality in human subjects caused by titanium exposure. Therefore, it is important to look at how nanotitania can interact with human cell lines under in vitro or simulated conditions. A critical evaluation of its interaction at the molecular level in the cellular environment may give some insight into its effects on chronic conditions, such as inflammatory bowel disease or neurodegenerative conditions. Although TiO_2 NPs may affect many cell types and organs, this chapter examines effects particularly on gastrointestinal, liver, kidney, lung, brain, and heart cell lines. Toxicity parameters examined include production of reactive oxygen species (ROS) and oxidative stress, genotoxicity, cytotoxicity, accumulation of nanotitania, and alterations in cell morphology. While NPs are classically defined as being less than 100 nm, some studies employing slightly larger sizes are also included in this review to reflect the commonly used food-grade titanium particles. Because TiO_2 nanoparticles tend to agglomerate, the effects of agglomerates greater than 100 nm have been considered in this chapter.

1.1 Gastrointestinal Tract Toxicity In Vitro

Since epithelial cells are a major component of the gastrointestinal tract (GIT), cell lines such as AGS, Caco-2, and HT-29 are often employed in toxicity studies. Table 12.1 illustrates the results of studies reviewed in this section. In vitro studies on gastric cell lines focusing on toxicity, as opposed to only absorption and accumulation, are more abundantly represented than in vivo research. Absorption of TiO_2 NPs across the GIT barrier is somewhat difficult to characterize and results have varied with different experimental procedures. It has been shown that TiO_2 NPs can be absorbed in vitro by Caco-2 cells (De Angelis et al. 2012; Ruiz et al. 2017). However, it was reported that TiO_2 NPs were absorbed in very low doses or not at all by GIT monolayer models (Janer et al. 2014; Jones et al. 2015; MacNicoll et al. 2015; Song et al. 2015). Contrasting results, such as these, are most likely due to variation in experimental design, and more importantly the physicochemical property differences of the TiO_2 NPs used in each study. Factors such as size, shape, agglomeration, hydrodynamic diameter, surface area, and structure can greatly affect how TiO_2 NPs behave both in vitro and in vivo. Since there is no standardization for toxicity and accumulation testing, it is difficult to ascertain as to which of these factors are influencing toxicity results. In addition to physicochemical properties of NPs, differences in responses to TiO_2 NPs in vitro could be due to the types of cell line and titanium crystal used. Song et al. (2015) demonstrated undifferentiated Caco-2 cells were much more sensitive to TiO_2 NPs than differentiated Caco-2 cells and Caco-2 monolayer. Undifferentiated cells in this study showed significant increases in NP uptake and ROS generation when compared to the differentiated and monolayer cells, although no cytotoxicity was observed.

An elegant study, based on combined in vivo and in vitro conditions of exposure, demonstrated a significant uptake of Ti nanoparticles by human intestinal epithelial cells (IECs) and macrophages in culture. The uptake was observed to trigger NLRP3-ASC-caspase-1 assembly, caspase-1 cleavage, and the release of NLRP3-associated interleukin (IL)-1β and IL-18. TiO_2 also induced reactive oxygen species generation and increased epithelial permeability in IEC monolayers (Ruiz et al. 2017). This study clearly suggests that larger TiO_2 NP size range, such as those found in food-grade TiO_2 NP, are likely to be absorbed by the gastrointestinal tract. However, Jones et al. (2015) reported no significant penetration of TiO_2 NPs into the basal chamber of Caco-2 cells at higher concentrations. The difference may be attributed to the size of particles used in this study (100 nm and smaller). Janer et al. (2014) indicated TiO_2 NPs did not affect Caco-2 membrane integrity at concentrations of 100 µg/ml but did demonstrate that a minute number of NPs were able to cross the monolayer membrane at just above the detection rate of 0.4%. This is consistent

Table 12.1 Effects of TiO$_2$ NPs on gastrointestinal tract in vitro

Cell type	NP primary size	Hydrodynamic/agglomerate size	NP structure	Exposure parameters	Results	Reference
Caco-2	TiO$_2$-F (fine): 40-300 nm TiO$_2$: 20–80 nm TiO$_2$-HSA: <10 nm	Not specified	Anatase Anatase/rutile Anatase modified	0–80 ug/cm^2 particles for up to 24 h	Cytotoxicity induced by nano-TiO$_2$; reduced metabolic activity in nano-TiO$_2$ at 20 ug/cm^2; no DNA breaks or oxidative lesions	Gerloff et al. (2009)
Caco-2	TF$_A$ (fine): 215 ± 2 nm TUF$_{A/R}$I: 25.20 ± 0.20 nm TUF$_A$II: 21.90 ± 0.30 nm TUF$_A$I: 6.7 ± 1.3 nm TUF$_A$II: 3.94 ± 0.05 nm	374.0 ± 5.5 nm 214.5 ± 2.6 nm 327.5 ± 27.6 nm 455.2 ± 74.4 nm 291.1 ± 56.6 nm	Anatase Anatase/rutile Anatase/rutile Anatase Anatase	0–80 ug/cm^2 particles for up to 24 h	Decreased viability after 24 h in all except fine TiO$_2$; low genotoxicity in TUF$_{A/R}$I; no significant oxidative damage or ROS observed	Gerloff et al. (2012)
Caco-2	20–60 nm	Serum-free medium: 771.90 ± 110.00 nm Complete medium: 1080.00 ± 190.50 nm	Anatase Anatase	0–20 ug/cm^2 for up to 24 h	No significant cytotoxicity; significant ROS production after 6 hr. but declined up to 24 h; no increase in IL-8	De Angelis et al. (2012)
Simulated digestive system; Caco-2; GES-1	Gum16: 40–300 nm	20–300 nm	Gum-2–6: Mainly anatase Gum-1: Anatase with some rutile	0–200 μg/ml for 24 h	95% of TiO2 in gum comes while chewing and is digested; slight oxidative stress (ROS) in both cell lines induced; no cell viability loss	Chen et al. (2013)

(continued)

Table 12.1 (continued)

Cell type	NP primary size	Hydrodynamic/agglomerate size	NP structure	Exposure parameters	Results	Reference
AGS	21 nm <25 nm	160.5 nm 420.7 nm	Anatase/rutile Oxide-anatase	0–150 µg/ml for up to 24 h	Increased proliferation, genotoxicity, and oxidative stress; decreased apoptosis; tumor-like phenotype possibly induced by oxidative stress	Botelho et al. (2014)
Caco-2, Caco2/HT29 co-culture, Caco-2/RajiB co-culture	12 ± 3 nm	H_2O: 132.0 ± 0.8 nm Simulated gastric fluid: 218.4 ± 2.9 nm Simulated intestinal fluid: >1000 nm	Anatase Anatase Anatase	0–200 µg/ml for up to 48 h	NPs translocated through Caco-2/RajiB co-culture; no translocation in Caco-2 and Caco-2/HT29; modulation of cell junction maintenance genes occurred; no overt cytotoxicity or apoptosis;	Brun et al. (2014)
Caco-2_BBe1	Food grade: 122 ± 48 nm Gum grade: 141 ± 56 nm	220 ± 9 nm 250 ± 10 nm	Anatase Anatase	0–350 µg/ml for unspecified time	Both NP types caused limp and reduced microvilli; gum NPs caused retracted and non-erect microvilli; food NPs caused significant loss of microvilli and disorganization	Faust et al. (2014)
Caco-2	Bulk: 103.2 ± 16 nm P25: 22.8 ± 0.64 nm Anatase: 16.4 ± 2.4 nm Rutile: 30.8 ± 2.5	179.3 ± 13.7 7.1 ± 4.1 142.3 ± 14.4 88.3 ± 34.1	Anatase/rutile Anatase/rutile Anatase Rutile	0–10 mg l^{-1} for 24 h	Good cell viability, low LDH leakage, normal morphology; uptake of Ti observed with increases in Ca^{2+}	Gitrowski et al. (2014)
Caco-2	18 ± 8 nm	270 nm	Anatase	100 µg/ml for 24 and 48 h	Membrane integrity not affected; Ti uptake detected just above minimum detection level	Janer et al. (2014)
Caco-2	15 nm 100 nm < 5 µm	Up to 1 µm 1 µm-sized Few µm-sized	Anatase Rutile Rutile	0–100 µg/ml for unspecified time	No NPs observed in basal chamber of cells; simulated gastric fluid showed more agglomeration of NPs than water	Jones et al. (2015)

Cell type	NP primary size	Hydrodynamic/agglomerate size	NP structure	Exposure parameters	Results	Reference
Caco-2, Caco-2/M-cell co-culture, Caco-2/RajiB co-culture	Sigma 15 nm, Nanocomposix 25 nm, Sigma 40–50 nm, Sigma <5 μm	250–400 nm, ~125 nm, Not specified, Not specified	Anatase, Anatase/rutile, Rutile, Rutile	0–250 μg/ml for 24 h	All NPs except Nanocomposix decreased cell viability at larger sizes, indicating effects were not related to nano size; lack of translocation across epithelium model	MacNicoll et al. (2015)
Caco-2 (differentiated, undifferentiated, monolayer)	T1: 99 ± 30 nm, T2: 26 ± 12 nm	H_2O: 233 nm, Medium: 719 ± 56 nm, H_2O: 497 nm, Medium: 1785 ± 237 nm	Anatase, Anatase	0–200 μg/ml for up to 24 h	No toxicity observed in differentiated and undifferentiated cells; differentiated cells showed lower ability to take up NPs; no significant crossing of NPs in monolayer; significant ROS generation in undifferentiated cells	Song et al. (2015)
Caco-2/THP-1 and Caco-2/MUTZ3 co-culture	7–10 nm	896 ± 133 nm	Not specified	24 h exposure, concentration not specified	No significant toxicity or inflammation observed	Susewind et al. (2016)
Caco-2	20–60 nm	220 ± 68 nm	Anatase	0–38 μg/ml for up to 24 h	Cytotoxicity observed only at 24 h for all concentrations	Tassinari et al. (2015)
Caco-2	20–60 nm	1080 ± 190 nm	Anatase	0–128.0 μg/ml for up to 24 h	No significant micronuclei despite increased ROS production; slight oxidative DNA damage	Zijno et al. (2015)
Caco-2, HT-29	0.36 μm	Not specified	Anatase	0–100 μg/ml for up to 24 h	Intracellular accumulation by endocytosis in both cell types; significant increase in ROS; increased permeability in Caco-2	Ruiz et al. (2017)
Caco-2	A50: 50 nm, A100: 100 nm, R50: 50 nm, R250: 250 nm, P25: 21 nm	205.30 ± 4.88 nm, 262.10 ± 4.66 nm, 193.28 ± 1.37 nm, 439.13 ± 8.665 nm, 181.55 ± 1.10 nm	Anatase, Anatase, Rutile, Rutile, Anatase/rutile	0–50 μg/ml for up to 72 h	At 24 h no reduced viability; at 72 h all but R250 reduced viability; significant oxidative stress induced by all NPs except P25; A50 increased IL-8 expression	Tada-Oikawa et al. (2016)

with the observation of Chen et al. (2013), which indicated significant aggregates of TiO$_2$ NPs in a human digestive model. The observed aggregation was due to the character of the stomach fluid and its effects on the NPs themselves. This interaction with stomach fluid could therefore lend cause as to why TiO$_2$ NPs are unable to cross the GIT barrier. In addition, Chen et al. (2013) also demonstrated inability of TiO$_2$ NPs to induce cytotoxicity in GES-1 and Caco-2 cells, while noted slight increases in ROS and oxidative stress. Song et al. (2015) confirmed similar results of stomach fluid influencing ability of TiO$_2$ NPs to infiltrate Caco-2 cells. Other specific effects of TiO$_2$ NPs have been recently documented in case of gastric cell cultures. Treatment of human AGS cells with TiO$_2$ NPs induced an increase in proliferation, oxidative stress, genotoxicity, and a decrease in the frequency of apoptosis (Botelho et al. 2014). The authors suggest that TiO$_2$ NP effect on the cell cycle and oxidative stress via ROS generation may be responsible for high growth rates and uncontrolled proliferation, respectively; however, the TiO$_2$ mechanism of action in this study was not fully elucidated. Conversely, Gitrowski et al. (2014) had observed good cell viability and negligible cytotoxicity in Caco-2 monolayers at higher concentrations of TiO$_2$ NP treatment along with intact cell morphology. It is interesting to note, however, that Ti uptake increased at a significant rate as concentration of NPs increased in these cells.

Generation of oxidative stress as a result of TiO$_2$ NP exposure is also controversial in recent research. Older studies have shown that TiO$_2$ NPs induce cell death in undifferentiated Caco-2 cells but do not cause oxidative stress (Gerloff et al. 2009, 2012). While the 2012 study by Gerloff et al. demonstrated genotoxic effects after TiO$_2$ NP treatment in the form of DNA strand breaks, these results were not attributed to oxidative stress. ROS in Caco-2 cells was shown to increase after 6 hours of TiO$_2$ NP treatment, but then decrease at 24 h (De Angelis et al. 2012; Zijno et al. 2015). This suggests cells may be able to repair themselves after initial NP shock. However, Zijno et al. (2015) did present evidence of genotoxicity at some test points, although no time- or concentration-dependent patterns could be established. Significant oxidative stress has been demonstrated in both rutile and anatase TiO$_2$ NP treatments at concentrations as little as 50 µg/ml (Tada-Oikawa et al. 2016). Tassinari et al. (2015) also demonstrated decreases in the enzyme lactate dehydrogenase (LDH) after TiO$_2$ treatment in Caco-2 cells, but it was not identified if this decrease was related to ROS or oxidative stress levels. Previously, it was shown that TiO$_2$ NPs can localize within and in between gut epithelial cells, but complete translocation was not observed in Caco-2 monolayer and Caco-2/HT29 coculture and no cytotoxicity or apoptosis was observed (De Angelis et al. 2012; Brun et al. 2014). More recently, Ruiz et al. (2017) have demonstrated oxidative stress and NP accumulation in Caco-2 and HT-29 cell lines in a dose-dependent manner, along with increased epithelial permeability at doses as low as 20 µg/ml TiO$_2$ NPs. MacNicoll et al. (2015) demonstrated that cell viability decreased as NP concentration increased, but it was shown that this effect was more apparent for larger size NPs which indicated viability was related to physical effects of particles as opposed to the NP size. Similar cytotoxic effects (after 72 hours) and increases in inflammation markers (IL-8) were also demonstrated by Tada-Oikawa et al. (2016), but relation to particle size and agglomeration was not specifically addressed. The largest size of 250 nm did not produce significant cytotoxic effects. Earlier studies indicated that TiO_2 NPs were not able to induce cytotoxicity or inflammation (IL-8) under inflamed and non-inflamed conditions in Caco-2 cells (De Angelis et al. 2012), but well-designed studies by MacNicoll et al. (2015) and Tada-Oikawa et al. (2016) led to a different conclusion showing the potential of large TiO_2 NPs agglomerates in causing cytotoxicity. Overall, from the studies reviewed here, it is clear that although TiO_2 NPs may be able to accumulate or permeate between cells, the cytotoxic effects produced is not very high. The ability of these NPs to increase epithelial permeability should be further explored in the GIT, since nanotitania are widely used in many food-manufacturing processes.

Other in vitro studies have focused on the physical effects and toxicity of TiO_2 NPs on the gastrointestinal tract. Microvilli disruption from food and gum-grade TiO_2 NP exposure in Caco-2_{BBe1} cells has been observed at doses as low as 350 ng/mL (Faust et al. 2014). The human gastrointestinal epithelium is lined with microvilli, which help increase surface area and nutrient absorption. Faust et al. (2014) estimated 42% microvilli loss at the 350 ng/mL dose in their study and characterized this change as a unique biological response to TiO_2 treatment. The food grade nanoparticles used for this study were not consistently less than 100 nm due to aggregation from experimental treatment, although primary particles less than 100 nm were present. It is worth noting that the TiO_2 NPs used for this study are the same as those used in worldwide food and gum manufacturing, and so their effects illustrate possible outcomes of overconsumption of TiO_2 NPs. One interesting connection to the Brun et al. (2014) study is that they observed more TiO_2 NPs taken up by Caco-2/HT29 cocultured cells that lacked microvilli. If TiO_2 NP uptake does decrease microvilli as indicated by Faust et al. (2014), then it could be facilitating TiO_2 NP uptake in the gastrointestinal tract based on the results demonstrated by Brun et al. (2014). This should be investigated more given there are some conflicting results regarding TiO_2 NP translocation across the gut epithelium.

In an interesting study, Dudefoi et al. (2017) examined a defined human model of intestinal bacterial community in vitro. Molecular phylogeny based on DNA profiles indicated effects of nano-Ti on the bacterial community, showing a modest decrease in the relative abundance of the dominant *Bacteroides ovatus* in favor of *Clostridium cocleatum*. Such shifts in the treated consortia may appear minor in single or limited doses, but their cumulative effect might be very significant due to chronic TiO_2 NP ingestion. This aspect needs further investigations to understand the interactions between food-grade Ti and the increased intestinal permeability in humans.

1.2   Liver Toxicity In Vitro

The liver is a site of filtration and detoxification for the body, which makes it an important site of observation when studying effects of TiO_2 NPs. Table 12.2 presents details of the studies examined; however, there have not been an overwhelming number of in vitro liver cell studies. Kermanizadeh et al. (2012) demonstrated low cytotoxicity and oxidative stress in C3A hepatocytes at concentrations of 0–256 µg/ml, but observed some genotoxicity, IL-8 induction, and increases in ROS production. It is interesting that although overall toxicity was low, TiO_2 NPs were still able to induce the aforementioned effects. In the same manner, Tobergte and Curtis (2013) observed low cytotoxicity and no change in C3A hepatocyte morphology at doses up to of 625 µg/cm². However, it was shown that glutathione (GSH) levels decreased at a dose of 250 µg/cm², indicating oxidative stress, and indicators of inflammation (IL-8 and TNF- α) also increased at lower exposures of 64 µg/cm². Although overall hepatocyte homeostasis was unaffected, these results indicate TiO_2 NPs carry potential for causing other possible damaging effects. Additionally, these results suggest that chronic uptake of nanotitania has potential to interfere with Cytochrome P450 drug metabolism in the liver. Considering these results, it is clear TiO_2 NPs are capable of inducing stressful effects on these liver cell types, which have been investigated further in vivo to confirm deleterious outcomes of TiO_2 NP exposure to the liver. In addition, cell type may play a critical role in sensitivity, as demonstrated by Sha et al. (2011) who characterized differences in cytotoxicity based on liver cell type examined. Although overall cytotoxicity was exhibited in a time- and dose-dependent manner for all cell types, the concentrations which showed increases in ROS and decreases in GSH varied between each cell type.

El-Said et al. (2014) have confirmed significant increases in oxidative stress and ROS generation in HEPG2 liver cells after TiO_2 NP exposure, which resulted in apoptosis and morphological changes such as chromatin condensation and nuclear fragmentation. This occurred at low concentrations of 10 µg/ml, but it was not specified how much of the TiO_2 NP solution was incubated with the cell cultures. In the same manner, Natarajan et al. (2015) also indicated significant increase in ROS production in cultured rat primary hepatocytes as well as higher cytotoxicity as concentration increased, but no change in cell morphology was observed. What is particularly interesting about this study is that it examined direct effects of TiO_2 NPs on actual liver function by examining urea and albumin synthesis after treatments. It was shown that in both cases, there was a loss of function after TiO_2 exposure. Mitochondrial function was also altered with loss of fibers and increased instances of fragmentation. This is consistent with the observed increase in ROS production. A separate study has confirmed that oxidative cellular conditions contribute to increasing TiO_2 NP toxicity in BRL-3A rat liver cells (Sha et al. 2014). It was found that TiO_2 NPs alone induced abnormal G0/G1 and S phase transition; however, under oxidative cellular conditions, cells were arrested at G2/M. Significant increases in cytotoxicity were observed when hydrogen peroxide (H_2O_2) was built during the oxidative stress induced by nanoparticles. This indicates TiO_2 NP may increase liver toxicity based on cellular oxidative conditions.

Table 12.2 Effects of TiO$_2$ NPs on liver in vitro

Cell type	NP primary size	NP hydrodynamic size	NP structure	Exposure parameters	Results	Reference
BRL3A	40 nm	Not specified	Not specified	0–250 µg/ml for 24 h	Mitochondrial function unaffected, low LDH leakage, and low cytotoxicity up to 250 µg	Hussain et al. (2005)
SMMC-7721, HL-7702, CBRH-7919, and BRL-3A	3.768 ± 0.242 nm	Not specified	Not specified	0–100 µg/ml for up to 48 h	All cell lines exhibited cytotoxicity, morphological changes, increased ROS, and decreased GSH	Sha et al. (2011)
C3A	NM101: 4–8 nm NRCWE001: 80–400 nm NRCWE002: 80–400 nm NRCWE003: 80–400 nm NRCWE004: Up to µm	185 nm 203 nm 287 nm 240, 1487 nm 339 nm	Anatase Rutile Rutile, pos. Charge Rutile, neg. Charge Rutile	0–256 µg/ml for up to 24 h	Low cytotoxicity; genotoxicity evident in NM101 and NRCWE002; no significant GSH depletion; increased ROS; IL-8 induction	Kermanizadeh et al. (2012)
C3A	NM101: 4–8 nm ST-01: 7 nm	In medium: 185, 742 nm In H$_2$O: 75.66 nm	Anatase Anatase	0–625 µg/cm^2 for up to 24 h	Low cytotoxicity; no effect on hepatocyte homeostasis; oxidative stress at 250 µg; no changes to cell morphology, but NPs found in cytoplasm; IL-8 and TNF- α increased	Tobergte and Curtis (2013)
HEPG2	Not specified	216 ± 70 nm	Not specified	0–10 µg/ml for 48 h	Increase in oxidative stress resulting in apoptosis and ROS; chromatin condensation, nuclear fragmentation	El-Said et al. (2014)
BRL3A	12–18 nm (diameter), 40–80 nm (length);	Not specified	Rutile	0–100 µg/ml for up to 24 h	Under oxidative stress conditions cytotoxicity increased; G0/G1 arrest exhibited	Sha et al. (2014)

(continued)

Table 12.2 (continued)

Cell type	NP primary size	NP hydrodynamic size	NP structure	Exposure parameters	Results	Reference
Rat primary hepatocytes	P25: 21 nm 50 nm 50 nm	800 nm 380 nm 700 nm	Anatase/rutile Anatase Rutile	0–1000 ppm for up to 72 h	No change in cellular morphology; cytotoxicity exhibited in dose-dependent manner; loss of function for urea and albumin synthesis; increased ROS; mitochondrial fragmentation	Natarajan et al. (2015)

A careful perusal of Table 12.2 shows some degree of cytotoxicity in several in vitro studies. It is apparent that TiO$_2$ NPs do have potential to cause liver toxicity under the right cellular conditions. Increased oxidative stress and ROS seem to be hallmarks of toxic NP effects. HepG2 cells appear to be particularly sensitive to TiO$_2$ NPs, showing toxic effects at concentrations as low as 10 µg/ml. One aspect to consider, however, is that in vitro studies focusing on TiO$_2$ NP toxicity are somewhat sparse in recent years. Most of the studies reviewed here used TiO$_2$ NP concentrations much lower than average limits of daily exposure for humans. However, these levels caused cellular stress and toxicity, indicating a need to explore in vitro studies more thoroughly in conjunction with animal models.

1.3 Kidney Toxicity In Vitro

Table 12.3 illustrates recent studies on TiO$_2$ NP toxicity in kidney cell lines. These investigations again present contrasting cytotoxicity effects depending on their experimental designs. Previously it was reported that human IP15 mesangial cells did not experience cytotoxicity after exposure to TiO$_2$ NPs, but LLC-PK1 cells showed more sensitivity and slightly increased cytotoxicity after NP contact (Bhowmick et al. 2010). Similar results of zero to low cytotoxicity after TiO$_2$ contact have also been demonstrated in HK-2, canine MDCK, and normal rat kidney (NRK) cells (Pujalté et al. 2011; Halamoda et al. 2013; Pujalté et al. 2015; Schoelermann et al. 2015). It is worth noting that both Bhowmick et al. (2010) and Pujalté et al. (2011) noted accumulation of NPs in kidney cells, but with contrasting histological and morphological results. While Bhowmick et al. (2010) reported uptake with changes in cell structure, such as cell shrinkage and detachment, Pujalté et al. (2011) observed no apparent loss of adhesion or morphological changes. In addition, Halamoda et al. (2013) demonstrated that LLC-PK1 cells were less sensitive than MDCK cells and had no cytotoxicity. MDCK cells only exhibited cytotoxicity at the highest concentrations used (50 and 235 µg/ml), whereas LLC-PK1 had low cytotoxicity at these concentrations. Furthermore, Meena et al. (2012) showed cytotoxicity with significant viability decreases in HEK-293 cells in a dose-dependent manner. Therefore, cell type may play a role in effects of TiO$_2$ NPs, much like the aforementioned Caco-2 cells in the gastrointestinal tract.

Surprisingly, several of the previously mentioned studies indicated increases in ROS and/or oxidative stress without cytotoxic effects (Bhowmick et al. 2010; Pujalté et al. 2011; Halamoda et al. 2013; Pujalté et al. 2015). This is particularly striking since oxidative stress and ROS generation are often associated with cytotoxic effects on the cell. Pujalté et al. (2011) demonstrated no reduction in GSH levels after exposure in both IP15 and HK-2 cell lines, even though increased ROS levels were detected. In a similar fashion, a later study by Pujalté et al. (2015) demonstrated induction of antioxidant systems via Nrf2 translocation, significant cellular oxidative damage by lipid peroxidation, and increases in key ROS genes (CAT, heme oxygenase [HO-1], glutathione peroxidase [GPx]) after TiO$_2$ NP exposure, but no

Table 12.3 Effects of TiO$_2$ NPs on kidney in vitro

Cell type	NP primary size	NP hydrodynamic/ agglomerate size	NP structure	Exposure parameters	Results	Reference
IP15, LLC-PK$_1$	15 nm 25–75 nm	481 ± 48 nm 1005 ± 270	Anatase/ rutile Anatase/ rutile	0–160 ug/cm^2 for up to 24 h	Cytotoxicity in LLC-PK$_1$ but not in IP15; cell shrinkage and detachment in both cell types; uptake of 15 nm NPs in LLC-PK$_1$; no significant increased ROS	Bhowmick et al. (2010)
IP15, HK-2	12 ± 2 nm	449 ± 393	Anatase/ rutile	5–20 ug/cm^2 for up to 24 h	No loss of adhesion or morphological changes; IP15 showed slight cytotoxicity; ROS generation in both cell types but no reduction in GSH	Pujalté et al. (2011)
HEK-293	<25 nm	In H$_2$O: 72 nm In culture medium: 165.6 nm	Anatase	0–200 μg/ml (cell viability assay) and 0–200 mg/l (cytotoxicity and LDH assays) for up to 72 h	Decreased viability; genotoxicity and cytotoxicity observed; increased lipid peroxidation; morphological changes; upregulated p53, Bax, and caspase-3	Meena et al. (2012)
HEK-293	2.3 nm	2.3 ± 0.5 nm	Not specified	0–100 μg/ml for 3 h	Significant genotoxicity at 100 μg	Demir et al. (2013)
MDCK, LLC-PK	21 nm	Not specified	Not specified	0–235 μg/ml for up to 72 h	In MDCK: Decreased mitochondrial metabolism, ROS production; in LLC-PK: ROS production	Halamoda et al. (2013)
HK-2	NM101: 4–8 nm NRCWE001: 80–400 nm NRCWE002: 80–400 nm NRCWE003: 80–400 nm NRCWE004: Up to 2 μm	358 nm 337.5 nm 378.8 nm 423.6 nm 482.6 nm	Anatase Rutile Rutile, pos. Charge Rutile, neg. Charge Rutile	0–256 μg/ml for up to 72 h	Reduced cell viability but overall low toxicity; all but NM 101 showed increased IL-6 and IL-8; no TNF-α changes observed; slight increase in oxidative stress for 4/5 NPs; genotoxicity at 4 hr. in 2/5 NPs (NRCWE002 and 004)	Kermanizadeh et al. (2013)

Cell type	NP primary size	NP hydrodynamic/agglomerate size	NP structure	Exposure parameters	Results	Reference
HK-2	12.2 ± 2.2	Not specified	Anatase	0–160 µg/cm² for 24 h	Lipid oxidative damage was observed; Nrf2 translocation, which indicated induction of anti-oxidant systems, but no NF-κB translocation was observed; increased, CAT, HO, and GSH peroxidase	Pujalté et al. (2015)
NRK, HeLa, CHO	5 nm 40 nm	In H₂O: 563.5 nm In medium: 162.5 nm In H₂O: 360 nm In medium: 163 nm	Anatase Anatase	5 mg/l for 16 h	No effect on cellular viability, but observed modified communication with surrounding cells; transfer of NPs between cells depends on and scales with cell contact	Schoelermann et al. (2015)
HK-2	21 nm	<50 nm	Not specified	0–300 µg/ml	Increased ROS production and protein expression of TGF-β	Huang et al. (2015)

cytotoxicity was observed. In contrast, Nrf2 has been shown in lung tissue to have a protective effect against TiO2 NP toxicity (Delgado-Buenrostro et al. 2009). Of the in vitro kidney studies reviewed, those that observed NF-κB activity showed no change, which is interesting given that NF-κB is often induced after NP exposure, leading to oxidative and inflammation effects. As mentioned previously, Halamoda et al. (2013) demonstrated cytotoxic effects at the highest concentrations of TiO_2 NPs in MDCK cells, and this was accompanied by significant increases in ROS. Although increases in ROS were demonstrated in LLC-PK cells in this study as well, treatment of cells with TiO_2 NPs failed to induce any changes in DNA synthesis. Kermanizadeh et al. (2013) revealed interesting results using five different types of TiO_2 NPs, showing that cell viability and genotoxicity varied depending on NP characteristics. In addition, four out of five (NM101 being the exception) TiO_2 NPs tested showed a dose-dependent increase in inflammatory markers IL-6 and IL-8. Overall cytotoxicity was evaluated to be low in HK-2 cells exposed to TiO_2 NPs; however, significant increases in genotoxicity and oxidative stress were noted. Holistically looking at the findings on cytotoxicity across various types of in vitro cell models, cell type sensitivity may be a characteristic to consider when assessing toxicity of TiO_2 NPs in the kidney.

Although most in vitro studies have shown overall low toxicity of TiO_2 NPs, some select few have demonstrated significant increases in apoptosis and genotoxicity after NP exposure in HEK-293 cells (Meena et al. 2012; Demir et al. 2013). Interestingly, Demir et al. (2013) have reported that the ionic form of TiO_2 did not demonstrate toxicity on cells, which was similarly observed with other types of metallic NPs. Meena et al.'s (2012) observation of upregulated p53, Bax, and caspase-3 correlated well with similar observations in other cell types and tissues (Marquez-Ramirez et al. 2012; Park et al. 2014; Lucie et al. 2015; Chen et al. 2016; De Simone et al. 2016). Schoelermann et al. (2015) had demonstrated that TiO_2 NP altered cell-cell communication, and that NP transfer between cells depends on and scales with cell contact. This may indicate some interactions of TiO_2 NPs with signaling pathways, which may not result in increased toxicity, but rather lead to other changes within the cell mentioned above. Overall, cell culture studies related to the effect of TiO_2 NP are indicative of their impact on the kidneys, but the findings are not as robust as animal model-based investigations.

1.3.1 Lung Toxicity In Vitro

Toxicity studies on lung cell types in vitro are quite numerous, since inhalation is a typical exposure route to TiO_2 NPs in humans. Table 12.4 shows the results of many cell culture studies conducted recently. Typical lung cell lines such as A549, alveolar macrophages (AMs), MH-S, BEAS-2B, and WI-38 have all shown increased uptake and/or accumulation of Ti after TiO_2 NP exposure (Liu et al. 2013; Tang et al. 2013; Park et al. 2014; Medina-Reyes et al. 2015b; Vales et al. 2015; Armand et al. 2016; Chen et al. 2016). Liu et al. (2013) demonstrated TiO_2 NPs were more easily phagocytized by AMs when compared to other NP types, and Park et al.

Table 12.4 Effects of TiO₂ NPs on lung in vitro

Cell type	NP primary size	NP hydrodynamic/ agglomerate size	NP structure	Exposure parameters	Results	Reference
MH-S	21 ± 3 nm 98 ± 20 nm 12 ± 2 nm 148 ± 35 nm	Not specified	Anatase Anatase Anatase Rutile	0–600 µg/ml for 24 h	No cytotoxicity from rutile; Anatase forms had little effect on viability until 600 µg	Zhang et al. (2012)
V79	MT15: 5.9 nm P25: 34.1 nm Nanofilament: 15.5 nm Bulk: 169.4 nm Coated: 1–10 nm	460 nm 400 nm 420 nm 365 nm 600 nm	Anatase Anatase, rutile Rutile Anatase Rutile	0–100 mg/l for up to 48 h	Nano-NPs taken up more than micron-sized NPs and induced cytotoxicity and genotoxicity; coated NPs much less genotoxic and cytotoxic;	Hamzeh and Sunahara (2013)
Alveolar macrophage (AMs)	22.82 ± 5.30 nm	Not specified	Not specified	0–100 µg/ml for 24 h	TiO₂ NPs more readily phagocytized than other NP types; decreased viability; nuclear condensation and vacuoles	Liu et al. (2013)
A549	20–50 nm	Not specified	Anatase	0–300 µg/ml for 44 h	Decreased viability and mitochondrial function; accumulation; mitochondria nearest NPs damaged more seriously;	Tang et al. (2013)

(continued)

Table 12.4 (continued)

Cell type	NP primary size	NP hydrodynamic/ agglomerate size	NP structure	Exposure parameters	Results	Reference
MH-S	TNS: 50 nm P25:	Vehicle: 106.7 ± 30.8 nm Media: 223.6 ± 21.5 nm H_2O: 483.7 ± 25.9 nm Media: 955.9 ± 349.9 nm	Anatase	0–20 µg/ml for 24 h	Accumulation and decreased viability in both NP types; decreased ATP production in TNS; P25 induced ROS generating cells; TNS induced NO secretion and pro-inflammatory cytokines, but not of ROS; TNS increased BAX/Bcl-2, SOD1 protein, and caspase-8	Park et al. (2014)
A-549 and BEAS-2B	13–90.6 nm	H_2O: 139.8 ± 4 nm RPMI: 197.6 ± 6.7 nm BEGM: 224.6 ± 7.5 nm	Anatase	0–40 µg/ml for up to 24 h	A549: Exhibited no significantly decreased viability, but genotoxicity was induced at 40 µg; increased IL-6 BEAS-2B: Significantly decreased viability; membrane damage; increased IL-6 (slight) and TNF-α	Ursini et al. (2014)
A549	4–8 nm	H_2O: 106.7 ± 8.0 nm Medium: 23.28 ± 2.0 nm	Anatase	0–200 µg/ml for up to 48 h	Genotoxicity attributed to oxidative stress and ROS; cell cycle arrest at G2/M phase; increased apoptosis at 24 h and necrosis at 48 h; micronuclei induced	Kansara et al. (2015)
A549	25 ± 7 nm	H_2O: 44 ± 25 nm Medium: 342 ± 15 nm	Anatase/ rutile	0, 2.5, or 50 µg/ml for up to 2 months	Modification of proteins involved in mitochondrial activity, intra/extra-cellular trafficking, glucose metabolism, and activation of p53 pathway	Lucie et al. (2015)

Cell type	NP primary size	NP hydrodynamic/ agglomerate size	NP structure	Exposure parameters	Results	Reference
A549	<25 nm	Medium: 588.3 ± 43.11 nm	Not specified	0–10 µg/cm^2 for 7 days	Synchronizing cell cycle after NP treatment showed increased proliferation	Medina-Reyes et al. (2015a)
A549	TiO$_2$-S: <25 nm TiO$_2$-B: Not specified	Medium 1: 588.3 ± 43.1 Medium 2: 310.4 ± 0.1 Medium 1: 454 ± 1.5 Medium 2: 602.2 ± 2	Anatase Not specified	0–10 µg/cm^2 or 0–200 µg/cm^2 (cell viability) For 7 days	Accumulation; both NP types caused decreased cell size and induced pro-inflammatory cytokine release—TiO2-S returned to basal levels after 7 days but TiO$_2$-B induced down regulation	Medina-Reyes et al. (2015b)
WI-38	Not specified	30–250 nm	Anatase/ rutile	0–400,200 µg/cm^2 for up to 48 h	Acute cellular toxicity; increased ROS and oxidative stress; morphology changes; increased amount of cells in G2/M phase and decreased G0/G1	Periasamy et al. (2015)
TT1	Not specified Not specified	H$_2$O: 6–20 nm Medium: 5–20 nm H$_2$O: 5–12 nm Medium: 2–12 nm	Anatase Rutile	0–100 µg/ml for up to 24 h	No significant changes in viability for either NP; increased IL-6 and Il-8; generation of O^2- for both NP types—Anatase treated cells reduced O^2- at 24 h but rutile treated cells did not	Sweeney et al. (2015)
BEAS-2B	20.99 ± 6.4 nm	Medium: 575.9 ± 8 nm	Anatase	0–20 µg/ml for up to 4 weeks	Accumulation; significant increase in IL-8 expression and IL-6 (slight)	Vales et al. (2015)

(continued)

Table 12.4 (continued)

Cell type	NP primary size	NP hydrodynamic/ agglomerate size	NP structure	Exposure parameters	Results	Reference
16HBE14o-	Not specified	H_2O: 220.4 nm Medium: 295.9 nm	Anatase/ rutile	0–100 µg/ml for up to 72 h	Increased ER stress and disruption of MAMs; morphological changes; decreased ATP; abnormal autophagy	Yu et al. (2015)
A549 Calu-3	Nanofibers: 9.9 ± 5.8 length $\times 0.3 \pm 0.1$ diameter P25: 21 nm	Not specified	Anatase Anatase/ rutile	0–80 µg/cm² and 1–4 mg/ml (hemolytic activity) for up to 72 h	Nanofibers: Significantly reduced viability and more hemolytic than P25 in A549; lowered TEER in Calu-3 P25: No effects on viability in A549 or TEER in Calu-3	Allegri et al. (2016)
A549	24 ± 6 nm	H_2O: 44 ± 25 nm Medium: 342 ± 15 nm	Anatase/ rutile	0–50 µg/ml for up to 2 months	Accumulation remained up to 2 months but no NPs found in nucleus or mitochondria; proliferation decreased; increased ROS but no changes in GSH, GSH, CAT, and SOD expression; genotoxicity observed up to 2 months	Armand et al. (2016)
BEAS WI-38	M212: 28 ± 8.4 nm l 10 ± 1.4 W AFDC: 142 ± 22.7 nm AFDC300: 214 ± 46.1 nm cNRs: 117 ± 12.5 nm l $\times 39 \pm 7.8$ nm W	Not specified	Rod shape Anatase Anatase Spindle shape	100 µg/ml (uptake, cell cycle, apoptosis assays), 0–800 µg/ml (viability assay), and 1000 µg/ml (caspase assay) for up to 72 h	All but cNRs aggregated in cytoplasm of both cell types; cytotoxicity more significant in BEAS cells; apoptosis induced in BEAS via caspase 3/7 activation	Chen et al. (2016)

Cell type	NP primary size	NP hydrodynamic/ agglomerate size	NP structure	Exposure parameters	Results	Reference
AM-like THP-1, A549, HPMEC-ST1.6R	Not specified	400 nm	Not specified	0–800 µg/ml for up to 24 h	All 3 cell types showed increased ROS and different profiles of activated signaling related to DNA damage; cytotoxicity observed in THP-1 and HPMEC; HPMEC showed cell cycle arrest at S phase and loss of cells in G2	Hanot-Roy et al. (2016)
BEAS-2B	P25: 30 ± 4 nm, 50 ± 6 nm AA: 7 (length) × 2 (dia) nm AB: 7.0 ± 0.9 nm, 5.6 ± 0.7 nm AO: 33.3 ± 12.5 nm	Medium: 344.0 ± 32.8 Medium: 188.8 ± 4.4 nm Medium: 222.7 ± 9.2 nm Medium: 232.4 ± 0.3 nm	Anatase/ rutile Anatase Anatase Anatase	0–50 µg/ml for up to 72 h	P25 and AB induced cytotoxicity; increased TNF-α in P25, AB, and AO groups; all NP types induced significant NO production; overall synthesize NPs were less toxic	Vergaro et al. (2016)
A549	Not specified	Not specified	Not specified	0–200 µg/cm² for 24 h	Changes to gene expression relating to inflammation, protein metabolism, cell migration, proliferation, death, and survival; decreased ATP production at highest dose	Vuong et al. (2016)

(2014) confirmed presence of TiO$_2$ NPs in phagosomes in MH-S cells as well. In addition, uptake of TiO$_2$ NPs in lung cell models has been shown to increase in a dose- and time-dependent manner (Tang et al. 2013; Vales et al. 2015). Interestingly, Armand et al. (2016) showed that accumulation of TiO$_2$ NPs in A549 cells decreased after 1 month, although levels remained significant. This group also found TiO$_2$ NPs in vacuole-like compartments within cells, similar to the uptake by phagosomes mentioned previously. This indicates uptake of TiO$_2$ NPs in lung cell types may depend on phagocytosis, rather than solely crossing the cell membrane barrier due to small size alone. In vitro studies on uptake and accumulation echo those of in vivo results and are discussed further in the next section.

Cytotoxicity and cell viability are the factors that remain unsettled when discussing TiO$_2$ NP toxicity on lung cell types. A wide variety of cell types has shown sensitivity to TiO$_2$ NPs concerning cytotoxicity and viability. These include hamster lung fibroblasts (V79), AMs, MH-S, BEAS-2B, A-549, alveolar macrophage-like THP-1, and human pulmonary microvascular endothelial cells (HPMEC-ST1.6R). Studies have shown these cell lines to decrease in viability, increase in apoptosis, and/or increase in necrosis after exposure to TiO$_2$ NPs (Hamzeh and Sunahara 2013; Liu et al. 2013; Park et al. 2014; Ursini et al. 2014; Kansara et al. 2015; Allegri et al. 2016; Chen et al. 2016; Hanot-Roy et al. 2016; Vergaro et al. 2016). Both Liu et al. (2013) and Ursini et al. (2014) also demonstrated accompanying decreases in LDH, which is indicative of cytotoxicity. In addition, Chen et al. (2016) had reported that apoptosis was attributed to the induction of caspase-3 and caspase-7. Caspase pathways were shown to be activated after TiO$_2$ NP exposure in many other studies of different cell types as well (Marquez-Ramirez et al. 2012; Meena et al. 2012; Park et al. 2014; Meena et al. 2012; De Simone et al. 2016). As with other cell types, lung cell toxicity in vitro has been shown to be dependent upon NP shape. Hamzeh and Sunahara (2013) demonstrated more significant decreases in viability with nanofilaments when compared to NPs, while Allegri et al. (2016) showed nanofibers to be more cytotoxic than NPs. In addition, Vergaro et al. (2016) showed low cytotoxicity of synthesized NPs when compared to commercial TiO$_2$ NP.

However, these results are not without disparity. For example, Zhang et al. (2012) demonstrated no cytotoxicity after exposure to rutile TiO$_2$ NPs. In addition, only high concentrations of 5 and 25 nm anatase TiO$_2$ NPs over 400 µg/ml were found to be cytotoxic. LDH assay mirrored these results, showing low leakage overall. Studies on A-549 cells are particularly contrasting, with several pointing to low cytotoxicity and/or apoptosis after TiO$_2$ NP exposure (Ursini et al. 2014; Armand et al. 2016; Hanot-Roy et al. 2016; Vuong et al. 2016). It should be noted that although Armand et al. (2016) did not observe significant cytotoxicity, cell proliferation was significantly decreased in cells after NP treatment. However, Vuong et al. (2016) observed the opposite; with no decreases in proliferation or LDH levels after TiO$_2$ NP treatment. Based on these results it is clear that variations in NP characteristics and experimental design can critically influence overall outcomes when examining nanotoxicity. BEAS-2B have also been shown to be resistant to changes in viability induced by TiO$_2$ with certain types of synthesized NPs, most likely due to differences in NP characteristics and their fabrication methods (Vergaro et al. 2016).

Other cell types which have shown no cytotoxic effects include human alveolar type-I-like epithelial cells (TT1) and Calu-3 cell lines, thus indicating a need to investigate these more thoroughly with regards to viability.

Several cell types have exhibited increases in ROS after TiO_2 NPs. Hamzeh and Sunahara (2013) reported increased ROS in V79 cells at 24-hours post-exposure using three types of NPs. ROS generation in A549 cells has been investigated extensively due to their wide use in lung toxicity studies. It has been demonstrated that TiO_2 NPs induce dose-dependent and significant increases in ROS generation and oxidative stress in this cell type (Kansara et al. 2015; Armand et al. 2016; Hanot-Roy et al. 2016). Armand et al. (2016) observed these effects in a long-term study, indicating possible damage from low dose, prolonged exposure. However, there were no significant changes in GSH, CAT, or SOD. Tang et al. (2013) also demonstrated mitochondrial damage in a dose-dependent and proximity-dependent manner in A549, as well as increased lipid peroxidation. WI-38, THP-1, and HPMEC cell types have also shown concentration- and/or time-dependent increases in ROS after TiO_2 NP exposure (Periasamy et al. 2015; Hanot-Roy et al. 2016). Interestingly MH-S cells have exhibited selective effects concerning ROS generation after TiO_2 NP exposure, showing sensitivity to P25 TiO_2 NPs but not to sheet-type titania (Park et al. 2014). Vergaro et al. (2016) used various types of TiO_2 NPs on BEAS-2B cells and found all types induced significant ROS increases in a short-term study, but no significant lipid peroxidation or changes in GSH. This may indicate BEAS-2B cells are more sensitive to TiO_2 NPs when compared to other cell types with respect to ROS generation. However, at similar concentrations Vales et al. (2015) observed no ROS induction in BEAS-2B cells after TiO_2 NP exposure in a long-term study, which could suggest initial shock of TiO_2 NP exposure is recoverable for these cells. This was followed by observation of no genotoxic effects, indicating low cytotoxicity. Similar results for TT1 cells were observed, with increases initially in ROS using anatase NPs, but then levels reduced after 24 h (Sweeney et al. 2015). Interestingly, in the same study, rutile NPs increase ROS significantly and levels remained high, illustrating once more that NP characteristics play a vital role in toxicity effects.

Along with increases in ROS, TiO_2 NPs have demonstrated the ability to induce inflammation in certain lung cell types in vitro. Ursini et al. (2014) reported significant increases in IL-6 in A549 after TiO_2 exposure at 2 h, but this decreased over time to nonsignificant levels. In the same manner, Medina-Reyes et al. (2015b) showed significant increases in several types of inflammatory cytokines in A549 cells, including IL-6 and TNF-α, but many of these returned to nonsignificant levels within 7 days. BEAS-2B cells have also demonstrated release of inflammatory cytokines after TiO_2 NP contact. Ursini et al. (2014) and Vales et al. (2015) found comparable behaviors of IL-6, IL-8, and TNF-α release at similar concentrations of TiO_2 NP exposure in both short- and long-term studies, but variations in significance for each cytokine were reported. Vergaro et al. (2016) also reported significant increases in TNF-α at 72 hours of TiO_2 NP contact in BEAS-2B cells using many NP types with different characteristics. Significant increases in IL-6 and IL-8 have also been demonstrated in TT1 cells (Sweeney et al. 2015). Since these biomarkers seem to be sensitive to TiO_2 NP exposure, they should be monitored in other cell types as well. This would also prove to be useful for in vivo studies, which would provide more insight into inhaled TiO_2 NP effects in humans.

Genotoxicity is of major concern after NP exposure in any cell type. Genotoxic effects have been reported in A549, THP-1, and HPMEC cell types. A549 cells in particular have been used frequently, showing oxidative DNA damage, concentration-dependent DNA damage, double-strand breaks, and changes in signaling pathways relating to DNA damage (Ursini et al. 2014; Kansara et al. 2015; Armand et al. 2016; Hanot-Roy et al. 2016). THP-1 and HPMEC cell lines have also shown changes in DNA damage signaling pathways, indicating DNA-TiO_2 NP interactions that can affect gene regulation (Hanot-Roy et al. 2016). The vast majority of in vitro studies used for this review that observed genotoxicity all indicated some level of DNA effects, indicating that this factor is a major concern when considering human TiO_2 NP inhalation risk. In addition, changes to cell cycle regulation, such as accumulation and loss of cells in certain phases and arrest in A549 and HPMEC cells, have been reported (Kansara et al. 2015; Medina-Reyes et al. 2015a; Periasamy et al. 2015; Hanot-Roy et al. 2016). Studies on A549 cells have also demonstrated changes in expression of genes relating to a wide variety of functions, such as mitochondrial activity, p53 signaling, inflammation, cell migration, proliferation, death, and survival (Lucie et al. 2015; Vuong et al. 2016). Given the data from recent research on TiO_2 NPs and DNA interaction, it is clear there is an inherent risk to these NPs. However, more cell types need to be tested in terms of genotoxicity specifically with standardized dosages in order to understand this mechanism more fully.

Another aspect of lung toxicity in vitro to consider is that of physiological and morphological changes post-NP exposure. In MH-S cells, changes such as increases in vacuole number, dilated mitochondria, and swelling of endoplasmic reticulum (ER) have been reported (Park et al. 2014). The widely used A549 cell line has shown significant increases in micronuclei and decreased cell size after TiO_2 NP exposure (Kansara et al. 2015; Medina-Reyes et al. 2015b). Periasamy et al. (2015) observed intracellular changes such as cell nuclear collapse and disruption of colony morphology in WI-38 cells. In addition, human bronchial epithelial cells (16HBE14o-) have shown disruption of inner and outer membranes and mitochondria-associated ER membranes (MAMs) after TiO_2 NP contact (Yu et al. 2015). This study also indicates disruption of calcium ions, which are acquired by mitochondria and act as important secondary messengers within the cell. The aforementioned three cell types also exhibited decreased levels of cellular ATP, indicating mitochondrial disruption and effects on cell viability (Park et al. 2014; Yu et al. 2015; Vuong et al. 2016). Several in vitro studies reviewed here did not include studies on the cell morphology, which deserves attention in order to gain better understanding of the interaction.

1.3.2   Brain Toxicity In Vitro

In vitro studies focusing on accumulation of TiO$_2$ NPs in the brain have shown that their uptake is common across many cell types. As shown in Table 12.5, U373, C6, SH-SY5Y, D384, BV-2, ALT, N2a, primary cultured hippocampal neurons, dorsal root ganglion (DRG), and satellite glial cells (SGC) have all demonstrated ability to uptake TiO$_2$ NPs after exposure. However, most of these studies utilized short exposure time (up to 24 hr). Experiments utilizing longer >24 h would possibly yield better insight for in vivo studies representing daily human intake. Many of these investigations showed that Ti uptake was limited to the cytoplasm lysosomes, without presence in other organelles or the nucleus (Marquez-Ramirez et al. 2012; Mao et al. 2015; De Simone et al. 2016). One particularly interesting finding by Hsiao et al. (2016) demonstrated BV-2 cells pretreated with bacterial lipopolysaccharide (LPS) took up more TiO$_2$ NPs than BV-2 cells without LPS treatment. This indicates that pre-existing infections may affect TiO$_2$ NP uptake in brain tissue. In addition, this group also showed that ALT and BV-2 cells were more sensitive to TiO$_2$ NPs concerning uptake when compared to N2a cells, which may indicate some brain cell types are more sensitive to NPs (Hsiao et al. 2016). Hong et al. (2015) demonstrated presence of TiO$_2$ NPs not only in the cytoplasm but also in the nuclei of primary cultured hippocampal cells. Genotoxicity was not observed for this particular experiment, but future studies should focus on NP presence in brain cell nuclei and possible effects.

Cytotoxicity was demonstrated in all cell types examined here in the forms of reduced cell viability, reduced cell proliferation, and increased apoptosis. Marquez-Ramirez et al. (2012) indicated proliferation of U373 and C6 cells to be inhibited in a dose-dependent manner using TiO$_2$ NPs, while Rihane et al. (2016) demonstrated a slight inhibition of growth in BV-2 cells after exposure. Reduction in cell viability after TiO$_2$ NP contact has been shown in D384, SH-SY5Y, primary hippocampal neurons, cortical rat astrocytes, ALT, and BV-2 cells (Coccini et al. 2015; Hong et al. 2015; Sheng et al. 2015; Wilson et al. 2015; Hsiao et al. 2016). After 24 h of exposure, Sheng et al. (2015) indicated severe cytotoxicity at all concentrations tested, indicating that low doses such as 5 µg/ml were enough to induce cell death. Coccini et al. (2015) also showed decreased mitochondrial activity over 7 days, as well as reduced colony size for D384 and SH-SY5Y starting at low doses of 1.5 and 0.1 µg/ml, respectively. These results at such low doses indicate TiO$_2$ NPs pose a significant risk to brain cell function in the cell environment. However, it has been reported in other studies that SH-SY5Y cells do not exhibit decreases in viability after TiO$_2$ NP exposure (Valdiglesias et al. 2013; Mao et al. 2015). Apoptosis and increase in apoptotic rates have also been demonstrated in U373, C6, SH-SY5Y, DRG, SGC, D384, and primary hippocampal neurons (Marquez-Ramirez et al. 2012; Valdiglesias et al. 2013; Erriquez et al. 2015; Sheng et al. 2015; De Simone et al. 2016). De Simone et al. (2016) specifically demonstrated increases in expression of apoptotic genes, such as p53, p21, Bax, Bcl-2, and activated caspase-3. Interestingly, Erriquez et al. (2015) indicated that the rutile form of TiO$_2$ NPs did not induce apoptosis, whereas an anatase/rutile mix initiated cell death. This lends

Table 12.5 Effects of TiO$_2$ NPs on brain in vitro

Cell type	NP primary size	NP hydrodynamic/ agglomerate size	NP structure	Exposure Parameters	Results	Reference
U373, C6	40–200 nm	421 ± 4.35 nm	Anatase/ rutile	0–40 µg/cm^2 for up to 96 h	Both cell types showed: Inhibited proliferation, morphological changes, uptake of NPs into cytoplasm, apoptosis, increased caspase-3, and fragmented nuclei	Marquez-Ramirez et al. (2012)
SH-SY5Y	TiO$_2$-S: 25 nm TiO$_2$-D: 25 nm	H$_2$O: 447.9 nm Medium: 504.5 nm H$_2$O: 160.5 nm Medium: 228.3 nm	Anatase Anatase/ rutile	0–150 µg/ml for up to 24 h	No change in viability or morphology induced by either NP type; NP uptake of both types observed; cytotoxic and genotoxic effects observed, but not related to ROS or DSB	Valdiglesias et al. (2013)
U373, C6	<50 nm	481 nm	Not specified	0 or 20 µg/ cm^2 for up to 24 h	Oxidative stress in both cell types due to lipid peroxidation and increased GPx, CAT, and SOD2; mitochondrial damage; cytotoxic effects	Huerta-García et al. (2014)
D384, SH-SY5Y	15 nm	Stock: 52 nm Medium: 299 nm, 356 nm	Anatase	0–250 µg/ml for up to 7 days	In both cell types: Decreased mitochondrial activity, loss of shape, decreased viability; SH-SY5Y showed more sensitivity to NPs	Coccini et al. (2015)
Dorsal root ganglion (DRG), satellite glial cells (SGC)	TiO$_2$-A/R: TiO$_2$-R: Data in Supp. File (unable to find)	Data in Supp File (unable to find)	Anatase/ rutile Rutile	0–5 µg/ml for 24 h	TiO$_2$A/R induced uptake, apoptosis, and ROS production in both cell types; increased pro-inflammatory cytokines (IL-1β) in DRG by A/R mix; morphological changes	Erriquez et al. (2015)
Primary hippocampal neurons	5–6 nm	Medium: 200–420 nm	Anatase	0–30 µg/ml for 24 h	Decreased viability; NPs found in cytoplasm and nuclei; decreased neurite length; changes in glutamic acid metabolism; increased ROS	Hong et al. (2015)
SH-SY5Y	20.90 ± 3.57 nm	H$_2$O: 314.0 ± 5.8 Medium: 110.0 ± 72.9 nm	Spherical	0–100 µg/ml for 24 h	Microtubule morphology disruption; uptake of NPs in lysosomes and cytoplasm; ROS present; interaction with tau proteins which led to microtubule instability	Mao et al. (2015)

Cell type	NP primary size	NP hydrodynamic/ agglomerate size	NP structure	Exposure Parameters	Results	Reference
Rat brain mitochondria	32.34 ± 2.37 nm	>100 nm	Anatase	0–50 µg/ml for 1 h	Increased lipid peroxidation, protein oxidation, GPx, GSR, and Mn-SOD; decreased GSH; cellular respiration enzymatic activity decreased	Nalika and Parvez 2015
Primary hippocampal neurons	5.5 nm	200–420 nm	Anatase	0–50 µg/ml for up to 48 h	Decreased viability; cytotoxicity observed; increased apoptosis linked to mitochondria and ER-mediated signaling pathway; changes to mitochondria and ER morphology	Sheng et al. (2015)
Primary rat cortical astrocytes	50 nm / 50 nm / P25: 21 nm	Medium: 360 nm / 540 nm / 360 nm	Rutile / Anatase / Anatase/rutile	0–100 ppm for 24 h	All three types induced loss of glutamate uptake, decreased viability, and morphological changes; P25 and anatase increased ROS and decreased mitochondrial membrane potential; P25 and anatase produced more severe effects on cells	Wilson et al. (2015)
D384	15 nm	Stock: 52 nm / Medium: 356 nm	Anatase	0–125 µg/ml for 24 h	Increased ROS production; increased expression of apoptotic genes p53, p21, Bax, Bcl-2, and activate caspase-3; accumulation of NPs in cytoplasm	De Simone et al. (2016)
BV-2	20–30 nm	150–200 nm	Not specified	0–200 µg/ml for 24 h	Some cytotoxicity with slight inhibition of cell growth; morphological changes; increased O^{2-}; accumulation of NPs in cells	Rihane et al. (2016)
ALT, BV-2, N2a	5.8 nm	H_2O: 44.4 nm / Medium: 92 nm	Anatase	0–100 µg/ml for up to 48 h	ALT and BV-2 showed more NP uptake and lower viability when compared to N2a; increased IL-1β in all 3 cell types; increased IL-6 in N2a; NPs mediated ROS and/or cytokines release from microglia, but not from astrocytes	Hsiao et al. (2016)

support to the common idea that NP characteristics can greatly influence toxicity. One particular cell type that should be investigated further is BV-2. Rihane et al. (2016) reported some cytotoxicity in this cell line, but no apoptosis was observed. Since there are few recent studies, which examine this particular cell type in terms of apoptosis, it may be worthwhile to further explore.

ROS production and oxidative stress, as with other cell types, are highly exhibited throughout many cell line studies. Huerta-García et al. (2014) demonstrated strong oxidative stress in both C6 and U373 cells, which led to lipid peroxidation and increases in GPx, CAT, and SOD2. Similar results showing increase in lipid and protein peroxidation, as well as changes in GSH, GPx, and glutathione reductase (GSR), have been reported in brain tissue mitochondria from rats (Nalika and Parvez 2015). Perhaps even more striking is that Nalika and Parvez (2015) reported significant decreases in NADH dehydrogenase and succinate dehydrogenase, two key enzymes required for cellular respiration. Decreases in these enzymes would cause drastic effects on ATP synthesis, which may contribute to the mechanism of overall toxicity of TiO_2 NPs. D384 and BV-2 cells have shown increased oxidative stress with increasing ROS levels and overproduction of O^{2-}, respectively (De Simone et al. 2016; Rihane et al. 2016). Interestingly, no genotoxicity was observed in BV-2 cells, even though slight cytotoxicity was observed. However, instances of direct genotoxicity have been observed in the forms of condensed and fragmented nuclei, as well as apoptotic bodies in U373 and C6 cells by DAPI staining (Marquez-Ramirez et al. 2012). Hong et al. (2015) observed significant increases in nitric oxide (NO) and nitric oxide synthetase (NOS) in primary cultured hippocampal cells. Inflammatory responses have been demonstrated in DRG cells in the form of significantly elevated IL-1β mRNA expression after anatase/rutile mix TiO_2 NP exposure (Erriquez et al. 2015). As with other toxicity indicators from this study, inflammatory response was not observed with rutile TiO_2 NPs. ALT, BV-2, and N2a cells have also shown increased IL-1β after exposure to anatase TiO_2 NPs, with N2a cells also demonstrating elevated levels of IL-6 (Hsiao et al. 2016).

Anatase TiO_2 NPs have also been shown to cause ROS production and decrease mitochondrial membrane potential in DRG and rat cortical astrocytes, whereas rutile NPs did not show these types of effects (Erriquez et al. 2015; Wilson et al. 2015). Hsiao et al. (2016) demonstrated significant ROS production in ALT and BV-2 cells, but not in N2a cells, after 24 h of TiO_2 NP exposure. Interestingly, however, BV-2 cells exposed first to bacterial LPS and then to TiO_2 NPs showed more ROS generation in comparison to untreated BV-2 cells. As with results for Ti uptake, this indicates that BV-2 cells that have sustained some injury may be more susceptible to TiO_2 NP toxic effects. One exception to the ROS generated by TiO_2 NP exposure is that of SH-SY5Y cells. Valdiglesias et al. (2013) observed no oxidative stress using anatase and anatase/rutile mix TiO_2 NPs in SH-SY5Y cells. Although genotoxic effects were observed, they were not related to double-strand breaks that are normally associated with oxidative DNA damage. Mao et al. (2015) confirmed ROS formation in SH-SY5Y cells, but there was no decrease in viability and no associated morphological changes. Instead, it was observed that NPs interacted with tau proteins, which are found largely in neurons and help stabilize microtubules.

Mao et al. (2015) indicated this finding as the source of the observed neurotoxicity demonstrated by SH-SY5Y cells. Taken together, these results suggest SH-SY5Y cells may not undergo the usual route of ROS- and mitochondrial damage-induced toxicity, but instead may experience a different molecular mechanism in response to TiO_2 NPs.

Notably, the areas most investigated relate to morphological and histological changes in cells after TiO_2 NP contact. Markers of toxicity such as damage to mitochondria and changes to mitochondrial membrane potential have been documented in many brain cell types (Huerta-García et al. 2014; Sheng et al. 2015; Wilson et al. 2015; Rihane et al. 2016). Other morphological change reported include reduced cell size and volume, membrane blebbing, loss of star shape, thinning neurites, microtubule disruption, inhibition of cell adhesion, enhanced cytoplasmic membrane permeability, apoptosis, nucleus shrinkage, and dilation of ER (Huerta-García et al. 2014; Coccini et al. 2015; Mao et al. 2015; Rihane et al. 2016; Sheng et al. 2015). In addition, changes to the cell cycle that inhibit progression from S phase have been demonstrated in SH-SY5Y cells, although no morphological changes were observed (Mao et al. 2015). This makes sense, given that the same study also reported microtubule disruption after TiO_2 NP exposure. Disruption to the microtubule system would undoubtedly have effects such as halting the cell cycle at S phase as reported in SH-SY5Y cells. For those cell types in which ROS generation and mitochondrial changes have been observed, it is most likely that a mitochondrial and/or ER-mediated pathway is being utilized to result in toxicity. Indeed, Ca^{2+} elevation has been reported in primary hippocampal neurons after TiO_2 NP exposure, which is an indicator of mitochondrial damage (Sheng et al. 2015). Loss of glutamate uptake and changes in glutamic acid metabolism, both of which are required for proper motor neuron function, has also been reported in rat cortical astrocytes and hippocampal cells after TiO_2 NP contact (Hong et al. 2015b; Wilson et al. 2015). Changes in mitochondrial Ca^{2+} and glutamate are clinically present in amyotrophic lateral sclerosis (ALS), a neurodegenerative disease (Kawamata and Manfredi 2010), thus indicating that TiO_2 NPs could potentially exacerbate such conditions.

1.4   Cardiotoxicity

Multiple in vitro and in vivo studies in the cardiovascular system have demonstrated myocardial damage, oxidative stress, inflammatory responses, and atherosclerosis in mice exposed to TiO_2 NPs. Various experiments have been performed to elucidate the underlying mechanisms. Savi et al. (2014) conducted functional studies on isolated adult rat cardiomyocytes by exposing them to TiO_2. Interestingly, they also administered a single dose of 2 mg/kg TiO_2 NPs via the trachea in healthy rats. Transmission electron microscopy was used to verify the TiO_2 nanoparticles within cardiac tissue, and toxicological assays were used to assess lipid peroxidation and DNA alteration. An in silico method was used to model the effect on action potential.

The investigators demonstrated that TiO_2 NPs can reach the heart via the respiratory system, enter ventricular cardiomyocytes, and create arrhythmic conditions by shortening of repolarization time and by increasing cardiac excitability.

Huerta-García et al. (2018) evaluated the toxicity of TiO_2 NPs on H9c2 rat cardiomyoblasts. They assessed internalization of TiO_2 NPs and their effect on cell proliferation, viability, oxidative stress, cell cycle alterations and cell death. It was found that TiO_2 NPs reduced metabolic activity and cell proliferation. A significantly increased H2DCFDA oxidation suggested the occurrence of increased oxidative stress. TiO_2 NPs disrupted the plasmatic membrane integrity and decreased the mitochondrial membrane potential. These effects correlated with changes in the distribution of cell cycle phases resulting in necrotic death and autophagy. Oxidative stress and cell death in cardiac myocytes induced by TiO_2 NPs represent a potential health risk, particularly in the development of cardiovascular disease.

In multiple in vivo studies, TiO_2 NPs were shown affecting different functional cardiovascular aspects, such as cytotoxicity of cardiomyocytes, altering hemodynamic parameters, such as systolic blood pressure (SBP) and heart rate (HR) variability, induction of thrombosis, promotion of arrhythmogenesis, and the changes in vasomotor responses (Baranowska-Wójcik et al. 2020). The inflammatory response triggered by TiO_2NPs is viewed as one of the main causes for cardiovascular system malfunction. Increased expression of inflammatory cytokines such as TNF-α, INF-g, and IL- 8 in blood after intake of TiO_2 NPs has been reported previously. Earlier, Wang et al. (2007) reported that intragastric administration of TiO_2 in different sizes (25 nm, and 80 nm) had increased lactate dehydrogenase (LDH) and alpha-hydroxybutyrate dehydrogenase (HBDH), compared to controls. Nano Ti treatment also induces an increased level of creatine kinase (CK). Some of these especially LDH and CK are used as markers of myocardial obstructive disease. However, it is unclear as to what kinds of LDH isoforms were elevated and that makes it difficult to assess a clear correlation between these elevations and myocardial ischemia/injury. Increases in CK and LDH levels after intragastric administration of TiO_2 in mice were subsequently confirmed by Liu et al. (2009) and Bu et al. (2010). Furthermore, Nemmar et al. (2011) instilled rutile Fe-doped TiO_2 nanorods intrathecally in rats and found a significant increase in SBP and HR in treated animals. This hemodynamic response could have been a result of the systemic inflammation induced by TiO_2.

The currently available data regarding the TiO_2 NP prothrombotic effect is somewhat unclear. Chen et al. (2009) had reported thrombosis in the pulmonary vascular system of mice treated via intraperitoneal injection with anatase TiO_2 NPs measuring 80–110 nm. However, Bihari et al. (2010) failed to confirm thrombosis induction in either the mesenteric or the cremasteric circulation of mice who were given intravenous TiO_2 measuring 10 nm. Activation of platelets and induction of intrinsic pathway of plasmatic coagulation are the common mechanisms for increased clotting induced by NPs. This mainly pro-coagulate action of NPs could be particularly dangerous for individuals suffering from hyper-coagulate states linked to common diseases, such as diabetes mellitus, arteriosclerosis, cancer, and obstructive pulmonary disease (Fröhlich 2016). More research is needed to further confirm the thrombotic nature and assess the mechanism of TiO_2.

The vasomotor response of TiO_2 probably has the most evidence among other cardiovascular patho-physiologic responses. The arteriolar dilatation in the bed of the spino-trapezius muscle arterioles was impaired by P25 anatase-rutile TiO_2 (21 nm) inhalation (LeBlanc et al. 2010) in rats. This vascular impairment was due to a dose-dependent reduction in NO endothelium production induced by microvascular oxidative stress. Reactive oxygen species produced by TiO_2 could deplete endothelium derived NO or adversely affect NO endogenous production that in turn would lead to vascular dysfunction. The loss of microvascular vasodilator capacity can significantly influence the normal homeostasis in any organ, causing impairment of tissue perfusion and creating ischemic conditions. The ex vivo contractile response to prostaglandin F2α and KCl and the relaxant response to Ach were not altered by TiO_2 NP treatment (LeBlanc et al. 2010). However, without appropriate particle characterization, it is difficult to compare findings between toxicological studies. Future studies are needed to provide more detailed knowledge regarding the toxic effects of TiO_2 NPs on the cardiovascular system.

2   Conclusion

Perusal of in vitro studies discussed above clearly indicates that the food-grade TiO_2 NPs are acquired, intracellularly distributed and accumulated in the cell and its organelles by the majority of cell lines derived from human organs. These nanoparticles increase the cell membrane permeability, but rarely affect the membrane integrity. However, changes in overall cell morphologies, for example, loss of microvilli in intestinal cells; cell shrinkage and detachment in kidney cells; dilated mitochondria, swelled endoplasmic reticulum, and increased number of vacuoles in lung cells; and membrane blebbing, thinning neurites, and microtubule disruption in brain cells were commonly observed. While the uptake of TiO_2 NPs by different cells was shown adequately, few studies delineated the method of cellular transport across the membrane: whether just gradient-based passive absorption, facilitated endocytosis, or phagocytosis. However, some studies have reported the presence of phagosomes and involvement of phagocytosis in Ti uptake in lung cells. While scrutinizing the effects of Ti nanoparticles, the range of deleterious effects include, but not limited to, decreased cell viability or cell division, increased apoptosis or necrosis, presence of micronuclei, mitochondrial damage, and increased proinflammatory cytokines. In most of the cases, occurrence of heightened oxidative stress and ROS production was common. Studies also demonstrate the changes in expression of genes relating to a wide variety of functions, such as mitochondrial activity, p53 signaling, inflammation, cell migration, proliferation, death, and survival. Demonstrated genotoxicity effects include oxidative DNA damage, concentration-dependent DNA damage, double-strand breaks, and changes in signaling pathways relating to DNA damage. The observed effects were largely dependent on the size, shape, concentration, crystal or surface characteristics of the nanoparticles used in the exposure experiments. TiO_2 NPs–anatase crystal type was found more toxic that the other crystal form, rutile. Data from human cell lines are compelling for envisioning the harmful effects of TiO$_2$ NPs on optimal organ functions in humans. We feel an urgent need for devising novel methods of investigation into the potential adverse effects of TiO$_2$ NPs in humans. While researchers in this field make progress in laying out the role of these nanomaterials on overall human health, we feel that the governing bodies should consider mandating the disclosure of the quantity and nature of TiO$_2$ NPs present in various consumer goods. This will allow the consumers to make an informed decision on whether to use these products or not.

References

Allegri, M., Bianchi, M., Chiu, M., Varet, J., Costa, A., Ortelli, S., Blosi, M., Bussolati, O., Poland, C., & Bergamaschi, E. (2016). Shape-related toxicity of titanium dioxide Nanofibres. *PLoS One, 11*(3), e0151365. https://doi.org/10.1371/journal.pone.0151365.

Armand, L., Tarantini, A., Beal, D., Biola-Clier, M., Bobyk, L., Sorieul, S., Pernet-Gallay, K., Marie-Desvergne, C., Lynch, I., Herlin-Boime, N., & Carriere, M. (2016). Long-term exposure of A549 cells to titanium dioxide nanoparticles induces DNA damage and sensitizes cells towards genotoxic agents. *Nanotoxicology, 10*(7), 913–923. https://doi.org/10.3109/17435390.2016.1141338.

Baranowska-Wójcik, E., Szwajgier, D., Oleszczuk, P., et al. (2020). Effects of titanium dioxide nanoparticles exposure on human health—A review. *Biological Trace Element Research, 193*, 118–129. https://doi.org/10.1007/s12011-019-01706-6.

Bhowmick, T., Yoon, D., Patel, M., Fisher, J., & Ehrman, S. (2010). In vitro effects of nanoparticles on renal cells. *Journal of Nanoparticle Research, 12*(80), 2757–2770. https://doi.org/10.1007/s11051-010-9849-x.

Bihari, P., Holzer, M., Praetner, M., Fent, J., Lerchenberger, M., Reichel, C. A., Rehberg, M., Lakatos, S., & Krombach, F. (2010). Single-walled carbon nanotubes activate platelets and accelerate thrombus formation in the microcirculation. *Toxicology, 269*(2–3), 148–154.

Botelho, M., Costa, C., Silva, S., Costa, S., Dhawan, A., Oliveira, P., & Teixeira, J. (2014). Effects of titanium dioxide nanoparticles in human gastric epithelial cells in vitro. *Biomedicine & Pharmacotherapy, 68*(1), 59–64. https://doi.org/10.1016/j.biopha.2013.08.006.

Brun, E., Barreau, F., Veronesi, G., Fayard, B., Sorieul, S., Chanéac, C., Carapito, C., Rabilloud, T., Mabondzo, A., Herlin-Boime, N., & Carrière, M. (2014). Titanium dioxide nanoparticle impact and translocation through ex vivo, in vivo and in vitro gut epithelia. *Particle and Fibre Toxicology, 11*, 16. https://doi.org/10.1186/1743-8977-11-13.

Bu, Q., Yan, G., Deng, P., Peng, F., Lin, H., Xu, Y., Cao, Z., Zhou, T., Xue, A., Wang, Y., Cen, X., & Zhao, Y. L. (2010). NMR-based metabonomic study of the sub-acute toxicity of titanium dioxide nanoparticles in rats after oral administration. *Nanotechnology, 21*(12), 125105.

Chen, C., Huang, J., Lai, T., Jan, Y., Hsiao, M., Chen, C., Hwu, Y., & Liu, R. (2016). Evaluation of the intracellular uptake and cytotoxicity effect of TiO$_2$ nanostructures for various human oral and lung cells under dark conditions. *Toxicology Research, 5*, 303–311. https://doi.org/10.1039/c5tx00312a.

Chen, J., Dong, X., Zhao, J., & Tang, G. (2009). In vivo acute toxicity of titanium dioxide nanoparticles to mice after intraperitioneal injection. *Journal of Applied Toxicology, 29*(4), 330–337.

Chen, X., Cheng, B., Yang, Y., Cao, A., Liu, J., Du, L., Liu, Y., Zhao, Y., & Wang, H. (2013). Characterization and preliminary toxicity assay of nano-titanium dioxide additive in sugar-coated chewing gum. *Small, 9*(9–10), 1765–1774. https://doi.org/10.1002/smll.201201506.

Coccini, T., Grandi, S., Lonati, D., Locatelli, C., & De Simone, U. (2015). Comparative cellular toxicity of titanium dioxide nanoparticles on human astrocyte and neuronal cells after acute and prolonged exposure. *Neurotoxicology, 44*, 77–89. https://doi.org/10.1016/j.neuro.2015.03.006.

De Angelis, I., Barone, F., Zijno, A., Bizzarri, L., Russo, M., Pozzi, R., Franchini, F., Giudetti, G., Uboldi, C., Ponti, J., Rossi, F., & De Berardis, B. (2012). Comparative study of ZnO and TiO 2 nanoparticles: Physicochemical characterisation and toxicological effects on human colon carcinoma cells. *Nanotoxicology, 7*(8), 1361–1372. https://doi.org/10.3109/17435390.2012.741724.

De Simone, U., Lonati, D., Ronchi, A., & Coccini, T. (2016). Brief exposure to nanosized and bulk titanium dioxide forms induces subtle changes in human D384 astrocytes. *Toxicology Letters, 254*, 8–21. https://doi.org/10.1016/j.toxlet.2016.05.006.888.

Delgado-Buenrostro, N., Medina-Reyes, E., Lastres-Becker, I., Freyre-Fonseca, V., Ji, Z., Hernandez-Pando, R., Marquina, B., Pedraza-Chaverri, J., Espada, S., Cuardrado, A., & Chirino, Y. (2009). Nrf2 protects the lung against inflammation induced by titanium dioxide nanoparticles: A positive regulator role of Nrf2 on cytokine release. *Environmental Toxicology, 30*(7), 782–792. https://doi.org/10.1002/tox.2195.

Demir, E., Burgucu, D., Turna, F., Aksakal, S., & Kaya, B. (2013). Determination of TiO2, ZrO2, and Al2O3 nanoparticles on genotoxic responses in human peripheral blood lymphocytes and cultured embryonic kidney cells. *Journal of Toxicology and Environmental Health Part A, 76*(16), 990–1002. https://doi.org/10.1080/15287394.2013.830584.

Dudefoi, W., Moniz, K., Allen-Vercoe, E., Ropers, M.-H., & Walker, V. (2017). Impact of food grade and nano-TiO2 particles on a human intestinal community. *Food and Chemical Toxicology, 106*, 242–249. https://doi.org/10.1016/j.fct.2017.05.050.

El-Said, K., Ali, E., Kanehira, K., & Taniguchi, A. (2014). Molecular mechanism of DNA damage induced by titanium dioxide nanoparticles in toll-like receptor 3 or 4 expressing human hepatocarcinoma cell lines. *Journal of Nanobiotechnology, 12*, 48. https://doi.org/10.1186/s12951-014-0048-2.

Erriquez, J., Bolis, V., Morel, S., Fenoglio, I., Fubini, B., Quagliotto, P., & Distasi, C. (2015). Nanosized TiO2 is internalized by dorsal root ganglion cells and causes damage via apoptosis. *Nanomedicine: Nanotechnology, Biology, and Medicine, 11*(6), 1309–1319. https://doi.org/10.1016/j.nano.2015.04.003.

Faust, J., Doudrick, K., Yang, Y., Westerhoff, P., & Capco, D. (2014). Food grade titanium dioxide disrupts intestinal brush border microvilli in vitro independent of sedimentation. *Cell Biology and Toxicology, 30*(3), 169–188. https://doi.org/10.1007/s10565-014-9278-1.

Fröhlich, E. (2016). Action of nanoparticles on platelet activation and plasmatic coagulation. *Current Medicinal Chemistry, 23*(5), 408–430. https://doi.org/10.2174/0929867323666160106151428.

Gerloff, K., Albrecht, C., Boots, A., Forster, I., & Schins, R. (2009). Cytotoxicity and oxidative DNA damage by nanoparticles in human intestinal Caco-2 cells. *Nanotoxicology, 3*(4), 355–364. https://doi.org/10.3109/17435390903276933.

Gerloff, K., Fenoglio, I., Carella, E., Kolling, J., Albrecht, C., Boots, A., Forster, I., & Schins, R. (2012). Distinctive toxicity of TiO2 rutile/Anatase mixed phase nanoparticles on Caco-2 cells. *Chemical Research in Toxicology, 25*(3), 646–655. https://doi.org/10.1021/tx200334k.

Gitrowski, C., Al-Jubory, A., & Handy, R. (2014). Uptake of different crystal structures of TiO2 nanoparticles by Caco-2 intestinal cells. *Toxicology Letters, 226*(3), 264–276. https://doi.org/10.1016/j.toxlet.2014.02.014.

Halamoda, K. B., Chapuis Bernasconi, C., & Juillerat-Jeanneret, L. (2013). Stress reaction of kidney epithelial cells to inorganic solid-core nanoparticles. *Cell Biology and Toxicology, 29*(1), 39–58. https://doi.org/10.1007/s10565-012-9236-8.

Hamzeh, M., & Sunahara, G. (2013). In vitro cytotoxicity and genotoxicity studies of titanium dioxide (TiO2) nanoparticles in Chinese hamster lung fibroblast cells. *Toxicology In Vitro, 27*(2), 864–873. https://doi.org/10.1016/j.tiv.2012.12.018.

Hanot-Roy, M., Tubeuf, E., Guilbert, A., Bado-Nilles, A., Vigneron, P., Trouiller, B., Braun, A., & Lacroix, G. (2016). Oxidative stress pathways involved in cytotoxicity and genotoxicity of titanium dioxide (TiO2) nanoparticles on cells constitutive of alveolo-capillary barrier in vitro. *Toxicology In Vitro, 33,* 125–135. https://doi.org/10.1016/j.tiv.2016.01.013.

Hong, F., Sheng, L., Ze, Y., Hong, J., Zhou, Y., Wang, L., Liu, D., Yu, X., Xu, B., Zhao, X., & Ze, X. (2015). Suppression of neurite outgrowth of primary cultured hippocampal neurons is involved in impairment of glutamate metabolism and NMDA receptor function caused by nanoparticulate TiO2. *Biomaterials, 53,* 76–85. https://doi.org/10.1016/j.biomaterials.2015.02.067.

Hsiao, I., Chang, C., Wu, C., Hsieh, Y., Chuang, C., Wang, C., & Huang, Y. (2016). Indirect effects of TiO$_2$ nanoparticle on neuron-glial cell interactions. *Chemico-Biological Interactions, 254,* 34–44. https://doi.org/10.1016/j.cbi.2016.05.024.

Huang, K., Wu, C., Huang, K., Lin, W., Chen, C., Guan, S., Chiang, C., & Liu, S. (2015). Titanium nanoparticle inhalation induces renal fibrosis in mice via an oxidative stress upregulated transforming growth factor-β Pathway. *Chemical Research in Toxicology, 28*(3), 354–364. https://doi.org/10.1021/tx500287f.

Huerta-García, E., Pérez-Arizti, J., Márquez-Ramírez, S., Delgado-Buenrostro, N., Chirino, Y., Iglesias, G., & López-Marure, R. (2014). Titanium dioxide nanoparticles induce strong oxidative stress and mitochondrial damage in glial cells. *Free Radical Biology and Medicine, 73C,* 84–94. https://doi.org/10.1016/j.freeradbiomed.2014.04.026.

Huerta-García, E., Zepeda-Quiroz, I., Sánchez-Barrera, H., Colín-Val, Z., Alfaro-Moreno, E., Ramos-Godinez, M. P., & López-Marure, R. (2018). Internalization of titanium dioxide nanoparticles is cytotoxic for H9c2 rat cardiomyoblasts. *Molecules, 2018*(23), 1955. https://doi.org/10.3390/molecules23081955.

Hussain, S., Hess, K., Gearhart, J., Geiss, K., Schlager, J. (2005). In vitro toxicity of nanoparticles in BRL 3A rat liver cells. *Toxicology in Vitro, 19*(7): 975–83. https://doi.org/10.1016/j.tiv.2005.06.034.

Janer, G., Mas del Molino, E., Fernandez-Rosas, E., Fernandez, A., & Vazquez-Campos, S. (2014). Cell uptake and oral absorption of titanium dioxide nanoparticles. *Toxicology Letters, 228*(2), 103–110. https://doi.org/10.1016/j.toxlet.2014.04.014.

Jones, K., Morton, J., Smith, I., Jurkschat, H. K., Helen, A., & Evans, G. (2015). Human in vivo and in vitro studies on gastrointestinal absorption of titanium dioxide nanoparticles. *Toxicology Letters, 233*(2), 95–101. https://doi.org/10.1016/j.toxlet.2014.12.005.

Kansara, K., Patel, P., Shah, D., Shukla, R., Singh, S., Kumar, A., & Dhawan, A. (2015). TiO2 nanoparticles induce DNA double strand breaks and cell cycle arrest in human alveolar cells. *Environmental and Molecular Mutagenesis, 56*(2), 204–217. https://doi.org/10.1002/em.21925.

Kawamata, H., & Manfredi, G. (2010). Mitochondrial dysfunction and intracellular calcium dysregulation in ALS. *Mechanisms of Ageing and Development, 131*(7–8), 517–526. https://doi.org/10.1016/j.mad.2010.05.003.

Kermanizadeh, A., Gaiser, B., Hutchison, G., & Stone, V. (2012). An in vitro liver model-assessing oxidative stress and genotoxicity following exposure of hepatocytes to a panel of engineered nanomaterials. *Particle and Fibre Toxicology, 9,* 28. https://doi.org/10.1186/1743-8977-9-28.

Kermanizadeh, A., Vranic, S., Boland, S., et al. (2013). An *in vitro* assessment of panel of engineered nanomaterials using a human renal cell line: Cytotoxicity, pro-inflammatory response, oxidative stress and genotoxicity. *BMC Nephrology, 14,* 96. https://doi.org/10.1186/1471-2369-14-96.

Kessler, R. (2011). Engineered nanoparticles in consumer products: Understanding a new ingredient. *Environmental Health Perspectives, 119*(3), 120–125. https://doi.org/10.1289/ehp.119-a120.

LeBlanc, A. J., Cumpston, J. L., Chen, B. T., Frazer, D., Castranova, V., & Nurkiewicz, T. R. (2010). Nanoparticle inhalation impairs coronary microvascular reactivity via a local reactive oxygen species-dependent mechanism. *Cardiovascular Toxicology, 10*(1), 27–36.

Liu, H., Ma, L., Zhao, J., Liu, J., Yan, J., Ruan, J., & Hong, F. (2009). Biochemical toxicity of nano-anatase TiO2 particles in mice. *Biological Trace Element Research, 129,* 170–180.

Liu, H., Yang, D., Yang, H., Zhang, H., Zhang, W., Fang, Y., Lin, Z., Tian, L., Lin, B., Yan, J., & Xi, Z. (2013). Comparative study of respiratory tract immune toxicity induced by three sterilization nanoparticles: Silver, zinc oxide and titanium dioxide. *Journal of Hazardous Materials, 248-249*(1), 478–486. https://doi.org/10.1016/j.jhazmat.2013.01.046.

Lucie, A., Mathilde, B., Laure, B., Véronique, C., Hélène, D., Jean-Marc, S., Sarah, C., Alain, V., Nathalie, H., Thierry, R., & Marie, C. (2015). Molecular responses of alveolar epithelial A549 cells to chronic exposure to titanium dioxide nanoparticles: A proteomic view. *Journal of Proteomics, 134*, 163–173. https://doi.org/10.1016/j.jprot.2015.08.006.

MacNicoll, A., Kelly, M., Aksoy, H., Kramer, E., Bouwmeester, H., & Chaudhry, Q. (2015). A study of the uptake and biodistribution of nano-titanium dioxide using in vitro and in vivo models of oral intake. *Journal of Nanoparticle Research, 17*(2), 20. https://doi.org/10.1007/s11051-015-2862-3.

Mao, Z., Xu, B., Ji, X., Zhou, K., Zhang, X., Chen, M., Han, X., Tang, Q., Wang, X., & Xia, Y. (2015). Titanium dioxide nanoparticles alter cellular morphology via disturbing the microtubule dynamics. *Nanoscale, 7*(18), 8466–8475. https://doi.org/10.1039/C5NR01448D.

Marquez-Ramirez, S., Delgado-Buenrostro, N., Chirino, Y., Iglesias, G., & Lopez-Marure, R. (2012). Titanium dioxide nanoparticles inhibit proliferation and induce morphological changes and apoptosis in glial cells. *Toxicology, 302*(2–3), 146–156. https://doi.org/10.1016/j.tox.2012.09.005.

Medina-Reyes, E., Bucio-López, L., Freyre-Fonseca, V., Sánchez-Pérez, Y., García-Cuéllar, C., Morales-Bárcenas, R., Pedraza-Chaverri, J., & Chirino, Y. (2015a). Cell cycle synchronization reveals greater G2/M-phase accumulation of lung epithelial cells exposed to titanium dioxide nanoparticles. *Environmental Science and Pollution Research, 22*(5), 3976–3982. https://doi.org/10.1007/s11356-014-3871-y.

Medina-Reyes, E., Deciga-Alcaraz, A., Freyre-Fonseca, V., Delgado-Buenrostro, N., Flores-Flores, J., Gutierrez-Lopez, G., Sanchez-Perez, Y., Garcia-Cuellar, C., Pedraza-Chaverri, J., & Chirino, Y. (2015b). Titanium dioxide nanoparticles induce an adaptive inflammatory response and invasion and proliferation of lung epithelial cells in chorioallantoic membrane. *Environmental Research, 136*, 424–434. https://doi.org/10.1016/j.envres.2014.10.016.

Meena, R., Rani, M., Pal, R., & Rajamani, P. (2012). Nano-TiO2-induced apoptosis by oxidative stress-mediated DNA damage and activation of p53 in human embryonic kidney cells. *Applied Biochemistry and Biotechnology, 167*(4), 791–808. https://doi.org/10.1007/s12010-012-9699-3.

Nalika, N., & Parvez, S. (2015). Mitochondrial dysfunction in titanium dioxide nanoparticle-induced neurotoxicity. *Toxicology Mechanisms and Functions, 25*(5), 355–363. https://doi.org/10.3109/15376516.2015.1020183.

Natarajan, V., Wilson, C., Hayward, S., & Kidambi, S. (2015). Titanium dioxide nanoparticles trigger loss of function and perturbation of mitochondrial dynamics in primary hepatocytes. *PLoS One, 10*(8), 1–19. https://doi.org/10.1371/journal.pone.0134541.

Nemmar, A., Melghit, K., Al-Salam, S., Zia, S., Dhanasekaran, S., Attoub, S., Al-Amri, I., & Ali, B. H. (2011). Acute respiratory and systemic toxicity of pulmonary exposure to rutile Fe-doped TiO2 nanorods. *Toxicology, 279*(1–3, 11), 167–175.

Park, E., Lee, S., Lee, G., Kim, D., Kim, Y., Cho, M., & Kim, J. (2014). Sheet-type titania, but not P25, induced paraptosis accompanying apoptosis in murine alveolar macrophage cells. *Toxicology Letters, 230*(1), 69–79. https://doi.org/10.1016/j.toxlet.2014.07.027.

Periasamy, V., Athinarayanan, J., Al-Hadi, A., Juhaimi, F., & Alshatwi, A. (2015). Effects of titanium dioxide nanoparticles isolated from confectionery products on the metabolic stress pathway in human lung fibroblast cells. *Archives of Environmental Contamination and Toxicology, 68*(3), 521–533. https://doi.org/10.1007/s00244-014-0109-4.

Pujalté, I., Passagne, I., Brouillaud, B., Tréguer, M., Durand, E., Ohayon-Courtès, C., & L'Azou, B. (2011). Cytotoxicity and oxidative stress induced by different metallic nanoparticles on human kidney cells. *Particle and Fibre Toxicology, 8*(1), 10. https://doi.org/10.1186/1743-8977-8-10.

Pujalté, I., Passagne, I., Daculsi, R., de Portal, C., Ohayon-Courtès, C., & L'Azou, B. (2015). Cytotoxic effects and cellular oxidative mechanisms of metallic nanoparticles on renal tubular cells: Impact of particle solubility. *Toxicology Research, 4*(2), 409–422. https://doi.org/10.1039/C4TX00184B.

Rihane, N., Nury, T., M'rad, I., El Mir, L., Sakly, M., Amara, S., & Lizard, G. (2016). Microglial cells (BV-2) internalize titanium dioxide (TiO2) nanoparticles: Toxicity and cellular responses. *Environmental Science and Pollution Research, 23*(10), 9690–9699. https://doi.org/10.1007/s11356-016-6190-7.

Ruiz, P., Moron, B., Becker, H., Lang, S., Atrott, K., Spalinger, M., Scharl, M., Wojtal, K., Fischbeck-Terhalle, A., Frey-Wagner, H. M., Kraemer, T., & Rogler, G. (2017). Titanium dioxide nanoparticles exacerbate DSS-induced colitis: Role of the NLRP3 inflammasome. *Gut, 66*, 1216–1224.

Savi, M., Rossi, S., Bocchi, L., et al. (2014). Titanium dioxide nanoparticles promote arrhythmias via a direct interaction with rat cardiac tissue. *Particle and Fibre Toxicology, 11*, 63. https://doi.org/10.1186/s12989-014-0063-3.

Schoelermann, J., Burtey, A., Allouni, Z., Gerdes, H., & Cimpan, M. (2015). Contact-dependent transfer of TiO2 nanoparticles between mammalian cells. *Nanotoxicology, 10*(2), 204–215. https://doi.org/10.3109/17435390.2015.1048322.

Sha, B., Gao, W., Wang, S., Xu, F., & Lu, T. (2011). Cytotoxicity of titanium dioxide nanoparticles differs in four liver cells from human and rat. *Composites Part B: Engineering, 42*(8), 2136–2144. https://doi.org/10.1016/j.compositesb.2011.05.009.

Sha, B., Gao, W., Wang, S., Gou, X., Li, W., Liang, X., Qu, Z., Xu, F., & Lu, T. (2014). Oxidative stress increased hepatotoxicity induced by nano-titanium dioxide in BRL-3A cells and Sprague—Dawley rats. *Journal of Applied Toxicology, 34*(4), 345–356. https://doi.org/10.1002/jat.2900.

Sheng, L., Ze, Y., Wang, L., Yu, X., Hong, J., Zhao, X., Ze, X., Liu, D., Xu, B., Zhu, Y., Long, Y., Lin, A., Zhang, C., Zhao, Y., & Hong, F. (2015). Mechanisms of TiO2 nanoparticle-induced neuronal apoptosis in rat primary cultured hippocampal neurons. *Journal of Biomedical Materials Research – Part A, 103*(3), 1141–1149. https://doi.org/10.1002/jbm.a.35263.

Song, Z., Chen, N., Liu, J., Tang, H., Deng, X., Xi, W., Han, K., Cao, A., Liu, Y., & Wang, H. (2015). Biological effect of food additive titanium dioxide nanoparticles on intestine: An in vitro study. *Journal of Applied Toxicology, 35*(10), 1169–1178. https://doi.org/10.1002/jat.3171.

Susewind, J., Carvalho-Wodarz, C., Repnik, U., Collnot, E-M., Schneider-Daum, N., Griffiths, G., Lehr, C-M. (2016). A 3D co-culture of three human cell lines to model the inflamed intestinal mucosa for safety testing of nanomaterials. *Nanotoxicology, 10*(1). https://doi.org/10.3109/17435390.2015.1008065.

Sweeney, S., Berhanu, D., Ruenraroengsak, P., Thorley, A., Valsami-Jones, E., & Tetley, T. (2015). Nano-titanium dioxide bioreactivity with human alveolar type-I-like epithelial cells: Investigating crystalline phase as a critical determinant. *Nanotoxicology, 9*(4), 1743–5404. https://doi.org/10.3109/17435390.2014.948518.

Tada-Oikawa, S., Ichihara, G., Fukatsu, H., Shimanuki, Y., Tanaka, N., Watanabe, E., Suzuki, Y., Murakami, M., Izuoka, K., Chang, J., Wu, W., Yamada, Y., & Ichihara, S. (2016). Titanium dioxide particle type and concentration influence the inflammatory response in Caco-2 cells. *International Journal of Molecular Sciences, 17*, 4. https://doi.org/10.3390/ijms17040576.

Tang, Y., Wang, F., Jin, C., Liang, H., Zhong, X., & Yang, Y. (2013). Mitochondrial injury induced by nanosized titanium dioxide in A549 cells and rats. *Environmental Toxicology and Pharmacology, 36*(1), 66–72. https://doi.org/10.1016/j.etap.2013.03.006.

Tassinari, R., La Rocca, C., Stecca, L., Tait, S., De Berardis, B., Ammendolia, M., Iosi, F., Di Virgilio, A., Martinelli, A., & Maranghi, F. (2015). In vivo and in vitro toxicological effects of titanium dioxide nanoparticles on small intestine. *AIP Conference Proceedings*. Melville, NY: AIP Publishing. doi: https://doi.org/10.1063/1.4922572.

Tibau, A. V., Grube, B. D., Velez, B. J., Vega, V. M., & Mutter, J. (2019). Titanium exposure and human health. *Oral Science International, 16*, 15–24. https://doi.org/10.1002/osi2.1001.

Tobergte, D., & Curtis, S. (2013). Effects of titanium dioxide nanoparticles on the liver and hepatocytes in vitro. *Journal of Chemical Information and Modeling, 53*(9), 1689–1699. https://doi.org/10.1017/CBO9781107415324.004.

United States of America, Foreign Agricultural Service (USDA-FAS), Washington, DC, The 2019 U.S. Agricultural Export Yearbook, https://www.fas.usda.gov/data/2019-united-states-agricultural-export-yearbook.

Ursini, C., Cavallo, D., Fresegna, A., Ciervo, A., Maiello, R., Tassone, P., Buresti, G., Casciardi, S., & Iavicoli, S. (2014). Evaluation of cytotoxic, genotoxic and inflammatory response in human alveolar and bronchial epithelial cells exposed to titanium dioxide nanoparticles. *Journal of Applied Toxicology, 34*(11), 1209–1219. https://doi.org/10.1002/jat.3038.

Valdiglesias, V., Costa, C., Sharma, V., Kilic, G., Pasaro, E., Teixeira, J., Dhawan, A., & Laffon, B. (2013). Comparative study on effects of two different types of titanium dioxide nanoparticles on human neuronal cells. *Food and Chemical Toxicology, 57*, 352–361. https://doi.org/10.1016/j.fct.2013.04.010.

Vales, G., Rubio, L., & Marcos, R. (2015). Long-term exposures to low doses of titanium dioxide nanoparticles induce cell transformation, but not genotoxic damage in BEAS-2B cells. *Nanotoxicology, 9*(5), 568–578. https://doi.org/10.3109/17435390.2014.957252.

Vergaro, V., Aldieri, E., Fenoglio, I., Marucco, A., Carlucci, C., & Ciccarella, G. (2016). Surface reactivity and in vitro toxicity on human bronchial epithelial cells (BEAS-2B) of nanomaterials intermediates of the production of titania-based composites. *Toxicology In Vitro, 34*, 171–178. https://doi.org/10.1016/j.tiv.2016.04.003.

Vuong, N., Goegan, P., Mohottalage, S., Breznan, D., Ariganello, M., Williams, A., Elisma, F., Karthikeyan, S., Vincent, R., & Kumarathasan, P. (2016). Proteomic changes in human lung epithelial cells (A549) in response to carbon black and titanium dioxide exposures. *Journal of Proteomics*. Advance online publication. https://doi.org/10.1016/j.jprot.2016.03.046.

Wang, J., Zhou, G., Chen, C., Yuab, H., Wang, T., Mad, Y., Jia, G., Gao, Y., Li, B., Sun, J., Li, F., Jiao, F., Zhao, Y., & Chai, Z. (2007). Acute toxicity and biodistribution of different sized titanium dioxide particles in mice after oral administration. *Toxicology Letters, 168*(2), 176–185.

Weir, A., Westerhoff, P., Fabricius, L., Hristovski, K., & von Goetz, N. (2012). Titanium dioxide nanoparticles in food and personal care products. Environmental Science and Technology, 46(4), 2242–2250. doi: https://doi.org/10.1021/es204168d.

Wilson, C., Natarajan, V., Hayward, S., Khalimonchuk, O., & Kidambi, S. (2015). Mitochondrial dysfunction and loss of glutamate uptake in primary astrocytes exposed to titanium dioxide nanoparticles. *Nanoscale, 7*(44), 18477–18488. https://doi.org/10.1039/c5nr03646a.

Yu, K., Chang, S., Park, S., Lim, J., Lee, J., Yoon, T., Kim, J., & Cho, M. (2015). Titanium dioxide nanoparticles induce endoplasmic reticulum stress-mediated autophagic cell death via mitochondria-associated endoplasmic reticulum membrane disruption in normal lung cells. *PLoS One, 10*(6). https://doi.org/10.1371/journal.pone.0131208.

Zhang, J., Song, G., Zhang, J., Sun, Z., Li, L., Ding, F., & Gao, M. (2012). Cytotoxicity of different sized TiO2 nanoparticles in mouse macrophages. *Toxicology and Industrial Health, 29*(6), 523–533. https://doi.org/10.1177/0748233712442708.

Zijno, A., De Angelis, I., De Berardis, B., Andreoli, C., Russo, M., Pietraforte, D., Scorza, G., Degan, P., Ponti, J., Rossi, F., & Barone, F. (2015). Different mechanisms are involved in oxidative DNA damage and genotoxicity induction by ZnO and TiO2 nanoparticles in human colon carcinoma cells. *Toxicology In Vitro, 29*(7), 1503–1512. https://doi.org/10.1016/j.tiv.2015.06.009.

Chapter 13
Green Engineering of Silver Nanoparticles Using *Leucas aspera* Extract: Cytotoxic Efficacy in HeLa Cell Line

P. Venkatachalam, T. Bhuvaneswari, and N. Geetha

Contents

1 Introduction

Currently, the nanomaterial-based drugs are playing a significant role in the field of pharmaceutical biotechnology and also provide the tools for the investigation of biological systems (Mohamed and van der Walle 2008). Nanoparticles exhibit different optical, electronic, and crystal properties than those of the bulk materials, and these novel properties are depending on the shape, size, and surface characteristics of a particular class of nanomaterials. Metal nanoparticles, especially silver and gold, have drawn great attention of scientists because of their extensive biomedical and therapeutic applications besides their roles in the development of new technologies related to electronics and material sciences at the nanoscale (Magudapathy et al. 2001; Kohler et al. 2001). Silver nanoparticles (AgNPs) display different

P. Venkatachalam (✉) · T. Bhuvaneswari
Department of Biotechnology, Periyar University, Salem, Tamil Nadu, India

N. Geetha
Department of Botany, Bharathiar University, Coimbatore, TN, India

© Springer Nature Switzerland AG 2021
N. Sharma, S. Sahi (eds.), *Nanomaterial Biointeractions at the Cellular, Organismal and System Levels*, Nanotechnology in the Life Sciences,
https://doi.org/10.1007/978-3-030-65792-5_13

physiochemical and biological characteristics from those of bulk silver, such as increased optical, electromagnetic, catalytic properties and antimicrobial activity (Choi et al. 2009; Venkatachalam et al., 2015; Jinu et al., 2016). AgNPs are used in a wide range of medical applications, including in wound dressings, contraceptive devices, surgical instruments, and bone prostheses that are coated or implanted with the silver nanoparticles (Cheng et al. 2004; Chen et al. 2006; Zhang and Sun 2007). Nanoparticles can be synthesized by several physical, chemical, and biological methods. Chemical methods of nanoparticle synthesis cause accumulation of toxic and non-eco-friendly by-products.

The development of biological method for synthesis of nanoparticles is crucial for the production of environmental nontoxic nanoparticles. Several biological systems including plants, bacteria, fungi, and algae have been used for the synthesis of nanoparticles (Kalimuthu et al. 2008; Kowshik et al. 2003; Shahverdi et al. 2007; Sangi and Verma 2009). Among biological methods, plant extract-induced synthesis of nanoparticles is a common, feasible alternative to available conventional chemical and physical methods (Sreeram et al. 2008; Willner et al. 2006). Earlier studies suggest that bioactive compounds from medicinal herbs are found to be cost-effective and served as potential chemopreventive agents for inducing apoptosis in different types of cancer cell lines in the recent past (Kalaiarasi et al., 2015; Kayalvizhi et al., 2016; Jinu et al., 2017). The formation of gold and silver nanoparticles was reported, for the first time, by using living plants (Gardea-Torresdey et al. 2002; Jose-yacamann et al. 2003). The plant material-based production of silver nanoparticles has found applications in medical industries (Becker 1999; Kayalvizhi et al., 2016; Jinu et al., 2017; Prasannaraj and Venkatachalam, 2017a, b, c; Venkatachalam et al. 2017).

Medicinal plants provide a rich resource for natural drug research and development. *Leucas aspera* (Willd) Link is a species of annual medicinal herb that belongs to the family Lamiaceae (Labiatae). It grows abundantly in the high land crop fields, roadsides, and fallow lands of the wide area of South Asia (India, Bangladesh, and Nepal), Malaysia, and Mauritius (Shrestha and Sutton 2000). *Leucas aspera* exhibits biological activities like analgesic-antipyretic, antirheumatic, anti-inflammatory, antibacterial, and antifungal treatment (Gani 2003). *Leucas aspera* is also used traditionally as a medicine for coughs, colds, painful swellings, chronic skin eruptions (Chopra et al. 2002), and cytotoxic activities (Prajapati et al. 2010). Due to their immense medicinal importance, many scientific studies have been carried out on the phytochemical and pharmacological values of *Leucas aspera*. The selection of the plant species for the present study was principally based on the traditional uses of *Leucas aspera* for the treatment of various diseases such as ulcer, skin infection, and cancer.

Oxidative stress reactions and generation of free radicles are associated with a plethora of chronic conditions including cancer (Pourmorad et al. 2006). A large number of phytochemicals are being today used worldwide for the management of various cancers and chronic conditions (Moongkarndi et al. 2004). Silver nanoparticles (AgNPs) have demonstrated a unique potential in cancer management because it is selectively involved in disruption of the mitochondrial respiratory chain that leads to the production of ROS and interruption of ATP synthesis, which, in turn, causes DNA damage (AshaRani et al. 2009; Morones et al. 2005). Similarly, AgNPs were found having a great promise for the treatment of retinal neovascularization, cancer, hepatitis B, acquired human immunodeficiency syndrome (AIDS), and diabetes (Bhattacharya and Mukherjee 2008). Biological methods are considered to be the most powerful technique for the synthesis of AgNPs, as nanoparticles produced have a longer shelf life and constant stability as natural capping takes place. The bioengineering of nanoparticles has cost-effective, eco-friendly, and simple downstream processing, and obtained by effective purification and extracellular synthesis methods.

In view of the above, the present experiment was aimed for effective phytosynthesis of silver nanoparticles using aqueous leaf extract of *Leucas aspera* and characterization of synthesized AgNPs. Further, the synthesized silver nanoparticles were evaluated for their cytotoxic effect in HeLa cancer cell line as well as in normal cell line, Vero.

2   Materials and Methods

2.1   Preparation of Leaf Extract

Fresh leaves of *Leucas aspera* were collected from the Botanical Garden, Periyar University, and surface sterilized with running tap water, followed by distilled water to remove the dust particles adhered on the leaves. Biosynthesis of silver nanoparticles was carried out by microwave irradiation method.

2.2   Synthesis of Silver Nanoparticles by Microwave Irradiation Method

Fresh leaves of *Leucas aspera* (10 g) were finely chopped and mixed with 50 ml of deionized water. The mixture was boiled in microwave oven for 10 min. After boiling, the samples were allowed to cool at room temperature and the extracts filtered with Whatman No.1 filter paper. About 25 ml of plant aqueous extract prepared by microwave irradiation method was added into the 225 ml aqueous silver nitrate (AgNO$_3$) solution in the Erlenmeyer flask. Then the extract was mixed with AgNO$_3$ to make the final volume concentration of 1 mM solution. The reaction mixture was placed on shaker and mixed thoroughly and kept in dark room condition until the color change was noticed. The reaction progress for the formation of AgNPs was monitored by visual color change and UV–Vis spectral scanning in the range of 300–700 nm.

2.3   Characterization of Synthesized Silver Nanoparticles

Synthesized silver nanoparticles were confirmed by sampling the reaction mixture at regular intervals and absorption maxima was scanned by UV–vis spectra, at the wavelength of 300–700 nm in Systronics 2350 double beam spectrophotometer. In X-ray diffraction analysis, crystalline metallic silver was examined by coating dried nanoparticles on XRD grid using Rigaku miniflex II. X-ray diffraction (XRD) measurement of film of the biologically synthesized silver nanoparticles was performed at voltage 30 Kv with Cu k (α) radiation of 1.54187 nm wavelength. The scanning region of 2θ ranges from 20° and 80°. Purified AgNPs in the form of powder were analyzed using FT-IR spectral measurements. The measurements were carried out on a Bruker tensor 27 instrument. The samples were mixed with KBr to make a pellet and it was placed into the sample holder. The spectrum was recorded at a resolution of 4 cm^{-1}. The shape and size of synthesized AgNP was examined using field emission scanning electron microscopy (FESEM, CARL ZEISS) and these images were operated at 5 kV. The presence of elemental silver was determined in order to check the surface inter-atomic distribution using energy-dispersive X-ray (EDX).

2.4   Cytotoxic Activity of Silver Nanoparticles

2.4.1   Culture of Cell Lines

HeLa (Cervical cancer) cell lines were obtained from the National Centre for Cell Sciences (NCCS), Pune, India. The cell lines were maintained in Eagle's minimal essential medium (EMEM) supplemented with glutamine (2 mM), fetal bovine serum (10% V/V), 1.5 g/l sodium bicarbonate, nonessential amino acids (0.1 mM), and penicillin and streptomycin (1 IU/1 µg). The cell lines were maintained in 25 cm^2 culture flasks at 37 °C in a humidified CO_2 incubator containing 5% CO_2.

2.4.2   In Vitro Cytotoxicity Assay and Determination of IC50

The inhibitory concentration (IC50) value was determined by using MTT assay. Cells were cultured and seeded into 96-well culture plates approximately as 1×10^4 in each plate incubated for 48 h. HeLa cells and Vero cells (normal cells) were treated with series of 10–100 µg/ml concentrations of phytosynthesized AgNPs along with plant extracts. The treated cells were incubated for 48 hours to study cytotoxicity effects by MTT assay. MTT assay was based on the measurement of the mitochondrial activity of viable cells by the reduction of the tetrazolium salt MTT (3-(4,5-dimethyathiazol-2-yl)-2,5-diphenyl tetrazolium bromide) to form a blue water-insoluble product, formazan. The stock concentration (5 mg/ml) of MTT (Sigma, USA), a yellow tetrazole, was prepared and 100 µL of MTT was added into

AgNPs and plant extract treated wells and incubated for 4 h at 37 °C. After incubation, the medium was removed and DMSO (100 μL) was added to dissolve the formazan crystals resulting from the reduction of the tetrazolium salt only by metabolically active cells (Mosmann 1983; Kleihues and Ohgaki 2000; Kleihues et al. 2002). The absorbance of dissolved formazan was measured at 620 nm using a multiwell ELISA plate reader (Thermo Multiskan EX, USA). Since the absorbance was directly correlated with the number of viable cells, the percentage of cell viability was calculated by using the following formula:

$$\text{Percentage of Cell Viability} = \frac{\text{OD value of experimental NPs treated sample}}{\text{OD value of experimental untreated sample}} \times 100$$

3 Results and Discussion

In this investigation, the leaf extract was prepared by microwave irradiation method and the leaf extract was used for fabrication of silver nanoparticles. Microwave irradiation mediated synthesis of nanoparticles holds several advantages over traditional methods as its controlled temperature operation provides enhanced reaction kinetic properties, fast initial heating eventually leading to increased reaction rates with higher yields of nanoparticles (Nadagouda et al. 2011). Earlier studies have been documented that facile synthesis of nanoparticles using microwave irradiation (Nadagouda et al. 2011; Priyadharshini et al., 2014), and nowadays it has become an efficient alternative method for synthesis of various nanoparticles including silver nanoparticles.

3.1 Fabrication and Characterization of Silver Nanoparticles

The addition of *Leucas aspera* leaf extracts to silver nitrate solution resulted in color change of reaction mixture from yellow to brown due to the production of silver nanoparticles after 4 h of incubation in the dark room condition (Fig. 13.1). The formation of silver nanoparticles was recorded by UV–vis spectrum using the mixture of *Leucas aspera* leaf extracts and 1 mM silver nitrate solution at various time intervals. The color changes arise because of the excitation of surface plasmon resonance spectrum for the synthesized silver nanoparticles (Mulvaney, 1996). The reaction mixture of silver nitrate and leaf extracts exhibits the plasmon resonance peak at 400 nm. Bioactive molecules present in the leaf extracts served as capping and reducing agents for bioreduction of silver ions. This result indicates the formation of silver nanoparticles in the reaction mixture. The silver nanoparticles were

Fig. 13.1 Visual observations of reaction mixtures by color changes: a- plant extract, b- synthesized AgNO₃ (microwave method) c- AgNO₃

collected by centrifugation and washed thoroughly to remove impurities and the dried nanoparticles were used for further characterization and cytotoxicity studies.

The FTIR spectroscopy measurements were carried out to identify the possible bioactive molecules from the leaf extract that bound specifically on the silver ion surface (Fig. 13.2). The biologically synthesized silver nanoparticles obtained via microwave irradiation method were mixed with potassium bromide to make a pellet. The IR spectrum of biologically synthesized $AgNO_3$ nanoparticles obtained from microwave method showed the absorption peaks located at 3786.11, 3700.37, 3662.17, 3637.40, 2360.96, 1665.17, 1591.13, 1528.75, 1484.51, 1325, and 671.02 cm^{-1}. The bands appeared at 2360.96, 1325, and 671.02 cm^{-1} are due to phosphines (P-H), amines (C-N), and Alky halides (C-Br) stretching, respectively. The bands at 1650–1550 cm^{-1} are characteristic of amide groups (Caruso et al., 1998). The amide band arises either as a result of stretch mode of the carbonyl group coupled to the amide linkage or N-H stretching mode of vibration in the amide linkage. The proteins can easily bind to silver nanoparticles through either free amine groups or cysteine residues in the proteins which ultimately stabilizes the silver nanoparticles (Gole et al., 2001). Biological components are known to interact with metal salts via these functional groups and mediate their reduction to nanoparticles. The IR spectroscopic studies confirmed that phosphines and amine groups possess strong binding ability and acting as capping agent to provide stability to the synthesized silver nanoparticles.

X-ray diffraction analysis was performed to confirm the crystalline structure of synthesized silver nanoparticles (Fig. 13.3). XRD results indicate that the silver nanoparticles obtained by microwave method revealed diffraction peaks at 27.52° (210), 32.25° (122), 38.21° (111), 46.40° (231), 54.74° (142), 57.42° (241), and 64.47° (220) (Fig. 13.5). The average crystal size of the silver nanoparticles formed

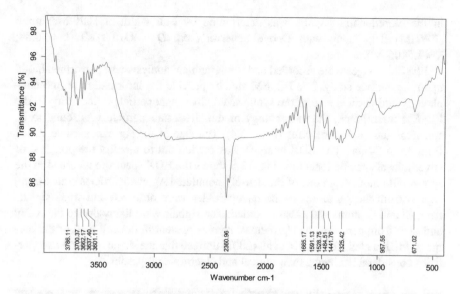

Fig. 13.2 FTIR spectra pattern of biosynthesized silver nanoparticles using leaf extracts of *Leucas aspera*

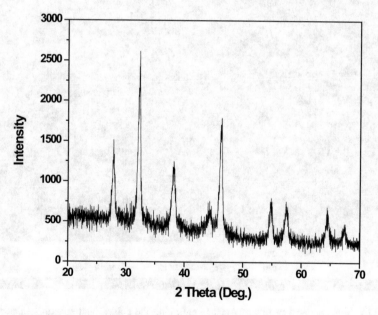

Fig. 13.3 XRD pattern of silver nanoparticles synthesized using *Leucas aspera* leaf extracts

in the bioreduction process was determined by Full width at Half Maximum (FWHM) data along with Debye Scherrer's eq. $D = k\lambda/\beta \, \cos\theta$ ($k = 0.89$; $\lambda = 1.5406 \, A°$).

FESEM images were recorded and topographical analysis was performed based upon the surface study. The FESEM studies provide the information on the morphology and particle size of the synthesized silver nanoparticles. According to the FESEM micrograph, the morphology of the silver nanoparticle was found to be spherical and cubic structures (Fig. 13.4). The size of the Ag nanoparticle varied from 35 to 54 nm. The EDX analysis was carried out to identify the presence of silver ions at specific locations. Fig. 13.5 shows the EDX spectrum recorded in the spot-profile mode from one of the densely populated Ag nanoparticles area. Strong signals from the Ag atoms in the nanoparticles were observed and weak signals from Cl and O atoms were also recorded. The signals were likely due to the X-ray emission from carbohydrates/proteins/enzymes present in the cell wall of the biomass (Mishra et al., 2010). Xu et al. (2002) noticed that the shape of metal nanoparticles considerably change their optical and electronic properties.

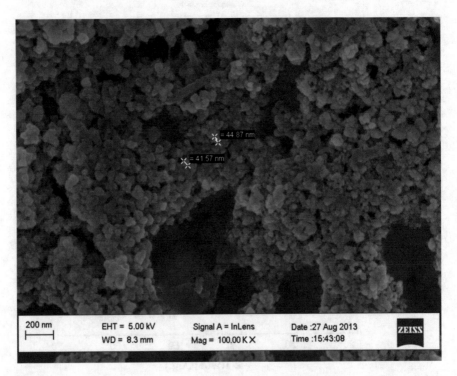

Fig. 13.4 Topographical results of AgNps confirming the cubical and spherical shaped from FE-SEM analysis

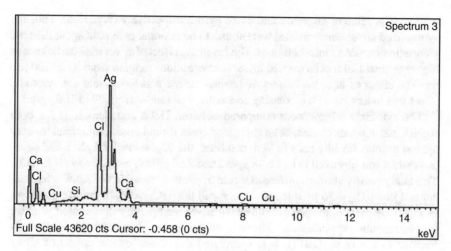

Fig. 13.5 Energy-dispersive X-ray spectrum (EDX) of metallic AgNPs

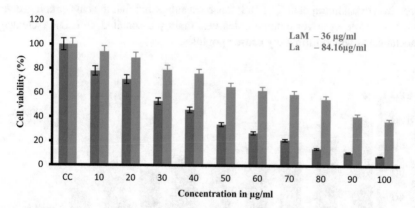

Fig. 13.6 MTT assay results confirming the in vitro cytotoxicity effect of silver nanoparticles and *Leucas aspera* plant extracts against HeLa cell lines

3.2   Evaluation of Silver Nanoparticle-Induced Cytotoxic Activity

The cytotoxicity of silver nanoparticles at different concentrations was evaluated using HeLa cancer cell line and Vero normal cell line by MTT assay. The dose-dependent cytotoxicity was observed in AgNPs and plant extract treated HeLa and Vero cells and the increase in concentration of AgNPs and plant extract demonstrated increased cytotoxicity in HeLa cells. The percentage values of cell viability obtained with continuous exposure for 48 h are depicted in Fig. 13.6. The in vitro screening of the AgNPs displayed maximum cytotoxic activity, compared to the plant extract, against both HeLa and Vero cell lines. The result displayed that HeLa cells proliferation were significantly inhibited by both, AgNPs and plant extract, with an IC_{50} value of 36 µg/ml and 84.16 µg/ml, respectively (Fig. 13.6). Thus the synthesized silver nanoparticles were found to be stronger cytotoxic agent than the plant extract against HeLa cell lines. The cytotoxic effect of silver nanoparticles was higher against HeLa cells even at lower concentrations ranging from 37 to 100 µg/ml. The effect of aqueous extract of *Leucas aspera* was lower, and the cytotoxic effect was increased with increasing concentrations ranges from 80 to 100 µg/ml.

The cytotoxic effects were compared between HeLa and Vero cells for both AgNPs and aqueous extract, as an antitumor agent should produce minimal toxicity against normal, healthy cells In Vero cell lines, the IC_{50} values of AgNPs and aqueous extract was observed to be 66.48 µg/ml and 72.4 µg/ml, respectively (Fig. 13.7). This study clearly shows a differential role of plant extract-derived AgNPs for cancer cell line (IC_{50} = 36 µg/ml) and normal cell line (IC_{50} = 66.48 µg/ml), as shown in Fig. 13.8. Hence, the phytosynthesized silver nanoparticles hold good promise for therapeutic applications. Similar results were also reported earlier by Venkatachalam et al. (2017). It is speculated that the cytotoxic impact of silver nanoparticles is due to the presence of various bioactive molecules coated silver atom interactions with the cell macromolecules such as nucleic acids and functional proteins (Prasannaraj et al. 2017). It has been suggested that the cancer cell growth-inhibitory role of silver nanoparticles was mainly associated with DNA-cleavage mechanisms which ultimately cause apoptosis.

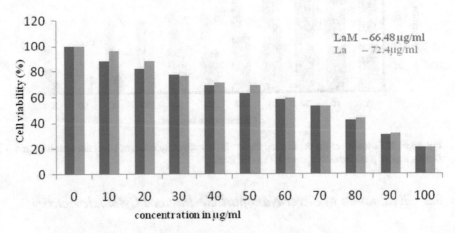

Fig. 13.7 MTT assay results confirming the in vitro cytotoxicity effect of *Leucas aspera* silver nanoparticles and aqueous leaf extract against the normal cell lines (Vero cells)

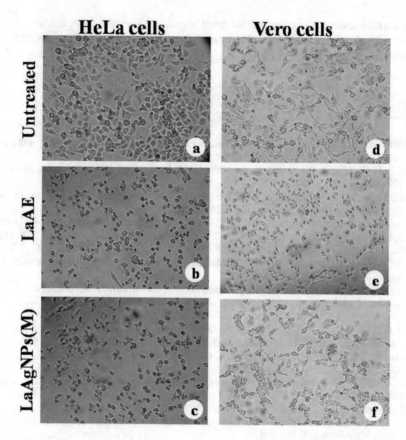

Fig. 13.8 Inverted microscopy images of LaLE and LaAgNPs treated cancer HeLa and normal Vero cell lines. (**a**) Untreated HeLa cells (**b**) *Leucas aspera* aqueous extract treated HeLa cells (**c**) *Leucas aspera* Silver nanoparticle treated Hela cells (**d**) Untreated Vero cells (**e**) *Leucas aspera* aqueous extract treated Vero cells (**f**) *Leucas aspera* Silver nanoparticle treated Vero cell

4   Conclusion

The present study was focused to develop an efficient protocol for fabrication of silver nanoparticles using cell-free aqueous leaf extracts of *Leucas aspera* by microwave irradiation method. *Leucas aspera* leaf extract is environmentally benign and renewable source which serves as reducing and capping agents for synthesis of nanoparticles. The synthesized silver nanoparticles were characterized and confirmed the presence of silver ions on nanoparticles. The presence of various bioactive compounds that are involved in reduction of silver ions was confirmed by FT-IR analysis. Mostly the synthesized nanoparticles were spherical in shape with average size of 55 nm. The impact of the silver nanoparticles on cell cytotoxicity was also investigated using HeLa cell lines. The cytotoxic effect of silver nanoparticles was positively correlated with dose on the cell line examined. It is interesting to note that the silver nanoparticles showed highest anticancer activity against HeLa cell lines. Therefore, the nanoparticles synthesized using *Leucas aspera* plant extracts hold therapeutic potential and deserve further detailed studies.

References

AshaRani, P. V., Kah Mun, G. L., Hande, M. P., & Valiyaveettil, S. (2009). Cytotoxicity and genotoxicity of silver nanoparticles in human cells. *ACS Nano, 3*, 279–290.

Becker, R. O. (1999). Silver ions in the treatment of local infections. *Metal-Based Drugs, 6*, 297–300.

Bhattacharya, R., & Mukherjee, P. (2008). Biological properties of naked metal nanoparticles. *Advanced Drug Delivery Reviews, 60*, 1289–1306.

Caruso, F., Caruso, A., & Helmuth, M. (1998). Nanoengineering of Inorganic and Hybrid Hollow Spheres by Colloidal Templating. *Science, 282.* https://doi.org/10.1126/science.282.5391.1111.

Chen, H. W., Su, S. F., Chien, C. T., Lin, W. H., Yu, S. L., Chou, C. C., Chen, J. J., & Yang, P. C. (2006). Titanium dioxide nanoparticles induce emphysema-like lung injury in mice. *The FASEB Journal, 20*, 2393–2395.

Cheng, D., Yang, J., & Zhao, Y. (2004). Antibacterial materials of silver nanoparticles application in medical appliances and appliances for daily use. *Chinese Medical Equipment Journal, 4*, 26–32.

Choi, O., Clevenger, T. E., Deng, B., Surampalli, R. Y., Ross, L., Jr., & Hu, Z. (2009). Role of sulfide and ligand strength in controlling nanosilver toxicity. *Water Research, 43*(7), 1879–1886.

Chopra, R. N., Nayar, S. L. and Chopra, I. C.. (2002). Glossary of Indian medicinal plants. NISCAIR, CSIR, New Delhi, pp. 153.

Gani, A. (2003). *Medicinal plants of Bangladesh: Chemical constituents and uses* (p. 215). Asiatic Society of Bangladesh.

Gardea-Torresdey, J. L., Parsons, J. G., Dokken, K., Peralta-Videa, J., Troiani, H. E., Santiago, P., & Jose-yacamann, M. (2002). Formation and growth of Au nanoparticles inside line alfalfa plants. *Nano Letters, 2*, 397–401.

Gole, A., Dash, C., Ramachandran, V., Sainkar, S., Mandale, A.,Rao, M., & Sastry, M. (2001). Pepsin-gold colloid conjugates: prepa-ration, characterization, and enzymatic activity. *Langmuir, 17*, 1674–1679.

Jinu, U., Jayalakshmi, N., Sujima Anbu, A., Mahendran, D., Sahi, S. V., & Venkatachalm, P. (2016). Biofabrication of Cubic Phase Silver Nanoparticles Loaded with Phytochemicals from *Solanum nigrum* Leaf Extract for Potential Antibacterial, Antibiofilm and Antioxidant Activities Against MDR Human Pathogens. *J Clust Sci, 28*, 489–505.

Jinu, U., Gomathi, M., Saiqa, I., Geetha, N., Benelli, G., & Venkatachalam, P. (2017). Green engineered biomolecule-capped silver and copper nanohybrids using *Prosopis cineraria* leaf extract: Enhanced antibacterial activity against microbial pathogens of public health relevance and cytotoxicity on human breast cancer cells (MCF-7). *Microbial Pathog, 105*, 86–95.

Jose-yacamann, M., Gardea-Torresdey, J. L., Gomez, E., Peralta-Videa, J., Parsons, J. G., & Troiani, H. E. (2003). Alfalfa sprout: A natural source for the synthesis of silver nanoparticles. *Langmuir, 19*, 1357–1361.

kalaiarasi, K., Prasannaraj, G., Sahi, S. V., & Venkatachalam P. (2015). Phytofabrication of biomolecule coated metallic silver nanoparticles using leaf extracts of in vitro-raised bamboo species & its anticancer activity against human PC3 cell lines. *J of biology, 39*, 223–232.

Kalimuthu, K., Babu, R. S., Venkataraman, D., Bilal, M., & Gurunathan, S. (2008). Biosynthesis of silver nanocrystals by bacillus licheniformis. *Colloids and Surface B, 65*, 150–153.

Kayalvizhi, T., Ravikumar, S., & Venkatachalam, P. (2016). Green Synthesis of Metallic Silver Nanoparticles Using Curculigo orchioides Rhizome Extracts and Evaluation of Its Antibacterial, Larvicidal, and Anticancer Activity. *J Environ Eng, 142*, C4016002. https://doi.org/10.1061/(asceee.1943-7870.0001098).

Kleihues, P., Louis, D. N., Scheithauer, B. W., Rorke, L. B., Eifenberger, G., Burger, P. C., & Cavenee, W. K. (2002). The WHO classification of tumors of the nervous system. *Journal of Neuropathology and Experimental Neurology, 61*, 215–225.

Kleihues, P., & Ohgaki, H. (2000). Phenotype vs genotype in the evolution of astrocytic brain tumors. *Toxicologic Pathology, 28*, 164–170.

Kohler, J. M., Csaki, A., Reichert, J., Moller, R., Straube, W., & Fritzsche, W. (2001). Selective labeling of oligonucleotide monolayers by metallic nanobeads for fast optical readout of DNA chips. *Sensors and Actuators B Chemical, 76*(1–3), 166–172.

Kowshik, M., Ashtaputure, S., Kharazi, S., Vogel, N., Urban, J., Kullkarni, S. K., & Panknikar, K. M. (2003). Extracellular synthesis of silver nanoparticles by a silver-tolerant yeast strain MKY3. *Nanotechnology, 14*, 95–100.

Magudapathy, P., Gangopadhyay, P., Panigrahi, B.K., Nair, K.G.M. and Dhara, S. 2001. Electrical transport studies of ag nanoclusters embedded in glass matrix. Physica B 299(1–2):142–146.

Mishra, A. N., Bhadauria, S., Gaur, M. S., Pasricha. P., & Kushwah, B. S. (2010). Synthesis of Gold Nanoparticles by Leaves of Zero-Calorie Sweetener Herb (*Stevia rebaudiana*) and their Nanoscopic Characterization by Spectroscopy and Microscopy. *InterJ Green Nanotech Physics Chemistry, 1*(2), 118–124.

Mohamed, F., & van der Walle, C. F. (2008). Engineering biodegradable polyester particles with specific drug targeting and drug release properties. *Journal of Pharmaceutical Sciences, 97*, 71–87.

Moongkarndi, P., Kosem, N., Luanratana, O., Jongsomboonkusol, S., & Pongpan, N. (2004). Antiproliferative activity of the medicinal plant extracts on human breast adenocarcinoma cell line. *Fitoterapia, 75*, 375–377.

Morones, J. R., Elechiguerra, L. J., Camacho, A., Holt, K., Kouri, B. J., Ramirez, T. J., et al. (2005). The bactericidal effect of silver nanoparticles. *Nanotechnology, 16*, 2346–2353.

Mosmann, T. (1983). Rapid colorimetric assay for cellular growth and survival: Application to proliferation and cytotoxicity assays. *Journal of Immunological Methods, 65*, 55–63.

Mulvaney, P. (1996). Surface Plasmon Spectroscopy of Nanosized Metal Particles. *Langmuir, 12*(3), 788–800.

Nadagouda, N., Speth, F., & Varma, S. (2011). Microwave-Assisted Green Synthesis of Silver Nanostructures. *Acc. Chem. Res, 44*(7), 469–478.

Pourmorad, F., Hosseinimehr, S. J., & Shahabimajd, N. (2006). Anti-oxidant activity, phenol and flavonoid contents of some selected Iranian medicinal plants. *African Journal of Biotechnology, 5*(11), 1142–1145.

Prajapati, M. S., Patel, J. B., Modi, K., & Shah, M. B. (2010). *Leucas aspera*: A review. *Pharmacognosy Reviews, 4*, 85–87.

Prasannaraj, G., Sahi, S. V., Benelli, G., & Venkatachalam, P. (2017a). Coating with active phytomolecules enhances anticancer activity of bio-engineered Ag nanocomplex. *J Clust Sci, 28*, 2349. https://doi.org/10.1007/s10876-017-1227-8.

Prasannaraj, G., & Venkatachalam, P. (2017b). Hepatoprotective effect of engineered silver nanoparticles coated bioactive compounds against diethylnitrosamine induced hepatocarcinogenesis in experimental mice. *J Photochem Photobiol b, 167*, 309–320.

Prasannaraj, G., & Venkatachalam, P. (2017c). Green engineering of biomolecule-coated metallic silver nanoparticles and their potential cytotoxic activity against cancer cell lines. *Adv. Nat. Sci. Nanosci. Nanotechnol, 8*, 025001

Priyadharshini, R., Prasannaraj, G., Geetha, N., & Venkatachalam, P. (2014). Microwave-Mediated Extracellular Synthesis of Metallic Silver and Zinc Oxide Nanoparticles Using Macro-Algae (*Gracilaria edulis*) Extracts and Its Anticancer Activity Against Human PC3 Cell Lines. *Appl Biochem Biotechnol, 174*, 2777–2790.

Rai, V., Agarwal, M., Agnihotri, A. K., Khatoon, S., Rawat, A. K., & Mehrotra, S. (2005). Pharmacognostical evaluation of *Leucas aspera*. *Natural Product Sciences, 11*, 109–114.

Sangi, R., & Verma, P. (2009). Biometric synthesis and characterization of protein capped silver nanoparticles. *Bioresource Technology, 100*, 501–504.

Shahverdi, A. R., Minacian, S., Shahverdi, H. R., Jamalifar, H., & Nohi, A. A. (2007). Rapid synthesis of silver nanoparticles using culture supernatants of enterobacteria: A novel biological approach. *Process Biochemistry, 42*, 919–923.

Shrestha, K. K. and Sutton, D. A. (2000). Annotated checklist of the flowering plants of Nepal. Press JR, The Natural History Museum.

Sreeram, K. J., Nidhin, M., & Nair, B. U. (2008). Microwave assisted template synthesis of silver nanoparticles. *Bulletin of Materials Science, 31*, 937–942.

Venkatachalam, P., Sangeetha, P., Geetha, N., & Sahi, S. V. (2015). Phytofabrication of Bioactive Molecules Encapsulated Metallic Silver Nanoparticles from *Cucumis sativus* L. and Its Enhanced Wound Healing potential in Rat Model. *J. Nanomat.* Article Id 753193, 9. https://doi.org/10.1155/2015/753193.

Venkatachalam, P., Kayalvizhi, T., Jinu, U., Benelli, G., & Geetha, N. (2017). Enhanced antibacterial and cytotoxic activity of phytochemical loaded-silver nanoparticles using *Curculigo orchioides* leaf extracts with different extraction techniques. *Journal of Cluster Science, 28*, 607–619.

Willner, I., Baron, R., & Willner, B. (2006). Growing metal nanoparticles by enzymes. *Advanced Materials, 18*, 1109–1120.

Xu, W., Gao, Y., & Liu., H. (2002). The Preparation, Characterization and their Photocatalytic Activities of Rare-Earth-Doped TiO_2 Nanoparticles. *J Catal, 207*(2), 151–157.

Zhang, Y., & Sun, J. (2007). A study on the bio-safety for nano-silver as anti-bacterial materials. *Chinese Journal of Medical Instruments, 31*, 35–38.

Chapter 14
Nanoparticles-Mediated Interventions to Prevent Herpes Simplex Virus (HSV) Entry into Susceptible Hosts

Jacob Beer, Hardik Majmudar, Yogendra Mishra, Deepak Shukla, and Vaibhav Tiwari

Contents

J. Beer · H. Majmudar · V. Tiwari (✉)
Department of Microbiology & Immunology, Midwestern University,
Downers Grove, IL, USA
e-mail: vtiwar@midwestern.edu

Y. Mishra
University of Southern Denmark, Mads Clausen Institute, NanoSYD, Sønderborg, Denmark

D. Shukla
Department of Ophthalmology & Visual Sciences, University of Illinois at Chicago,
Chicago, IL, USA

© Springer Nature Switzerland AG 2021
N. Sharma, S. Sahi (eds.), *Nanomaterial Biointeractions at the Cellular,
Organismal and System Levels*, Nanotechnology in the Life Sciences,
https://doi.org/10.1007/978-3-030-65792-5_14

1 Background

Herpesviruses are larger family members of viruses, which are known to infect wide variety organisms ranging from oysters to humans. They consist of a large linear double-stranded DNA genome, ranging in size from about 100,000 to 250,000 base pairs; an icosahedral capsid, which packages the genome and has 162 morphological units; and a layer of proteins called the tegument, which surrounds the capsid and is in turn enclosed within a lipid-containing envelope (Roizman and Sears 1996; Whitley et al. 1998). To replicate, herpesviruses kill the infected cells via lytic cycle. In addition, these viruses also have the capacity to establish nonlytic, latent infections in cells that, under certain conditions, do not support viral replication. In latently infected cells, the viral genome is stably associated with the cell nucleus and expresses few if any viral proteins. There is little or no pathology associated with latent infections, but activation of the latent viral genomes may occur due to changes in the state of the cell, resulting in new rounds of virus replication and new episodes of disease.

Herpesviruses have been divided into three subfamilies containing eight different type of herpesviruses, differing markedly in their biology (Whitley and Roizman 2001). These include herpes simplex virus: types 1 and 2 (HSV-1, HSV-2), varicella-zoster virus (VZV), cytomegalovirus (CMV), Epstein-Barr virus (EBV), human herpesvirus 6 (HHV-6), HHV-7, and HHV-8, or Kaposi sarcoma-associated herpesvirus. Based on genomic analysis and other biologic characteristics, the herpesviruses are classified into three subfamilies. The alphaherpesviruses subfamily contains HSV-1, HSV-2, and VZV, which are neurotropic cytolytic viruses. These viruses can establish asymptomatic latent infections in neurons of the peripheral nervous system. The betaherpesviruses subfamily includes cytomegalovirus (CMV), so named because infected cells become massively enlarged or "cytomegalic." This subfamily also includes HHV-6 and HHV-7. Members of the gammaherpesvirus subfamily are lymphotropic and include EBV and HHV-8. Both HHV-8 and EBV are considered important cofactors in malignancies (Shukla and Spear 2001).

1.1 A Wide Spectrum of Clinical Manifestations in Herpesvirus Diseases in Humans

Herpes simplex virus type-1 (HSV-1) infections are extremely widespread in the human population. The virus causes a broad range of diseases ranging from labial herpes, ocular keratitis, genital disease, and encephalitis (Fatahzadeh and Schwartz 2007; Anzivino et al. 2009). The herpetic infection is a major cause of morbidity especially in immunocompromised patients. Following initial infection in epithelial cells, HSV establishes latency in the host sensory nerve ganglia (Whitley and Roizman 2001). The virus emerges sporadically from latency and causes lesions on mucosal epithelium, skin, and the cornea, among other locations. Prolonged or multiple recurrent episodes of corneal infections can result in vision impairment or blindness, due to the development of herpetic stromal keratitis (HSK) (Farooq et al. 2011). HSK accounts for 20–48% of all recurrent ocular HSV infections leading to significant vision loss (Wilhelmus 1998). HSV infection may also lead to other diseases including retinitis, meningitis, and encephalitis [3].

Infections with each of the eight human herpesviruses, if one includes latent infections, are highly prevalent in all human populations. The clinical manifestations of infection can be divided into two categories: those evident after primary infection (primary disease) and those resulting from activation of latent virus (reactivation disease) (Whitley and Roizman 2001; Nicoll et al. 2012) The importance of cell-mediated immunity in controlling herpesvirus replication and in limiting reactivation of latent virus in vivo is apparent from the dramatic increase in severity and frequency of disease in immunodeficient persons. The transmission of herpesviruses usually requires intimate contact between persons, such as kissing or sexual intercourse. Often, the transmitting person is an asymptomatic shedder of infectious virus. Among the human herpesviruses, HSV has the broadest host range. Although HSV is normally isolated only from humans, many animal species can be experimentally infected, and many types and species of cultured cells will support HSV replication (Wertheim et al. 2014). In contrast, the other human herpesviruses have a much more limited host range, are fastidious about the cell types in which they will replicate, and are often much more difficult to propagate in cell culture (Shukla and Spear 2001). For all the herpesviruses, the ability to establish latent infections and to reactivate from latency are key to their ability to persist for the lifetime of the infected host, to maintain a high prevalence in most human populations, and to cause significant disease long after the initial infection and primary disease, if any.

The most common form of disease caused by HSV in humans is manifested as mucocutaneous lesions, which occur usually in or near the mouth (cold sores or fever blisters), on the cornea (keratitis), or on genital tissues (Roizman et al. 2007). Because the virus that causes the primary lesions establishes latent infections in sensory or autonomic peripheral neurons and can be reactivated by appropriate stimuli, periodic recurrences of herpetic lesions are common and present one of the troublesome aspects of infections with HSV. Less frequently, HSV can also cause life-threatening disease affecting vital organs, including encephalitis in apparently normal adults and disseminated disease in infants and immunocompromised individuals. Virus reactivation and replication may occur periodically in asymptomatic persons. Although HSV-1 and HSV-2 can infect the same body sites and cause indistinguishable lesions, the viruses are mostly associated with different sets of diseases. The viruses isolated from cases of adult encephalitis, keratitis, and facial lesions are usually HSV-1, whereas those isolated from cases of neonatal disease, adult meningitis, and genital lesions are usually HSV-2 (Fatahzadeh and Schwartz 2007).

Like HSV, VZV can replicate in the epidermis and establish latent infections in neurons. In contrast to HSV, VZV is transmitted in an epidemic fashion, via aerosols and the respiratory route, and causes a systemic rather than local primary disease. Varicella or chicken pox has the hallmarks of an acute viral disease affecting the entire lymphoreticular system with a generalized vesicular eruption of the skin and mucosal membranes. VZV establishes latent infections in sensory nerves. Reactivation of virus is infrequent and occurs usually in older individuals, resulting in zoster or shingles, an eruption of skin lesions that is limited to a single dermatome and often causes severe pain (Shukla and Spear 2001).

CMV infections are typically asymptomatic. The most common clinical manifestation is a self-limited mononucleosis, although disease can be life threatening in immunodeficient individuals. CMV can cause hepatitis, chorioretinitis, pneumonitis, colitis, and congenital cytomegalic inclusion disease and can be a contributing factor to failure of organ transplantation. The cell types that harbor latent virus are probably of the monocytic lineage. The major means of CMV transmission are by the congenital, oral, and sexual routes and by blood transfusion or tissue transplantation (Britt 2008). The other two human members of the betaherpesvirus family, HHV-6 and HHV-7, are T lymphotropic and are both capable of causing a childhood illness called sixth disease or roseola infantum. Most children become infected with these viruses, but very few show clinical manifestations of disease (Shukla and Spear 2001).

Primary infections of children with EBV are usually asymptomatic, but, in young adults, the consequence can be infectious mononucleosis. The virus establishes latent infections in B lymphocytes and can induce proliferation of B cells through expression of a limited number of viral genes. A competent immune system keeps these latently infected B cells in check, but abrogation of cell-mediated immunity can result in lymphoproliferative disease. The other human member of the gammaherpesvirus subfamily, HHV-8, is the cause of Kaposi sarcoma, a vascular tumor of mixed cell composition, and body cavity-based lymphoma found in AIDS patients. It is thus evident that the human herpesviruses differ markedly in their ability to infect various cell types and to establish productive or latent infections in these cell types (Shukla and Spear 2001).

1.2   Herpesvirus Entry into the Host Cells: A Dynamic Event

Enveloped viruses enter cells by inducing fusion between the viral envelope and a cell membrane. This membrane fusion can be triggered in at least two ways, resulting in different pathways of entry. For example, the binding of a virus to a cell may induce endocytosis of the virus, followed by acidification of the endosome, which can trigger fusion between the viral envelope and endosome membrane. Alternatively, the binding of a virus to a cell may result in multiple receptor-ligand interactions at the cell surface that can trigger fusion between the viral envelope and the plasma membrane. Most herpesviruses apparently enter cells via the latter pathway. They can fuse directly with the cell plasma membrane. Fusion might also occur after endocytosis of the virus particle in an early endosome, but may not be dependent on the acidification of endosomes. In addition, a novel mode of HSV entry via phagocytic uptake has been suggested by our group using primary cultures derived from human eye donors (Clement et al. 2006).

The herpesvirus envelope is a lipid bilayer derived from a host cell membrane in which most cell proteins have been displaced by viral membrane proteins. As many as a dozen or more integral membrane proteins and glycoproteins encoded by the virus become incorporated into the viral envelope. For most herpesviruses, only four or five of the envelope proteins may participate in the process of viral entry (Shukla and Spear 2001). Rigorous proof that a viral envelope protein has a role in viral entry comes from characterizing the specific infectivity of a viral mutant that produces virions devoid of the protein in question. A viral mutant that cannot produce a protein required for viral entry will not be propagated. Such mutants can be propagated, however, in a cell line transformed to express the missing protein (a complementing cell line) so that the protein is supplied in trans for viral assembly. Passage of the virus once through a noncomplementing cell line results in production of virus particles missing the protein in question, and these particles can be assessed for their ability to enter cells.

By constructing and characterizing many such HSV-1 mutants, it has been possible to identify five viral glycoproteins (gB, gC, gD, gH, and gL) that contribute to viral entry (Spear 1993). Four of them, gB, gD, gH, and gL, are essential for viral entry. In the absence of any one of these glycoproteins, the mutant virions can bind to cells but fail to enter them. Absence of gC results in reduced binding of virus to cells, although the virus that binds can enter cells and initiate infection (Herold et al. 1991). Absence of both gB and gC severely reduces the binding of HSV-1 to cells (Herold et al. 1994). These findings, and others discussed below, indicate that either gC or gB can mediate the binding of virus to cells, whereas gB, gD, gH, and gL are all required for the membrane fusion that leads to viral entry (Eisenberg et al. 2012).

1.3 Current Limitations with Herpes Simplex Virus (HSV) Therapy

Herpes simplex virus (HSV) is among the ubiquitous human infections and persists lifelong as a latent infection in their host. The latency has potential to reactivate, thus causing a multitude of symptoms, the most common being painful blisters or ulcers at the site of infection. In immunocompromised people, HSV-1 can also lead to severe complications such as encephalitis or keratitis. The development of novel strategies to eradicate HSV is a global public health priority (Wilson et al. 2009; Schulte et al. 2010; Schepers et al. 2014; Pan et al. 2014). While Acyclovir and related nucleoside analogs provide successful modalities for treatment and suppression, HSV remains highly prevalent worldwide (Wilson et al. 2009). The current focus of antiviral drugs solely on HSV replication, emergence of Acyclovir-resistant strains, and the ability of HSV to uniformly establish latency coupled with adverse effects of anti-herpetic drugs, demand search for new and more effective antiviral agents to replace or synergistically enhance efficacy of the current treatment methods (Superti et al. 2008). In addition, Acyclovir fails to prevent latency and the virus can reactivate and cause a clinical disease. Long-term use of drugs to treat ocular herpes often results in cataract and elevated intraocular pressure. Therefore, there is always a need to develop new interventions to prevent virus infectivity and control drug resistance.

1.4   Novel Approach to Fight HSV Infections with Broader Applications

The best way to control the HSV epidemic is prevention (Wilson et al. 2009). A novel strategy is to design the drugs that prevent viral entry into target cells. In this regard, cell-associated HS represents an attractive target. Heparin sulfate (HS) is endowed with the remarkable ability to bind numerous viral proteins including HSV-1 glycoproteins (Liu and Thorp 2002). The significance of HS in viral diseases is also evident from recent reports that polysulfated compounds and HS-mimetic are very effective in blocking viral infections (Copland et al. 2008; Tiwari et al. 2009a, b; Ekblad et al. 2010; Thakkar et al. 2010; Gangji et al. 2018; Majmudar et al. 2019). In addition, the cells that are defective in HS exhibit a dramatic reduction in susceptibility to HSV-1 infection (Campadelli-Fume and Menotti 2007). It is worth noting that interaction with cell surface HS has been found to be a common pathway for attachment by several other human and animal viruses (Liu and Thorp 2002). Our recently published data indicates that the zinc oxide-based nanoparticles (ZnO-NPs) capped with negatively charged nanospikes can trap HSV-1 virion and thereby reduce the viral infectivity, suggesting the strong possibility of developing NPs-based enveloped viral infection antagonists (Mishra et al. 2011; Antoine et al. 2012). In recent years NPs have gained wide popularity in therapeutics, drug delivery, biological imaging, and biosensors (Eugenii and Willner 2004; De et al. 2008).

2   Virostatic Potential of Zinc Oxide Nanoparticles (ZnO NPs) on Herpes Simplex Virus Type-1 Entry and Spread

Multivalent interactions are important for many biological processes involving protein-protein interactions (Varga et al. 2014; Waldmann et al. 2014; Connolly et al. 2011). However, efficient blockade of a multivalent process may not always be achievable by monovalent inhibitors. Such is the case for viral entry into host cell, which is mediated by multiple interactions between virus envelope proteins and host cell surface receptors. As shown (Fig. 14.1a), three major steps (I-III) are involved in HSV-1 entry. (I). Positive (+) residues in HSV-1 gB bind to negatively (−) charged cell surface heparan sulfate (HS) (Shukla and Spear 2001), (II) HSV-1 gD then binds to its receptors to initiate virus-cell membrane fusion or endocytosis (Connolly et al. 2011). (III) Viral capsid is transported to the nucleus for replication

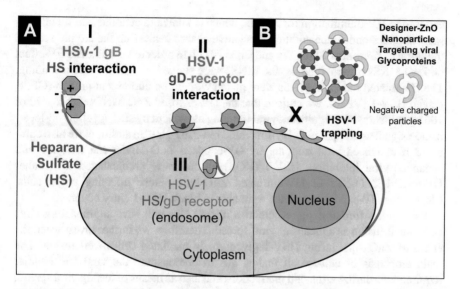

Fig. 14.1 A model depicting the significance of ZnO-micro-nano structures (MNs) in blocking HSV-1 entry by interfering multivalent interactions between the virus and the host cell. The panel **a** shows the three major steps (I-III) involved in HSV-1 entry. (I). Positive residues in HSV-1 glycoprotein B (gB) binds to negatively charged cell surface heparan sulfate (HS), (II) HSV-1 glycoprotein D (gD) then binds to its receptors to initiate virus-cell membrane fusion or endocytosis, (III) Viral capsid is transported to the nucleus for replication. Viral infectivity can be blocked by using zinc oxide nanoparticles (ZnO NPs) bearing multiple negatively charged nanospikes (panel **b**), as they competitively bind multiple entry-related positively charged glycoproteins such as gB, gC, and gD. Thus, ZnO-NPs already have proven to interfere and block the multivalent interactions (Mishra et al. 2011 *Antiviral Res*)

(Spear and Longnecker 2003). Viral infectivity can be blocked by ZnO-NPs bearing multiple negatively charged nanospikes (Fig. 14.1b), as they competitively bind multiple entry-related positively charged glycoproteins such as gB, gC, and gD (Mishra et al. 2011). For example, using cell culture model we tested the effect of ZnO nanoparticles on HSV-1 entry into the target cells. HSV-1 entry into a cell was determined by using β-galactosidase expressing HSV-1 reporter virus (gL86) into wild-type Chinese hamster ovary (CHO-K1) cells expressing gD receptor nectin-1. It was clear that HSV-1 preincubation with ZnO-NPs significantly blocked the viral entry in a dose-dependent manner in CHO-K1 cells expressing gD receptors. The control cells treated with 1× PBS (untreated) showed HSV-1 entry. The blocking activity of nanoparticles was pronounced even at low concentrations (0.01 mg/ml or 100 μg/ml). We further confirmed the blocking activity of ZnO-nanoparticles on HSV-1 entry, using human corneal fibroblasts (CF)—a natural target for HSV-1 infection. The CF expresses HVEM and 3-OST-3 as gD receptors (Tiwari et al. 2006). Similar results were found in HeLa cells that express all the known gD-receptors. In all cases, the mock-treated control cells showed HSV-1 entry.

We further rationalized that viral entry inhibition property of ZnO nanoparticle is due to its partial negatively charged oxygen vacancies. Therefore, ZnO-NPs were exposed to UV illumination for 30 min, which is known to generate additional oxygen vacancies and hence additional negative charge centers on the atomic scale at the surface (Kong et al. 2008; Wu and Chen 2011). In order to visualize the UV-effect on ZnO MNSs to viral binding, the MNSs were stained red via phalloidin staining. The UV-treated red-ZnO-MNSs were mixed with green fluorescent protein (GFP)-tagged HSV-1 (VP26). We noticed that the UV-exposed ZnO-MNSs (0.1 mg/ml or 100 µg/ml) showed a significant viral trapping as evident by strong yellow co-localization signal as compared to UV-untreated red-ZnO-Nps. In addition, we also compared if enhanced viral trapping by UV-exposed ZnO-MNSs can translate into enhanced viral inhibition. HSV-1 (KOS) virions were preincubated with either UV-treated ZnO-NPs or UV-untreated ZnO-NPs before infecting target cells. Clearly, the UV-exposed particles were able to block HSV-1 entry better.

Our studies further explored the idea if ZnO NPs will have an impact on virus cell-to-cell fusion as a means of viral spread. Therefore, we further investigated the effect of ZnO-Nps during HSV-1 glycoprotein-mediated cell-to-cell fusion. The main emphasis of cell-to-cell fusion was to demonstrate the viral and cellular requirements during virus-cell interactions and also as means of testing viral spread. We sought to determine whether ZnO NPs interaction with HSV-1 envelope glycoproteins essential for viral entry affects cell-to-cell fusion. Surprisingly, effector cells expressing HSV-1 glycoproteins treated with ZnO NPs (0.01 mg/ ml) impaired the cell-to-cell fusion in CHO-K1 cells expressing gD receptor nectin-1. In parallel, the control-untreated effector cells cocultured with target cells showed expected fusion. This response was further confirmed when polykaryocytes formation was estimated. ZnO NP-treated effector cells failed to form polykaryons when cocultured with target cells. The control-untreated effector cells efficiently showed larger polykaryons. Our results indicate that the presence of ZnO NPs significantly reduces viral penetration. We, therefore, propose that ZnO NPs can possibly disrupt the viral envelope glycoproteins binding to cell surface HS, thereby preventing the virus attachment, surfing, and fusion processes.

Since our studies with ZnO-NPs involved the lab strain of HSV-1, we thus decided to test the ability of NPs to block viral entry in different clinically relevant strains of HSV (F, G, and MP) (Dean et al. 1994). Here we used nectin-1 expressing CHO Ig8 cells that express β-galactosidase upon viral entry (Montgomery et al. 1996). The virulent strains were preincubated with MNSs and then used for infecting the cells. The results from this experiment again showed that MNSs blocked entry of additional HSV strains as evident by using reporter-based ONPG assay. Finally, we asked whether an in vivo significance of NPs could be demonstrated in an animal model. For this, we chose embryos from zebrafish, which provide a quick and easy model for testing HSV-1 infection in vivo (Burgos et al. 2008). Our results showed that NPs were able to prophylactically block infection of the zebrafish embryos as well (Tiwari, Unpublished results). This result reaffirms that NPs hold a strong promise for development as an effective anti-HSV prophylactic agent.

Certainly, nanotechnology offers an opportunity to re-explore biological properties of known antimicrobial compounds by manipulation of their sizes (Travan et al. 2011). ZnO has long been known for its antibacterial and antifungal properties including the recent report for selective destruction of tumor cells by ZnO nanoparticles and its potential in the development of anticancer agents (Rasmussen et al. 2010). In addition, the use of ZnO nanoparticles in sunscreens is one of the most common uses of nanotechnology in consumer products (Donathan and Meyer 2010). The uses of ZnO nanostructures have been suggested in nonresonant nonlinear optical microscopy in biology and medicine (Aliaksandr et al. 2008). Cell surface HS is involved in viral pathogenesis including viral binding, transport, and membrane fusion aiding to viral spread (Shukla and Spear 2001; Jihan and Shukla 2009). Therefore, the use of ZnO NPs, in present case, represents a unique approach to antiviral therapy forcing a competition for the HS chain used by HSV-1 and, possibly, many other medically important herpesviruses that bind to the cell surface HS (Liu and Thorp 2002). Our efforts to synthesize structurally defined ZnO NPs bearing nanoscopic filopodia-like spikes was based on the speculation that it will expand the surface area of partially negatively charged molecules, thereby mimicking the overall negative charge present on HS structures. Our initial quantitative viral entry assay revealed that pretreatment of HSV-1 with ZnO NPs significantly affected the viral entry at nontoxic concentrations. UV-irradiated ZnO MNSs were even more potent in blocking HSV-1 entry and spread. In addition, the fluorescent-based imaging experiment further confirmed the quantitative viral entry data that UV-treated ZnO NPs neutralized the viral infectivity by "viral-trapping" or "virostatic activity," which was evident from the enhanced accumulation of GFP-tagged virus around NPs. The viral trapping activity of NPs was expected as UV exposure to ZnO spikes enhanced the distribution of negative charge by oxygen vacancies and thereby allowing more viruses to bind. The use of functionalized nanoparticles to develop antivirals that act by interfering with viral infection, in particular attachment and entry, are gaining wide popularity (Tallury et al. 2010; Bowman et al. 2008; Lara et al. 2010; Vig et al. 2008). Similar studies with nanoparticles are known to inhibit cell-to-cell spread of HSV (Baram-Pinto et al. 2010).

The major advantage of ZnO NPs is their effectiveness at lower concentrations (µg), the low cost of their synthesis, molecular specificity to viral envelope protein without affecting the expression of native HS chain, and ease in designing nanoparticle capsules coated with additional anti-HSV-1 agents, including envelop glycoprotein B (gB) and D (gD)-based peptides, to block HSV-1 entry receptors, while keeping the HSV-1 virions trapped to MNSs. ZnO is an integral component of skin, face, and lip creams where HSV-1 infection or reactivation leads to painful blisters. Therefore, ZnO-NPs exhibit strong potentials to develop anti-HSV medication for cold sore in the form of protective gel or cream, which may be further activated by the UV part of the sun light. In addition, such MNSs will become the bench tool to create additional antiviral agents against many other viruses with the conjugation of peptides against specific virus envelope glycoproteins. Furthermore, they can also be used to deliver antiviral peptides with minimal pharmacokinetic problems together with enhanced activity of drug for the treatment of HSV infection.

Taken together our previously published findings support the model by which partially negatively charged ZnO NPs trap the HSV-1 to prevent virus-cell interaction, which are key steps for successful viral infection of the host cells and, therefore, MNSs-based compounds present a useful therapeutic approach. This is further supported by our observation that MNSs also block infection in vivo, in a zebrafish model of HSV-1 infection.

2.1 Unique "Tetrapod" Symmetry of ZnO-NPs

While screening different types of ZnO NPs, we discovered that ZnO with unique tetrapod symmetry can efficiently trap HSV at multiple steps of infection and provide preventive as well as significant therapeutic benefits by reducing the number of plaques formed. Keeping this key structural feature in mind, we plan to conjugate anti-HSV peptides to the tetrapod form of ZnO. ZnO-NPs with tetrapod symmetry can be commercially synthesized in large quantities (>kilograms). As indicated in Fig. 14.2 ZnO nano-micro scale tetrapod structures can be synthesized by flame transport approach (panels a: Mishra et al. 2011) Glass bottle shows the large amount of ZnO tetrapod structures which were synthesized in just one run. Figure 14.2 (panels b–e) show the scanning electron microscopy images of different type of tetrapod structures from the ZnO powder shown in panel a (Fig. 14.2).

2.2 Precise Engineering of ZnO "Nanospikes" to Trap Multiple Strains of HSV and Associated Coinfections

Developing a "broad-spectrum inhibitor" against multiple strains of HSV and other viruses is ideal and possible because of our ability to create negative charges on ZnO spikes and add peptides against specific viral envelop proteins from one or multiple viruses (Di Gianvincenzo et al. 2012). For instance, it has been shown that HSV reactivation is associated with increased replication of HIV on mucosal surfaces (Rollenhagen et al. 2014; Meque et al. 2014; McConville et al. 2014); hence the ZnO-NPs that trap both HSV-2 and HIV will be of high value. Conjugation of anti-HSV-2 gB/gD together with anti-HIV gp41 peptides will specifically inhibit both HSV-2 and HIV, and therefore can be useful for future microbicide development (Whaley et al. 2010).

2.3 Unique Ability of Micro–Nano Spikes of ZnO to Trap HSV-1

Our study further identified ZnO-nanostructures bearing nanospikes (Fig. 14.2), which trapped HSV virions and blocked both viral entry and cell-to-cell spread (Figs. 14.3 and 14.4). Interestingly HSV entry blocking activities of ZnO-NPs were

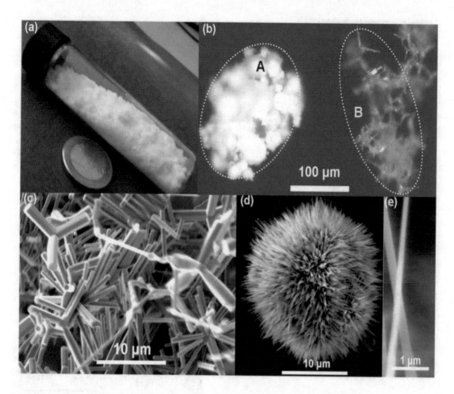

Fig. 14.2 Zinc oxide-based micro-nano structures. ZnO-MNs with unique tetrapod symmetry which efficiently traps HSV at multiple steps of infection and provide preventive as well as significant therapeutic benefits by reducing entry and the number of plaques formed during viral spread. The panels a-e show the details of the micro nano structures. (**a**) Synthesis of the ZnO material can be done in large quantities, please note the 23 mm diameter coin. (**b**) Microscopic image, comparison between a standard powder (A) and the material synthesized here (B). (**c**) Electron micrograph showing the complex geometries. (**d**) The powder contains a larger quantity of filopodia-like structures, which have spikes down to the nanoscale (**e**) (Mishra et al. 2011, *Antiviral Res*)

significantly enhanced by the addition of extra oxygen vacancies induced by UV-light illumination. We further provided the evidence for the broader significance of UV-treated ZnO as an anti-HSV agent affecting entry of clinical isolates of HSV (F, G, and MP), cell-to-cell fusion, and polykaryocyte formation examined by previously described methods (Mishra et al. 2011).

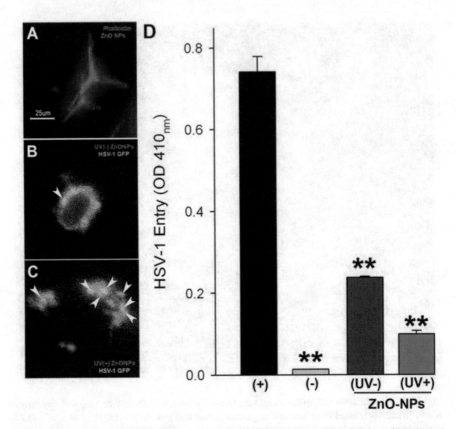

Fig. 14.3 Zinc oxide-based micro-nano structures (ZnO-MNs) trap HSV-1 virions and block viral entry. Panels (**a–c**). UV-illumination on ZnO MNSs significantly enhances HSV-1 binding. ZnO-MNSs were exposed to UV illumination for 30 min. MNSs were stained as red via phalloidin treatment (panel **a**). UV-untreated (panel **b**) and UV-treated (panel **c**) ZnO MNSs were mixed with green fluorescent protein (GFP)-tagged HSV-1 (VP26). The UV-exposed ZnO-MNSs showing significant HSV-1 trapping as indicated by strong yellow co-localization signal (highlighted by arrows) compared to UV-untreated red-ZnO-MNSs. Panel **d** shows that the preincubation of UV-treated ZnO MNSs with HSV-1 significantly block viral entry. In this experiment, β-galactosidase-expressing recombinant virus HSV-1 (KOS) gL86 (25 pfu/cell) was preincubated for 90 min with the UV-pretreated (+) or -untreated (−) ZnO-MPs at 0.1 mg/ml. HSV-1 KOS gL86 mock-incubated with 1 × phosphate buffer saline (PBS; black bar) was used as positive control. The uninfected cells were used as negative control (gray bar). After 90 min the soup was challenged to CF. After 6 h, the cells were washed, permeabilized, and incubated with ONPG substrate (3.0 mg/ml) for quantitation of β-galactosidase activity expressed from the input viral genome. The enzymatic activity was measured at an optical density of 410 nm (OD_{410}). The value shown is the mean of three or more determinations (±SD) (Mishra et al. 2011; *Antiviral Res*)

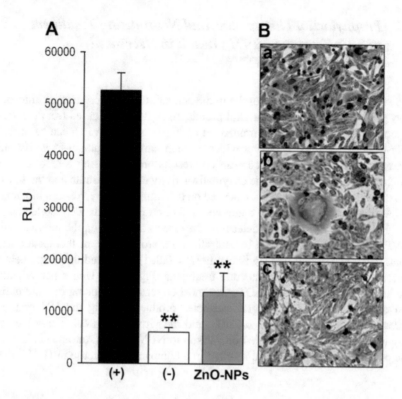

Fig. 14.4 Negatively charged UV-treated zinc-oxide-based micro-nano structures (ZnO-MNs) negatively impairs HSV-1 glycoprotein-mediated cell-to-cell fusion and polykaryocyte formation. Panel **a**: In this experiment, a reporter-based cell-to-cell fusion assay was used. Briefly, the "effector CHO-K1 cells" expressing HSV-1 glycoproteins (gB, gD, gH–gL) along with T7 plasmid were preincubated with 100 μg/ml UV-treated ZnO-MPs or with 1 × PBS for 90 min. The two pools of effector cells (ZnO-MNSs treated and PBS treated) were mixed with target CHO-K1 cells expressing luciferase gene along with specific gD receptor nectin-1. Membrane fusion as a means of viral spread was detected by monitoring luciferase activity. Relative luciferase units (RLUs) determined using a Sirius luminometer (Berthold detection systems). Black bars and gray bars represent 1× PBS-treated and ZnO-MNSs-treated cells, respectively. The effector cells devoid of HSV-1 glycoprotein mixed with target CHO-K1 nectin-1 expressing cells was used a negative control (white bar). Error bars represent standard deviations. *$P < 0.05$, one-way ANOVA. Panel **b**: Microscopic visualization of polykaryocyte impairments by ZnO-MNSs. In this experiment effector CHO-K1 cells expressing four essential HSV-1 glycoproteins (gB, gD, gH–gL) were either preincubated with ZnO-MNSs or with 1 × PBS for 90 min before they were cocultured in 1:1 ratio with target nectin-1 expressing CHO-K1 cells for 24 hrs. The cells were fixed (2% formaldehyde and 0.2% glutaraldehyde) for 20 min. and then stained with Gimesa stain (Fluka) for 20 min. Shown are photographs of representative cells (Zeiss Axiovert 200) pictured under microscope at 40 × objective. The upper a panel shows no polykaryocytes formation in the absence of HSV-1 glycoprotein (negative control), middle panel b shows significant inhibition of polykaryocytes formation in the presence of HSV-1 glycoprotein in effector cells fused with target nectin-1 CHO-K1 cells. Lower panel c shows no polykaryocytes formation in the presence of ZnO-MNSs during coculture of HSV-1 glycoprotein expressing cells with target nectin-1 expressing CHO-K1 cells (Mishra et al. 2011; *Antiviral Res*)

2.4 Prophylactic, Therapeutic, and Neutralizing Treatment of ZnO-Tetrapod (ZnOT) Result in Decreased Internalization of HSV

The virus entry into the cells results in the activation of transcription of immediate early gene. In case of HSV-2, a viral protein 16 (VP16) is released from the virion to activate the transcription of immediate early genes. Since VP16 can be immediately detected in cells, it also provides an alternative mechanism to verify HSV internalization. Therefore, to confirm our results obtained using a reporter virus, which could be subjective to the enzymatic activity of the substrate and the translation of the reporter gene, we also focused on detecting VP16 by Western blot analysis (Antoine et al. 2012). Two hours post-infection HeLa cells were collected and total lysates were analyzed to determine the effects of ZnOT on the internalization of VP16. Cells were subjected to neutralization, prophylaxis, or therapeutic treatment. VP16 expression was significantly ($P < 0.001$) decreased following neutralization, prophylaxis, and therapeutic treatment (Fig. 14.5). Under neutralization condition, the effect of UV ZnOTs was most effective in reducing the internalization of the viruses. The oxygen vacancies produced during the UV treatment increases the attraction between ZnOT and the viruses, thus enhancing the viral trapping ability of UV ZnOT in comparison to NUV ZnOT (Antoine et al. 2012). The prophylaxis treatment also resulted in a comparable decrease in HSV-2 (333)

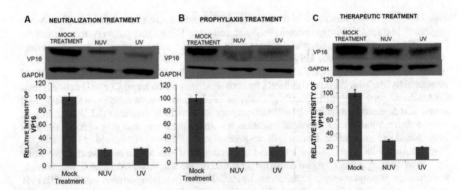

Fig. 14.5 Future utilization of ZnO-based tetrapod structures (ZnOT) as a prevention and a therapy against genital herpes (HSV-2) infections. Using cell culture model with a reporter based HSV-2 virus, treatment with ZnOT resulted decrease in virus internalization suggesting neutralization, prophylaxis, or therapeutic options with nanoparticles (NPs). Western blot analysis of VP16 expression was performed to determine the effect of ZnOT on HSV-2 internalization. As indicated, VP16 protein expression was determined for neutralization, prophylaxis, therapeutic, or mock-treated cells infected with wild-type HSV-2(333). The cell lysates were prepared at 2 h post-infection and Western blots were performed. VP 16 expression and relative protein intensity are shown. (a) Neutralization, (b) prophylactic, and (c) therapeutic treatment. GAPDH was measured as a loading control. Results are representative of three independent experiments (Adapted from Antoine et al. 2012; *Antiviral Res*)

internalization in both UV and NUV ZnOTs conditions (Fig. 14.5). The decreased internalization, however, supports the preventative function of ZnOT against the ability of viruses to enter into susceptible cells. Lastly, the therapeutical usage of ZnOTs was found to decrease the viral internalization as tegument protein VP16 expression was significantly decreased following UV and NUV ZnOT treatments (Fig. 14.5) (Antoine et al. 2012).

2.5 ZnOT Reduces Infectious Cell Cluster and Syncytia Formation

The ZnO NPs have also been sued to investigate if it has inhibitory effect on the syncytia formation. Using a reporter-based HSV-2(333) GFP virus we determine the effect of ZnOT on viral replication and cell-to-cell spread. Overall, we have found that the cellular expression of GFP and its distribution mark the ability of HSV-2 to form clusters of infected cells under various treatment conditions. Reduced clusters were noted in all conditions where ZnOT was present. Using a confocal microscope syncytia formation was recorded 24 h post-mixing suggesting the potent effect of NPs on viral spread (Antoine et al. 2012).

3 Effect of ZnO NPs in Corneal HSV Infection

HSV-1 is a leading cause of infectious blindness in developed countries, which causes keratitis, resulting in loss of vision. Although the infections are usually treatable, corneal epithelial keratitis can become devastating causing dendritic ulceration, corneal neovascularization, scarring, and eventual blindness if left untreated or infection with the drug-resistant strains. Since the virus has ability to cause recurrent infection, and therefore there is also a need to develop a model system that not only facilitates testing of novel drugs in a timely manner but also accurately replicates the human host environment to study both viral cycles. Therefore, cornea based on ex vivo model was developed to visualize the spread of HSV-1 infection. Using a β-galactosidase reporter virus infection in the cornea using x-gal staining represents a highly specific and rapid method for confirming the active infection of corneal tissue with HSV-1 by visual inspection (Mockli and Auerbach 2004; Li et al. 2012). Additionally, β-galactosidase is not expressed during the latent viral phase but is reexpressed during viral reactivation in secondary infected cells, providing an avenue for studying the latency establishment and reactivation mechanisms of HSV-1 (Lachmann et al. 1999; Shimeld et al. 2001; Summers et al. 2001). This was an important development in studying the ability of HSV-1 to penetrate corneal tissue, as use of reliable ex vivo tissue models is often more efficient and ethical in comparison to in vivo models. Given this, ex vivo models have been shown to be promising for developing next-generation methods of drug delivery to combat lytic and latent HSV-1 infection

(Agarwal and Rupenthal 2016). UV-treated zinc oxide tetrapods (ZnOT) were used to test antiviral properties against the HSV-1(KOS)tk12 strain virus. Neutralization treatment, in which virus and the ZnOTs were incubated before infection, statistically significantly reduced the amount of β-galactosidase expression, as evidenced by the decreased presence of blue staining in the pig cornea epithelium and the HCE cells. Samples infected with the zinc oxide/ HSV-1(KOS)tk12 virus showed statistically significantly less blue coloring in the in vitro and ex vivo samples compared with the samples infected with HSV-1(KOS)tk12 alone. Quantification of X-gal via the amount of blue present in these samples was completed by using MetaMorph software to compare the infection visualized infected versus the infected plus ZnOT-treated samples. This represented a more objective method for determining the ability of zinc oxide to combat HSV-1 insult than simple visual assessment. To assess the antiviral effect of the UV zinc oxide tetrapods on proteins necessary for the spread of HSV-1, quantification of viral glycoproteins, gB (HCE cells) and gD (corneas), was undertaken. The infected samples were compared with the infected plus zinc oxide tetrapod-treated samples to assess the levels relative to GAPDH. Consistent with the decrease seen in β-galactosidase expression, these glycoproteins were significantly reduced in the in vitro (HCE cells) and ex vivo (cornea tissue) neutralization treatment samples confirming the antiviral potential of ZnOT.

4   Broad-Spectrum Antiviral Effects of Zinc Oxide Tetrapod (ZnOT) Structures against Herpes Simplex Virus Type-2 (HSV-2) Infection

HSV-2 is one of the most frequent sexually transmitted infections worldwide with global estimates of 536 million infected people and an annual incidence of 23.6 million cases (Tronstein et al. 2011). HSV-2 is the prototype of the neurotropic alpha-herpesviruses, all of which cause latency in sacral root ganglion (Avitabile et al. 2007). The initial experiments were conducted with using both UV-treated ZnOT and (n)UV ZnOT treatment on the viability of HeLa and human vaginal epithelial (VK2/E6) cells using MTS cytotoxicity assay. A concentration range from 0.1 to 1.5 mg/ml was examined and the highest ZnOT concentration resulted in up to 40 percent loss of viability after the 24-h treatment. Viability assays also showed comparable effects on by both UV and NUV ZnOT. The concentration of ZnOT during the remainder of experiments was kept at a nontoxic concentration of 0.1 mg/ml throughout the study to preserve cell health during the investigation. The structures of ZnOT were synthesized by flame transport approach and verified using correspond SEM images at different magnifications. SEM images show the tetrapod morphologies and homogeneity of synthesized structures. The diameters of tetrapod arms vary in the range from 200 nm to 2 μm and their lengths vary from 5 to 25 μm which have been successfully reproduced several times as per requirements.

To establish the broad spectrum usage of ZnOT structures during different stages of infection, HeLa and V2K/E6 cells were infected at a concentration of 0.1 mg/ml under three different conditions: neutralization, prophylactic and therapeutic treatment. Each group was contained with UV ZnOTs and NUV ZnOTs. Since UV-treated ZnO structures have been reported to show the enhanced antiviral properties (Wiesenthal et al. 2011), both UV and NUV ZnOTs were examined to determine their antiviral properties against HSV-2. Following either neutralization treatment in which the virus and ZnOTs were incubated first, prophylaxis treatment (pretreatment of cells with ZnOTs prior to infection), therapeutic treatment (viral infection prior to the addition of ZnOTs), or mock treatment (infection in the absence of ZnOTs), entry of a β-galactosidase-expressing reporter virus, HSV-2(333) gJ−, was determined. The experiment with the neutralization treatment resulted in a significant decrease in β-galactosidase expression in HeLa cells and vaginal epithelial cells when treated with both UV and NUV ZnOTs. UV neutralization however resulted in more pronounced decrease of β-galactosidase expression. The prophylactic treatment effect on reporter virus strain HSV-2(333) gJ- also resulted in a significant decrease in HSV-2 entry in UV ZnOT-treated cells suggesting an enhanced effect of the UV-treated ZnOT on the virus. Lastly, the effectiveness of ZnOT in the presence of an active infection was studied through the therapeutic treatment. This condition allowed us to study the inhibiting effects of ZnOTs after viral entry. It has already been shown that HSV is able to enter the cells within the first 10 min of infection at 37 °C (Cheshenko et al. 2003); therefore, we allowed the infection to persist for 30 min prior to the addition ZnOTs. Therapeutic treatment with ZnOTs resulted in a decrease in viral entry as expression of β-galactosidase was reduced in UV and NUV ZnOT-treated cells; reduction in entry was not as significant as seen in other treatment groups.

The other mechanism that contributes to the pathogenesis of HSV-2 is its ability to infect the neighboring cells without diffusing through the extracellular environment; this is known as cell-to-cell fusion. Through this process, the virus spreads rapidly while evading detection by the immune system (Fischer et al. 2001; Sattentau 2008). The cell-to-cell spread of viruses is mediated through the coordinated efforts of surface-exposed glycoproteins from infected cells that contact their specific receptors on neighboring uninfected cells. Through the specific glycoprotein-receptor interactions, the cells fuse and generate multinucleated cells. Since ZnOTs treatment resulted in a decreased viral entry and internalization, we decided to investigate its effect on virus-free cell-to-cell fusion. To create this scenario in vitro, CHO-K1 cells were split into two populations: target cells and effector cells (Tiwari et al. 2009a, b). Target cells were transfected with the gD receptor Nectin-1 and the luciferase reporter gene under the control of the T7 promoter. Effector cells were transfected HSV-2 glycoproteins essential for cell fusion and T7 polymerase. The fusion between both populations of cells allows the T7-polymerase to bind its promoter, thus initiating the synthesis of the luciferase gene. By addition of the substrate, firefly luciferase, we can analyze the amount of fusion that has occurred between the two populations of cells.

5   Zinc Oxide Nanoparticles: A Platform for Future Live Virus Vaccine

Since the zinc oxide-based NPs show potent activity against virus entry and virus neutralization (Mishra et al. 2011; Antoine et al. 2012), the studies were further undertaken to address if they can be exploited to trigger the localized immune response for the future development of a live virus vaccine, which in turn may reduce infection without compromising local immune responses. Using a mouse model of HSV-2 infection, it was clearly visible that HSV-2 treated with ZnO NPs had comparable morphology and thickness in the vaginal epithelium with respect to the mock mice. Interestingly, ZnONPs/HSV-2-infected mice exhibited significantly larger draining lymph nodes than mock infected, comparable to HSV-2-infected mice without treatment, providing an indirect evidence that the presence of ZnO NPs mediates an immune response similar to non-treated infection. The elicited immune response by ZnO NPs/HSV-2 also decrease local cell infiltration and inflammation and, therefore, results in a global decrease of pathogenesis (Alex Agelidis et al. 2019). A trend of decreased infiltration of CD45+, Gr-1+, and F4/80+ cells was observed in the vaginal tissue upon ZnO NPs/HSV-2 genital infection. ZnO NPs also restored basal levels of CD49b + and CD11c + cells. It is understood that neutrophils (expressing Gr-1) are a major component of the innate inflammatory infiltrate at the primary site of herpes infection (Wang et al. 2012). The observed trends of decreased CD45+ and Gr-1+ infiltrating cells in the vaginal epithelium in addition to lower levels of proinflammatory IL-1β transcripts further demonstrates the decreased local inflammation observed in ZnO NPs/HSV-2-infected mice. Therefore, it is clear that ZnO NPs treatment in mice activates immune response, which in turn suppresses clinical manifestation in the mice. Since NPs also responds well against both the HSV-1 and HSV-2 viruses, a major future application of NPs may herald the development of a potent microbicide as well as the strategy with the live virus vaccine.

6   High Impact of Emerging Designer NPs in the Treatment of Human Viral Infections

The use of NP dendrimers represents nanostructures with great promise against HIV (Domenech et al. 2010; Tyssen et al. 2010). Emulsified NPs containing inactivated influenza virus and CpG nucleotides are being tested for the development of prophylactic vaccines (Huang et al. 2010). Use of gold NPs to deliver ssRNA in the form of nanoplex to inhibit H1N1 influenza virus replication has been demonstrated (Chakravarthy et al. 2010). Recently, biodegradable NPs carrying HIV gp120 showed protective effects in rhesus macaques (Himeno et al. 2010). NP-based vaginal drug delivery system has also been proposed for HIV prevention (Mallipeddi and Rohan 2010). Gold or Silver NPs against HIV (Elechiguerra et al. 2005;

Bowman et al. 2008) or modified Ag-NPs against monkey pox virus, hepatitis B virus, and HSV-1 have all shown strong inhibitory effects (Baram-Ointo et al. 2009; Rogers et al. 2008; Liu et al. 2008). Recently, a novel cellular nanosponge was reported to be an effective medical countermeasure to the SARS-CoV-2 virus (Zhang et al. 2020). These nanosponges display the same protein receptors, both identified and unidentified, required by SARS-CoV-2 for cellular entry. Therefore, a neutralized effect was visible in the presence of nanosponges for future possible future antiviral applications (Zhang et al. 2020).

7 Future Innovations: Construction of NPs Ligated with Anti-HSV-1 gB/gD Peptides to Develop "Virus-Specific Capturing" Properties of ZnO

ZnO-NP in concept is similar to a vacuum cleaner that traps dust particles on the basis of charge distribution. Our results show that the anti-HSV-1 virostatic activity was further enhanced by the addition of extra negative charge by UV-light treatment by creating oxygen vacancies within the material (Mishra et al. 2011; Antoine et al. 2012). One NP with negatively charged nanospikes can potentially trap multiple viruses. In addition, further structural variations introduced into NPs can fine-tune antiviral and cell survival activities. For instance, each HSV-1 virion contains 600–750 glycoprotein moieties with variable packing densities (Grunewald et al. 2003). Conjugation of anti-HSV-1 gB/gD peptide to ZnO-NPs will make them more potent and highly specific by targeting multiple critical sites embedded in HSV gB and gD, which are involved in cell interaction during viral entry.

8 Conclusions

Viruses are obligate intracellular parasites whose interactions with host cells often comprise a variety of receptor-ligand interactions. The intrinsic characteristics of viral disease, which include complexities in life cycles, different stages of replication in different sub-cellular compartments or organelles, differences in replication dynamics, the possibility of latent infection in inaccessible biological compartments, and the development of drug resistance, all result in unique requirements for drug design. Our past work has demonstrated that nanoparticles provide an attractive platform for drug design to prevent viral entry and cell-to-cell spread.

Acknowledgments The authors sincerely thank the University of Illinois at Chicago (UIC), University of Southern Denmark, and the Midwestern University, Downers Grove, (IL) for facilitating a joint collaborative research project.

References

Agarwal, P., & Rupenthal, I. D. (2016). In vitro and ex vivo corneal penetration and absorption models. *Drug Delivery and Translational Research, 6,* 634–647.

Agelidis, A., Koujah, L., Suryawanshi, R., Yadavalli, T., Mishra, Y. K., Adelung, R., & Shukla, D. (2019). An intra-vaginal zinc oxide tetrapod nanoparticles (ZOTEN) and genital herpesvirus cocktail can provide a novel platform for live virus vaccine. *Frontiers in Immunology, 10,* 500.

Aliaksandr, V. K., Kuzmin, A. N., Nyk, M., Roy, I., & Prasad, P. N. (2008). Zinc oxide nanocrystals for non-resonant nonlinear optical microscopy in biology and medicine. *The Journal of Physical Chemistry. C, Nanomaterials and Interfaces, 112*(29), 10721–10724.

Antoine, T. E., Mishra, Y. K., Trigilio, J., Tiwari, V., Adelung, R., & Shukla, D. (2012). Prophylactic, therapeutic and neutralizing effects of zinc oxide tetrapod structures against herpes simplex virus type-2 infection. *Antiviral Research, 96*(3), 363–375.

Anzivino, E., Fioriti, D., Mischitelli, M., et al. (2009). HSV infection in pregnancy and in neonate: Status of art of epidemiology, diagnosis, therapy and prevention. *Virology Journal, 6*(1), 40.

Avitabile, E., Forghieri, C., & Campadelli-Fiume, G. (2007). Complexes between herpes simplex virus glycoproteins gD, gB, and gH detected in cells by complementation of split enhanced green fluorescent protein. *Journal of Virology, 81,* 11532–11537.

Baram-Ointo, D., Shukla, S., Perkas, N., Gedanken, A., & Sarid, R. (2009). Inhibition of herpes simplex virus type-1 infection by sliver nanoparticles capped with mercaptoethane sulfornate. *Biconjugate Chemistry, 20,* 1497–1502.

Baram-Pinto, D., Shukla, S., Gedanken, A., & Sarid, R. (2010). Inhibition of HSV-1 attachment, entry, and cell-to-cell spread by functionalized multivalent gold nanoparticles. *Small, 6*(9), 1044–1050.

Bowman, M. C., Ballard, T. E., Ackerson, C. J., Feldheim, D. L., Margolis, D. M., & Melander, C. (2008). Inhibition of HIV fusion with multivalent gold nanoparticles. *Journal of the American Chemical Society, 130*(22), 6896–6897.

Britt, W. (2008). Manifestations of human cytomegalovirus infection: Proposed mechanisms of acute and chronic disease. In T. Shenk & M. Stinski (Eds.), *Current topics in microbiology and immunology: Human cytomegalovirus* (Vol. 325, pp. 417–470). Berlin and Heidelberg: Springer-Verlag.

Burgos, J. S., Ripoll-Gomez, J., Alfaro, J. M., Sastre, I., & Valdivieso, F. (2008). Zebrafish as a new model for herpes simplex virus type 1 infection. *Zebrafish, 5*(4), 323–333.

Campadelli-Fume, G., & Menotti, L. (2007). In A. Arvin, G. Campadelli-Fumi, E. Mocarski, P. S. Morre, B. Roizman, R. Whitley, & K. Yamanishi (Eds.), Chapter 777 *Entry of alphaherpesviruses into cell, in human herpepesviruses: Biology, therpay, and immunoprophylaxis* (pp. 93–111). New York: Cambridge.

Chakravarthy, K. V., Bonoiu, A. C., Davis, W. G., Ranjan, P., Ding, H., Hu, R., Bowzard, J. B., Bergey, E. J., Katz, J. M., Knight, P. R., Sambhara, S., & Prasad, P. N. (2010). Gold nanorod delivery of an ssRNA immune activator inhibits pandemic H1N1 influenza viral replication. *Proceedings of the National Academy of Sciences, 107*(22), 10172–10177.

Cheshenko, N., Del Rosario, B., Woda, C., Marcellino, D., Satlin, L. M., & Herold, B. C. (2003). Herpes simplex virus triggers activation of calcium-signaling pathways. *The Journal of Cell Biology, 163,* 283–293.

Clement, C., Tiwari, V., Scanlon, P., Vali-Nagy, T., Yue, B. Y. J. T., & Shukla, D. (2006). A novel for phagocytosis-like uptake in herpes simplex virus entry. *Journal of Cell Biology, 174,* 1009–1021.

Connolly, S. A., Jackson, J. O., Jardetzky, T. S., & Longnecker, R. (2011). Fusing structure and function: A structural view of the herpesvirus entry machinery. *Nature Reviews. Microbiology, 9*(5), 369–381. https://doi.org/10.1038/nrmicro2548. Epub 2011 . Apr 11.

Copland, R., Balasubramaniam, A., Tiwari, V., Zhang, F., Bridges, A., Linhardt, R., Shukla, D., & Liu, J. (2008). Using a 3-O-sulfated heparin octasaccharide to inhibit the entry of herpes simplex virus-1. *Biochemistry, 47,* 5774–5783.

De, M., Ghosh, P. S., & Rotello, V. M. (2008). Applications of nanoparticles in biology. *Advanced Materials, 20*, 4225–4241.

Dean, H. J., Terhune, S., Shieh, M. T., Susmarski, N., & Spear, P. G. (1994). Single amino acid substitutions in gD of herpes simplex virus 1 confer resistant to gD-mediated interference and cause cell type-dependent alterations in infectivity. *Virology, 199*, 67–80.

Di Gianvincenzo, P., Chiodo, F., Marradi, M., & Penadés, S. (2012). Gold manno-glyconanoparticles for intervening in HIV gp120 carbohydrate-mediated processes. *Methods in Enzymology, 509*, 21–40.

Domenech, R., Abian, O., Bocanegra, R., Correa, J., Sousa-Herves, A., Riguera, R., Mateu, M. G., Fernandez-Megia, E., Velazquez-Campoy, A., & Neira, J. L. (2010). Dendrimers as potential inhibitors of the dimerization of the capsid protein of HIV-1. *Biomacromolecules, 11*, 2069–2078.

Donathan, G. B., & Meyer, T. A. (2010). Characterization of the UVA protection provided by avobenzone, zinc oxide, and titanium dioxide in broad-spectrum sunscreen products. *American Journal of Clinical Dermatology, 11*(6), 413–421.

Eisenberg, R. J., Atanasiu, D., Cairns, T. M., Gallagher, J. R., Krummenacher, C., & Cohen, G. H. (2012). Herpes virus fusion and entry: A story with many characters. *Viruses, 4*(5), 800–832.

Ekblad, M., Adamiak, B., Bergstrom, T., Johnstone, K. D., Karoli, T., Liu, L., Ferro, V., & Trybala, E. (2010). A highly lipophilic sulfated tetrasaccharide glycoside related to muparfostat (PI-88) exhibits virucidal activity against herpes simplex virus. *Antiviral Research, 86*(2), 196–203.

Elechiguerra, J. L., Burt, J. L., Morones, J. R., Camacho-Bragado, A., Gao, X., Lara, H. H., & Yacaman, M. J. (2005). Interaction of silver nanoparticles with HIV-1. *Journal of Nanbiotechnology, 3*, 6.

Eugenii, K., & Willner, I. (2004). Integrated nanoparticles-biomolecule hybrid systems: Synthesis, properties, and applications. *Angewandte Chemie, International Edition, 43*, 6042–6108.

Farooq, A. V., Valyi-Nagy, T., & Shukla, D. (2011). Mediators and mechanisms of herpes simplex virus entry into ocular cells. *Current Eye Research, 1*. Author manuscript; available in PMC.

Fatahzadeh, M., & Schwartz, R. A. (2007). Human HSV infections: Epidemiology, pathogenesis, symptomatology, diagnosis, and management. *Journal of the American Academy of Dermatology, 57*(5), 737–763.

Fischer, N. O., Mbuy, G. N., & Woodruff, R. I. (2001). HSV-2 disrupts gap junctional intercellular communication between mammalian cells in vitro. *Journal of Virological Methods, 91*, 157–166.

Gangji, R. N., Sankaranarayanan, N. V., Elste, J., Al-Horani, R. A., Afosah, D. K., Joshi, R., Tiwari, V., & Desai, U. R. (2018). Inhibition of herpes simplex Virus-1 entry into human cells by non-saccharide glycosaminoglycan mimetics. *ACS Medicinal Chemistry Letters, 9*(8), 797–802.

Grunewald, K., Desai, P., Winkler, D. C., Heyman, J. B., Belnap, D. M., Baumeister, W., & Steven, A. C. (2003). Three dimensional structure of herpes simplex virus from cryo electron microscopy. *Science, 302*, 1396–1398.

Herold, B. C., Visalli, R. J., Susmarski, N., Brandt, C. R., & Spear, P. G. (1994). Glycoprotein C-independent binding of herpes simplex virus to cells requires cell surface heparan sulphate and glycoprotein B. *The Journal of General Virology, 75*, 1211–1222.

Herold, B. C., WuDunn, D., Soltys, N., & Spear, P. G. (1991). Glycoprotein C of herpes simplex virus type 1 plays a principal role in the adsorption of virus to cells and in infectivity. *Journal of Virology, 65*, 1090–1098.

Himeno, A., Akagi, T., Uti, T., Wang, X., Baba, M., Ibuki, K., Matsuyama, M., Horiike, M., Igarashi, T., Miura, T., & Akashi, M. (2010). Evaluation of the immune response and protective effects of the rhesus macaques vaccinated with biodegradable nanoparticles carrying gp120 of human immunodeficiency virus. *Vaccine, 28*(32), 5377–5385.

Huang, M. H., Lin, S. C., Hsiao, C. H., Chao, H. J., Yang, H. R., Liao, C. C., Chuang, P. W., Wu, H. P., Huang, C. Y., Leng, C. H., Liu, S. J., Chen, H. W., Chou, A. H., Hu, A. Y., & Chong, P. (2010). Emulsified nanoparticles containing inactivated influenza virus and CpG oligodeoxyneculeotides critically influences the host immune response in mice. *PLoS One, 5*, e12279.

Jihan, A., & Shukla, D. (2009). Viral entry mechanisms: Cellular and viral mediators of herpes simplex virus entry. *The FEBS Journal, 276*(24), 7228–7236.

Kong, J., Chu, S., Olmedo, M., Li, L., Yang, Z., & Liu, J. (2008). Dominant ultraviolet light emissions in packed ZnO columnar homojunction diodes Appl. *Physics Letters, 93*, 132113.

Lachmann, R. H., Sadarangani, M., Atkinson, H. R., & Efstathiou, S. (1999). An analysis of herpes simplex virus gene expression during latency establishment and reactivation. *The Journal of General Virology, 80*, 1271–1282.

Lara, H. H., Ayala-Nuñez, N. V., Ixtepan-Turrent, L., & Rodriguez-Padilla, C. (2010). Mode of antiviral action of silver nanoparticles against HIV-1. *Journal of Nanobiotechnology, 20*(8), 1.

Li, W., Zhao, X., Zou, S., Ma, Y., Zhang, K., & Zhang, M. (2012). Scanning assay of beta-galactosidase activity. *Prikladnaia Biokhimiia i Mikrobiologiia, 48*, 668–672.

Liu, J., & Thorp, S. C. (2002). Cell surface heparan sulfate and its roles in assisting viral infections. *Medicinal Research Reviews, 22*, 1–25.

Liu, L., Sun, R. W., Chen, R., Hui, C. K., Hui, C. M., Ho, C. M., Luk, J. M., Lau, G. K., & Che, C. M. (2008). Silver nanoparticles inhibits hepatitis B virus replication. *Antiviral Theraphy, 13*, 253–262.

Majmudar, H., Hao, M., Sankaranarayanan, N. V., Zanotti, B., Volin, M. V., Desai, U. R., & Tiwari, V. (2019). A synthetic glycosaminoglycan mimetic blocks HSV-1 infection in human iris stromal cells. *Antiviral Research, 161*, 154–162.

Mallipeddi, R., & Rohan, L. C. (2010). Nanoparticle-based vaginal drug delivery system for HIV prevention. *Expert Opinion on Drug Delivery, 7*(1), 37–48.

McConville, C., Boyd, P., & Major, I. (2014). Efficacy of Tenofovir 1% vaginal gel in reducing the risk of HIV-1 and HSV-2 infection. *Clinical Medicine Insights Womens Health, 7*, 1–8. eCollection 2014. Review. PMID: 24741339.

Meque, I., Dubé, K., Feldblum, P. J., Clements, A. C., Zango, A., Cumbe, F., Chen, P. L., Ferro, J. J., & van de Wijgert, J. H. (2014). Prevalence, incidence and determinants of herpes simplex virus type 2 infection among HIV-seronegative women at high-risk of HIV infection: A prospective study in Beira, Mozambique. *PLoS One, 9*(2), e89705. https://doi.org/10.1371/journal.pone.0089705. eCollection 2014. PMID: 24586973.

Mishra, Y. K., Adelung, R., Röhl, C., Shukla, D., Spors, F., & Tiwari, V. (2011). Virostatic potential of micro-nano filopodia-like ZnO structures against herpes simplex virus-1. *Antiviral Research, 92*(2), 305–312.

Mockli, N., & Auerbach, D. (2004). Quantitative beta-galactosidase assay suitable for high-throughput applications in the yeast two-hybrid system. *BioTechniques, 36*, 872–876.

Montgomery, R. I., Warner, M. S., Lum, B. J., & Spear, P. G. (1996). Herpes simplex virus-1 entry into cells mediated by a novel member of the TNF/NGF receptor family. *Cell, 87*(3), 427–436.

Nicoll, M. P., Proença, J. T., & Efstathiou, S. (2012). The molecular basis of herpes simplex virus latency. *FEMS Microbiology Reviews, 36*(3), 684–705.

Pan, D., Kaye, S. B., Hopkins, M., Kirwan, R., Hart, I. J., & Coen, D. M. (2014). Common and new acyclovir resistant herpes simplex virus-1 mutants causing bilateral recurrent herpetic keratitis in an immunocompetent patient. *The Journal of Infectious Diseases, 209*(3), 345–349. https://doi.org/10.1093/infdis/jit437. Epub 2013 Aug 14. PMID: 23945375.

Rasmussen, J. W., Martinez, E., Louka, P., & Wingett, D. G. (2010). Zinc oxide nanoparticles for selective destruction of tumor cells and potential for drug delivery applications. *Expert Opinion on Drug Delivery, 7*(9), 1063–1077.

Rogers, J. V., Parkinson, C. V., Choi, Y. W., Speshock, J. L., & Hussain, S. M. (2008). A prelimary assessment of silver nanoparticle inhibition of monkey pox virus plaque formation. *Nanoscale Research Letters, 3*, 129–133.

Roizman, B., Knipe, D. M., & Whitley, R. J. (2007). Herpes simplex viruses. In D. M. Knipe & P. M. Howley (Eds.), *Fields Virology* (5th ed., pp. 2503–2602). Baltimore, MD: Lippincott Williams & Wilkins.

Roizman, B., & Sears, A. E. (1996). Herpes simplex viruses and their replication. In B. N. Fields, D. M. Knipe, R. M. Chanock, M. S. Hirsch, J. L. Melnick, T. P. Monath, & B. Roizman (Eds.), *Virology. 3* (Vol. 2, pp. 2231–2295). Philadelphia, PA: Lippincott-Raven Publishers.

Rollenhagen, C., Lathrop, M. J., Macura, S. L., Doncel, G. F., & Asin, S. N. (2014). Herpes simplex virus type-2 stimulates HIV-1 replication in cervical tissues: Implications for HIV-1 transmission and efficacy of anti-HIV-1 microbicides. *Mucosal Immunology*. https://doi.org/10.1038/mi.2014.3. [Epub ahead of print] PMID: 24496317.

Sattentau, Q. (2008). Avoiding the void: Cell-to-cell spread of human viruses. Nature reviews. *Microbiology, 6*, 815–826.

Schepers, K., Hernandez, A., Andrei, G., Gillemot, S., Fiten, P., Opdenakker, G., Bier, J. C., David, P., Delforge, M. L., Jacobs, F., & Snoeck, R. (2014). Acyclovir-resistant herpes simplex encephalitis in a patient treated with anti-tumor necrosis factor-α monoclonal antibodies. *Journal of Clinical Virology, 59*(1), 67–70. https://doi.org/10.1016/j.jcv.2013.10.025. Epub 2013 Oct 31. PMID: 24257111.

Schulte, E. C., Sauerbrei, A., Hoffmann, D., Zimmer, C., Hemmer, B., & Mühlau, M. (2010). Acyclovir resistance in herpes simplex encephalitis. *Annals of Neurology, 67*(6), 830–833.

Shimeld, C., Efstathiou, S., & Hill, T. (2001). Tracking the spread of a lacZ-tagged herpes simplex virus type 1 between the eye and the nervous system of the mouse: Comparison of primary and recurrent infection. *Journal of Virology, 75*, 5252–5262.

Shukla, D., & Spear, P. G. (2001). Herpesviruses and heparan sulfate: An intimate relationship in aid of viral entry. *The Journal of Clinical Investigation, 108*(4), 503–510.

Spear, P. G. (1993). Entry of alphaherpesviruses into cells. *Seminars in Virology, 4*, 167–180.

Spear, P. G., & Longnecker, R. (2003). Herpesvirus entry: An update. *Journal of Virology, 77*(19), 10179–10185.

Summers, B. C., Margolis, T. P., & Leib, D. A. (2001). Herpes simplex virus type 1 corneal infection results in periocular disease by zosteriform spread. *Journal of Virology, 75*, 5069–5075.

Superti, F., Ammendolia, M. G., & Marchetti, M. (2008). New advances in anti-HSV chemotherapy. *Current Medicinal Chemistry, 15*(9), 900–911.

Tallury, P., Malhotra, A., Byrne, L. M., & Santra, S. (2010). Nanobioimaging and sensing of infectious diseases. *Advanced Drug Delivery Reviews, 62*(4–5), 424–437.

Thakkar, J. N., Tiwari, V., & Desai, U. R. (2010). Nonsulfated, cinnamic acid-based lignins are potent antagonists of HSV-1 entry into cells. *Biomacromolecules, 11*(5), 1412–1416.

Tiwari, V., Clement, C., Xu, D., Valyi-Nagy, T., Yue, B. Y., Liu, J., & Shukla, D. (2006). Role for 3-O-sulfated heparan sulfate as the receptor for herpes simplex virus type 1 entry into primary human corneal fibroblasts. *Journal of Virology, 80*(18), 8970–8980. https://doi.org/10.1128/JVI.00296-06.

Tiwari, V., Darmani, N. A., Thrush, G. R., & Shukla, D. (2009a). An unusual dependence of human herpesvirus-8 glycoproteins-induced cell-to-cell fusion on heparan sulfate. *Biochemical and Biophysical Research Communications, 390*, 382–387.

Tiwari, V., Shukla, S. Y., & Shukla, D. (2009b). A sugar binding protein cyanovirin-N blocks herpes simplex virus type-1 entry and cell fusion. *Antiviral Research, 84*(1), 67–75. Epub 2009 Aug 7.

Travan, A., Marsich, E., Donati, I., Benincasa, M., Giazzon, M., Felisari, L., & Paoletti, S. (2011). Silver-polysaccharide nanocomposite antimicrobial coatings for methacrylic thermosets. *Acta Biomaterialia, 7*(1), 337–346.

Tronstein, E., Johnston, C., Huang, M. L., Selke, S., Magaret, A., Warren, T., Corey, L., & Wald, A. (2011). Genital shedding of herpes simplex virus among symptomatic and asymptomatic persons with HSV-2 infection. *JAMA, 305*, 1441–1449.

Tyssen, D., Henderson, S. A., Johnson, A., Sterjovski, J., Moore, K., La, J., Zanin, M., Sonza, S., Karellas, P., Giannis, M. P., Krippner, G., Wesselingh, S., McCarthy, T., Gorry, P. R., Ramsland, P. A., Cone, R., Paull, J. R., Lewis, G. R., & Tachedjian, G. (2010). Structure activity relationship of dendrimer microbicides with dual action antiviral activity. *PLoS One, 5*(8) [Epub ahead of print].

Varga, N., Sutkeviciute, I., Ribeiro-Viana, R., Berzi, A., Ramdasi, R., Daghetti, A., Vettoretti, G., Amara, A., Clerici, M., Rojo, J., Fieschi, F., & Bernardi, A. (2014). A multivalent inhibitor of the DC-SIGN dependent uptake of HIV-1 and dengue virus. *Biomaterials, 35*(13), 4175–4184. https://doi.org/10.1016/j.biomaterials.2014.01.014. Epub 2014 Feb 6.

Vig, K., Boyoglu, S., Rangari, V., Sun, L., Singh, A., Pillai, S., & Singh, S. R. (2008). Use of nanoparticles as therapy for respiratory syncytial virus inhibition. *Nanotechnology.* Life Sciences, Medicine & Bio Materials - Technical Proceedings of the 2008 NSTI Nanotechnology Conference and Trade Show; 2008. pp. 543 – 546.

Waldmann, M., Jirmann, R., Hoelscher, K., Wienke, M., Niemeyer, F. C., Rehders, D., & Meyer, B. (2014). A nanomolar multivalent ligand as entry inhibitor of the hemagglutinin of avian influenza. *Journal of the American Chemical Society, 136*(2), 783–788. https://doi.org/10.1021/ja410918a. Epub 2014 Jan 7.PMID: 24377426.

Wang, J. P., Bowen, G. N., Zhou, S., Cerny, A., Zacharia, A., Knipe, D. M., et al. (2012). Role of specific innate immune responses in herpes simplex virus infection of the central nervous system. *Journal of Virology, 86*, 2273. https://doi.org/10.1128/JVI.06010-11.

Wertheim, J. O., Smith, M. D., Smith, D. M., Scheffler, K., & Pond, S. L. K. (2014). Evolutionary origins of human herpes simplex viruses 1 and 2. *Molecular Biology and Evolution, 31*(9), 2356–2364.

Whaley, K. J., Hanes, J., Shattock, R., Cone, R. A., & Friend, D. R. (2010). Novel approaches to vaginal delivery and safety of microbicides: Biopharmaceuticals, nanoparticles, and vaccines. *Antiviral Research, 88*(Suppl 1), S55–S66. https://doi.org/10.1016/j.antiviral.2010.09.006. PMID: 21109069.

Whitley, R. J., Kimberlin, D. W., & Roizman, B. (1998). Herpes simplex viruses. *Clinical Infectious Diseases, 26*, 541–553.

Whitley, R. J., & Roizman, B. (2001). HSV infections. *Lancet, 357*(9267), 1513–1518.

Wiesenthal, A., Hunter, L., Wang, S., Wickliffe, J., & Wilkerson, M. (2011). Nanoparticles: Small and mighty. *International Journal of Dermatology, 50*, 247–254.

Wilhelmus, K. R. (1998). Epidemiology of ocular infections. In W. Tasman & E. A. Jaeger (Eds.), *Duane's foundations of clinical ophthalmology* (pp. 1–46). Philadelphia, PA: Lippincott Williams & Wilkins.

Wilson, S. S., Faakioglu, E., & Herold, B. C. (2009). Novel approaches in fighting herpes simplex virus infections. *Expert Review of Anti Infective Therapy, 7*(5), 559–568.

Wu, J. M., & Chen, Y.-R. (2011). Ultraviolet-light-assisted formation of ZnO nanowires in ambient air: Comparison of photoresponsive and photocatalytic activities in zinc hydroxide. *The Journal of Physical Chemistry, 115*(5), 2235–2243.

Zhang, Q., Honko, A., Zhou, J., Gong, H., Downs, S. N., Vasquez, J. H., Fang, R. H., Gao, W., Griffiths, A., & Zhang, L. (2020). Cellular Nanosponges inhibit SARS-CoV-2 infectivity. *Nano Letters*. acs.nanolett.0c02278.

Chapter 15
Nanoparticle Biosynthesis and Interaction with the Microbial Cell, Antimicrobial and Antibiofilm Effects, and Environmental Impact

Rajesh Prabhu Balaraman, Jovinna Mendel, Lauren Flores, and Madhusudan Choudhary

Contents

R. Prabhu Balaraman · J. Mendel · L. Flores · M. Choudhary (✉)
Department of Biological Sciences, Sam Houston State University, Huntsville, TX, USA
e-mail: mxc017@SHSU.EDU

© Springer Nature Switzerland AG 2021   371
N. Sharma, S. Sahi (eds.), *Nanomaterial Biointeractions at the Cellular, Organismal and System Levels*, Nanotechnology in the Life Sciences,
https://doi.org/10.1007/978-3-030-65792-5_15

1 Introduction

Nanotechnology is emerging as a cutting-edge technology in several disciplines of biology and materials science research. The concept of nanotechnology was first presented by Richard Feynmann, and the word nanotechnology was introduced by Prof. Norio Taniguchi. The prefix "nano" refers to the Greek word *nanos*, which means "dwarf" (Hulkoti and Taranath 2014). Nanotechnology involves the synthesis and development of various nanoscale systems, including nanoparticles (NPs). Nanoparticles refers to objects or engineered materials that range in sizes from 1 to 100 nm and exist in different shapes (Hasan 2015). NPs are currently being used widely in cancer therapy and drug delivery treatments (Zhang et al. 2008; Dobson 2006), food emulsions (Dickinson 2012), solar cells (Fahr et al. 2009), water treatment (Pradeep 2009), and cosmetics industries (Patel et al. 2011). Synthesis of NPs has gained interest because of their electronic, magnetic, optical, thermal, chemical, dielectric, and photo-electrochemical properties. These unique physiochemical properties are attributed to their nanosize of <100 nm, crystallinity, solubility, surface morphology, reactivity, chemical composition, purity, and shapes (Gatoo et al. 2014). Recently, metal NPs made of gold, silver, copper, zinc, titanium, platinum, aluminum, iron, magnesium, cerium, and palladium have gained interest because of their technological importance (Kulkarni and Muddapur 2014).

Nanoparticles are commonly synthesized through two main strategies: bottom-up and top-down approaches. The top-down approach involves the breakdown of bulk materials, and the bottom-up approach comprises the assembly of atoms or molecules into nanosized materials (Vollath 2008). NPs are generally synthesized through chemical and physical methods. However, chemicals absorbed on the surface of the NPs make it undesirable for medical applications due to its toxicity. To overcome these drawbacks, NP synthesis through eco-friendly biological methods using microorganisms, enzymes, fungus, algae, actinomycetes, and plants has gained attention in the areas of green nanotechnology (Gahlawat and Choudhury 2019; Kulkarni and Muddapur 2014; Hulkoti and Taranath 2014). Microbes serve as potential nanofactories for green synthesis of metal NPs and metal oxide NPs in different shapes and forms such as nanowires, nanoparticles, nanotubes, nanoconjugates, and nanorods (Gahlawat and Choudhury 2019). The synthesis of such nanoforms is inexpensive and has shown promising anticancer and antimicrobial properties for biomedical applications (Thanh and Green 2010; Nune et al. 2009).

2 Mechanisms of NP Synthesis Using Microbes

Microorganisms synthesize NPs through intracellular and extracellular mechanisms based on different biological agents. The intracellular synthesis of NPs involves the transportation of a particular ion through the microbial cell wall. The mechanism involves the attachment of the negative charge of the microbial cell wall with the positive charge of the NPs through electrostatic attractions, and the ion reduces in the cell walls by the enzymes present (Hulkoti and Taranath 2014). The mechanisms of producing NPs vary with different microbial systems. The fungal plant pathogen *Verticillium* sp. follows the mechanism of trapping, bioreduction, and capping to synthesize NPs (Mukherjee et al. 2001). In *Lactobacillus* sp., the construction of nanoclusters occurs by the electrostatic interactions between the bacterial cell and nucleated nanoclusters during the early steps (Nair and Pradeep 2002). The extracellular synthesis of metal NPs (particularly silver) using microorganisms generally involves the nitrate reductase mediated mechanism. Several studies have reported that the bioreduction of metal ions occurs through the enzyme nitrate reductase secreted by the microbes (Ingle et al. 2008). The mechanism was proven evident through nitrate reductase assay that indicated the reaction of nitrate with 2,3-diaminophthalene. Fluorescence emission intensity peaks at 405 nm and 490 nm demonstrated the maximum emission of nitrate, and the presence of enzyme nitrate reductase with 0.1% KNO_3 solution has confirmed the reduction of silver in fungi (Kumar et al. 2007; Ingle et al. 2008). A similar mechanism was observed in the bacterium *Rhodopseudomonas capsulata* with the synthesis of gold nanoparticles, where the cofactors NADH and NADH-dependent enzymes secreted by the bacterium assisted with the bioreduction of gold ions for NP synthesis (He et al. 2007; Hulkoti and Taranath 2014).

3   Synthesis of Different Nanoparticles Using Bacteria

Bacteria are considered an ideal candidate for the synthesis of NPs because of its unique properties in reducing metal ions into NPs, possessing high growth rates, and enabling easy handling in laboratory environments (Table 15.1). Compared to other microbes, bacteria can be easily molded and genetically modified for the biomineralization of metal ions (Gahlawat and Choudhury 2019). Besides, bacteria have proven to withstand harsh and toxic environments consisting of high concentrations of heavy metal ions (e.g., *Pseudomonas stutzeri* and *Pseudomonas aeruginosa*) in their surroundings (Kumar and Mamidyala 2011). Bacteria have evolved with defense mechanisms to combat such lethal environments through intracellular sequestration, efflux pumps, extracellular precipitation, and altering metal ions concentration (Iravani 2014). Bacteria exhibit both intracellular and extracellular mechanisms to synthesize NPs. The deposition of gold NPs extracellularly was first reported (Beveridge and Murray 1980) in the unfixed cell wall of *Bacillus subtilis* with the suspension of gold chloride solution. The intracellular synthesis of silver NPs using bacteria was identified in *Pseudomonas stutzeri,* resulting in NPs with a size less than 200 nm. The NADH-dependent reductase enzyme enables the transfer and supply of electrons by oxidizing itself to NAD^+ necessary for reducing silver ions to silver NPs (Klaus-Joerger et al. 2001; Haefeli et al. 1984). The capability of *Pseudomonas aeruginosa* to intracellularly synthesize a wide variety of NPs such as Pd, Ag, Rh, Ni, Fe, Co, Pt, and Li in the absence of external stabilizing agents,

Table 15.1 Biosynthesis of metal nanoparticles using bacteria

Bacteria	Size and shape of NPs	References	Extracellular or intracellular
Gold nanoparticles (AuNPs)			
Bacillus subtilis 168	5–25 nm; octahedral	Beveridge and Murray (1980)	Both
Marinobacter pelagius sp.	20 nm; spherical	Joerger et al. (2000)	Intracellular
Lactobacillus sp.	20–50 nm; hexagonal	Nair and Pradeep (2002)	Intracellular
Pseudomonas aeruginosa	15–30 nm	Husseiny et al. (2007)	Extracellular
Rhodopseudomonas capsulata	10–20 nm; spherical	He et al. (2007)	Extracellular
Escherichia coli	20–25 nm	Deplanche and Macaskie (2008)	Intracellular
Klebsiella pneumoniae	35–65 nm; spherical	Nangia et al. (2009)	Intracellular
Pseudomonas fluorescens	50–70 nm; spherical	Rajasree and Suman (2012)	Extracellular
Stenotrophomonas maltophilia	40 nm; spherical	Sharma et al. (2012)	–
Geobacillus sp.	5–50 nm; quasi-hexagonal	Correa-Llantén et al. (2013)	Intracellular
Silver nanoparticles (AgNPs)			
Acinetobacter calcoaceticus	8–12 nm; spherical	Singh et al. (2013)	–
Aeromonas sp.	6.4 nm	Mouxing et al. (2006)	Both
Bordetella sp.	63–90 nm	Thomas et al. (2012)	Extracellular
Enterobacter aerogenes	25–35 nm; spherical	Karthik and Radha (2012)	Extracellular
Escherichia coli	42–90 nm; spherical	Gurunathan et al. (2009)	Extracellular
Gluconobacter roseus	10 nm	Krishnaraj and Berchmans (2013)	Extracellular
Idiomarina sp.	25 nm	Seshadri et al. (2012)	Intracellular
Klebsiella pneumoniae	15–37 nm; spherical	Kalpana and Lee (2013)	Intracellular
Morganella sp.	10–40 nm; quasispherical	Parikh et al. (2008)	Extracellular
Proteus mirabilis	10–20 nm; spherical	Samadi et al. (2009)	Both
Rhodobacter sphaeroides	3–15 nm; spherical	Bai et al. (2011)	Extracellular
Rhodopseudomonas palustris	5–20 nm; spherical	Chai and Bai (2010)	Intracellular
Shewanella oneidensis	2–16 nm; spherical	Debabov et al. (2013)	–
Xanthomonas oryzae	15 nm; spherical, triangular and rod-shaped	Narayanan and Sakthivel (2013)	Extracellular
Bacillus sp.	5–15 nm	Pugazhenthiran et al. (2009)	Intracellular

(continued)

Table 15.1 (continued)

Bacteria	Size and shape of NPs	References	Extracellular or intracellular
Bacillus flexus	12 and 65 nm; spherical and triangular	Priyadarshini et al. (2013)	Extracellular
Bacillus licheniformis	19–63 nm; spherical	Shanthi et al. (2016)	Extracellular
Bacillus safensis	5–30 nm; spherical	Lateef et al. (2015)	Extracellular
Bacillus methylotrophicus	10–30 nm; spherical	Wang et al. (2016)	Extracellular
Bacillus thuringiensis	44–143 nm; spherical	Banu et al. (2014)	Extracellular
Brevibacterium casei	10–50 nm; spherical	Kalishwaralal et al. (2010)	Extracellular
Exiguobacterium sp.	5–50 nm; spherical	Tamboli and Lee (2013)	Extracellular
Geobacillus stearothermophilus	5–35 nm; spherical	Fayaz et al. (2011)	Extracellular
Lactobacillus mindensis	2–20 nm; spherical	Dhoondia and Chakraborty (2012)	–
Rhodococcus sp.	10–15 nm; spherical	Otari et al. (2014)	Extracellular
Thermoactinomyces sp.	20–40 nm; spherical	Deepa et al. (2013)	Extracellular
Ureibacillus thermosphaericus	10–100 nm; spherical	Juibari et al. (2011)	Extracellular
Pseudomonas meridian	2–21.5 nm; spherical	Shivaji et al. (2011)	Extracellular
Pseudomonas proteolytica	3–23 nm; spherical	Shivaji et al. (2011)	Extracellular
Zinc oxide nanoparticles (ZnO NPs)			
Aeromonas hydrophila	57.72 nm; spherical, oval	Jayaseelan et al. (2012)	–
Lactobacillus sporogenes	5–15 nm; hexagonal	Prasad and Jha (2009)	–
Pseudomonas aeruginosa	35–80 nm; spherical	Singh et al. (2014)	–
Rhodococcus pyridinivorans	100–120 nm; hexagonal	Kundu et al. (2014)	Extracellular
Bacillus licheniformis	400 × 40 nm; nanoflowers	Tripathi et al. (2014)	–
Serratia ureilytica	170–250 nm; spherical to nanoflowers	Dhandapani et al. (2014)	–
Bacillus megaterium	45–95 nm; rod and cubic	Saravanan et al. (2018)	–
Halomonas elongata	10–27 nm; multiform	Taran et al. (2018)	Extracellular
Lactobacillus johnsonii	4–9 nm; spherical	Al-Zahrani et al. (2018)	Extracellular
Lactobacillus paracasei	1100–3000 nm; spherical	Król et al. (2018)	Intracellular
Lactobacillus plantarum	7–19 nm; spherical	Selvarajan and Mohanasrinivasan (2013)	–

(continued)

Table 15.1 (continued)

Bacteria	Size and shape of NPs	References	Extracellular or intracellular
Sphingobacterium thalpophilum	40 nm; triangle	Rajabairavi et al. (2017)	Extracellular
Rhodococcus pyridinivorans	100–120 nm; spherical	Kundu et al. (2014)	Extracellular
Staphylococcus aureus	10–50 nm; acicular	Rauf et al. (2017)	Intracellular
Streptomyces sp.	20–50 nm; spherical	Balraj et al. (2017)	–
Iron-based nanoparticles (Fe₃O₄)			
Thermoanaerobacter ethanolicus	35–65 nm	Yeary et al. (2005)	–
Magnetospirillum magnetotacticum	50–100 nm	Prozorov (2015)	Intracellular
Aquaspirillum magnetotacticum	40–50 nm; octahedral prism	Mann (1985)	Extracellular
Magnetospirillum gryphiswaldense	35–120 nm	Lang and Schüler (2006)	Intracellular
Geobacter metallireducens	10–50 nm	Lovley et al. (1987)	–
Actinobacter sp.	10–40 nm	Bharde et al. (2005)	
Titanium oxide nanoparticles (TiO₂ NPs)			
Bacillus subtilis	66–77 nm; spherical and oval	Kirthi et al. (2011)	Extracellular
Bacillus amyloliquefaciens	22.11–97.28 nm; spherical	Khan and Fulekar (2016)	–
Lactobacillus crispatus	92.65 and 112.26 nm; spherical and oval	Abdulsattar (2014)	–

electron donors, and preventing modifying pH step during biomineralization have also been investigated (Srivastava and Constanti 2012). Different types of NPs synthesized using bacteria with their sizes, shapes, and mechanisms are summarized in Table 15.1 and described in the following paragraphs.

4   Silver Nanoparticles (AgNPs)

In recent years, AgNPs have gained enormous interest because of its wide range of applications in electronics, nanomedicine, energy, biosensors, catalysis, and antimicrobial activities (Srikar et al. 2016). The green synthesis (using algae, fungi, bacteria, and plants) of AgNPs is preferred over chemical (microemulsion technique, Tollens's method, pyrolysis, electrochemical method, and microwave-assisted technique) and physical processes (irradiation, ball milling, laser ablation, and evaporation-condensation) due to its cost-effectiveness, safety, nontoxicity, and eco-friendly approach (Singh et al. 2015; Iravani et al. 2014). Compared to other biological agents, the use of bacteria in synthesizing AgNPs is widely accepted because of its biocompatibility, sustainability, nontoxicity, and mass production. The extracellular synthesis of AgNPs using bacteria (e.g., *Bacillus licheniformis, Bacillus pumilus,* and *Bacillus persicus* produces 72–92 nm) involves the biogenic reduction of Ag^+ to Ag^o, with the activity of proteins or small secreted enzymes present on the bacterial cell wall. The extracellular mechanisms produce NPs of varied shapes and sizes, such as hexagonal, spherical, triangular, circular, disc, and cuboidal (Nanda and Saravanan 2009; Pugazhenthiran et al. 2009). The size of AgNPs produced using bacteria ranges from 1 to 200 nm and greatly depends on the reducing agents and the type of bacterial species. However, nanoscale synthesis of AgNPs is preferred because of its high surface-to-volume ratio, low melting point, superconductivity, transition temperature, optical property, and high reactivity (Bogunia-Kubik and Sugisaka 2002; Baker et al. 2005). Extracellularly synthesized AgNPs are generally extracted through ultracentrifugation (at 10,000 to 12,000 rpm), forming a pellet that can be resuspended in the desired solvent (Javaid et al. 2018). Intracellular synthesis of AgNPs by bacteria involves the transportation of metal ions inside the bacterial cell facilitated by membrane proteins. The silver restraining property of some bacterial species reduce Ag^+ to Ag^o, and the resulting NPs accumulate either on the cell wall or periplasmic space (Narayanan and Sakthivel 2010). Bacterial species *Pseudomonas stutzeri* AG259 produces AgNPs (size ~200 nm) and a small fraction of the product Ag sulfide acanthite by reducing $AgNO_3$ solution (Klaus et al. 1999). Similarly, *Corynebacterium* sp. synthesizes AgNPs ranging in size from 10 to 15 nm, forming a diamine Ag complex on the cell wall (Zhang et al. 2005). The AgNPs produced through intracellular mechanism require additional extraction steps such as cell lysis by ultrasonication, heat treatment through autoclaving, and use of chemical salts and detergents for the release of NPs accumulated inside the bacterial cell (Iravani et al. 2014; Javaid et al. 2018).

Several research studies have investigated the synthesis of AgNPs through extracellular and intracellular methods in the literature (Table 15.1) (Iravani 2014; Kulkarni and Muddapur 2014; Hulkoti and Taranath 2014). Green synthesis of AgNPs using culture supernatant of *Bacillus subtilis* has yielded monodispersed AgNPs in size range of ~5–50 nm and was combined with microwave irradiation to prevent the aggregation of AgNPs, while increasing the rate of reaction (Saifuddin et al. 2009). Monodispersed AgNPs (~50 nm) synthesized extracellularly using *Bacillus licheniformis* by bioreduction of Ag ion have been reported (Kalimuthu et al. 2008; Kalishwaralal et al. 2008). The biosynthesis of monodispersed spherical, circular, and triangular-shaped AgNPs with efficient antimicrobial activity has been reported with *Bacillus flexus* and *Bacillus amyloliquefaciens* with size ranging from 12–65 nm and 14.5 nm (Wei et al. 2012; Priyadarshini et al. 2013). Studies have also reported the biosynthesis of stable AgNPs (8 months stability in the dark) with cell-free culture extracts of psychrophilic bacteria (such as *Phaeocystis antarctica, Pseudomonas proteolytica, Pseudomonas meridiana, Arthrobacter kerguelensis,* and *Arthrobacter gangotriensis*) and mesophilic bacteria (*Bacillus indicus and Bacillus cecembensis*). The synthesized AgNPs range in size 6–13 nm, and its stability varied with pH, temperature, and bacterial species involved in the process. AgNPs synthesized using cell-free culture supernatants varied with bacterial species, i.e., *A. kerguelensis* supernatant could not synthesize AgNPs at the same temperature as *P. antarctica* that produced the AgNPs (Shivaji et al. 2011; Iravani 2014).

5   Gold Nanoparticles (AuNPs)

AuNPs are commonly used in the field of bio-nanotechnology because of its unique properties and multiple surface functionalities. The simplicity in functionalizing AuNPs surface makes it a versatile tool for biological assemblies with biomolecules and the design of novel biomaterials for investigation of biological systems (Yeh et al. 2012). Synthesis of spherical AuNPs is of great interest because of its opto-electronic properties, large surface-to-volume ratio, high biocompatibility, and less toxicity. Besides these properties, AuNPs have unique properties such as surface plasmon resonance (SPR), surface-enhanced Raman scattering (SERS), fluores-cence quenching, and redox activity. These properties make AuNPs exhibit broad applications in electronic devices, electrochemical sensing, imaging and sensing, photothermal therapy, sensor fabrication, and materials science (Eustis and El-Sayed 2006). Spherical AuNPs changes colors with the variation in the core size of NPs from 1 to 100 nm in aqueous solutions, and it is generally observed with the size-relative absorption peak from 500 to 550 nm (Link and El-Sayed 1999; Jain et al. 2006). The biosynthesis of AuNPs in bacteria occurs with the reduction of Au (III) ion into Au atoms and adheres to the bacterial cell wall that further aggregates to form AuNPs. AuNPs with different shapes and sizes produced by diverse microor-ganisms are widely used in medical fields as anti-angiogenesis, anti-arthritic, and antimalarial agents (Moshfegh et al. 2011).

Several bacterial species have shown the ability to reduce Au^{3+} ions into different shaped gold NPs. *Bacillus subtilis* 168 shown to reduce Au^{3+} ions into octahedral gold NPs with size ranging from 5 to 25 nm using gold chloride solution under ambient temperature and pressure conditions (Beveridge and Murray 1980). Biosynthesis of AuNPs occurs in both extracellular and intracellular mechanisms with the reduction of chloroaurate and silver ions using *Bacillus subtilis*, in contrast to AgNPs predominantly synthesized through extracellular mechanism (Iravani 2014; Reddy et al. 2010). Biosynthesis of AuNPs using *Pseudomonas aeruginosa* and *Rhodopseudomonas capsulata* has shown to produce spherical AuNPs in the size range of 10–20 nm (pH 7) and nanoplates at pH 4. The study involved the incu-bation of *R. capsulata* biomass and aqueous $HAuCl_4$ solution at varying pH values ranging from 7 to 4, which is an important parameter for controlling the size and shapes of AuNPs (Singh and Kundu 2014). *Bacillus megaterium* has shown promis-ing results with the production of spherical AuNPs (~1.5–2.5 nm) with a self-assem-bled layer of thiol. These synthesized monodispersed spherical AuNPs stayed stable over several weeks (Ahmad et al. 2003). Similarly, *Lactobacillus* sp. produced AuNPs when exposed to gold ions resulting in the nucleation of AuNPs on the bac-terial cell wall and further movement of Au nuclei into the cell for aggregation of Au atoms into large-sized AuNPs (Nair and Pradeep 2002). *Escherichia coli* DH5α showed the ability to produce spherical AuNPs (~17–33 nm) with mixed amounts of triangular and quasi-hexagonal shaped AuNPs (Du et al. 2007). *E. coli* MC4100 and *Desulfovibrio desulfuricans* ATCC 29577 produced spherical AuNPs at acidic pH, and varied sizes (~10 nm and ~ 50 nm) of well-defined triangle-, hexagon-, and rod-shaped AuNPs at pH 7–9 (Deplanche and Macaskie 2008).

6   Zinc Oxide Nanoparticles (ZnO NPs)

Zinc is an essential nutrient for living organisms and is observed to have wide biological applications due to its antimicrobial activity (Sirelkhatim et al. 2015). Zinc is a trace mineral that plays a vital role in many physiological functions of the body (Vallee and Falchuk 1993) and also effectively inhibits the growth of a broad-spectrum of pathogens (Saravanan et al. 2018). Due to the increased absorption of zinc in the body, it is widely investigated for health and productivity. It has already shown potential use in the poultry and livestock industries as a supplement feed for animals (Swain et al. 2016). The application of zinc includes improving growth performances, enhancing antioxidative property, improving immune response, and increasing the bioavailability of zinc (Swain et al. 2016; Li et al. 2016). ZnO NPs use in the cosmetic and sunscreen industry has shown potential use because of its transparency and ability to reflect, scatter, and absorb UV radiation (Jayaseelan et al. 2012).

The synthesis of ZnO NPs using chemical and physical methods (vapor condensation, interferometric lithography, physical fragmentation, sol-gel process, solvent evaporation, and microemulsion method) has drawn disadvantage due to its low biocompatibility, high cost, and is toxic, making it unsuitable for clinical and biomedical applications. Biosynthesis of ZnO NPs using microbes is commonly preferred due to its cost-effectiveness and eco-friendly technique (Mirzaei and Darroudi 2017; Ahmed et al. 2017). Compared to other microbes, bacteria have shown promising results with the production of ZnO NPs because of its ease of handling and manipulative genetic attributes compared to other eukaryotic organisms (Table 15.1) (Velusamy et al. 2016). Lactic acid bacteria (e.g., *Lactobacillus plantarum, Lactobacillus sporogenes*) are facultative anaerobic that contain negative electrokinetic potential making it an ideal candidate to absorb metal ions for the reduction/oxidation process (Selvarajan and Mohanasrinivasan 2013; Mishra et al. 2013). Additionally, lactic acid bacteria are gram-negative bacteria containing a thick cell layer made of peptidoglycan, teichoic acid, lipoteichoic acid, protein, and polysaccharides that assists with the biosorption and bioreduction of metal ions (Chapot-Chartier and Kulakauskas 2014). Exopolysaccharides that exist in such bacteria help in combatting the metal ions, thereby acting as an additional site (Korbekandi et al. 2012). A reproducible bacteria *Aeromonas hydrophila* as a reducing and capping agent showed the synthesis of ZnO NPs through an inexpensive and easy procedure. The presence of ZnO NPs was confirmed with a peak at 374 nm on the UV-Vis spectrum. The size of the ZnO NPs was ~57 nm and shaped in spherical and oval forms (Jayaseelan et al. 2012).

7   Magnetic Nanoparticles

Nanocrystalline iron-based particles (such as magnetite and maghemite) possess a high surface-to-volume ratio with magnetic properties, unlike bulk materials. Such magnetic nanoparticles are highly important in scientific and technological applications due to their biocompatibility (Revati and Pandey 2011). Magnetite (Fe_3O_4) is a ferric material consisting of a cubic inverse spinel structure with unique electric and magnetic properties that depend on the transfer of electrons from Fe^{2+} to Fe^{3+} in the octahedral sites. These magnetic NPs have great applications in storage devices, sensors, separation processes, imaging, environmental remediation, targeted cancer treatment, drug delivery, and gene therapy (Bharde et al. 2005). Biosynthesis of magnetite crystals in culture solution follows the biologically induced biomineralization process. The process depends on parameters such as pH, pO_2, pCO_2, redox, potential, and temperature. The general process involves the release of metabolites by the microbes into the surrounding medium. The metabolites further react with specific ions or compounds in the solution or the cell wall surface, resulting in mineral particle formation (Revati and Pandey 2011). Fe(III) reducing bacteria such as *Geobacter metallireducens* and *Shewanella putrefaciens* are common microbes that produce magnetite in the solution as a by-product. The secretion of Fe(II) in the surrounding medium occurs when iron-reducing bacteria respire oxidized Fe(III) compound in the form of Fe(III) oxyhydroxide under anaerobic conditions. The released Fe(II) crystals unite with the abundant ferric hydroxide grain resulting in magnetite formation, the process which is generally favored at high pH (Bazylinski and Schübbe 2007; Bazylinski et al. 2007). Biosynthesis of magnetite through sulfate-reducing bacteria such as *Desulfuromonas* strains, releasing H_2S under anaerobic conditions, has been reported (Mandal et al. 2006; Revati and Pandey 2011).

Other ways of magnetite NPs formation involve the biologically controlled biomineralization process. The process involves the synthesis of magnetite crystals intracellularly within the cytoplasm or cell wall, which is kept encapsulated as a geochemical environment. The sequestered specific ions from the organic matrix get transferred to an isolated compartment within the cell, where the nucleation occurs and results in the growth of highly ordered magnetite crystals. Bacteria such as *Magnetospirillum magnetotacticum, M. gryphiswaldense,* and *Desulfovibrio magneticus* follow the biologically controlled biomineralization mode to synthesize well-ordered crystalline, consistent morphology, and nearly uniform-sized magnetite NPs (Table 15.1). These magnetotactic bacteria (Yan et al. 2012) are gram-negative and microaerophilic, generally found in freshwater swamps and ponds, salt marsh ponds, and sediments of freshwater and marine, and places containing high iron content with neutral pH and less oxygenated (Bazylinski and Schübbe 2007; Bazylinski et al. 2007; Sparks et al. 1989). These species come under the heterogeneous group of aquatic microorganisms that aligns with the magnetic field lines (a phenomenon known as magnetotaxis). This orientation is due to the presence of magnetosomes that consist of magnetic crystals such as magnetite or greigite encapsulated intracellularly within a membrane (Faivre and Schuler 2008; Blakemore 1975; DeLong et al. 1993).

8   Titanium Dioxide Nanoparticles (TiO$_2$ NPs)

TiO$_2$ NPs have been widely studied due to their interesting properties of catalysis, photocatalysis, and antibacterial activity (Haider et al. 2019). TiO$_2$ generally exists in rutile, anatase, and brookite form. TiO$_2$ NPs possess unique properties of high specific surface area, optimum electronic band structure, high quantum efficiency, chemical innerness, and stability. Due to its extensive applications, the synthesis of TiO$_2$ NPs through biological methods has gained importance. Several bacterial species (e.g., *Lactobacillus* sp., *Bacillus subtilis*) have successfully synthesized TiO$_2$ NPs through biological processes. *Lactobacillus* sp. has reported producing TiO$_2$ NPs biosynthetically in smaller sizes of 5–9 nm and larger particles of 60–80 nm (Al-Zahrani et al. 2018; Prasad et al. 2007). The synthesis of TiO$_2$ NPs observed to be dependent on the energy source, pH, and overall oxidation-reduction potential. *Lactobacillus crispatus* showed the synthesis of TiO$_2$ NPs, resulting in spherical- and oval-shaped NPs ranging from 70 to 115 nm (Abdulsattar 2014). Furthermore, studies have shown that TiO$_2$ NPs produced from *L. crispatus* has the ability to reduce biofilm formation, hemolysin, and also urease that are responsible for developing multidrug resistance in pathogens (Ibrahem et al. 2014). In *Lactobacillus* sp., the formation of TiO$_2$ NPs occurs due to the electrostatic interactions between the bacterial species and metal clusters (Nair and Pradeep 2002). Studies involving the synthesis of TiO$_2$ NPs using *Bacillus subtilis* (NPs size 66–77 nm) and *Bacillus amyloliquefaciens* (NPs size 15–90 nm) have also been reported (Khan and Fulekar 2016; Kirthi et al. 2011).

9   Interaction of Nanoparticles with Extra- and Intracellular Structures

Advancement of NP synthesis and biosynthesis practices has greatly increased the interest in utilizing them in biomedical applications. Implementing metal NPs for medicinal purposes is of particular interest as they have shown to interact with bacterial extra- and intracellular components (Fig. 15.1), eliciting differing types and degrees of toxicity in gram-negative and gram-positive bacteria. NPs exhibit antibacterial properties by predominantly engaging in cell-surface attachment and disruption, alteration of membrane potential, enzyme and protein inhibition, DNA damage, and induction of reactive oxygen species (ROS), and subsequently oxidative stress (Sondi

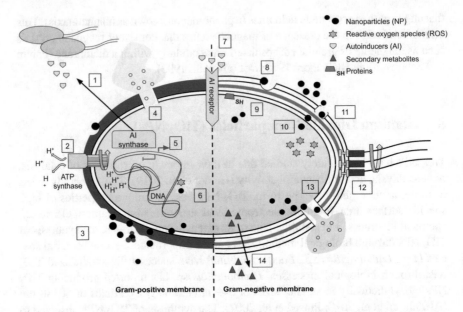

Fig. 15.1 Effects of nanoparticles on the microbial cell structure and function. This illustration shows the many ways in which nanoparticles can interact with the bacterial cell via intracellularly or extracellularly. 1. NPs can interfere with autoinducer synthesis and secretion. 2. NPs can alter ATP synthesis via inhibition of ATP synthase or alteration of membrane potential. 3. NPs disrupt the thick peptidoglycan layer of gram-positive cells. 4. NPs can cause membrane damage, resulting in spillage of intracellular contents. 5. NPs influence up- and downregulation of various genes. 6. NPs generate reactive oxygen species (ROS), which can lead to DNA damage. 7. NPs can influence the efficacy of autoinducer signaling via the receptor. 8. NPs can disrupt the outer membrane in gram-negative bacteria, resulting in pit formation. 9. NPs interact with thiol groups of proteins causing inhibition. 10. NPs generate reactive oxygen species (ROS) via oxidative stress of the cell. 11. NPs can pass through porins found in the outer membrane of gram-negative bacteria. 12. NPs alter the function of bacterial swarming via flagellar movement. 13. NPs can disrupt all layers of gram-negative bacteria, resulting in NP entry to the cell and spillage of intracellular contents. 14. NPs interfere with secondary metabolite secretion

and Salopek-Sondi 2004; Chang et al. 2012; Park et al. 2009; Dibrov et al. 2002; Ramalingam et al. 2016). Differing metal NPs elicit varying bactericidal properties; however, cell membrane disruption in tandem with oxidative stress damage is believed to play the largest role in cell toxicity (Morones et al. 2005).

The toxicity of NPs is primarily influenced by physicochemical properties such as size, shape, and surface charge, in addition to NPs concentration (Arakha et al. 2015; Raza et al. 2016). Spherical AgNPs ≤ 10 nm (10 nm, 7 nm) have shown to possess greater bactericidal effect against *Pseudomonas fluorescens*, *Staphylococcus aureus*, and *Escherichia coli* than corresponding 20 nm – 89 nm NPs (Ivask et al. 2014; Martínez-Castañon et al. 2008). Likewise, in comparison to 0.3 and 0.6 μm MgO NPs, 20 nm MgO NPs expressed greater bactericidal effect against *Xanthomonas perforans* at half the concentration (50 μg/ml) (Liao et al. 2019). It is also observed that the reduction of size permits easy bacterial penetration by the

NPs (Manuela et al. 2013). A series of four bacteria (*E. coli, Pseudomonas aeruginosa, Bacillus megaterium*, and *Staphylococcus epidermidis*) treated with fluorescent 1.5 nm AgNPs exhibited uniform antimicrobial activity (MIC 1.95 μg/ml), wherein the most plausible explanation was membrane penetration (Bera et al. 2014). Some studies have observed contradicting outcomes. One study showed that spherical 25 nm AuNPs penetrated the membrane of *Corynebacterium pseudotuberculosis*, while in a similar study spherical 16 nm AuNPs were incapable of penetrating *Salmonella typhimurium*, implicating the influence of other physicochemical characteristics of NPs and bacteria (Mohamed et al. 2017; Wang et al. 2011). High concentrations of metal NPs generally elicit bactericidal effects, although result diversity exists when comparing responses of gram-negative and gram-positive species against concentrations. Upon exposure of *E. coli* and *S. aureus* to varying ZnO NPs concentrations (0.125 ml/ml–16 mg/ml), it has been observed that the former resisted the bactericidal effect until 16 mg/ml, whereas the latter resisted up to 8 mg/ml (Emami-Karvani and Chehrazi 2011). Moreover, exposure of negative bacteria *E. coli, P. aeruginosa*, and *Vibrio cholera* to 25–100 μg/ml AgNPs showed bactericidal effects surfacing at 75 μg/ml AgNPs (Morones et al. 2005). The surface charge of NPs is a main determinant in initial attachment to bacteria cell surfaces. As gram-negative and gram-positive cell surfaces generally express a negative net charge, cationic NPs will have stronger electrostatic attraction towards these bacteria (Dickson and Koohmaraie 1989). Cationic NPs have shown increased affinity towards gram-negative bacteria however, as their negative charge is greater due to physiological differences in membrane structure (Mandal et al. 2016).

10 Interaction of Nanoparticles with Gram-Positive and Gram-Negative Bacteria

Gram-negative bacteria such as *E. coli* and *P. aeruginosa* are composed of a three-layer cell wall structure: an inner plasma membrane (PM), a thin peptidoglycan layer, and an outer membrane (OM), respectively (Fig. 15.1). The NPs-cell surface electrostatic attraction observed in gram-negative bacteria interactions is attributable to the lipopolysaccharide layer of the OM (Arakha et al. 2015). Embedded in the OM are porins, transmembrane proteins which mediate the passive diffusion of molecules into the periplasmic space of gram-negative bacteria, and exist as either specific or nonspecific types, with the latter serving as the gateway of acquiring antimicrobial resistance (Iyer et al. 2017; Smani et al. 2014). Comparatively, gram-positive bacteria such as *S. aureus* and *S. epidermidis* comprise a two-layer structure: an inner plasma membrane and a thick outer peptidoglycan layer embedded with teichoic and lipoteichoic acids, which give the cell rigidity, its overall negative net charge, and act as a chelating agent for metal NPs (Wickham et al. 2009). While OM protection partly attributes to the antibacterial resistance of gram-negative bacteria, the overall physiological composition of gram-negative membranes is more permeable, and thus more compromised to invasion than gram-positive bacteria. Spherical 10 and 35 nm AgNPs were observed to have stronger antibacterial activity against gram-negative bacteria *E. coli* and *P. aeruginosa* compared to gram-positive bacteria *S. aureus*, in which the AgNPs were approximately only half as effective as the latter (Kaviya et al. 2011). The effectiveness of NPs penetration is more substantial following NPs cell-surface interactions in which membranes become disrupted and highly permeable. Observation of bacteria-NP interactions requires consideration of NPs and bacterial physiochemical properties as studies have shown gram-negative species such as *E. coli, P. aeruginosa*, and *Klebsiella pseudomonas* to be equally or more resistant to toxicity in the presence of ZnO and Ag (Ag_2S, Ag_2Se) NPs than gram-positive counterparts (*S. aureus, Streptococcus agalactiae*); results which appear to indicate NP capability of overcoming gram-positive bacterias' antibacterial resistance mechanisms (Fig. 15.1) (Cakić et al. 2016; Emami-Karvani and Chehrazi 2011; Delgado-Beleño et al. 2018).

11 Nanoparticle Effects on Membrane Disruption and Reorganization

The electrostatic attractions heavily influence initial attachment of NPs to gram-negative and gram-positive cell surfaces (Fig. 15.1). Adherence of NPs to cell surfaces primarily results in physical disruption and degradation of lipid membranes, and leads to further intracellular damages upon penetration of NPs, including disruption of permeability (Xie et al. 2011). Some NPs-cell membrane interactions—as seen with graphene nanosheets and *E. coli*—pierce and incise lipid membranes or extract phospholipid molecules, effectively removing and thinning sections of the membrane (Tu et al. 2013). Similar interactions see the release of lipopolysaccharides from the membrane, creating "pits" in the cell surface (Sondi and Salopek-Sondi 2004; Mandal et al. 2016). Direct physical damage to cell membranes leads to leakage of intracellular components as the membrane gradually comes apart and lyses (Brayner et al. 2006). Such a mechanism occurred in gram-negative and gram-positive bacteria *Proteus vulgaris* and *Enterococcus faecalis,* where the species experienced both degrees of pit formation, leakage, and membrane fragmentation when treated with AgNPs (Mandal et al. 2016). Although adherence of NPs to cell surfaces to cause extra- and intracellular damages is a major antibacterial mechanism, mere presence does not inherently indicate toxicity. Rather, accumulation (i.e., concentration) of NPs in and near intracellular spaces are regarded as a clearer indicator (Simon-Deckers et al. 2009).

12   Interaction of Nanoparticles with Enzymes and DNAs

At the intracellular level, NPs are capable of altering membrane potential (Fig. 15.1) and inhibiting enzymes with roles in DNA synthesis, membrane synthesis, and protein folding (Sohm et al. 2015; Wigginton et al. 2010). Much of metal NPs toxicity to intracellular components is due to their affinity for thiols (-SH), a functional group found in respiratory and cell wall synthesis enzymes, and cysteine, an amino acid essential for protein folding (Slavin et al. 2017). Furthermore, the prevention of DNA unwinding may also occur upon binding of DNA by NPs; AgNPs induced such a reaction when paired with *P. aeruginosa* (Batarseh 2004). An *E.coli* proteome analysis following TiO$_2$ NPs treatment revealed significant downregulation of 152 genes, including genes involved in cell structure, protein folding, energy metabolism, transcription, and translation (Sohm et al. 2015). Likewise, another proteome analysis of *E. coli* following AgNPs treatment identified membrane integrity proteins such as those associated with porins, chaperonins, lipopolysaccharide assembly, and OM assembly as targets for AgNPs binding (Wigginton et al. 2010). It is also reported that metal NPs may act as an antivirulence property by inducing virulence factor inhibition of pathogens, as biofilm disruption, modulation of swarming motility, and quorum sensing inhibition of *P. aeruginosa* occurred upon treatment with ZnO NPs and FeOOH NPs (Pham et al. 2019; Saleh et al. 2019).

Moreover, the inactivation of plasma membrane respiratory proteins upon chemical group binding may induce reactive oxygen species (ROS) accumulation near and within bacterial cells. ROS are produced endogenously in prokaryotes as a by-product of oxygen metabolism and occur in the forms of highly reactive free radicals, peroxides, and superoxides, all of which are ultimately toxic to bacterial cells in superfluous amounts (Park et al. 2009). Bacteria have become capable of responding and maintaining ROS levels below the lethal threshold; however, upon prolonged exposure to stressors such as UV radiation and NPs, endogenous ROS production increases above the threshold and induces oxidative stress, damaging DNA, proteins, and lipids, and may subsequently result in cell death (Stohs and Bagchi 1995).

13   Nanoparticle Effects on Bacterial Metabolism and Signaling Systems

The use of nanoparticles as antimicrobial agents significantly impacts different aspects of bacterial metabolism, such as ATP synthesis, secondary metabolite synthesis, and redox reactions (Slavin et al. 2017). Additionally, impacts on common cellular signaling pathways such as quorum sensing and chemotaxis are observed with different nanoparticles in groups of bacteria. The effects of such nanoparticles on microbial cellular processes is highly dependent on the size, shape, and concentration of the nanoparticle as well as length of exposure (Azam et al. 2012;

Gómez-Gómez et al. 2019). The magnitude of impact that the nanoparticle exhibits on the cell is highly species-specific, although some bacteria exhibit similar effects (Baek and An 2011).

Nanoparticles influence the secretion of secondary metabolites from bacterial cells, resulting in positive and negative impacts on the surrounding environment. This phenomenon is commonly important for agricultural and soil studies due to its specificity with species and nanoparticles. Two different nanoparticles (Ag and SiO_2) have shown to induce expression of pyrrolnitrin, an antifungal compound produced in the soil bacterium *Pseudomonas protegens* CHA0, leading to inhibition of the common plant fungal pathogen *C. albicans* (Khan et al. 2018). Zinc oxide and titanium oxide nanoparticles exhibited negative effects on the secretion of beneficial secondary metabolites such as siderophore and indole acetic acid (IAA) molecules by three different soil bacteria (*Pseudomonas aeruginosa, P. fluorescens*, and *Bacillus amyloliquefaciens*) (Haris and Ahmad 2017).

14 ATP Synthesis and Cellular Respiration

Gold nanoparticles alter the activity of adenosine triphosphate (ATP) synthase in four different enteric pathogens by adjusting the membrane potential of the cells, resulting in a decrease of ATP levels within the bacteria (Shamaila et al. 2016; Cui et al. 2012). *D. vulgaris*, a sulfur-reducing bacterium which is responsible for utilizing sulfur in energy synthesis, was shown to have a decrease in sulfur reduction upon exposure to copper oxide nanoparticles (CuO-NPs). Genes involved in electron transfer and cellular respiration were downregulated; however, upregulation in genes involved in ATP synthesis was present, suggesting that the toxic nanoparticle exposure generated a stress response in the cell leading to the utilization of a different mechanism of synthesizing ATP in the bacterium (Chen et al. 2019).

Iron oxide nanoparticles (Fe_3O_4-NPs) were observed to reduce proton movement across both *Escherichia coli* and *Escherichia hirae* membranes, with *E. coli* having a larger fold reduction when compared to *E. hirae*. The decrease in proton transport correlated with a decrease in activity of the F_0F_1-ATPase (Gabrielyan et al. 2019), which is responsible for utilizing the transmembrane proton gradient to synthesize ATP molecules from ADP precursors (Kagawa 1978). However, impacts were much more severe for *E. hirae* than *E. coli*, indicating a species-specific mechanism for ATPase inhibition by the nanoparticles (Gabrielyan et al. 2019).

15 Quorum Sensing and Chemotaxis

Quorum sensing is a method in which bacteria interact with their surrounding environment, particularly with other bacterial cells, via secretion of secondary molecules known as autoinducers (Fig. 15.1). Once the bacteria interact with a particular concentration of autoinducers, networks of genes will become transcriptionally active to alter the bacterial cell for adaptation to the environment (Hagen 2014). The purpose of quorum sensing in bacteria is important for synchronizing gene activation among bacterial cell communities, which can lead to events such as biofilm formation and increased pathogenicity (Schuster et al. 2003; Xiong et al. 2020). In the bacterium *Chromobacterium violaceum*, metallic nanoparticles such as ZnO NPs, TiO_2 NP, and Ag NPs were shown to negatively impact different steps within the quorum-sensing pathway via quorum quenching. Ag-NPs and TiO_2-NPs disrupted the synthesis of *N*-acyl-L-homoserine lactones (AHL), an autoinducer of *C. violaceum*. Additionally, exposure to ZnO NPs led to a decrease in the bacterium's ability to recognize changing concentrations of autoinducers in the surrounding environment (Gómez-Gómez et al. 2019). Green synthesized silver nanoparticles (AgNPs) were observed to negatively impact the quorum sensing ability of *Serratia marcescens* and also downregulate virulence genes mediated by quorum sensing of the bacterium (Ravindran et al. 2018). A decrease in pyocyanin (a molecule mediated by AHL synthesis) demonstrated inhibition of quorum sensing in *P. aeruginosa* upon exposure to increasing concentrations of gold nanoparticles (Samanta et al. 2017).

Chemotactic signaling in bacteria is responsible for the directional movement of the cell towards a chemically favorable or away from a chemically unfavorable environment. In chemotactic signaling, certain gene expressions lead to activation or inactivation of flagellar movement. Unfavorable environments generate a swarming response from the bacterium to move away from the chemical concentration, thus utilizing ATP for flagellar activation (Baker et al. 2006). Nanoparticles exhibit both positive and negative effects of chemotactic signaling in bacteria. Upon initial exposure to gold nanoparticles, *Escherichia coli* cells showed an increase in chemotactic gene transcription, resulting in an increased swarming motility. This phenomenon, which utilizes large amounts of ATP, was quickly diminished over time, presumably due to the negative impacts of the AuNPs on the bacterial metabolism (Cui et al. 2012). In contrast, the bacterium *Desulfovibrio vulgaris* was shown to have a decrease in gene expression of cell motility genes when exposed to copper oxide nanoparticles (CuO NPs), thus indicating a decline in chemotactic signaling and movement (Chen et al. 2019). Silver nanoparticles synthesized by the plant *Vetiveria zizanioides* were shown to inhibit the swarming motility of *S. marcescens*, thus impacting the efficacy of virulence by the bacterium (Ravindran et al. 2018).

16 Antimicrobial and Antibiofilm Activity of Nanoparticles

Microbial cells are densely crowded on solid surfaces to form biofilms and are naturally found as simple or composite bacterial communities in the abiotic and biotic environment. Microbial cells produce and secrete extracellular polymeric substances (EPS), such as proteins, polysaccharides, and DNA (Lasa 2006; Whitchurch et al. 2002), and these substances together form the matrix in which microbial cells are embedded. Altered metabolic activity of the microbial cells in the biofilm further increases the rates of EPS production, activation, or inhibition of specific genes associated with biofilm formation, and/or decrease microbial growth rate providing bacterial community an adaptation to these specific environments (Fulaz et al. 2019; Flemming et al. 2007).

Bacterial biofilms not only afford the protection to the microorganisms against altered environmental conditions, such as pH, osmolarity, nutrient availability, mechanical, and shear forces (Fux et al. 2005; Costerton et al. 1995), but they also protect bacterial cells from host-defense system, including antibodies and macrophages (Stewart and Costerton 2001; Costerton et al. 1999). Therefore, the increased resistance to bacteria to the harsh conditions and antimicrobial treatments leads to the emergence of multidrug resistance to the latent, persistent, and recurrent bacterial infections. Bacterial biofilms were observed in several types of intracellular microbial infections (Olsen 2015; Donlan and Costerton 2002), which cause diseases such as tuberculosis, salmonellosis, periodontitis, endocarditis, and chronic lung infection in cystic fibrosis (Davies 2003; Singh et al. 2002; Gómez and Prince 2007). Of the total chronic microbial infections, 80% microbial infections are biofilm-related, which are 10–1000 times more antibiotic resistant than the planktonic cells (Hall and Mah 2017). It has been previously shown that the two-third of the antibiotics are futile against biofilm-related intracellular pathogens (Stewart and Costerton 2001).

The structural and physiological complexity of the biofilms not only reduces the antibiotic diffusion, but also keep retention of antibiotics at low concentration in the cells. The low diffusion and retention of antibiotics in the biofilm are circumvented by naturally produced D-amino acids and polyamine by the microbial cells, and avert reformation of bacterial biofilms in *S. aureus* and *E. coli* (Tack and Sabath 1985; Schlessinger 1988). It has also been shown that the matrix digesting enzymes of *S. aureus, Vibrio cholerae*, and *Pseudomonas aeruginosa* promote the degradation of biofilms. Inhibitors of the quorum sensing genes also dysregulate the reorganization of biofilms and are used as an effective treatment of biofilm-related bacterial infections (Flemming et al. 2007).

In recent years, metal nanoparticles, organic nanoparticles, and green nanoparticles have been used as antimicrobial and antibiofilm agents to dismantle and eradicate several biofilm-related bacterial infections (Luo et al. 2016). However, synthesizing metal nanoparticles using microorganisms and plants has now been well recognized for rapid, ecofriendly, and easily scaled-up production of bionanoparticles (Singh et al. 2016). Recently, the employment of the nanoparticle-based antimicrobial therapies has allowed inhibition, disruption, and abolition of multidrug resistance bacterial biofilms (Koo et al. 2017). We have only discussed selected major types of nanoparticles, which have presented novel delivery applications as antibiofilm and antibacterial activities (Wu et al. 2015).

Metal oxide nanoparticles (MgO, CuO, CaO, ZnO, Fe_2O_3, and TiO_2) have important characteristics, such as small size and shape, high stability and catalytic activity, and significant antibacterial activity against bacterial biofilms (Dizaj et al. 2014). CuO NPs and Fe_2O_3 NPs were investigated for their antibacterial activity against methicillin-resistant *Staphylococcus aureus* (MRSA) and *E. coli* (Agarwala

et al. 2014), where CuO NPs were observed to be more toxic and antibiotic susceptibility than Fe_2O_3 NPs. TiO_2 NPs showed a breakdown of organic compounds and
form superoxide ions under nonlethal ultraviolet light exposure, and are highly efficient in inhibiting the growth of MRSA biofilm (Shah et al. 2008).

CaF_2 NPs especially suppress the effects of major virulence factors (*vicR, gtfC,
ftf, sapP, comDE*) in *Streptococcus mutans* by inactivating the enzyme activity associated with glucan synthesis, cell adhesion, acid production and tolerance, and
genes associated with the quorum sensing pathway (Kalia and Purohit 2011), which
drastically reduced the rate of biofilm reformation. NO-silica NPs have been
observed to release silica nanoparticles to destroy 99% biofilm-based microbial
cells of *Pseudomonas aeruginosa, Escherichia coli, Staphylococcus aureus,
Staphylococcus epidermis*, and *Candida albicans* (Falsetta et al. 2014; Hetrick et al.
2009). AgNPs are renowned for in vivo and in vitro antimicrobial activity with 95%
inhibition in biofilm formation of many bacterial species, including *P. aeruginosa*
and *Staphylococcus epidermis*. Results have indicated that the loosely bound EPSs
of the wastewater biofilms were invariably removed with Ag-NPs exposure, which
has exhibited various levels of susceptibility to different bacterial species (Sheng
and Liu 2011).

Phosphatidylcholine-associated gold nanoparticles (PA NPs) with gentamicin
have revealed that GPA NPs maintained their antibiotic activities and are more
effective in disrupting biofilms and inhibited biofilm reformation of pathogens,
including gram-positive and gram-negative bacteria. In addition, GPA-NPs were
observed to be nontoxic to the host cells and radially taken up by the macrophages,
facilitating the killing of intracellular bacteria in infected macrophages. These
results suggested GPA NPs might be a promising antibacterial agent for effective
treatment of intracellular biofilm-related chronic infections (reference). The differential toxicity of these metal-based nanoparticles are seen against multidrug resistance bacterial biofilms and have also been applied as indicative tools for surgical
devices (Adams et al. 2006; Sawai 2003).

Several synthetic nanoparticles and bio-nanoparticles, including the described
NPs, have currently emerged as alternatives to the traditional antibiotics for biofilm-
related microbial infections.

17   Environmental Impact of Nanoparticles

With the increasing applications of nanotechnology in medicine and the environmental sector, the role of engineered and biosynthesized NPs for clinical use and its
release in the environment makes it inevitable. Understanding the impact of these
released NPs into the environment further provides information on the interactions
between metal-NPs and the microorganisms, particularly bacteria that constitute the
majority of the biosphere over three billion years (Saif et al. 2016).

Nanotechnology products and by-products have become a predominant ingredient in industrial wastes that merge with the aquatic environment through waterways

such as drainage, ditches, lakes, and rivers despite effective risk assessment procedures (Moore 1990; Daughton 2004). Nanoscale particles are well understood to associate with larger biotic and non-biotic particles in both deposited and suspended sediments. The chemical contaminants from industrial wastes get transported through suspended sediments over long distances, and their distribution of NPs greatly depends on the hydrodynamic and morphological characteristics of water bodies (Smedes 1994; Bucheli and Gustafsson 2000). Uptake of these NPs by the aquatic biota through direct ingestion or epithelial ways such as gills, olfactory, and body walls is a major concern to the marine environment. Unlike eukaryotes, prokaryotes (particularly bacteria) are greatly protected from bulk transport of supramolecular and colloidal particles across the cell wall because of their defense mechanisms. In comparison, eukaryotes such as metazoans undergo endocytosis and phagocytosis mechanisms with the cellular internalization of nanoscale particles into the body. In invertebrate animals, NPs likely target the cellular immune system, hepatopancreas, and gut epithelium (Panyam et al. 2003). The liver is highly targeted in fishes after the endocytotic transport across the intestinal epithelium and endocytosis into hepatocytes. Fullerenes (coated and uncoated) have reported oxidative damage and lipophilicity in mammalian cells, and also inducing oxidative damage in the brains of largemouth bass (Oberdörster 2000). The release of such metal NPs from industrial effluents into waterways and aquatic systems affects human health through direct exposure by skin contact, inhalation of water aerosols, direct ingestion of NPs absorbed on foods, and drinking NPs contaminated water (Livingstone 2001). Indirect exposure of NPs to humans occurs through the ingestion of fish and shellfish, such as mollusks and crustaceans. Mollusks have the characteristics of accumulating suspended nanoparticles and sediments with conventional pollutants from environmental release (Livingstone et al. 1990; Moore 2006).

The toxic effect of AgNPs studied on the zebrafish model due to its fast development and transparent have reported the deposition of NPs on organs and severe developmental effects (Yeo and Kang 2008). A study involving in vivo imaging of single AgNPs with a diameter of ~13 nm showed the biocompatibility and toxicity of AgNPs inside embryos of zebrafish at each development stage. The study also demonstrated the transport in and out of embryos through chorion pore canals exhibiting Brownian diffusion (Lee et al. 2007). The antibacterial effect of AgNPs against E. coli analyzed through the proteomics method indicated that AgNPs destabilize the outer membrane resulting in depletion of intracellular ATP and breakdown of plasma membrane potential. Dissipation of protein motive force was seen with the accumulation of envelope protein precursors with AgNPs penetration (Abdelhamid and Wu 2015; Lok et al. 2006). The inhibitory effect of AgNPs in wastewater treatment has been evaluated with extant respirometry and automatic microtiter fluorescence assay techniques. The study showed that AgNPs inhibited the nitrifying bacteria involved in the nitrification process, which is critical for biological nutrient removal in wastewater treatment. This impact has initiated stringent regulations with the use of AgNPs in wastewater treatment plants (Choi et al. 2008; Sharma et al. 2009).

AgNPs also inactivate other microorganisms including fungi, viruses, and algae. AgNPs with a diameter size of ~3 nm have shown significant antibacterial effect on fungi *Candida albicans* by inhibiting mycelial formation, cell membrane disruption, inhibition of normal budding process, and hinder growth and biofilm formation in catheter coated AgNPs (Kim et al. 2009). Studies involving AgNPs on viruses have reported inhibition of hepatitis B virus replication, adhesion on HIV-1 surface and preventing its attachment with the host cells, and reduction of the Syncytial virus infection by 44% (Lu et al. 2008; Elechiguerra et al. 2005; Sun et al. 2008). In higher organisms, the toxicity effect of AgNPs is observed in *Diptera* species (*Drosophila melanogaster*) and mammalian cell lines such as fibroblasts of human and mice (Arora et al. 2009; AshaRani et al. 2009). The toxicity in these organisms occurs through cell leakage, reduction of mitochondrial function, cytotoxic, genotoxic, antiproliferative, and reduction in lung function and inflammatory lesions (Marambio-Jones and Hoek 2010).

Gold nanomaterials occur in different sizes (1–500 nm) and geometrical shapes such as rods, spheres, tubes, wires, and ribbons. Its small size and needle-like structure makes it an ideal candidate for penetrating inside the cells important for biomedical and biomolecular applications (Yah 2013). The needle-like form makes it easy for absorption, penetration, circulation, and distribution of AuNPs into cell biosystems without cell injury and toxicity (Connor et al. 2005; De Jong et al. 2008). The release of excess AuNPs into the aquatic environment was investigated in zebrafish *Danio rerio*, which showed various effects in fish including genome composition. The AuNPs exposure showed variations in oxidative stress, mitochondrial metabolism, detoxification, and DNA repair. Also, acetylcholine esterase activity increased in brains when exposed to sediments containing AuNPs showing altered neurotransmission (Dedeh et al. 2015). The impact of AuNPs on the activity of five extracellular enzymes important in nutrient cycling was investigated on agricultural soil with 50 nm citrate-coated and PVP-coated AuNPs. The soil enzyme activities and the bacterial community composition increased with the incorporation of surface coated AuNPs (Asadishad et al. 2017).

Though iron NPs have wide environmental applications, it presents a risk to the environment through improper waste management from industries, leakage, pollution remediation that has proven serious harm to the soil and groundwater. Zero-valent nano-iron particles used as a permeable reactive barrier for in situ treatment of groundwater undergo transformation in the presence of contaminants and the environment, resulting in toxic impacts on microorganisms and soil fauna (Saif et al. 2016). These iron NPs showed a high impact on soil microorganisms (such as *Bacillus cereus*) and change microbial biomass by inducing transcriptional and proteomic stress responses (Vittori Antisari et al. 2013; Saccà et al. 2014; Fajardo et al. 2013). The cytotoxic impact of nanoparticles (zero-valent, magnetite, and maghemite iron NPs) towards gram-negative bacteria *E. coli* showed that the toxicity depends on the oxidation state of the NPs generated from reactive oxygen species (ROS). ROS include highly unstable superoxide and hydroxyl radicals that get absorbed on the bacterial cell membrane and disrupt the functioning of the cell (Auffan et al. 2008). Zero-valent iron NPs are more toxic than other iron NPs and exhibit strong bactericidal activity under anaerobic conditions (Li et al. 2010; Lee et al. 2008), and also their toxicity varies based on age and surface modifications (Phenrat et al. 2009). Unlike engineered NPs, green nanomaterials show less toxicity towards microorganisms and human cell lines such as human keratinocyte cells. Highly toxic zero-valent iron NPs showed no or less toxicity when green synthesized using green tea (Nadagouda et al. 2010; Saif et al. 2016).

18 Conclusion

Nanotechnology is an emerging field in science that involves the synthesis and development of nanoparticles of various sizes, shapes, composition, and properties. Metal NPs are currently used in various fields such as medical imaging, water treatments, and cosmetics industries. Due to its diverse applications, synthesis of metal NPs using various methods such as physical and chemical methods has gained attention. Biogenic approaches with clean, nontoxic, and cost-effective processes are being carried out to overcome the drawbacks of high toxicity and high cost associated with such physical and chemical methods. Biosynthesis of metal NPs using bacteria through intracellular and extracellular mechanisms is a greener approach and eco-friendly technique. It is necessary to classify the types of metal NPs produced through biological methods, their interactions with the bacterial structures, and mechanisms involved in their antibacterial activity for therapeutic applications. Hence, we have provided a detailed review of the several metal NPs synthesized using different bacterial species through extracellular and intracellular mechanisms. Besides, we have highlighted the interactions of NPs at the bacterial cell membranes and their antibacterial activity. We believe this can provide a resourceful insight while developing a nontoxic metal NPs through biogenic approaches.

References

Abdelhamid, H. N., & Wu, H.-F. (2015). Proteomics analysis of the mode of antibacterial action of nanoparticles and their interactions with proteins. *TrAC Trends in Analytical Chemistry, 65*, 30–46.

Abdulsattar, J. (2014). Effect of culture media on biosynthesis of titanium dioxide nanoparticles using lactobacillus crispatus. *International Journal, 2*(5), 1014–1021.

Adams, L. K., Lyon, D. Y., & Alvarez, P. J. (2006). Comparative eco-toxicity of nanoscale TiO_2, SiO_2, and ZnO water suspensions. *Water Research, 40*(19), 3527–3532.

Agarwala, M., Choudhury, B., & Yadav, R. (2014). Comparative study of antibiofilm activity of copper oxide and iron oxide nanoparticles against multidrug resistant biofilm forming uropathogens. *Indian Journal of Microbiology, 54*(3), 365–368.

Ahmad, A., Senapati, S., Khan, M. I., Kumar, R., & Sastry, M. (2003). Extracellular biosynthesis of monodisperse gold nanoparticles by a novel extremophilic actinomycete, Thermomonospora sp. *Langmuir, 19*(8), 3550–3553.

Ahmed, S., Chaudhry, S. A., & Ikram, S. (2017). A review on biogenic synthesis of ZnO nanoparticles using plant extracts and microbes: A prospect towards green chemistry. *Journal of Photochemistry and Photobiology B: Biology, 166*, 272–284.

Al-Zahrani, H., El-Waseif, A., & El-Ghwas, D. (2018). Biosynthesis and evaluation of TiO_2 and ZnO nanoparticles from in vitro stimulation of lactobacillus johnsonii. *Journal of Innovations in Pharmaceutical and Biological Sciences, 5*, 16–20.

Arakha, M., Saleem, M., Mallick, B. C., & Jha, S. (2015). The effects of interfacial potential on antimicrobial propensity of ZnO nanoparticle. *Scientific Reports, 5*, 9578.

Arora, S., Jain, J., Rajwade, J., & Paknikar, K. (2009). Interactions of silver nanoparticles with primary mouse fibroblasts and liver cells. *Toxicology and Applied Pharmacology, 236*(3), 310–318.

Asadishad, B., Chahal, S., Cianciarelli, V., Zhou, K., & Tufenkji, N. (2017). Effect of gold nanoparticles on extracellular nutrient-cycling enzyme activity and bacterial community in soil slurries: Role of nanoparticle size and surface coating. *Environmental Science: Nano, 4*(4), 907–918.

AshaRani, P., Low Kah Mun, G., Hande, M. P., & Valiyaveettil, S. (2009). Cytotoxicity and genotoxicity of silver nanoparticles in human cells. *ACS Nano, 3*(2), 279–290.

Auffan, M., Achouak, W., Rose, J., Roncato, M.-A., Chaneac, C., Waite, D. T., Masion, A., Woicik, J. C., Wiesner, M. R., & Bottero, J.-Y. (2008). Relation between the redox state of iron-based nanoparticles and their cytotoxicity toward Escherichia coli. *Environmental Science & Technology, 42*(17), 6730–6735.

Azam, A., Ahmed, A. S., Oves, M., Khan, M., & Memic, A. (2012). Size-dependent antimicrobial properties of CuO nanoparticles against gram-positive and-negative bacterial strains. *International Journal of Nanomedicine, 7*, 3527.

Baek, Y.-W., & An, Y.-J. (2011). Microbial toxicity of metal oxide nanoparticles (CuO, NiO, ZnO, and Sb_2O_3) to Escherichia coli, Bacillus subtilis, and Streptococcus aureus. *Science of the Total Environment, 409*(8), 1603–1608.

Bai, H.-J., Yang, B.-S., Chai, C.-J., Yang, G.-E., Jia, W.-L., & Yi, Z.-B. (2011). Green synthesis of silver nanoparticles using Rhodobacter Sphaeroides. *World Journal of Microbiology and Biotechnology, 27*(11), 2723.

Baker, C., Pradhan, A., Pakstis, L., Pochan, D. J., & Shah, S. I. (2005). Synthesis and antibacterial properties of silver nanoparticles. *Journal of Nanoscience and Nanotechnology, 5*(2), 244–249.

Baker, M. D., Wolanin, P. M., & Stock, J. B. (2006). Signal transduction in bacterial chemotaxis. *BioEssays, 28*(1), 9–22.

Balraj, B., Senthilkumar, N., Siva, C., Krithikadevi, R., Julie, A., Potheher, I. V., & Arulmozhi, M. (2017). Synthesis and characterization of zinc oxide nanoparticles using marine Streptomyces sp. with its investigations on anticancer and antibacterial activity. *Research on Chemical Intermediates, 43*(4), 2367–2376.

Banu, A. N., Balasubramanian, C., & Moorthi, P. V. (2014). Biosynthesis of silver nanoparticles using Bacillus thuringiensis against dengue vector, Aedes aegypti (Diptera: Culicidae). *Parasitology Research, 113*(1), 311–316.

Batarseh, K. I. (2004). Anomaly and correlation of killing in the therapeutic properties of silver (I) chelation with glutamic and tartaric acids. *Journal of Antimicrobial Chemotherapy, 54*(2), 546–548.

Bazylinski, D. A., Frankel, R. B., & Konhauser, K. O. (2007). Modes of biomineralization of magnetite by microbes. *Geomicrobiology Journal, 24*(6), 465–475.

Bazylinski, D. A., & Schübbe, S. (2007). Controlled biomineralization by and applications of magnetotactic bacteria. *Advances in Applied Microbiology, 62*, 21–62.

Bera, R., Mandal, S., & Raj, C. R. (2014). Antimicrobial activity of fluorescent ag nanoparticles. *Letters in Applied Microbiology, 58*(6), 520–526.

Beveridge, T., & Murray, R. (1980). Sites of metal deposition in the cell wall of Bacillus subtilis. *Journal of Bacteriology, 141*(2), 876–887.

Bharde, A., Wani, A., Shouche, Y., Joy, P. A., Prasad, B. L., & Sastry, M. (2005). Bacterial aerobic synthesis of nanocrystalline magnetite. *Journal of the American Chemical Society, 127*(26), 9326–9327.

Blakemore, R. (1975). Magnetotactic bacteria. *Science, 190*(4212), 377–379.

Bogunia-Kubik, K., & Sugisaka, M. (2002). From molecular biology to nanotechnology and nanomedicine. *Biosystems, 65*(2–3), 123–138.

Brayner, R., Ferrari-Iliou, R., Brivois, N., Djediat, S., Benedetti, M. F., & Fiévet, F. (2006). Toxicological impact studies based on Escherichia coli bacteria in ultrafine ZnO nanoparticles colloidal medium. *Nano Letters, 6*(4), 866–870.

Bucheli, T. D., & Gustafsson, Ö. (2000). Quantification of the soot-water distribution coefficient of PAHs provides mechanistic basis for enhanced sorption observations. *Environmental Science & Technology, 34*(24), 5144–5151.

Cakić, M., Glišić, S., Nikolić, G., Nikolić, G. M., Cakić, K., & Cvetinov, M. (2016). Synthesis, characterization and antimicrobial activity of dextran sulphate stabilized silver nanoparticles. *Journal of Molecular Structure, 1110*, 156–161.

Chai, C.-J., & Bai, H.-J. (2010). Biosynthesis of silver nanoparticles using the phototrophic bacteria Rhodopseudomonas palustris and its antimicrobial activity against Escherichia coli and Staphylococcus aureus. *Microbiology/Weishengwuxue Tongbao, 37*(12), 1798–1804.

Chang, Y.-N., Zhang, M., Xia, L., Zhang, J., & Xing, G. (2012). The toxic effects and mechanisms of CuO and ZnO nanoparticles. *Materials, 5*(12), 2850–2871.

Chapot-Chartier, M.-P., & Kulakauskas, S. (2014). Cell wall structure and function in lactic acid bacteria. In *Microbial cell factories* (Vol. S1, p. S9). Springer.

Chen, Z., Gao, S.-h., Jin, M., Sun, S., Lu, J., Yang, P., Bond, P. L., Yuan, Z., & Guo, J. (2019). Physiological and transcriptomic analyses reveal CuO nanoparticle inhibition of anabolic and catabolic activities of sulfate-reducing bacterium. *Environment International, 125*, 65–74.

Choi, O., Deng, K. K., Kim, N.-J., Ross, L., Jr., Surampalli, R. Y., & Hu, Z. (2008). The inhibitory effects of silver nanoparticles, silver ions, and silver chloride colloids on microbial growth. *Water Research, 42*(12), 3066–3074.

Connor, E. E., Mwamuka, J., Gole, A., Murphy, C. J., & Wyatt, M. D. (2005). Gold nanoparticles are taken up by human cells but do not cause acute cytotoxicity. *Small, 1*(3), 325–327.

Correa-Llantén, D. N., Muñoz-Ibacache, S. A., Castro, M. E., Muñoz, P. A., & Blamey, J. M. (2013). Gold nanoparticles synthesized by Geobacillus sp. strain ID17 a thermophilic bacterium isolated from Deception Island, Antarctica. *Microbial Cell Factories, 12*(1), 1–6.

Costerton, J. W., Lewandowski, Z., Caldwell, D. E., Korber, D. R., & Lappin-Scott, H. M. (1995). Microbial biofilms. *Annual Review of Microbiology, 49*(1), 711–745.

Costerton, J. W., Stewart, P. S., & Greenberg, E. P. (1999). Bacterial biofilms: A common cause of persistent infections. *Science, 284*(5418), 1318–1322.

Cui, Y., Zhao, Y., Tian, Y., Zhang, W., Lü, X., & Jiang, X. (2012). The molecular mechanism of action of bactericidal gold nanoparticles on Escherichia coli. *Biomaterials, 33*(7), 2327–2333.

Daughton, C. G. (2004). Non-regulated water contaminants: Emerging research. *Environmental Impact Assessment Review, 24*(7–8), 711–732.

Davies, D. (2003). Understanding biofilm resistance to antibacterial agents. *Nature Reviews Drug Discovery, 2*(2), 114–122.

De Jong, W. H., Hagens, W. I., Krystek, P., Burger, M. C., Sips, A. J., & Geertsma, R. E. (2008). Particle size-dependent organ distribution of gold nanoparticles after intravenous administration. *Biomaterials, 29*(12), 1912–1919.

Debabov, V., Voeikova, T., Shebanova, A., Shaitan, K., Emel'yanova, L., Novikova, L., & Kirpichnikov, M. (2013). Bacterial synthesis of silver sulfide nanoparticles. *Nanotechnologies in Russia, 8*(3–4), 269–276.

Dedeh, A., Ciutat, A., Treguer-Delapierre, M., & Bourdineaud, J.-P. (2015). Impact of gold nanoparticles on zebrafish exposed to a spiked sediment. *Nanotoxicology, 9*(1), 71–80.

Deepa, S., Kanimozhi, K., & Panneerselvam, A. (2013). Antimicrobial activity of extracellularly synthesized silver nanoparticles from marine derived actinomycetes. *International Journal of Current Microbiology and Applied Sciences, 2*(2), 223–230.

Delgado-Beleño, Y., Martinez-Nuñez, C., Cortez-Valadez, M., Flores-López, N., & Flores-Acosta, M. (2018). Optical properties of silver, silver sulfide and silver selenide nanoparticles and antibacterial applications. *Materials Research Bulletin, 99*, 385–392.

DeLong, E. F., Frankel, R. B., & Bazylinski, D. A. (1993). Multiple evolutionary origins of magnetotaxis in bacteria. *Science, 259*(5096), 803–806.

Deplanche, K., & Macaskie, L. (2008). Biorecovery of gold by Escherichia coli and Desulfovibrio desulfuricans. *Biotechnology and Bioengineering, 99*(5), 1055–1064.

Dhandapani, P., Siddarth, A. S., Kamalasekaran, S., Maruthamuthu, S., & Rajagopal, G. (2014). Bio-approach: Ureolytic bacteria mediated synthesis of ZnO nanocrystals on cotton fabric and evaluation of their antibacterial properties. *Carbohydrate Polymers, 103*, 448–455.

Dhoondia, Z. H., & Chakraborty, H. (2012). Lactobacillus mediated synthesis of silver oxide nanoparticles. *Nanomaterials and Nanotechnology, 2*, 15.

Dibrov, P., Dzioba, J., Gosink, K. K., & Häse, C. C. (2002). Chemiosmotic mechanism of antimicrobial activity of ag+ in Vibrio cholerae. *Antimicrobial Agents and Chemotherapy, 46*(8), 2668–2670.

Dickinson, E. (2012). Use of nanoparticles and microparticles in the formation and stabilization of food emulsions. *Trends in Food Science & Technology, 24*(1), 4–12.

Dickson, J. S., & Koohmaraie, M. (1989). Cell surface charge characteristics and their relationship to bacterial attachment to meat surfaces. *Applied and Environmental Microbiology, 55*(4), 832–836.

Dizaj, S. M., Lotfipour, F., Barzegar-Jalali, M., Zarrintan, M. H., & Adibkia, K. (2014). Antimicrobial activity of the metals and metal oxide nanoparticles. *Materials Science and Engineering: C, 44*, 278–284.

Dobson, J. (2006). Magnetic nanoparticles for drug delivery. *Drug Development Research, 67*(1), 55–60.

Donlan, R. M., & Costerton, J. W. (2002). Biofilms: Survival mechanisms of clinically relevant microorganisms. *Clinical Microbiology Reviews, 15*(2), 167–193.

Du, L., Jiang, H., Liu, X., & Wang, E. (2007). Biosynthesis of gold nanoparticles assisted by Escherichia coli DH5α and its application on direct electrochemistry of hemoglobin. *Electrochemistry Communications, 9*(5), 1165–1170.

Elechiguerra, J. L., Burt, J. L., Morones, J. R., Camacho-Bragado, A., Gao, X., Lara, H. H., & Yacaman, M. J. (2005). Interaction of silver nanoparticles with HIV-1. *Journal of Nanobiotechnology, 3*(1), 1–10.

Emami-Karvani, Z., & Chehrazi, P. (2011). Antibacterial activity of ZnO nanoparticle on gram-positive and gram-negative bacteria. *African Journal of Microbiology Research, 5*(12), 1368–1373.

Eustis, S., & El-Sayed, M. A. (2006). Why gold nanoparticles are more precious than pretty gold: Noble metal surface plasmon resonance and its enhancement of the radiative and nonradiative properties of nanocrystals of different shapes. *Chemical Society Reviews, 35*(3), 209–217.

Fahr, S., Rockstuhl, C., & Lederer, F. (2009). Metallic nanoparticles as intermediate reflectors in tandem solar cells. *Applied Physics Letters, 95*(12), 121105.

Faivre, D., & Schuler, D. (2008). Magnetotactic bacteria and magnetosomes. *Chemical Reviews, 108*(11), 4875–4898.

Fajardo, C., Saccà, M., Martinez-Gomariz, M., Costa, G., Nande, M., & Martin, M. (2013). Transcriptional and proteomic stress responses of a soil bacterium Bacillus cereus to nanosized zero-valent iron (nZVI) particles. *Chemosphere, 93*(6), 1077–1083.

Falsetta, M. L., Klein, M. I., Colonne, P. M., Scott-Anne, K., Gregoire, S., Pai, C.-H., Gonzalez-Begne, M., Watson, G., Krysan, D. J., & Bowen, W. H. (2014). Symbiotic relationship between Streptococcus mutans and Candida albicans synergizes virulence of plaque biofilms in vivo. *Infection and Immunity, 82*(5), 1968–1981.

Fayaz, A. M., Girilal, M., Rahman, M., Venkatesan, R., & Kalaichelvan, P. (2011). Biosynthesis of silver and gold nanoparticles using thermophilic bacterium Geobacillus stearothermophilus. *Process Biochemistry, 46*(10), 1958–1962.

Flemming, H.-C., Neu, T. R., & Wozniak, D. J. (2007). The EPS matrix: The "house of biofilm cells". *Journal of Bacteriology, 189*(22), 7945–7947.

Fulaz, S., Vitale, S., Quinn, L., & Casey, E. (2019). Nanoparticle–biofilm interactions: The role of the EPS matrix. *Trends in Microbiology, 27*(11), 915–926.

Fux, C. A., Costerton, J. W., Stewart, P. S., & Stoodley, P. (2005). Survival strategies of infectious biofilms. *Trends in Microbiology, 13*(1), 34–40.

Gabrielyan, L., Hovhannisyan, A., Gevorgyan, V., Ananyan, M., & Trchounian, A. (2019). Antibacterial effects of iron oxide (Fe_3O_4) nanoparticles: Distinguishing concentration-dependent effects with different bacterial cells growth and membrane-associated mechanisms. *Applied Microbiology and Biotechnology, 103*(6), 2773–2782.

Gahlawat, G., & Choudhury, A. R. (2019). A review on the biosynthesis of metal and metal salt nanoparticles by microbes. *RSC Advances, 9*(23), 12944–12967.

Gatoo, M. A., Naseem, S., Arfat, M. Y., Mahmood Dar, A., Qasim, K., & Zubair, S. (2014). Physicochemical properties of nanomaterials: Implication in associated toxic manifestations. *BioMed Research International, 2014*, 498420.

Gómez, M. I., & Prince, A. (2007). Opportunistic infections in lung disease: Pseudomonas infections in cystic fibrosis. *Current Opinion in Pharmacology, 7*(3), 244–251.

Gómez-Gómez, B., Arregui, L., Serrano, S., Santos, A., Pérez-Corona, T., & Madrid, Y. (2019). Unravelling mechanisms of bacterial quorum sensing disruption by metal-based nanoparticles. *Science of the Total Environment, 696*, 133869.

Gurunathan, S., Kalishwaralal, K., Vaidyanathan, R., Venkataraman, D., Pandian, S. R. K., Muniyandi, J., Hariharan, N., & Eom, S. H. (2009). Biosynthesis, purification and characterization of silver nanoparticles using Escherichia coli. *Colloids and Surfaces B: Biointerfaces, 74*(1), 328–335.

Haefeli, C., Franklin, C., & Hardy, K. (1984). Plasmid-determined silver resistance in Pseudomonas stutzeri isolated from a silver mine. *Journal of Bacteriology, 158*(1), 389–392.

Hagen, S. J. (2014). *The physical basis of bacterial quorum communication*. Springer.

Haider, A. J., Jameel, Z. N., & Al-Hussaini, I. H. (2019). Review on: Titanium dioxide applications. *Energy Procedia, 157*, 17–29.

Hall, C. W., & Mah, T.-F. (2017). Molecular mechanisms of biofilm-based antibiotic resistance and tolerance in pathogenic bacteria. *FEMS Microbiology Reviews, 41*(3), 276–301.

Haris, Z., & Ahmad, I. (2017). Impact of metal oxide nanoparticles on beneficial soil microorganisms and their secondary metabolites. *International Journal of Life Science Scientific Research, 3*(3), 1020–1030.

Hasan, S. (2015). A review on nanoparticles: Their synthesis and types. *Research Journal of Recent Sciences, 2277*, 2502.

He, S., Guo, Z., Zhang, Y., Zhang, S., Wang, J., & Gu, N. (2007). Biosynthesis of gold nanoparticles using the bacteria Rhodopseudomonas capsulata. *Materials Letters, 61*(18), 3984–3987.

Hetrick, E. M., Shin, J. H., Paul, H. S., & Schoenfisch, M. H. (2009). Anti-biofilm efficacy of nitric oxide-releasing silica nanoparticles. *Biomaterials, 30*(14), 2782–2789.

Hulkoti, N. I., & Taranath, T. (2014). Biosynthesis of nanoparticles using microbes—A review. *Colloids and Surfaces B: Biointerfaces, 121*, 474–483.

Husseiny, M., Abd El-Aziz, M., Badr, Y., & Mahmoud, M. (2007). Biosynthesis of gold nanoparticles using Pseudomonas aeruginosa. *Spectrochimica Acta Part A: Molecular and Biomolecular Spectroscopy, 67*(3–4), 1003–1006.

Ibrahem, K. H., Salman, J. A. S., & Ali, F. A. (2014). Effect of titanium nanoparticles biosynthesis by lactobacillus Crispatus on urease, Hemolysin& Biofilm Forming by some Bacteria causing recurrent UTI in Iraqi women. *European Scientific Journal, 10*(9).

Ingle, A., Gade, A., Pierrat, S., Sonnichsen, C., & Rai, M. (2008). Mycosynthesis of silver nanoparticles using the fungus Fusarium acuminatum and its activity against some human pathogenic bacteria. *Current Nanoscience, 4*(2), 141–144.

Iravani, S. (2014). Bacteria in nanoparticle synthesis: current status and future prospects. International scholarly research notices 2014.

Iravani, S., Korbekandi, H., Mirmohammadi, S. V., & Zolfaghari, B. (2014). Synthesis of silver nanoparticles: Chemical, physical and biological methods. *Research in Pharmaceutical Sciences, 9*(6), 385.

Ivask, A., Kurvet, I., Kasemets, K., Blinova, I., Aruoja, V., Suppi, S., Vija, H., Käkinen, A., Titma, T., & Heinlaan, M. (2014). Size-dependent toxicity of silver nanoparticles to bacteria, yeast, algae, crustaceans and mammalian cells in vitro. *PLoS One, 9*(7), e102108.

Iyer, R., Moussa, S. H., Durand-Reville, T. F., Tommasi, R., & Miller, A. (2017). Acinetobacter baumannii OmpA is a selective antibiotic permeant porin. *ACS Infectious Diseases, 4*(3), 373–381.

Jain, P. K., Lee, K. S., El-Sayed, I. H., & El-Sayed, M. A. (2006). Calculated absorption and scattering properties of gold nanoparticles of different size, shape, and composition: Applications in biological imaging and biomedicine. *The Journal of Physical Chemistry B, 110*(14), 7238–7248.

Javaid, A., Oloketuyi, S. F., Khan, M. M., & Khan, F. (2018). Diversity of bacterial synthesis of silver nanoparticles. *BioNanoScience, 8*(1), 43–59.

Jayaseelan, C., Rahuman, A. A., Kirthi, A. V., Marimuthu, S., Santhoshkumar, T., Bagavan, A., Gaurav, K., Karthik, L., & Rao, K. B. (2012). Novel microbial route to synthesize ZnO nanoparticles using Aeromonas hydrophila and their activity against pathogenic bacteria and fungi. *Spectrochimica Acta Part A: Molecular and Biomolecular Spectroscopy, 90*, 78–84.

Joerger, R., Klaus, T., & Granqvist, C. G. (2000). Biologically produced silver–carbon composite materials for optically functional thin-film coatings. *Advanced Materials, 12*(6), 407–409.

Juibari, M. M., Abbasalizadeh, S., Jouzani, G. S., & Noruzi, M. (2011). Intensified biosynthesis of silver nanoparticles using a native extremophilic Ureibacillus thermosphaericus strain. *Materials Letters, 65*(6), 1014–1017.

Kagawa, Y. (1978). Reconstitution of the energy transformer, gate and channel subunit reassembly, crystalline ATPase and ATP synthesis. *Biochimica et Biophysica Acta (BBA)—Reviews on Bioenergetics, 505*(1), 45–93.

Kalia, V. C., & Purohit, H. J. (2011). Quenching the quorum sensing system: Potential antibacterial drug targets. *Critical Reviews in Microbiology, 37*(2), 121–140.

Kalimuthu, K., Babu, R. S., Venkataraman, D., Bilal, M., & Gurunathan, S. (2008). Biosynthesis of silver nanocrystals by Bacillus licheniformis. *Colloids and Surfaces B: Biointerfaces, 65*(1), 150–153.

Kalishwaralal, K., Deepak, V., Pandian, S. R. K., Kottaisamy, M., BarathManiKanth, S., Kartikeyan, B., & Gurunathan, S. (2010). Biosynthesis of silver and gold nanoparticles using Brevibacterium casei. *Colloids and Surfaces B: Biointerfaces, 77*(2), 257–262.

Kalishwaralal, K., Deepak, V., Ramkumarpandian, S., Nellaiah, H., & Sangiliyandi, G. (2008). Extracellular biosynthesis of silver nanoparticles by the culture supernatant of Bacillus licheniformis. *Materials Letters, 62*(29), 4411–4413.

Kalpana, D., & Lee, Y. S. (2013). Synthesis and characterization of bactericidal silver nanoparticles using cultural filtrate of simulated microgravity grown Klebsiella pneumoniae. *Enzyme and Microbial Technology, 52*(3), 151–156.

Karthik, C., & Radha, K. (2012). Biosynthesis and characterization of silver nanoparticles using Enterobacter aerogenes: A kinetic approach. *Digest Journal of Nanomaterials and Biostructures, 7*(3), 1007–1014.

Kaviya, S., Santhanalakshmi, J., Viswanathan, B., Muthumary, J., & Srinivasan, K. (2011). Biosynthesis of silver nanoparticles using Citrus sinensis peel extract and its antibacterial activity. *Spectrochimica Acta Part A: Molecular and Biomolecular Spectroscopy, 79*(3), 594–598.

Khan, R., & Fulekar, M. (2016). Biosynthesis of titanium dioxide nanoparticles using Bacillus amyloliquefaciens culture and enhancement of its photocatalytic activity for the degradation of a sulfonated textile dye reactive red 31. *Journal of Colloid and Interface Science, 475,* 184–191.

Khan, S. T., Ahmad, J., Ahamed, M., & Jousset, A. (2018). Sub-lethal doses of widespread nanoparticles promote antifungal activity in Pseudomonas protegens CHA0. *Science of the Total Environment, 627,* 658–662.

Kim, K.-J., Sung, W. S., Suh, B. K., Moon, S.-K., Choi, J.-S., Kim, J. G., & Lee, D. G. (2009). Antifungal activity and mode of action of silver nano-particles on Candida albicans. *Biometals, 22*(2), 235–242.

Kirthi, A. V., Rahuman, A. A., Rajakumar, G., Marimuthu, S., Santhoshkumar, T., Jayaseelan, C., Elango, G., Zahir, A. A., Kamaraj, C., & Bagavan, A. (2011). Biosynthesis of titanium dioxide nanoparticles using bacterium Bacillus subtilis. *Materials Letters, 65*(17–18), 2745–2747.

Klaus, T., Joerger, R., Olsson, E., & Granqvist, C.-G. (1999). Silver-based crystalline nanoparticles, microbially fabricated. *Proceedings of the National Academy of Sciences, 96*(24), 13611–13614.

Klaus-Joerger, T., Joerger, R., Olsson, E., & Granqvist, C.-G. (2001). Bacteria as workers in the living factory: Metal-accumulating bacteria and their potential for materials science. *Trends in Biotechnology, 19*(1), 15–20.

Koo, H., Allan, R. N., Howlin, R. P., Stoodley, P., & Hall-Stoodley, L. (2017). Targeting microbial biofilms: Current and prospective therapeutic strategies. *Nature Reviews Microbiology, 15*(12), 740.

Korbekandi, H., Iravani, S., & Abbasi, S. (2012). Optimization of biological synthesis of silver nanoparticles using lactobacillus casei subsp. casei. *Journal of Chemical Technology & Biotechnology, 87*(7), 932–937.

Krishnaraj, R. N., & Berchmans, S. (2013). In vitro antiplatelet activity of silver nanoparticles synthesized using the microorganism Gluconobacter roseus: An AFM-based study. *RSC Advances, 3*(23), 8953–8959.

Król, A., Railean-Plugaru, V., Pomastowski, P., Złoch, M., & Buszewski, B. (2018). Mechanism study of intracellular zinc oxide nanocomposites formation. *Colloids and Surfaces A: Physicochemical and Engineering Aspects, 553,* 349–358.

Kulkarni, N., & Muddapur, U. (2014). Biosynthesis of metal nanoparticles: A review. *Journal of Nanotechnology, 2014,* 510246.

Kumar, C. G., & Mamidyala, S. K. (2011). Extracellular synthesis of silver nanoparticles using culture supernatant of Pseudomonas aeruginosa. *Colloids and Surfaces B: Biointerfaces, 84*(2), 462–466.

Kumar, S. A., Abyaneh, M. K., Gosavi, S., Kulkarni, S. K., Pasricha, R., Ahmad, A., & Khan, M. (2007). Nitrate reductase-mediated synthesis of silver nanoparticles from $AgNO_3$. *Biotechnology Letters, 29*(3), 439–445.

Kundu, D., Hazra, C., Chatterjee, A., Chaudhari, A., & Mishra, S. (2014). Extracellular biosynthesis of zinc oxide nanoparticles using Rhodococcus pyridinivorans NT2: Multifunctional textile finishing, biosafety evaluation and in vitro drug delivery in colon carcinoma. *Journal of Photochemistry and Photobiology B: Biology, 140,* 194–204.

Lang, C., & Schüler, D. (2006). Biogenic nanoparticles: Production, characterization, and application of bacterial magnetosomes. *Journal of Physics: Condensed Matter, 18*(38), S2815.

Lasa, I. (2006). Towards the identification of the common features of bacterial biofilm development. *International Microbiology, 9*(1), 21–28.

Lateef, A., Adelere, I., Gueguim-Kana, E., Asafa, T., & Beukes, L. (2015). Green synthesis of silver nanoparticles using keratinase obtained from a strain of Bacillus safensis LAU 13. *International Nano Letters, 5*(1), 29–35.

Lee, C., Kim, J. Y., Lee, W. I., Nelson, K. L., Yoon, J., & Sedlak, D. L. (2008). Bactericidal effect of zero-valent iron nanoparticles on Escherichia coli. *Environmental Science & Technology, 42*(13), 4927–4933.

Lee, K. J., Nallathamby, P. D., Browning, L. M., Osgood, C. J., & X-HN, X. (2007). In vivo imaging of transport and biocompatibility of single silver nanoparticles in early development of zebrafish embryos. *ACS Nano, 1*(2), 133–143.

Li, M. Z., Huang, J. T., Tsai, Y. H., Mao, S. Y., Fu, C. M., & Lien, T. F. (2016). Nanosize of zinc oxide and the effects on zinc digestibility, growth performances, immune response and serum parameters of weanling piglets. *Animal Science Journal, 87*(11), 1379–1385.

Li, Z., Greden, K., Alvarez, P. J., Gregory, K. B., & Lowry, G. V. (2010). Adsorbed polymer and NOM limits adhesion and toxicity of nano scale zerovalent iron to E. coli. *Environmental Science & Technology, 44*(9), 3462–3467.

Liao, Y., Strayer-Scherer, A., White, J., De La Torre-Roche, R., Ritchie, L., Colee, J., Vallad, G., Freeman, J., Jones, J., & Paret, M. (2019). Particle-size dependent bactericidal activity of magnesium oxide against Xanthomonas perforans and bacterial spot of tomato. *Scientific Reports, 9*(1), 1–10.

Link, S., & El-Sayed, M. A. (1999). Size and temperature dependence of the plasmon absorption of colloidal gold nanoparticles. *The Journal of Physical Chemistry B, 103*(21), 4212–4217.

Livingstone, D. (2001). Contaminant-stimulated reactive oxygen species production and oxidative damage in aquatic organisms. *Marine Pollution Bulletin, 42*(8), 656–666.

Livingstone, D., Martinez, P. G., Michel, X., Narbonne, J., O'hara, S., Ribera, D., & Winston, G. (1990). Oxyradical production as a pollution-mediated mechanism of toxicity in the common mussel, Mytilus edulis L., and other molluscs. *Functional Ecology, 4*, 415–424.

Lok, C.-N., Ho, C.-M., Chen, R., He, Q.-Y., Yu, W.-Y., Sun, H., Tam, P. K.-H., Chiu, J.-F., & Che, C.-M. (2006). Proteomic analysis of the mode of antibacterial action of silver nanoparticles. *Journal of Proteome Research, 5*(4), 916–924.

Lovley, D. R., Stolz, J. F., Nord, G. L., & Phillips, E. J. (1987). Anaerobic production of magnetite by a dissimilatory iron-reducing microorganism. *Nature, 330*(6145), 252–254.

Lu, L., Sun, R., Chen, R., Hui, C.-K., Ho, C.-M., Luk, J. M., Lau, G., & Che, C.-M. (2008). Silver nanoparticles inhibit hepatitis B virus replication. *Antiviral Therapy, 13*(2), 253.

Luo, X., Xu, S., Yang, Y., Li, L., Chen, S., Xu, A., & Wu, L. (2016). Insights into the ecotoxicity of silver nanoparticles transferred from Escherichia coli to Caenorhabditis elegans. *Scientific Reports, 6*, 36465.

Mandal, D., Bolander, M. E., Mukhopadhyay, D., Sarkar, G., & Mukherjee, P. (2006). The use of microorganisms for the formation of metal nanoparticles and their application. *Applied Microbiology and Biotechnology, 69*(5), 485–492.

Mandal, D., Dash, S. K., Das, B., Chattopadhyay, S., Ghosh, T., Das, D., & Roy, S. (2016). Biofabricated silver nanoparticles preferentially targets gram positive depending on cell surface charge. *Biomedicine & Pharmacotherapy, 83*, 548–558.

Mann, S. (1985). Structure, morphology, and crystal growth of bacterial magnetite. In *Magnetite biomineralization and magnetoreception in organisms* (pp. 311–332). Springer.

Manuela, V., Ingo, K., & Arno, K. (2013). Zinc oxide nanoparticles in bacterial growth medium: Optimized dispersion and growth inhibition of Pseudomonas putida. *Advances in Nanoparticles, 2*, 287.

Marambio-Jones, C., & Hoek, E. M. (2010). A review of the antibacterial effects of silver nanomaterials and potential implications for human health and the environment. *Journal of Nanoparticle Research, 12*(5), 1531–1551.

Martínez-Castañon, G.-A., Nino-Martinez, N., Martinez-Gutierrez, F., Martinez-Mendoza, J., & Ruiz, F. (2008). Synthesis and antibacterial activity of silver nanoparticles with different sizes. *Journal of Nanoparticle Research, 10*(8), 1343–1348.

Mirzaei, H., & Darroudi, M. (2017). Zinc oxide nanoparticles: Biological synthesis and biomedical applications. *Ceramics International, 43*(1), 907–914.

Mishra, M., Paliwal, J. S., Singh, S. K., Selvarajan, E., Subathradevi, C., & Mohanasrinivasan, V. (2013). Studies on the inhibitory activity of biologically synthesized and characterized zinc oxide nanoparticles using lactobacillus sporogens against Staphylococcus aureus. *Journal of Pure Applied Microbiology, 7*(2), 1–6.

Mohamed, M. M., Fouad, S. A., Elshoky, H. A., Mohammed, G. M., & Salaheldin, T. A. (2017). Antibacterial effect of gold nanoparticles against Corynebacterium pseudotuberculosis. *International Journal of Veterinary Science and Medicine, 5*(1), 23–29.

Moore, M. (2006). Do nanoparticles present ecotoxicological risks for the health of the aquatic environment? *Environment International, 32*(8), 967–976.

Moore, M. N. (1990). Lysosomal cytochemistry in marine environmental monitoring. *The Histochemical Journal, 22*(4), 187.

Morones, J. R., Elechiguerra, J. L., Camacho, A., Holt, K., Kouri, J. B., Ramírez, J. T., & Yacaman, M. J. (2005). The bactericidal effect of silver nanoparticles. *Nanotechnology, 16*(10), 2346.

Moshfegh, M., Forootanfar, H., Zare, B., Shahverdi, A., Zarrini, G., & Faramarzi, M. (2011). Biological synthesis of Au, Ag and Au-Ag bimetallic nanoparticles by α-amylase. *Digest Journal of Nanomaterial Biostructures, 6*, 1419–1426.

Mouxing, F., Qingbiao, L., Daohua, S., Yinghua, L., Ning, H., Xu, D., Huixuan, W., & Huang, J. (2006). Rapid preparation process of silver nanoparticles by bioreduction and their characterizations. *Chinese Journal of Chemical Engineering, 14*(1), 114–117.

Mukherjee, P., Ahmad, A., Mandal, D., Senapati, S., Sainkar, S. R., Khan, M. I., Ramani, R., Parischa, R., Ajayakumar, P., & Alam, M. (2001). Bioreduction of AuCl4− ions by the fungus, Verticillium sp. and surface trapping of the gold nanoparticles formed. *Angewandte Chemie International Edition, 40*(19), 3585–3588.

Nadagouda, M. N., Castle, A. B., Murdock, R. C., Hussain, S. M., & Varma, R. S. (2010). In vitro biocompatibility of nanoscale zerovalent iron particles (NZVI) synthesized using tea polyphenols. *Green Chemistry, 12*(1), 114–122.

Nair, B., & Pradeep, T. (2002). Coalescence of nanoclusters and formation of submicron crystallites assisted by lactobacillus strains. *Crystal Growth & Design, 2*(4), 293–298.

Nanda, A., & Saravanan, M. (2009). Biosynthesis of silver nanoparticles from Staphylococcus aureus and its antimicrobial activity against MRSA and MRSE. *Nanomedicine: Nanotechnology, Biology and Medicine, 5*(4), 452–456.

Nangia, Y., Wangoo, N., Goyal, N., Shekhawat, G., & Suri, C. R. (2009). A novel bacterial isolate Stenotrophomonas maltophilia as living factory for synthesis of gold nanoparticles. *Microbial Cell Factories, 8*(1), 39.

Narayanan, K. B., & Sakthivel, N. (2010). Biological synthesis of metal nanoparticles by microbes. *Advances in Colloid and Interface Science, 156*(1–2), 1–13.

Narayanan, K. B., & Sakthivel, N. (2013). Biosynthesis of silver nanoparticles by phytopathogen Xanthomonas oryzae pv. Oryzae strain BXO8. *Journal of Microbiology and Biotechnology, 23*(9), 1287–1292.

Nune, S. K., Gunda, P., Thallapally, P. K., Lin, Y.-Y., Laird Forrest, M., & Berkland, C. J. (2009). Nanoparticles for biomedical imaging. *Expert Opinion on Drug Delivery, 6*(11), 1175–1194.

Oberdörster, G. (2000). Toxicology of ultrafine particles: In vivo studies. *Phil Trans R Soc Lond A, 358*, 2719–2740. Find this article online.

Olsen, I. (2015). Biofilm-specific antibiotic tolerance and resistance. *European Journal of Clinical Microbiology & Infectious Diseases, 34*(5), 877–886.

Otari, S., Patil, R., Nadaf, N., Ghosh, S., & Pawar, S. (2014). Green synthesis of silver nanoparticles by microorganism using organic pollutant: Its antimicrobial and catalytic application. *Environmental Science and Pollution Research, 21*(2), 1503–1513.

Panyam, J., Sahoo, S. K., Prabha, S., Bargar, T., & Labhasetwar, V. (2003). Fluorescence and electron microscopy probes for cellular and tissue uptake of poly (D, L-lactide-co-glycolide) nanoparticles. *International Journal of Pharmaceutics, 262*(1–2), 1–11.

Parikh, R. Y., Singh, S., Prasad, B., Patole, M. S., Sastry, M., & Shouche, Y. S. (2008). Extracellular synthesis of crystalline silver nanoparticles and molecular evidence of silver resistance from Morganella sp.: Towards understanding biochemical synthesis mechanism. *Chembiochem, 9*(9), 1415–1422.

Park, H.-J., Kim, J. Y., Kim, J., Lee, J.-H., Hahn, J.-S., Gu, M. B., & Yoon, J. (2009). Silver-ion-mediated reactive oxygen species generation affecting bactericidal activity. *Water Research, 43*(4), 1027–1032.

Patel, A., Prajapati, P., & Boghra, R. (2011). Overview on application of nanoparticles in cosmetics. *Asian Journal of Pharmaceutical and Clinical Research, 1*, 40–55.

Pham, D. T. N., Khan, F., Phan, T. T. V., S-k, P., Manivasagan, P., Oh, J., & Kim, Y.-M. (2019). Biofilm inhibition, modulation of virulence and motility properties by FeOOH nanoparticle in Pseudomonas aeruginosa. *Brazilian Journal of Microbiology, 50*(3), 791–805.

Phenrat, T., Long, T. C., Lowry, G. V., & Veronesi, B. (2009). Partial oxidation ("aging") and surface modification decrease the toxicity of nanosized zerovalent iron. *Environmental Science & Technology, 43*(1), 195–200.

Pradeep, T. (2009). Noble metal nanoparticles for water purification: A critical review. *Thin Solid Films, 517*(24), 6441–6478.

Prasad, K., & Jha, A. K. (2009). ZnO nanoparticles: Synthesis and adsorption study. *Natural Science, 1*(02), 129.

Prasad, K., Jha, A. K., & Kulkarni, A. (2007). Lactobacillus assisted synthesis of titanium nanoparticles. *Nanoscale Research Letters, 2*(5), 248–250.

Priyadarshini, S., Gopinath, V., Priyadharsshini, N. M., MubarakAli, D., & Velusamy, P. (2013). Synthesis of anisotropic silver nanoparticles using novel strain, Bacillus flexus and its biomedical application. *Colloids and Surfaces B: Biointerfaces, 102*, 232–237.

Prozorov, T. (2015). Magnetic microbes: Bacterial magnetite biomineralization. In *Seminars in cell & developmental biology* (pp. 36–43). Elsevier.

Pugazhenthiran, N., Anandan, S., Kathiravan, G., Prakash, N. K. U., Crawford, S., & Ashokkumar, M. (2009). Microbial synthesis of silver nanoparticles by Bacillus sp. *Journal of Nanoparticle Research, 11*(7), 1811.

Rajabairavi, N., Raju, C. S., Karthikeyan, C., Varutharaju, K., Nethaji, S., Hameed, A. S. H., & Shajahan, A. (2017). Biosynthesis of novel zinc oxide nanoparticles (ZnO NPs) using endophytic bacteria Sphingobacterium thalpophilum. In *Recent trends in materials science and applications* (pp. 245–254). Springer.

Rajasree, S. R., & Suman, T. (2012). Extracellular biosynthesis of gold nanoparticles using a gram negative bacterium Pseudomonas fluorescens. *Asian Pacific Journal of Tropical Disease, 2*, S796–S799.

Ramalingam, B., Parandhaman, T., & Das, S. K. (2016). Antibacterial effects of biosynthesized silver nanoparticles on surface ultrastructure and nanomechanical properties of gram-negative bacteria viz. Escherichia coli and Pseudomonas aeruginosa. *ACS Applied Materials & Interfaces, 8*(7), 4963–4976.

Rauf, M. A., Owais, M., Rajpoot, R., Ahmad, F., Khan, N., & Zubair, S. (2017). Biomimetically synthesized ZnO nanoparticles attain potent antibacterial activity against less susceptible S. aureus skin infection in experimental animals. *RSC Advances, 7*(58), 36361–36373.

Ravindran, D., Ramanathan, S., Arunachalam, K., Jeyaraj, G., Shunmugiah, K., & Arumugam, V. (2018). Phytosynthesized silver nanoparticles as antiquorum sensing and antibiofilm agent against the nosocomial pathogen Serratia marcescens: An in vitro study. *Journal of Applied Microbiology, 124*(6), 1425–1440.

Raza, M. A., Kanwal, Z., Rauf, A., Sabri, A. N., Riaz, S., & Naseem, S. (2016). Size-and shape-dependent antibacterial studies of silver nanoparticles synthesized by wet chemical routes. *Nanomaterials, 6*(4), 74.

Reddy, A. S., Chen, C.-Y., Chen, C.-C., Jean, J.-S., Chen, H.-R., Tseng, M.-J., Fan, C.-W., & Wang, J.-C. (2010). Biological synthesis of gold and silver nanoparticles mediated by the bacteria Bacillus subtilis. *Journal of Nanoscience and Nanotechnology, 10*(10), 6567–6574.

Revati, K., & Pandey, B. (2011). Microbial synthesis of iron-based nanomaterials—A review. *Bulletin of Materials Science, 34*(2), 191–198.

Saccà, M. L., Fajardo, C., Costa, G., Lobo, C., Nande, M., & Martin, M. (2014). Integrating classical and molecular approaches to evaluate the impact of nanosized zero-valent iron (nZVI) on soil organisms. *Chemosphere, 104*, 184–189.

Saif, S., Tahir, A., & Chen, Y. (2016). Green synthesis of iron nanoparticles and their environmental applications and implications. *Nanomaterials, 6*(11), 209.

Saifuddin, N., Wong, C., & Yasumira, A. (2009). Rapid biosynthesis of silver nanoparticles using culture supernatant of bacteria with microwave irradiation. *Journal of Chemistry, 6*(1), 61–70.

Saleh, M. M., Refa't, A. S., Latif, H. K. A., Abbas, H. A., & Askoura, M. (2019). Zinc oxide nanoparticles inhibits quorum sensing and virulence in Pseudomonas aeruginosa. *African Health Sciences, 19*(2), 2043–2055.

Samadi, N., Golkaran, D., Eslamifar, A., Jamalifar, H., Fazeli, M. R., & Mohseni, F. A. (2009). Intra/extracellular biosynthesis of silver nanoparticles by an autochthonous strain of proteus mirabilis isolated fromphotographic waste. *Journal of Biomedical Nanotechnology, 5*(3), 247–253.

Samanta, S., Singh, B. R., & Adholeya, A. (2017). Intracellular synthesis of gold nanoparticles using an ectomycorrhizal strain EM-1083 of Laccaria fraterna and its nanoanti-quorum sensing potential against Pseudomonas aeruginosa. *Indian Journal of Microbiology, 57*(4), 448–460.

Saravanan, M., Gopinath, V., Chaurasia, M. K., Syed, A., Ameen, F., & Purushothaman, N. (2018). Green synthesis of anisotropic zinc oxide nanoparticles with antibacterial and cytofriendly properties. *Microbial Pathogenesis, 115*, 57–63.

Sawai, J. (2003). Quantitative evaluation of antibacterial activities of metallic oxide powders (ZnO, MgO and CaO) by conductimetric assay. *Journal of Microbiological Methods, 54*(2), 177–182.

Schlessinger, D. (1988). Failure of aminoglycoside antibiotics to kill anaerobic, low-pH, and resistant cultures. *Clinical Microbiology Reviews, 1*(1), 54–59.

Schuster, M., Lostroh, C. P., Ogi, T., & Greenberg, E. P. (2003). Identification, timing, and signal specificity of Pseudomonas aeruginosa quorum-controlled genes: A transcriptome analysis. *Journal of Bacteriology, 185*(7), 2066–2079.

Selvarajan, E., & Mohanasrinivasan, V. (2013). Biosynthesis and characterization of ZnO nanoparticles using lactobacillus plantarum VITES07. *Materials Letters, 112*, 180–182.

Seshadri, S., Prakash, A., & Kowshik, M. (2012). Biosynthesis of silver nanoparticles by marine bacterium, Idiomarina sp. PR58-8. *Bulletin of Materials Science, 35*(7), 1201–1205.

Shah, M. S. A. S., Nag, M., Kalagara, T., Singh, S., & Manorama, S. V. (2008). Silver on PEG-PU-TiO$_2$ polymer nanocomposite films: An excellent system for antibacterial applications. *Chemistry of Materials, 20*(7), 2455–2460.

Shamaila, S., Zafar, N., Riaz, S., Sharif, R., Nazir, J., & Naseem, S. (2016). Gold nanoparticles: An efficient antimicrobial agent against enteric bacterial human pathogen. *Nanomaterials, 6*(4), 71.

Shanthi, S., Jayaseelan, B. D., Velusamy, P., Vijayakumar, S., Chih, C. T., & Vaseeharan, B. (2016). Biosynthesis of silver nanoparticles using a probiotic Bacillus licheniformis Dahb1 and their antibiofilm activity and toxicity effects in Ceriodaphnia cornuta. *Microbial Pathogenesis, 93*, 70–77.

Sharma, N., Pinnaka, A. K., Raje, M., Ashish, F., Bhattacharyya, M. S., & Choudhury, A. R. (2012). Exploitation of marine bacteria for production of gold nanoparticles. *Microbial Cell Factories, 11*(1), 86.

Sharma, V. K., Yngard, R. A., & Lin, Y. (2009). Silver nanoparticles: Green synthesis and their antimicrobial activities. *Advances in Colloid and Interface Science, 145*(1–2), 83–96.

Sheng, Z., & Liu, Y. (2011). Effects of silver nanoparticles on wastewater biofilms. *Water Research, 45*(18), 6039–6050.

Shivaji, S., Madhu, S., & Singh, S. (2011). Extracellular synthesis of antibacterial silver nanoparticles using psychrophilic bacteria. *Process Biochemistry, 46*(9), 1800–1807.

Simon-Deckers, A., Loo, S., Mayne-L'hermite, M., Herlin-Boime, N., Menguy, N., Reynaud, C., Gouget, B., & Carriere, M. (2009). Size-, composition-and shape-dependent toxicological impact of metal oxide nanoparticles and carbon nanotubes toward bacteria. *Environmental Science & Technology, 43*(21), 8423–8429.

Singh, B. N., Rawat, A. K. S., Khan, W., Naqvi, A. H., & Singh, B. R. (2014). Biosynthesis of stable antioxidant ZnO nanoparticles by Pseudomonas aeruginosa rhamnolipids. *PLoS One, 9*(9), e106937.

Singh, P., Kim, Y.-J., Zhang, D., & Yang, D.-C. (2016). Biological synthesis of nanoparticles from plants and microorganisms. *Trends in Biotechnology, 34*(7), 588–599.

Singh, P. K., & Kundu, S. (2014). Biosynthesis of gold nanoparticles using bacteria. *Proceedings of the National Academy of Sciences, India Section B: Biological Sciences, 84*(2), 331–336.

Singh, P. K., Parsek, M. R., Greenberg, E. P., & Welsh, M. J. (2002). A component of innate immunity prevents bacterial biofilm development. *Nature, 417*(6888), 552–555.

Singh, R., Shedbalkar, U. U., Wadhwani, S. A., & Chopade, B. A. (2015). Bacteriagenic silver nanoparticles: Synthesis, mechanism, and applications. *Applied Microbiology and Biotechnology, 99*(11), 4579–4593.

Singh, R., Wagh, P., Wadhwani, S., Gaidhani, S., Kumbhar, A., Bellare, J., & Chopade, B. A. (2013). Synthesis, optimization, and characterization of silver nanoparticles from Acinetobacter calcoaceticus and their enhanced antibacterial activity when combined with antibiotics. *International Journal of Nanomedicine, 8*, 4277.

Sirelkhatim, A., Mahmud, S., Seeni, A., Kaus, N. H. M., Ann, L. C., Bakhori, S. K. M., Hasan, H., & Mohamad, D. (2015). Review on zinc oxide nanoparticles: Antibacterial activity and toxicity mechanism. *Nano-Micro Letters, 7*(3), 219–242.

Slavin, Y. N., Asnis, J., Häfeli, U. O., & Bach, H. (2017). Metal nanoparticles: Understanding the mechanisms behind antibacterial activity. *Journal of Nanobiotechnology, 15*(1), 1–20.

Smani, Y., Fàbrega, A., Roca, I., Sánchez-Encinales, V., Vila, J., & Pachón, J. (2014). Role of OmpA in the multidrug resistance phenotype of Acinetobacter baumannii. *Antimicrobial Agents and Chemotherapy, 58*(3), 1806–1808.

Smedes, F. (1994). Sampling and partition of neutral organic contaminants in surface waters with regard to legislation, environmental quality and flux estimations. *International Journal of Environmental Analytical Chemistry, 57*(3), 215–229.

Sohm, B., Immel, F., Bauda, P., & Pagnout, C. (2015). Insight into the primary mode of action of TiO_2 nanoparticles on Escherichia coli in the dark. *Proteomics, 15*(1), 98–113.

Sondi, I., & Salopek-Sondi, B. (2004). Silver nanoparticles as antimicrobial agent: A case study on E. coli as a model for gram-negative bacteria. *Journal of Colloid and Interface Science, 275*(1), 177–182.

Sparks, N., Lloyd, J., & Board, R. (1989). Saltmarsh ponds—A preferred habitat for magnetotactic bacteria? *Letters in Applied Microbiology, 8*(3), 109–111.

Srikar, S. K., Giri, D. D., Pal, D. B., Mishra, P. K., & Upadhyay, S. N. (2016). Green synthesis of silver nanoparticles: A review. *Green and Sustainable Chemistry, 6*(1), 34–56.

Srivastava, S. K., & Constanti, M. (2012). Room temperature biogenic synthesis of multiple nanoparticles (ag, Pd, Fe, Rh, Ni, Ru, Pt, co, and Li) by Pseudomonas aeruginosa SM1. *Journal of Nanoparticle Research, 14*(4), 831.

Stewart, P. S., & Costerton, J. W. (2001). Antibiotic resistance of bacteria in biofilms. *The Lancet, 358*(9276), 135–138.

Stohs, S. J., & Bagchi, D. (1995). Oxidative mechanisms in the toxicity of metal ions. *Free Radical Biology and Medicine, 18*(2), 321–336.

Sun, L., Singh, A. K., Vig, K., Pillai, S. R., & Singh, S. R. (2008). Silver nanoparticles inhibit replication of respiratory syncytial virus. *Journal of Biomedical Nanotechnology, 4*(2), 149–158.

Swain, P. S., Rao, S. B., Rajendran, D., Dominic, G., & Selvaraju, S. (2016). Nano zinc, an alternative to conventional zinc as animal feed supplement: A review. *Animal Nutrition, 2*(3), 134–141.

Tack, K. J., & Sabath, L. (1985). Increased minimum inhibitory concentrations with anaerobiasis for tobramycin, gentamicin, and amikacin, compared to latamoxef, piperacillin, chloramphenicol, and clindamycin. *Chemotherapy, 31*(3), 204–210.

Tamboli, D. P., & Lee, D. S. (2013). Mechanistic antimicrobial approach of extracellularly synthesized silver nanoparticles against gram positive and gram negative bacteria. *Journal of Hazardous Materials, 260*, 878–884.

Taran, M., Rad, M., & Alavi, M. (2018). Biosynthesis of TiO_2 and ZnO nanoparticles by Halomonas elongata IBRC-M 10214 in different conditions of medium. *BioImpacts: BI, 8*(2), 81.

Thanh, N. T., & Green, L. A. (2010). Functionalisation of nanoparticles for biomedical applications. *Nano Today, 5*(3), 213–230.

Thomas, R., Jasim, B., Mathew, J., & Radhakrishnan, E. (2012). Extracellular synthesis of silver nanoparticles by endophytic Bordetella sp. isolated from Piper nigrum and its antibacterial activity analysis. *Nano Biomedicine & Engineering, 4*(4).

Tripathi, R., Bhadwal, A. S., Gupta, R. K., Singh, P., Shrivastav, A., & Shrivastav, B. (2014). ZnO nanoflowers: Novel biogenic synthesis and enhanced photocatalytic activity. *Journal of Photochemistry and Photobiology B: Biology, 141*, 288–295.

Tu, Y., Lv, M., Xiu, P., Huynh, T., Zhang, M., Castelli, M., Liu, Z., Huang, Q., Fan, C., & Fang, H. (2013). Destructive extraction of phospholipids from Escherichia coli membranes by graphene nanosheets. *Nature Nanotechnology, 8*(8), 594.

Vallee, B. L., & Falchuk, K. H. (1993). The biochemical basis of zinc physiology. *Physiological Reviews, 73*(1), 79–118.

Velusamy, P., Kumar, G. V., Jeyanthi, V., Das, J., & Pachaiappan, R. (2016). Bio-inspired green nanoparticles: Synthesis, mechanism, and antibacterial application. *Toxicological Research, 32*(2), 95–102.

Vittori Antisari, L., Carbone, S., Gatti, A., Vianello, G., & Nannipieri, P. (2013). Toxicity of metal oxide (CeO$_2$, Fe$_3$O$_4$, SnO$_2$) engineered nanoparticles on soil microbial biomass and their distribution in soil. *Soil Biology & Biochemistry, 60*, 87.

Vollath, D. (2008). Nanomaterials an introduction to synthesis, properties and application. *Environmental Engineering and Management Journal, 7*(6), 865–870.

Wang, C., Kim, Y. J., Singh, P., Mathiyalagan, R., Jin, Y., & Yang, D. C. (2016). Green synthesis of silver nanoparticles by Bacillus methylotrophicus, and their antimicrobial activity. *Artificial Cells, Nanomedicine, and Biotechnology, 44*(4), 1127–1132.

Wang, S., Lawson, R., Ray, P. C., & Yu, H. (2011). Toxic effects of gold nanoparticles on Salmonella typhimurium bacteria. *Toxicology and Industrial Health, 27*(6), 547–554.

Wei, X., Luo, M., Li, W., Yang, L., Liang, X., Xu, L., Kong, P., & Liu, H. (2012). Synthesis of silver nanoparticles by solar irradiation of cell-free Bacillus amyloliquefaciens extracts and AgNO3. *Bioresource Technology, 103*(1), 273–278.

Whitchurch, C. B., Tolker-Nielsen, T., Ragas, P. C., & Mattick, J. S. (2002). Extracellular DNA required for bacterial biofilm formation. *Science, 295*(5559), 1487–1487.

Wickham, J. R., Halye, J. L., Kashtanov, S., Khandogin, J., & Rice, C. V. (2009). Revisiting magnesium chelation by teichoic acid with phosphorus solid-state NMR and theoretical calculations. *The Journal of Physical Chemistry B, 113*(7), 2177–2183.

Wigginton, N. S., Titta, A., Piccapietra, F., Dobias, J., Nesatyy, V. J., Suter, M. J., & Bernier-Latmani, R. (2010). Binding of silver nanoparticles to bacterial proteins depends on surface modifications and inhibits enzymatic activity. *Environmental Science & Technology, 44*(6), 2163–2168.

Wu, H., Moser, C., Wang, H.-Z., Høiby, N., & Song, Z.-J. (2015). Strategies for combating bacterial biofilm infections. *International Journal of Oral Science, 7*(1), 1–7.

Xie, Y., He, Y., Irwin, P. L., Jin, T., & Shi, X. (2011). Antibacterial activity and mechanism of action of zinc oxide nanoparticles against campylobacter jejuni. *Applied and Environmental Microbiology, 77*(7), 2325–2331.

Xiong, Q., Liu, D., Zhang, H., Dong, X., Zhang, G., Liu, Y., & Zhang, R. (2020). Quorum sensing signal autoinducer-2 promotes root colonization of Bacillus velezensis SQR9 by affecting biofilm formation and motility. *Applied Microbiology and Biotechnology, 104*(16), 7177–7185.

Yah, C. S. (2013). The toxicity of Gold Nanoparticles in relation to their physiochemical properties. *Biomedical Research, 24*(3), 400–413.

Yan, L., Zhang, S., Chen, P., Liu, H., Yin, H., & Li, H. (2012). Magnetotactic bacteria, magnetosomes and their application. *Microbiological Research, 167*(9), 507–519.

Yeary, L. W., Moon, J.-W., Love, L. J., Thompson, J. R., Rawn, C. J., & Phelps, T. J. (2005). Magnetic properties of biosynthesized magnetite nanoparticles. *IEEE Transactions on Magnetics, 41*(12), 4384–4389.

Yeh, Y.-C., Creran, B., & Rotello, V. M. (2012). Gold nanoparticles: Preparation, properties, and applications in bionanotechnology. *Nanoscale, 4*(6), 1871–1880.

Yeo, M.-K., & Kang, M.-S. (2008). Effects of nanometer sized silver materials on biological toxicity during zebrafish embryogenesis. *Bulletin of the Korean Chemical Society, 29*(6), 1179–1184.

Zhang, H., Li, Q., Lu, Y., Sun, D., Lin, X., Deng, X., He, N., & Zheng, S. (2005). Biosorption and bioreduction of diamine silver complex by Corynebacterium. *Journal of Chemical Technology & Biotechnology, 80*(3), 285–290.

Zhang, L., Gu, F., Chan, J., Wang, A., Langer, R., & Farokhzad, O. (2008). Nanoparticles in medicine: Therapeutic applications and developments. *Clinical Pharmacology & Therapeutics, 83*(5), 761–769.

Part IV
General Mechanisms of Interaction

Chapter 16
The Chemistry behind Nanotoxicological Processes in Living Systems

Guadalupe de la Rosa, Edgar Vázquez-Núñez, Pabel Cervantes, and Ma. Concepción García-Castañeda

Contents

G. de la Rosa (✉)
Departamento de Ingenierías Química, Electrónica y Biomédica,
División de Ciencias e Ingenierías, Universidad de Guanajuato, León, Guanajuato, Mexico

Center for Environmental Implications of Nanotechnology (UCCEIN),
University of California, Santa Barbara, CA, USA
e-mail: gdelarosa@fisica.ugto.mx

E. Vázquez-Núñez · M. C. García-Castañeda
Departamento de Ingenierías Química, Electrónica y Biomédica,
División de Ciencias e Ingenierías,Universidad de Guanajuato, León, Guanajuato, Mexico

P. Cervantes
Bren School of Environmental Science and Management, University of California,
Santa Barbara, CA, USA

Tecnológico de Monterrey, Escuela de Ingeniería y Ciencias,
Reserva Territorial Atlixcáyotl, Puebla, Pue, Mexico
e-mail: pabel.cervantes@tec.mx

© Springer Nature Switzerland AG 2021
N. Sharma, S. Sahi (eds.), *Nanomaterial Biointeractions at the Cellular,
Organismal and System Levels*, Nanotechnology in the Life Sciences,
https://doi.org/10.1007/978-3-030-65792-5_16

1   Introduction

Living organisms interact with nanomaterials (NMs) of natural and synthetic origin. Natural NMs include those from meteorites dust, seashells, smoke particles derived from combustion and volcano ashes, or others originated in different physical processes such as erosion. Engineered NMs (ENMs) are those produced by man either intentionally or unintentionally. ENMs are used in agriculture, cosmetics, medicine, energy, packaging, textiles, construction, electronic, optic, sensors, wastewater treatment, food, and environmental remediation, among others. Since NMs size is similar to that of biological macromolecules, and because of their antibacterial and odor-fighting properties, they are extensively used for a number of commercial products including wound dressing, detergents, or antimicrobial coatings.

During the last years, the synthesis of new ENMs has grown rapidly. Different international institutions including the National Nanotechnology Initiative (Roco 2004), Carbon Nanotechnology Research Institute (Dai 2006), and Woodrow Wilson International Center for Scholars' Project on Emerging Nanotechnologies (PEN; Maynard 2006) have estimated the products and residues containing NMs that could be released to global markets and the environment. Keller et al. (2013) predicted that the global distribution of ENMs released to the environment could be as follows: 63–91% to landfills, 8–28% to soils, 0.4–7% to water, and 0.2–1.5% to the atmosphere.

Soil, water, and air embrace different types of biota. Moreover, they display a profound interrelationship which at times makes it difficult to treat them as independent matrixes. Due to physical and chemical interactions, nanoparticles are able to cross the soil layers eventually permeating to groundwater and other water bodies. Similarly, NMs present in water may ultimately end in the soil matrix and sediments. It has been reported that most NMs imbedded in organic matter are physically stable (Lu et al. 2012); however, some of them may suffer modifications that can facilitate their incorporation into plants (Kumar et al. 2018).

Limited processes are reported for the release of NPs into the air, i.e., catalysis (Barge and Vaidya 2018; Davies et al. 2018; Fu et al. 2018), lubricants and fuel additives production and use (Ali et al. 2018), paint and coatings (Saber et al. 2018; Scifo et al. 2018), and the use of agrochemicals (Kah et al., 2018; Xin et al. 2018). It is expected that the aerial concentration of nanomaterials be significantly lower than those concentrations in soil and water (Loureiro et al. 2018; Meng et al. 2018). The most known routes by which nanoparticles dispersed in air can be in contact with living organisms are deposition either in dry and wet conditions (Fang et al. 2018) followed by their incorporation into the organisms.

In the last years, important efforts have been made to understand the mechanisms of interaction of NMs with living organisms. However, systemic effects are still unclear, and this includes toxicological processes. Toxicology is the study of adverse effects of chemicals on living organisms (Klaassen 2001), and as such requires a good understanding of physical and chemical phenomena in order to properly explain the processes, and eventually prevent negative effects.

This chapter intends to summarize current knowledge on physical and chemical phenomena behind toxicological processes of metallic nanoparticles. This analysis would serve for decision-making purposes, mainly when ENMs are intended for biomedical purposes.

2 Classification and Properties of Nanomaterials

The word nanomaterial includes the prefix "nano," which is used to describe objects in the range of nanometer units. In the regulatory realm, a complete and constant consideration is that of materials in the size range of 1–100 nm. There are several

Table 16.1 Classification of nanostructured materials

Classification	Types	Examples
Dimensionality	Zero dimension—0D	Clusters and Q-dots
	One dimension—1D	Thin films and surface coatings
	Two dimensions—2D	Strands of fibers, nanostructured films, nanopore filters and free particles
	Three dimensions—3D	Free nanoparticles with several morphologies, colloids, fixed small nanostructures, membranes with nanopores
Morphology	High-aspect ratio	Wires (with shapes as helices or zig-Zag), tubes, belts and helices.
	Low-aspect ratio	Particles or powder in the form of suspension or colloids with spherical, oval, cubic, and prism shapes
Composition	Single constituent material	Can be synthetized by different methods. They can be compact or hollow particles and tubes
	Composite of several materials	Often found in nature as different agglomerated by mixed materials. Or can be synthetized to produce coated, encapsulated, barcode or grafted materials
Uniformity and agglomeration state	Isometric size	Based in their chemical and electromagnetic properties they can exist as dispersed aerosol, colloidal and suspension or in agglomerated state. This effect can have repercussions in their size and surface reactivity
	Inhomogeneous size	

classifications for NMs; Table 16.1 shows some of them as well as a brief description based on Buzea C. et al. (2007). NMs are classified according to their dimensionality, morphology, composition, and nature of agglomeration (Table 16.1). Other considerations may include the type of approach for their production, i.e., bottom-up or top-down. Given the diversity of ENMs, evaluation of toxicity becomes complex and several organisms are working towards providing regulations of these type of materials. In this context, understanding physical and chemical nature of the interactions of living organisms with ENMs becomes crucial.

Physicochemical behavior of NMs differs from their bulk counterparts mainly because surface area is extremely different in both materials. The smallest the particle, the highest the surface area (Zarschler et al. 2016), and the number of atoms at the surface. Since surface atoms are less tightly bound as compared to inner atoms, those are more likely to interact with the surrounding environment. Some of the consequences include a higher reactivity and changes in melting points, among others (Schmid 2008). These characteristics will also direct their interaction with living organisms. This will be discussed in further sections.

3 Characterization Techniques to Study Physical and Chemical Interaction of NMS and Living Organisms

Multiple techniques have been used to study the physical and chemical interactions between ENMs and living organisms. Numerous methodologies exist and usually they are used in combination. In addition, the choice of one technique over another will depend on the objective or goal to be pursued. Herein, we summarize some of the main techniques and approaches applied to study the interactions between ENMs and organisms.

3.1 Scanning and Transmission Electron Microscopy (SEM and TEM)

Electron microscopy uses a beam of electrons to create an image of a sample or specimen. In nanotoxicology studies, specimens are mainly cells, insects, leave, roots, and tissue samples. An electron beam is generated using a filament or "electron gun," which passes through a thin specimen in the case of transmission electron microscopy (TEM), or scans over the surface of the sample in the case of scanning electron microscopy (SEM). Although SEM and TEM are used for imaging, the information provided by both of them is different. SEM gives a detailed surface analysis of a given specimen and is a better option than TEM in terms of resolution and depth of field. However, the magnifying power is commonly higher in TEM (up to 5,000,000×) as compared to SEM (up to 200,000×), because of the efficiency in electrons passing through the specimen. An important consideration with the magnifying power is that TEM applies more energy (80–120 kV) than SEM (5–10 kV), which may damage the sample after long exposition to the electron beam. Biological sample preparation procedure has few differences for TEM and SEM. Biological samples prepared for TEM and SEM need to be dehydrated, then embedded in epoxy resin for TEM or coated with a conductive layer for SEM, commonly gold. In both cases, the embedding and coating of samples reduce the thermal damage and enhances sample imaging.

Nanotoxicology has taken advantage of electron microscopy in order to understand the physical interactions between ENMs and cells. The main interactions observed by TEM and SEM are the presence of ENMs in the cell membrane and in the cell wall for plant cells. TEM also provides cytoplasm images; therefore, ENMs can be observed in this space either unbound or bound to organelles. Some examples of application for SEM and TEM are the in vitro experiments, where samples are analyzed to observe the cell damage of specific strains such as *Escherichia coli*, *Staphylococcus aureus*, and *Bacillus subtilis*, among others. TEM has been also applied to observe the interactions between ENMs and complex culture of microorganisms in biological reactors (Cervantes-avilés et al. 2018). To avoid misinterpretation of artifacts with ENMs during TEM observation, it is important not to use staining agents such as uranyl acetate or lead citrate in sample preparation. Therefore, it is highly recommended to combine electron microscopy techniques with any analytical or diffraction technique, in order to verify the chemical composition of materials interacting with cells.

3.2   Energy Dispersive X-Ray Spectroscopy (EDS) and High-Angle Annual Dark Field (HAADF)

Some spectroscopic techniques such as energy dispersive X-ray spectroscopy (EDS) and high-angle annual dark field imaging provide compositional and chemical information of ENMs. These techniques are frequently coupled to SEM and TEM to analyze specific imaging fields. The purpose is to determine the chemical composition in selected fields by detecting the electrons released. In samples containing microorganisms and nanoparticles (NPs), EDS coupled to SEM or TEM is the most reproduced approach (Hondow et al. 2011). However, this technique considers only small areas of a cell. The HAADF-STEM imaging has been used to identify the chemical composition and perform elemental mapping of single bacteria (Cervantes-Avilés et al. 2017). A limitation for elemental mapping in HAADF-STEM is the high energy level applied in beam (200–300 kV), which may burn the sections of the biological samples after some minutes of exposition.

3.3 Cryo-Transmission Electron Microscopy (Cryo-TEM)

In cryo-transmission electron microscopy (Cryo-TEM), a beam of electrons is fired at a frozen hydrated sample. The emerging scattered electrons pass through a lens to create a magnified image of the structure (Callaway 2017). Thousands to hundreds of thousands of high-resolution 2D images are recorded and aligned to construct 3D images of organelles as well as molecules. However, the amount of radiation required to collect an image of a specimen in the Cryo-TEM is high enough to be a potential source of specimen damage for delicate structures (Wang et al. 2018). Cryo-TEM is complemented with several spectroscopic techniques to reveal some interactions between ENMs and organisms as well as molecules such as lipids and proteins. For example, the use of X-ray crystallography provides structural details (Wang and Wang 2017), while nuclear magnetic resonance (NMR) enables the study of specimens larger than 150 kDa (Perilla et al. 2017). Nowadays, the single particle Cryo-TEM is an increasingly popular technique to study interactions of ENMs with viruses, small organelles, as wells as molecular interactions in cells (Stewart 2017), which is due to its ability to solve structures at atomic resolution and affinity with X-ray crystallography.

3.4 Atomic Force Microscopy (AFM)

Atomic force microscopy (AFM) is a mechanical-optical instrument based on a physical probe, which simultaneously measures specimen surface morphology in three dimensions with high resolution. AFM has been employed to study the inter-action of positively charged dendrimer nanoparticles with lipid bilayers (Nievergelt et al. 2015). The limitations in AFM are the softness of biological samples, which are easily deformed or damaged by the AFM tip (Wang et al. 2018). AFM has been complemented with X-ray 2D imaging methods to differentiate NPs transported inside the cells with those adhered to the cell surface (Ma and Lin 2013).

3.5 Laser Scanning Confocal Microscopy (LSCM)

Confocal laser scanning microscopy (CLSM or LSCM) builds an image in a computer from collected points of light by a photomultiplier tube, which is positioned behind a confocal pinhole. LSCN is a valuable tool for obtaining high-resolution images and 3-D reconstructions (Paddock 2000). Confocal images are obtained in the XY and XZ confocal planes through optical sectioning (Z-stack mode). This technique has been used to track the permeation of the fluorescent chitosan nanoparticles through deeper skin layers (Abdel-Hafez et al. 2018). LSCN is well compatible with fluorescence labeling and imaging (Sulheim et al. 2016) that allows the targeting of ENMs and particular receptors in living organisms.

3.6 Enhanced Darkfield Microscopy (EDFN) and Hyperspectral Imaging System (HIS)

In enhanced darkfield microscopy (EDFM or EDM), darkfield-based illuminators emits highly collimated light to the sample. Scattered light from the specimen is collected to obtain improved contrast images with bright objects on a black background. Hyperspectral imaging system (HSI) uses continuous and contiguous ranges of wavelengths to obtain the spectrum for each pixel of an image, with the purpose of identifying objects. Spectral profiles of known materials can be generated and saved as reference libraries, in order to be compared with unknown samples. By using the different spectrum, HSI systems can generate 2D and 3D dimensional representation of spatial and spectral data, known as a plane and hypercube or data cube, respectively (Kiani et al. 2018). The simultaneous uses of EDFM and HSI have provided a solution for the characterization of NPs in histological samples (Roth et al. 2015), and in wastewater samples (Théoret and Wilkinson 2017). EDFM-HSI approach allows a faster image acquisition and analysis, which saves time and cost in comparison to more intensive conventional techniques such as TEM, SEM, and LSCM (Kiani et al. 2018). Additionally, sample preparation for EDFM-HSI is typically minimal as well as nondestructive (Roth et al. 2015). Validation and comparison of this approach with traditional techniques (TEM, SEM, etc.) in more applications fields will help to extend and intensify its application. Despite this, one of the major advantages of EDFM-HSI is the combination of imaging with spectroscopy, which allows determining the location and distribution of ENMs on in vivo or ex vivo experiments.

3.7 Fluorescence Lifetime Imaging Microscopy (FLIM) and Thermometry

Fluorescence lifetime imaging microscopy (FLIM) is based on the timely and intensity response of a fluorophore signal within a sample, which usually corresponds to cells from culture or tissues (Orte et al. 2013). To investigate interactions between ENMs and living organisms, some macromolecules including proteins and lipids, as well as ENMs, are labeled with a fluorophore and analyzed by photon detection (Sulheim et al. 2016). The employment of quantum dots (QDs) has allowed the detection of intracellular pH (Orte et al. 2013). One of the most advanced approaches of FLIM is the coupling with time-correlated single photon counting (TCSPC). This approach has been used in cellular thermometry for spatially resolved temperature measurements in living cells by tracking gold nanoclusters (Shang et al. 2013). Since TCSPC-FLIM has demonstrated the potential at spatial and temperature resolution, it can be used to target subcellular areas in biological samples exposed to ENMs.

3.8 Flow Cytometry

Flow cytometry is a laser-based technology employed for cell counting, where fluorescent, quantum dots, or isotopes are used as biomarkers. The capacity of many ENMs to reflect light is used as an advantage to visualize and quantify their presence inside cells to some extent. Recently, a standard operating procedure (SOP) for the quantification of nanoparticle uptake by flow cytometry has been developed by interlaboratory studies (Salvati et al. 2018). Absolute nanoparticle numbers consisted in calibrating the label signal in cells followed by relative measurements, which were carried out by comparing samples signals with the determined dose-response calibration curves or uptake kinetics (Kim et al. 2012). The application of flow cytometry is a suitable technique to identify the uptake and storage of ENMs within cells, which is one of the most critical interactions with living organisms.

3.9 Surface-Enhanced Raman Scattering Detection (SERS)

SERS is an advanced technique that incorporates nanoscale noble metal substrates (Ag, Au and Cu) into normal Raman spectroscopy (Sharma et al. 2012). Weak Raman signals can be improved by many orders of magnitude because the excitation of localized surface plasmon resonance (LSPR) on nanoscale-roughened surfaces can generate a large electromagnetic field, which increases the Raman cross section from the molecules adsorbed to noble metal nanostructures (Guo et al. 2016). Ag, Au, and Cu NPs have LSPRs that cover most of the visible and near-infrared wavelength range (400–1200 nm). SERS intensity is dependent on the inherent properties of the ligand molecules and NPs. Therefore, SERS was used to know the chemical and physical interactions of those NPs onto fresh spinach leaves and pond water samples by analyzing the ligand molecules (Guo et al. 2016). SERS has also been applied to detect ENMs inside cells due to the clear difference from the cell background. Although Raman signals are limited to the detectable LSPR, SERS seems to be appropriate to determine the physicochemical interactions between ENMs and organisms.

3.10 Fourier Transform Infrared Spectroscopy (FTIR): Surface Functionalization

Fourier transform infrared (FTIR) is a spectroscopy technique based on the infrared spectrum of a solid, liquid, or gas, and it is used for qualitative and quantitative analysis of mainly organic compounds. FTIR spectroscopy has become a popular technique in the field of cancer therapy with an ability to elucidate molecular interactions (Kalmodia et al. 2015). In agriculture, this technique has been used to study interactions between ENMs and cucumber fruit (Servin et al. 2013), as well as at molecular level to elucidate the extracellular polymeric substances binding to ENMs (Adeleye and Keller 2016). A new approach for infrared spectroscopy is the attenuated total reflectance (ATR) or ATR-FTIR, which is a useful and noninvasive method for the study of phospholipid structures in aqueous solutions exposed to magnetic nanoparticles (Kręcisz et al. 2016). The application of infrared techniques, either FTIR or ATR-FTIR, is useful to elucidate the chemical interactions along ENMs and cell macromolecules commonly attached to surface of the particles.

3.11   Techniques Based on Synchrotron Radiation (SR)

Synchrotron radiation (SR) is an advanced light source with a wide frequency range, from infrared up to the highest-energy X-rays. SR is an outstanding technique due to its brilliance and pulsed light emission. Brilliance is much higher than conventional sources, while pulse durations can be at or below one nanosecond. Also, SR is highly polarized, tunable, and collimated and can be focused over a small area with much more photons than a conventional source (Li et al. 2015). The radiation beam may interact with specimen by the absorption and the scattering phenomena. Probably the most used SR techniques to study the nanotoxicology on living organisms are X-ray absorption near edge structure (XANES) and extended X-ray absorption fine structure (EXAFS). The exposition of a specimen to X-ray induces the ejection of one or more electrons from inner levels of energy in the atoms. The releasing of electrons generates energy gradients between orbitals, which is detected as photons fluorescence. X-ray fluorescence spectrometry (XRF) measures the characteristic fluorescence to determine the elemental composition of a sample (Li et al. 2015). XRF has allowed a more direct observation of NPs crossing the nasal barrier to cerebral cortex and hippocampus region in mice (Wang et al. 2008).

The transmission X-ray microscopy (TXM) based on SR is another excellent method for studying the distribution of ENMs in biological samples, in both 2D and 3D. The combined application of TXM and XANES was able to determine the distribution and biotransformation of La_2O_3 NPs in cucumber (Ma et al. 2011). One of the most powerful SR approaches for nanotoxicological studies is the combination of XRF with XAFS. This strategy is useful to track ENMs inside organisms and to get information about both spatial distribution and chemical species from metals of interest (Li et al. 2015). This combined approach was useful to know the fate and biotransformation of ENMs in microorganisms (Qu et al. 2011), in fruits (Servin et al. 2013), and in biological reactors for wastewater treatment (Ma et al. 2014). Therefore, the use of single or multiple SR methods clarify both physical and chemical interactions between ENMs and living organisms. However, it is important to highlight that some of these approaches and techniques applied for biological and histopathological samples rely on statistical confidence and operator expertise.

Of course, other techniques exist and the selection of one over another will depend on the purpose of the study. For example, if the researcher intends to determine the impact of NPs on protein expression, biochemistry-based techniques would be needed. In any case, detection limit and quantification limit are important parameters to be taken into account.

4 Physicochemical Interactions of ENMs with Living Organisms

When living organisms encounter NPs, physical and chemical properties both dictate the way these entities interrelate. These interrelationships may be influenced by particle size, shape, chemical composition, and surface charge (Kamaly et al. 2012; Allegri et al. 2016). Chemistry cannot be separated from physics in this intent to explain how NPs interact in different media and/or with different organisms. For example, a given NP coated with distinctive functional groups will behave differently, resulting in a selected mechanism of absorption/adsorption (Lin et al. 2010). Figure 16.1 depicts physical interactions between ENMs and living organisms from terrestrial and aquatic ecosystems. In summary, these correspond to adsorption, ingestion, absorption, internalization, and translocation. This section examines general physicochemical interactions of ENMs with selected models including plants, human cells, and microorganisms. A more detailed discussion on the chemistry related to nanotoxicity processes will be provided lately.

4.1 Microorganisms

Bacteria are more abundant in soils than in water and air (Coleman et al. 2018). They, along with fungus, allow maintaining and enhancing crop production. Since they are ubiquitous in ecosystems and rivers and are vital for nutrient cycles, these organisms are an important target for studying the toxic effect of nanoparticles. In fact, bacterial diversity and richness are an easy and feasible bioindicator of environmental equilibrium and soil productivity (Cravo-Laureau et al. 2017; Asadishad et al. 2018; Huang et al. 2018).

Diverse studies have demonstrated that the microbial populations in soils are rapidly affected after exposition with NMs (Liu et al. 2017). Due to the economic importance of crops, the concerns are focused on the toxicity of nanomaterial on microorganisms that promote the plant growth and those that affect the nutrient cycling in soils, i.e., plant growth promoting rhizobacteria (PGPR) like *Pseudomonas aeruginosa, P. putida, P. fluorescens, Bacillus subtilis* and N cycle bacteria (Kashyap et al. 2017). Regarding to the N-cycling bacteria, some nitrifying and denitrifying bacteria have shown varying degrees of inhibition when exposed to ENPs in culture conditions or aqueous suspension (Mishra and Kumar 2009).

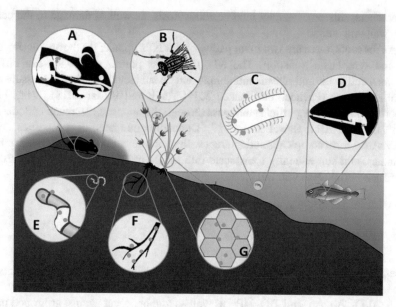

Fig. 16.1 Physical interactions between ENMs (orange dots) and living organisms from terrestrial and aquatic ecosystems. ENMs-organisms interface represents interactions corresponding to the processes of ingestion (**a, d**), absorption, and internalization (**c, e**), adsorption (**b, f**) and translocation (**g**)

It seems that antibacterial mechanisms of a variety of ENMs include the production of reactive oxygen species (ROS) that induces oxidative stress and lipid peroxidation, as well as non-ROS pathways (Tang and Lv 2014; Leung et al. 2014; Aurore et al. 2018). In some cases, membrane rupture caused by ENMs has been identified (Liu et al. 2018). Evidence of nanotoxicity and decrease in population growth has also been reported in freshwater and marine microorganisms (*Daphnia magna* and phytoplankton) (Keller et al. 2012).

4.2   Plants

Plants are all covered by a lipophilic cuticle (Meng et al. 2018), which protects them from uncontrolled water loss, pathogens, particles, and dirt (Barthlott et al. 2017; Tafolla-Arellano et al. 2018). Thus, cuticle is the first barrier for NPs interaction with these living organisms. NPs uptake by plants via gas transporters (stomata) has been reported. When CeO_2 NPs were applied to cucumber leaves, around 3% of the total cerium in the plant was found in the roots, demonstrating the leaf-root translocation process (Hong et al. 2014). Foliar application of TiO_2 NPs in lettuce resulted in their internalization into leaf tissue (Larue et al. 2014). According to these researchers, this process may involve endocytosis, as well as damage to the cellulose wall and cuticle.

As for roots, there are two main pathways for NPs uptake: (a) the apoplastic, and (b) the symplastic transport. The NPs that cross porous cell walls can diffuse between cell walls and plasma membrane and be subjected to osmotic pressure and capillary forces (Pacheco and Buzea 2017; Yanik and Vardar 2018). It has been reported that the nanoparticles that pass through the apoplast can reach endodermis (Tripathi et al. 2017; Rawat et al. 2017). Depending on materials' properties, some ENMs may form complexes and interact with membrane transporter proteins or root exudates, and subsequently translocate into the plant system (Avellan et al. 2017; Martínez-Fernández et al. 2017).

4.3   Terrestrial Animals

Insects and aquatic invertebrates are used as models to determine ENPs toxicity. A limited number of efforts have been carried out by some researchers to clarify how SiO_2, Al_2O_3, Ag, Au, and TiO_2 NPs, as well as graphene, act against arthropod pests and vectors (Benelli 2018). The most common interactions of those particles with insects are by attachment to the body as well as ingestion, which induce metabolic damage (Benelli 2018). More specific modes of action include modification of wax layer in cuticle resulting in insect dehydration (Stadler et al. 2017), and inhibition of enzymes. A very complete review on the interaction of NPs with insects was written by Benelli (2018).

In different in vivo an ex vivo experiments with mice, the main physical interactions of NPs were due to absorption, distribution, metabolism, and excretion in different organs such as lungs and gastrointestinal tract (Srivastava et al. 2015). Toxicological responses to the absorption of ENMs in mice organs has been the persistent inflammation and fibrosis (Shvedova et al. 2010). In rats, brain damage induced by Cu NPs has been reported (Trickler et al. 2012).

As for humans, there are three pathways to get into the body: (a) inhalation, (b) oral uptake, and (c) via the skin, called dermal uptake (Guleria et al. 2018). There is no doubt that the respiratory system is the most important exposure route for ENMs in the human occupational environment (Savolainen et al. 2010). Diverse studies in controlled conditions have evaluated the uptake of ENMs via inhalation, especially for cosmetics products. Nazarenko et al. (2012) determined the potential effect for human contact and inhalation exposure to nanomaterials when using nanotechnology-based cosmetic powders; in this study it was suggested that this exposure was mainly in the form of agglomerates or NMs attached to larger particles that would deposit in the upper airways of the human respiratory system rather than in the alveolar and tracheobronchial regions of the lung.

As previously mentioned, skin is the largest organ in the human body and an entrance pathway of ENMs in humans. Some evidences confirm that ENMs deposited in the skin may reach the gut lumen through hand-mouth contact (Pietroiusti et al. 2016). For epithelial layer uptake mechanisms, it has been reported that major routes are macropinocytosis, clathrin-mediated endocytosis and caveole-mediated and non-clathrin, and finally non-caveole-mediated uptake (Fröhlich and Roblegg 2012).

Concerning the orogastrointestinal pathway, NPs should reach different tissue structures including a cellular layer, an acellular structure (mucus; mucin of proteins), as well as epithelial layers, which represent the highest resistance against the passage of chemical compounds and NMs. Dietary intake of TiO_2 NPs has been estimated as 2.5 mg/individual/d (0.036 mg/kg for a person of 70 kg (Lomer et al. 2000)). Once ingested, ENMs are subjected to modifications due to the physico-chemical environment in the orogastrointestinal tract. For instance, the pH variations from acid to neutral in the oral cavity and intestine have a strong effect on surface charge of the particles and, as a consequence, on the agglomeration and cellular uptake, among other factors, that can affect bioavailability and absorption.

The mucus barrier is composed by glycosylated extracellular proteins with gel-forming properties. Mucus layers are formed after the saliva binds to mucins and these are deposited in the surface of the epithelium. The thickness of this layer may be around 70–100 µm, depending of its composition and location through the gastrointestinal section. Thus, in mucus, the interaction of NPs may result in change and variation of NPs permeability (Cone 2009). In addition, it has been proven that TiO_2 NPs induce mucus production (Chen et al. 2011).

4.4 Human Cells

The physical interaction of ENMs with different organs has been studied during in vitro assays by exposing histological samples to ENMs. It has been determined that the main physical interactions at cell level are the uptake, intracellular distribution, transformation, and expulsion of ENMs (Li et al. 2015; Srivastava et al. 2015). Receptor-ligand interactions, hydrophobic interactions, electrostatic attractions, and hydrogen bonds are often involved in the adsorption of ENMs into the cell membrane (Srivastava et al. 2015). Membrane fusion, endocytosis and passive movement through the clathrin and caveolae-independent endocytosis or caveolae-mediated endocytosis may occur during the internalization of ENMs (Zhao et al. 2011). Passive diffusion across cell membrane has been observed (Mühlfeld et al. 2008). Metal ions released from dissolvable metal oxide nanomaterials can be transported into the cells via membrane channels (He et al. 2015; Ma and Lin 2013).

Long-term effects studies are still needed and this information may help in elaborating proper regulations.

5 The Chemistry behind Nanotoxicological Processes in Living Organisms

Identifying chemical changes in living structures when they interact with ENMs is necessary to understand and evaluate their positive and negative systemic effects. This information is crucial for decision-making purposes, for example, in agro and biomedical areas. This section discusses selected chemical reactions that take place when various metallic ENMs interact with living organisms.

The chemistry of an ENM can be affected even before they enter in contact with a biosystem. In addition, NPs per se may be able to modify environmental conditions. For example, Vítková et al. (2015) demonstrated that nano-maghemite (γ-Fe_2O_3, NMa) decreased pH of water to 3.0, and that of a simulated root exudate solution from 6.42 to 4.93. In addition, NMa, as well as nano-hydroxyapatite and nano-hematite were able to reduce soluble As, Pb, and Sb in a mine soil (Arenas-Lago et al. 2019), which indicates possible adsorption of these elements to the ENMs. Because of hydration, maghemite surface displays Fe-OH and Fe-O($^-$) motifs (Liu et al. 2008), which may be the responsible for cation removal from waters. Also, at low pH, protonated maghemite is able to interact with anions (As and Sb). Similar processes remove mineral nutrients from growth media affecting their availability for living organisms, potentially causing nutrient deficiency. In this case, cation or anion exchange may be the processes behind NP toxicity.

Another source of nutrient deficiency in plants is membrane pores and xylem vessel obstruction by NPs. Dimkpa et al. (2013) reported that roots of wheat plants grown in ZnO and CuO NPs for fourteen days were surface covered with the NPs. Also, Chen et al. (2016) found that Y_2O_3 NPs nanotubes (31×236 nm in size) were uptaken by cabbage lateral roots; however, no translocation occurred. In addition, they observed NP aggregation and accumulation in external root epidermis.

Interaction of ENMs with cell membrane depends on NP surface charge, size, shape, topology, and composition, as well as on cell type. When a cell membrane meets NPs, Brownian collisions, wrapping, and internalization may occur, the internalization particularly through both active and passive transport. These phenomena are influenced by factors such as: (a) adhesion energy; (b) membrane bending modulus; and (c) membrane surface tension (Contini et al. 2018), which in turn depend on the chemical structure of the biomolecules. It has been proven that CuO NPs may cause membrane rupture (Liu et al. 2018), which may imply that energetic interactions of CuO NPs with cell membrane are higher than the forces maintaining membrane integrity. Also, production of ROS leads to membrane rupture.

Wang et al. (2012) found CuO NPs in epidermal cell wall, intercellular area, cortical cells cytoplasm, nuclei, xylem and phloem, and leaves in maize plants. These researchers hypothesized that untransformed CuO NPs may have entered the plant system through root apex, transported to the stele with subsequent mobilization to xylem and phloem. Dimkpa et al. (2013) demonstrated that, in wheat plants, CuO NPs were uptaken in the form of CuO NPs and Cu^{2+}. Moreover, Cu speciation indicated the presence of Cu(II) and Cu(I), this last mainly forming Cu-S bindings. Cu(I) and Cu(II) are able to react with H_2O_2 in Fenton-like reactions to produce ROS. Additionally, Pham et al. (2013) determined that Cu(I) also leads to the formation of the extremely oxidant Cu(III) species. In summary, it seems that physicochemical phenomena leading to CuO NPs toxicity includes (a) dissolution of CuO NPs to yield Cu^{2+}; (b) reduction of Cu(II) to Cu(I) by the action of reductases, ferredoxins, or sugars (Wang et al. 2012); (c) Cu(I) and Cu(II) reaction with H_2O_2 to produce reactive oxygen species ($O^{2•}$, •OH, and $^1O_2$); (d) Cu(I) reaction with H_2O_2 to yield the very reactive Cu(III); (e) Lipid peroxidation in cell membrane; (f) production of malondialdehyde (MDA) as a result of ROS reaction with unsaturated fatty acids and phospholipids (Halliwell and Gutteridge 1989); (g) modifications in cell membrane architecture, including potential rupture due to MDA, among others; (h) oxidative damage mediated by ROS in proteins and DNA; and (i) unspecific Cu binding to proteins. Other metallic NPs yielding specific ROS are TiO_2, CeO_2, ZnO, Fe_2O_3, and Ag, (Li et al. 2012; Oukarroum et al. 2013).

In some cases, it has been proven that ENMs affects not only DNA but also RNA. Sui et al. (2018) demonstrated that TiO_2 NPs promote cell apoptosis in mouse cells by affecting mi-RNA (miR-350), which are RNA fragments participating in the expression of proteins. A great deal of research has been dedicated to study the antibacterial properties of a variety of ENMs. Similar to what was observed in plants, it seems that main toxic mechanisms include the production of reactive oxygen species (ROS) that induces oxidative stress and lipid peroxidation, as well as non-ROS pathways (Tang and Lv 2014; Leung et al. 2014; Aurore et al. 2018).

Photocatalysis is another toxicity mechanism in living organisms. TiO_2 NPs interact with UV light to promote a valence electron to the conduction band. This leaves a vacancy that allows TiO_2 to extract an electron from H_2O or OH^-, yielding radicals OH•, which in turn reacts with biomolecules as previously explained (Zhang et al. 2015).

Regarding NP surface characteristics, Asati et al. (2010) demonstrated that CeO_2 NPs displaying dissimilar surface charges were differentially uptaken by normal (cardiomyocytes H9c2; human embryonic kidney cells, HEK293) and cancer cell lines (lung carcinoma A549; breast carcinoma, MCF-7). They also determined that endocytosis may be the mechanism of CeO_2 NP uptake by the studied cells.

Table 16.2 shows CeO_2 NPs distribution in the cells, as per the results of either dyeing with Lysotracker or tracking lysosome activity. Researchers also determined cell viability, which was negatively correlated with the presence of ceria in the lysosomes. Thus, CeO_2 NPs toxicity is associated with lysosomic activity which may lead to degradation of biological material causing cell death. These researchers concluded that positively charged CeO_2 NPs were the most toxic to the tested cell lines.

A similar conclusion was reached by Fröhlich (2012) who stated that a positive surface charge in some polymeric NPs results in a higher toxicity in nonphagocytes because of plasma membrane rupture and organelles dysfunction. However, phagocytes prefer negatively charged NPs. This researcher also concluded that Clarthin mediated the endocytic process.

Table 16.2 Distribution of polymer-coated CeO_2 NPs in different cell lines. Polymers provided positive (+), negative (−), or zero (0) surface charge (Asati et al. 2010

Cell line Cell compartment	Cardiomyocytes (H9c2)	Human embryonic kidney cells (HEK293)	Lung carcinoma cells (A549)	Breast carcinoma cells (MCF-7)
Cytoplasm	(+)	–	–	–
	(−)	(−)	-	–
	(0)	(0)	(0)	–
Lysosomes	(+)	(+)	(+)	–
	–	–	(−)	–
	–	–	–	–

Nutrient competition appears to be another toxic mechanism. Dimkpa et al. (2013) observed that the aerial plant tissue Zn was in the form of Zn phosphate in wheat plants fed with ZnO NPs. These researchers proved that the phosphate was formed inside plant tissue and not in the rhizosphere environment. Phosphate is an important constituent of nucleic acids and phospholipids. It is also essential for energy conversion in the cell. Since Zn phosphate is insoluble in water, this process may be involved in the toxicity, as no phosphate was available for the formation of the mentioned biomolecules. Further research is needed to confirm this hypothesis.

6 Conclusions

Different physicochemical interactions take place when a nanoparticle/material encounters living organisms. To study these phenomena, a variety of analytical techniques can provide different types of information. Even before entering biosystems, ENMs interact with different chemical species by adsorbing them, avoiding nutrient availability. First contact of ENMs with cells involves physicochemical and thermodynamic interactions where NMs surface charge, as well as membrane adhesion and surface tension dictate uptake and toxicity rates. Endocytic processes and ROS generation are the main events responsible for NPs internalization, and toxicity, respectively. ROS react with a variety of biomolecules to disrupt mechanical integrity or metabolic activity. Further kinetic models detailing metals and metal oxides reactions of ENMs with biomolecules will provide extremely useful information to evaluate and compare nanotoxicity of different materials.

References

Abdel-Hafez, S. M., Hathout, R. M., & Sammour, O. A. (2018). Tracking the transdermal penetration pathways of optimized curcumin-loaded chitosan nanoparticles via confocal laser scanning microscopy. *International Journal of Biological Macromolecules, 108*, 753–764.

Adeleye, A. S., & Keller, A. A. (2016). Interactions between algal extracellular polymeric substances and commercial TiO $_2$ nanoparticles in aqueous media. *Environmental Science & Technology, 50*(22), 12258–12265.

Ali, M. K. A., Fuming, P., Younus, H. A., Abdelkareem, M. A., Essa, F. A., Elagouz, A., & Xianjun, H. (2018). Fuel economy in gasoline engines using Al2O3/TiO2 nanomaterials as nanolubricant additives. *Applied Energy, 211*, 461–478.

Allegri, M., Bianchi, M. G., Chiu, M., Varet, J., Costa, A. L., Ortelli, S., Blosi, M., Bussolati, O., Poland, C. A., & Bergamaschi, E. (2016). Shape-related toxicity of titanium dioxide nanofibres. *PLoS One, 11*(3), e0151365.

Arenas-Lago, D., Abreu, M. M., Andrade Couce, L., & Veja, F. A. (2019). Is nanoremediation an effective tool to reduce the bioavailable as, Pb and Sb contents in mine soils from Iberian Pyrite Belt? *Catena, 176*, 362–371.

Asadishad, B., Chahal, S., Akbari, A., Cianciarelli, V., Azodi, M., Ghoshal, S., & Tufenkji, N. (2018). Amendment of agricultural soil with metal nanoparticles: Effects on soil enzyme activity and microbial community composition. *Environmental Science & Technology, 52*(4), 1908–1918.

Asati, A., Santra, S., Kaittanis, C., & Perez, J. M. (2010). Surface-charge-dependent cell localization and cytotoxicity of cerium oxide nanoparticles. *ACS Nano, 4*(9), 5321–5331.

Aurore, V., Caldana, F., Blanchard, M., Hess, S. K., Lannes, N., Mantel, P. Y., Filgueira, L., & Walch, M. (2018). Silver-nanoparticles increase bactericidal activity and radical oxygen responses against bacterial pathogens in human osteoclasts. *Nanomedicine: Nanotechnology, Biology and Medicine, 14*(2), 601–607.

Avellan, A., Schwab, F., Masion, A., Chaurand, P., Borschneck, D., Vidal, V., Rose, J., Santaella, C., & Levard, C. (2017). Nanoparticle uptake in plants: Gold nanomaterial localized in roots of Arabidopsis thaliana by X-ray computed nanotomography and hyperspectral imaging. *Environmental Science and Technology, 51*(15), 8682–8691.

Barge, A. S., & Vaidya, P. D. (2018). Wet air oxidation of cresylic spent caustic–a model compound study over graphene oxide (GO) and ruthenium/GO catalysts. *Journal of Environmental Management, 212*, 479–489.

Barthlott, W., Mail, M., Bhushan, B., & Koch, K. (2017). Plant surfaces: Structures and functions for biomimetic innovations. *Nano-Micro Letters, 9*(2), 23.

Benelli, G. (2018). Mode of action of nanoparticles against insects. *Environmental Science and Pollution Research, 25*(13), 12329–12341.

Buzea, C., Pacheco, I. I., & Robbie, K. (2007). Nanomaterials and nanoparticles: sources and toxicity. *Biointerphases, 2*(4), MR17–MR71.

Callaway, E. (2017). Molecular-imaging pioneers scoop Nobel. *Nature, 550*(7675), 167.

Cervantes-Avilés, P., Díaz Barriga-Castro, E., Palma-Tirado, L., & Cuevas-Rodríguez, G. (2017). Interactions and effects of metal oxide nanoparticles on microorganisms involved in biological wastewater treatment. *Microscopy Research and Technique, 80*(10), 1103–1112.

Cervantes-avilés, P., Ida, J., Toda, T., & Cuevas-Rodríguez, G. (2018). Effects and fate of TiO$_2$ nanoparticles in the anaerobic treatment of wastewater and waste sludge. *Journal of Environmental Management, 222*, 227–233.

Chen, E. Y., Garnica, M., Wang, Y. C., Chen, C. S., & Chin, W. C. (2011). Mucin secretion induced by titanium dioxide nanoparticles. *PLoS One, 6*, e16198.

Chen, Y., Sanchez, C., Yue, Y., de Almeida, M., González, J. M., Parkinson, D. Y., & Liang, H. (2016). Observation of yttrium oxide nanoparticles in cabbage (*Brassica oleracea*) through dual energy K-edge subtraction imaging. *Journal of Nanobiotechnology, 14*, 31.

Coleman, D. C., Callaham, M. A., Jr., & Crossley, D. A., Jr. (2018). *Fundamentals of soil ecology* (3rd ed.). London: Academic Press.

Cone, R. A. (2009). Barrier properties of mucus. *Advanced Drug Delivery Reviews, 61*, 75–85.

Contini, C., Schneemilch, M., Gaisford, S., & Quirke, N. (2018). Nanoparticle–membrane interactions. *Journal of Experimental Nanoscience, 13*(1), 62–81.

Cravo-Laureau, C., Lauga, B., Cagnon, C., & Duran, R. (2017). Microbial responses to pollution—Ecotoxicology: Introducing the different biological levels. In *Microbial ecotoxicology* (pp. 45–62). Cham: Springer.

Dai, L. (2006). From conventional technology to carbon nanotechnology: The fourth industrial revolution and the discovery of C60, carbon nanotube and nanodiamond. In L. Dai (Ed.), *Carbon nanotechnology: Recent developments in chemistry, physics, materials science and device applications* (pp. 3–12). The Netherlands: Elsevier.

Davies, D., Golunski, S., Johnston, P., Lalev, G., & Taylor, S. H. (2018). Dominant effect of support wettability on the reaction pathway for catalytic wet air oxidation over Pt and Ru nanoparticle catalysts. *ACS Catalysis, 8*(4), 2730–2734.

Dimkpa, C. O., Latta, D. E., McLean, J. E., Britt, D. W., Boyanov, M. I., & Anderson, A. J. (2013). Fate of CuO and ZnO Nano- and Microparticles in the plant environment. *Environmental Science and Technology, 47*(9), 4734–4742.

Fang, G. C., Zhuang, Y. J., & Huang, W. C. (2018). Seasonal ambient air particulates and metallic elements (Cr, Cu, Zn, Cd, Pb) pollutants dry depositions fluxes predictions and distributions with appropriate models at a farmland site. *Atmospheric Research, 210*, 58–65.

Fröhlich, E., & Roblegg, E. (2012). Models for oral uptake of nanoparticles in consumer products. *Toxicology, 291*(1–3), 10–17.

Fröhlich, E. (2012). The role of surface charge in cellular uptake and cytotoxicity of medical nanoparticles. *International Journal of Nanomedicine, 7*, 5577–5591.

Fu, J., Yue, Q., Guo, H., Ma, C., Wen, Y., Zhang, H., Zhang, N., Zheng, Y., Zheng, J., & Chen, B. H. (2018). Constructing Pd/CeO$_2$/C to achieve high leaching resistance and activity for catalytic wet air oxidation of aqueous amide. *ACS Catalysis, 8*(6), 4980–4985.

Guleria, P., Guleria, S., & Kumar, V. (2018). Effect of route of exposure on the toxicity behavior of nanomaterials. In V. Kumar, N. Dasgupta, & S. Ranjan (Eds.), *Nanotoxicology: Toxicity evaluation, risk assessment and management* (p. 81).

Guo, H., Xing, B., Hamlet, L. C., Chica, A., & He, L. (2016). Surface-enhanced Raman scattering detection of silver nanoparticles in environmental and biological samples. *Science of the Total Environment, 554–555*, 246–252.

Halliwell, B., & Gutteridge, J. M. C. (1989). *Free radicals in biology and medicine* (2nd ed.). Oxford: Oxford University Press.

He, X., Aker, W. G., Fu, P. P., & Hwang, H. M. (2015). Toxicity of engineered metal oxide nanomaterials mediated by nano-bio-eco-interactions: A review and perspective. *Environmental Science Nano, 2*(6), 564–582.

Hondow, N., Harrington, J., Brydson, R., Doak, S. H., Singh, N., Manshian, B., & Brown, A. (2011). STEM mode in the SEM: A practical tool for nanotoxicology. *Nanotoxicology, 5*(2), 215–227.

Hong, J., Peralta-Videa, J. R., Rico, C., Sahi, S., Viveros, M. N., Bartonjo, J., et al. (2014). Evidence of translocation and physiological impacts of foliar applied CeO2 nanoparticles on cucumber (*Cucumis sativus*) plants. *Environmental Science and Technology, 48*, 4376–4385.

Huang, J., Cao, C., Yan, C., Guan, W., & Liu, J. (2018). Comparison of Iris pseudacorus wetland systems with unplanted systems on pollutant removal and microbial community under nanosilver exposure. *Science of the Total Environment, 624*, 1336–1347.

Kah, M., Kookana, R. S., Gogos, A., & Bucheli, T. D. (2018). A critical evaluation of nanopesticides and nanofertilizers against their conventional analogues. *Nature Nanotechnology, 13*(8), 677–684.

Kalmodia, S., Parameswaran, S., Yang, W., Barrow, C. J., & Krishnakumar, S. (2015). Attenuated total reflectance Fourier transform infrared spectroscopy: An analytical technique to understand therapeutic responses at the molecular level. *Scientific Reports, 5*, 1–14.

Kamaly, N., Xiao, Z., Valencia, P. M., Radovic-Moreno, A. F., & Farokhzad, O. C. (2012). Targeted polymeric therapeutic nanoparticles: Design, development and clinical translation. *Chemical Society Reviews, 41*(7), 2971–3010.

Kashyap, A. S., Pandey, V. K., Manzar, N., Kannojia, P., Singh, U. B., & Sharma, P. K. (2017). Role of plant growth-promoting Rhizobacteria for improving crop productivity in sustainable agriculture. In *Plant-microbe interactions in agro-ecological perspectives* (pp. 673–693). Singapore: Springer.

Keller, A. A., Garner, K., Miller, R. J., & Lenihan, H. S. (2012). Toxicity of Nano-zero valent iron to freshwater and marine organisms. *PLoS One, 7*(8), e43983.

Keller, A. A., McFerran, S., Lazareva, A., & Suh, S. (2013). Global life cycle releases of engineered nanomaterials. *Journal of Nanoparticle Research, 15*(6), 1692–1694.

Kiani, S., van Ruth, S. M., Minaei, S., & Ghasemi-Varnamkhasti, M. (2018). Hyperspectral imaging, a non-destructive technique in medicinal and aromatic plant products industry: Current status and potential future applications. *Computers and Electronics in Agriculture, 152*, 9–18.

Kim, J. A., Aberg, C., Salvati, A., & Dawson, K. A. (2012). Role of cell cycle on the cellular uptake and dilution of nanoparticles in a cell population. *Nature Nanotechnology, 7*(1), 62–68.

Klaassen, C. D. (2001) Principles of toxicology and treatment of poisoning. In J. D. Hardman, L. E. Limbird & A. Goodman-Gilman (eds) Goodman & Gilman's the pharmaceutical basis of therapeutics, 10th edn. New York: McGraw-Hill; 67–80.

Kręcisz, M., Rybka, J. D., Strugała, A. J., Skalski, B., Figlerowicz, M., Kozak, M., & Giersig, M. (2016). Interactions between magnetic nanoparticles and model lipid bilayers—Fourier transformed infrared spectroscopy (FTIR) studies of the molecular basis of nanotoxicity. *Journal of Applied Physics, 120*(12), 124701.

Kumar, A., Joseph, S., Tsechansky, L., Privat, K., Schreiter, I. J., Schüth, C., & Graber, E. R. (2018). Biochar aging in contaminated soil promotes Zn immobilization due to changes in biochar surface structural and chemical properties. *Science of the Total Environment, 626*, 953–961.

Larue, C., Castillo-Michel, H., Sobanska, S., Trcera, N., Sorieul, S., Cécillon, L., Ouerdane, L., Legros, S., & Sarret, G. (2014). Fate of pristine TiO_2 nanoparticles and aged paint-containing TiO_2 nanoparticles in lettuce crop after foliar exposure. *Journal of Hazardous Materials, 273*, 17–26.

Leung, Y. H., Ng, A. M., Xu, X., Shen, Z., Gethings, L. A., & Wong, M. T. (2014). Mechanisms of antibacterial activity of MgO: Non-ROS mediated toxicity of MgO nanoparticles towards Escherichia coli. *Small, 10*(6), 1171–1183.

Li, Y., Zhang, W., Niu, J., & Chen, Y. (2012). Mechanism of photogenerated reactive oxygen species and correlation with the antibacterial properties of engineered metal-oxide nanoparticles. *ACS Nano, 6*(6), 5164–5173.

Li, Y. F., Zhao, J., Qu, Y., Gao, Y., Guo, Z., Liu, Z., Zhao, Y., & Chen, C. (2015). Synchrotron radiation techniques for nanotoxicology. *Nanomedicine: Nanotechnology, Biology, and Medicine, 11*(6), 1531–1549.

Lin, J., Zhang, H., Chen, Z., & Zheng, Y. (2010). Penetration of lipid membranes by gold nanoparticles: Insights into cellular uptake, cytotoxicity, and their relationships. *ACS Nano, 4*(9), 5421–5429.

Liu, G., Zhang, M., Jin, Y., Fan, X., Xu, J., Zhu, Y., Fu, Z., Pan, X., & Qian, H. (2017). The effects of low concentrations of silver nanoparticles on wheat growth, seed quality, and soil microbial communities. *Water, Air, & Soil Pollution, 228*(9), 348.

Liu, J. F., Zhao, Z. Z., & Jiang, G. B. (2008). Coating Fe_3O_4 magnetic nanoparticles with humic acid for high efficient removal of heavy metals in water. *Environmental Science and Technology, 42*(18), 6949–6954.

Liu, X., Tang, J., Wang, L., & Giesy, J. P. (2018). Mechanisms of oxidative stress caused by CuO nanoparticles to membranes of the bacterium Streptomyces coelicolor M145. *Ecotoxicology and Environmental Safety, 158*, 123–130.

Lomer, M. C., Thompson, R. P., Commisso, J., Keen, C. L., & Powell, J. J. (2000). Determination of titanium dioxide in foods using inductively coupled plasma optical emission spectrometry. *Analyst, 125*, 2339–2343.

Loureiro, S., Tourinho, P. S., Cornelis, G., Van Den Brink, N. W., Díez-Ortiz, M., Vázquez-Campos, S., Pomar-Portillo, V., Svedsen, C., & Van Gestel, C. A. (2018). Nanomaterials as soil pollutants. In A. C. Duarte, A. Cachada, & T. Rocha-Santos (Eds.), *Soil pollution: From monitoring to remediation* (pp. 161–190). London: Academic Press.

Lu, G., Li, S., Guo, Z., Farha, O. K., Hauser, B. G., Qi, X., Wang, Y., Wang, X., Han, S., Liu, X., DuChene, J. S., Zhang, H., Zhang, Q., Chen, X., Ma, J., Joachim Loo, S. C., Wei, W. D., Yang, Y., Hupp, J. T., & Huo, F. (2012). Imparting functionality to a metal–organic framework material by controlled nanoparticle encapsulation. *Nature Chemistry, 4*(4), 310–316.

Ma, R., Levard, C., Judy, J. D., Unrine, J. M., Martin, B., Jefferson, B., Lowry, G. V., & Durenkamp, M. (2014). Fate of zinc oxide and silver nanoparticles in a pilot waste water treatment plant and in processed biosolids fate of zinc oxide and silver nanoparticles in a pilot waste water treatment plant and in processed biosolids. *Environmental Science and Technology, 48*(1), 104–112.

Ma, S., & Lin, D. (2013). The biophysicochemical interactions at the interfaces between nanoparticles and aquatic organisms: Adsorption and internalization. *Environmental Sciences: Processes and Impacts, 15*(1), 145–160.

Ma, Y., He, X., Zhang, P., Zhang, Z., Guo, Z., Tai, R., Xu, Z., Zhang, L., Ding, Y., Zhao, Y., & Chai, Z. (2011). Phytotoxicity and biotransformation of La_2O_3 nanoparticles in a terrestrial plant cucumber (*Cucumis sativus*). *Nanotoxicology, 5*, 743–753.

Martínez-Fernández, D., Vítková, M., Michálková, Z., & Komárek, M. (2017). Engineered nanomaterials for phytoremediation of metal/metalloid-contaminated soils: Implications for plant physiology. In A. Ansari, S. Gill, R. Gill, G. Lanza, & L. Newman (Eds.), *Phytoremediation* (pp. 369–403). Cham: Springer.

Maynard, A. 2006. Woodrow Wilson International Center for Scholars Project on Emerging Nanotechnologies, Report. *Nanotechnology: A Research Strategy for Addressing Risk.*

Meng, B., Li, Y., Cui, W., Jiang, P., Liu, G., Wang, Y., Richards, J., Feng, X., & Cai, Y. (2018). Tracing the uptake, transport, and fate of mercury in sawgrass (*Cladium jamaicense*) in the Florida Everglades using a multi-isotope technique. *Environmental Science and Technology, 52*(6), 3384–3391.

Mishra, V. K., & Kumar, A. (2009). Impact of metal nanoparticles on the plant growth promoting rhizobacteria. *Digest Journal of Nanomaterials and Biostructures, 4*, 587–592.

Mühlfeld, C., Gehr, P., & Rothen-Rutishauser, B. (2008). Translocation and cellular entering mechanisms of nanoparticles in the respiratory tract. *Swiss Medical Weekly, 138*(27–28), 387–391.

Nazarenko, Y., Zhen, H., Han, T., Lioy, P. J., & Mainelis, G. (2012). Potential for inhalation exposure to engineered nanoparticles from nanotechnology-based cosmetic powders. *Environmental Health Perspectives, 120*(6), 885–892.

Nievergelt, A. P., Erickson, B. W., Hosseini, N., Adams, J. D., & Fantner, G. E. (2015). Studying biological membranes with extended range high-speed atomic force microscopy. *Nature Scientific Reports, 5*, 1–13.

Orte, A., Alvarez-Pez, J. M., & Ruedas-Rama, M. J. (2013). Fluorescence lifetime imaging microscopy for the detection of intracellular pH with quantum dot nanosensors. *ACS Nano, 7*, 6387–6395.

Oukarroum, A., Barhoumi, L., Pirastru, L., & Dewez, D. (2013). Silver nanoparticle toxicity effect on growth and cellular viability of the acquatic plant *Lemna gibba. Environmental Toxicology and Chemistry, 32*, 902–907.

Pacheco, I., & Buzea, C. (2017). Nanoparticle interaction with plants. In M. Ghorbanpour, K. Manika, & A. Varma (Eds.), *Nanoscience and plant–soil systems* (pp. 323–355). Cham: Springer.

Paddock, S. W. (2000). Principles and practices of laser scanning confocal microscopy. *Molecular Biotechnology, 16*(2), 127–149.

Perilla, J. R., Zhao, G., Lu, M., Ning, J., Hou, G., Byeon, I. J. L., Gronenborn, A. M., Polenova, T., & Zhang, P. (2017). CryoEM structure refinement by integrating NMR chemical shifts with molecular dynamics simulations. *The Journal of Physical Chemistry B, 121*(15), 3853–3863.

Pham, A. N., Xing, G., Miller, C. J., & Waite, T. D. (2013). Fenton-like copper redox chemistry revisited: Hydrogen peroxide and superoxide mediation of copper-catalyzed oxidant production. *Journal of Catalysis, 301*, 54–64.

Pietroiusti, A., Magrini, A., & Campagnolo, L. (2016). New frontiers in nanotoxicology: Gut microbiota/microbiome-mediated effects of engineered nanomaterials. *Toxicology and Applied Pharmacology, 299*, 90–95.

Qu, Y., Li, W., Zhou, Y., Liu, X., Zhang, L., Wang, L., Li, Y. F., Iida, A., Tang, Z., Zhao, Y., Chai, Z., & Chen, C. (2011). Full assessment of fate and physiological behavior of quantum dots utilizing *Caenorhabditis elegans* as a model organism. *Nano Letters, 11*(8), 3174–3183.

Rawat, S., Apodaca, S. A., Tan, W., Peralta-Videa, J. R., & Gardea-Torresdey, J. L. (2017). Terrestrial Nanotoxicology: Evaluating the Nano-biointeractions in vascular plants. In B. Yan, H. Zhou, & J. L. Gardea-Torresdey (Eds.), *Bioactivity of engineered nanoparticles* (pp. 21–42). Singapore: Springer.

Roco, M. C. (2004). The US national nanotechnology initiative after 3 years (2001–2003). *Journal of Nanoparticle Research, 6*(1), 1–10.

Roth, G. A., Peña, M. d. P. S., Neu-Baker, N. M., Tahiliani, S., & Brenner, S. A. (2015). Identification of metal oxide nanoparticles in histological samples by enhanced Darkfield microscopy and hyperspectral mapping. *Journal of Visualized Experiments, 106*, 53317.

Saber, A. T., Mortensen, A., Szarek, J., Jacobsen, N. R., Levin, M., Koponen, I. K., Jensen, K. A., Vogel, U., & Wallin, H. (2018). Toxicity of pristine and paint-embedded TiO₂ nanomaterials. *Human & Experimental Toxicology, 38*(1), 11–24.

Salvati, A., Nelissen, I., Haase, A., Åberg, C., Moya, S., Jacobs, A., Alnasser, F., Bewersdorff, T., Deville, S., Luch, A., & Dawson, K. A. (2018). Quantitative measurement of nanoparticle uptake by flow cytometry illustrated by an interlaboratory comparison of the uptake of labelled polystyrene nanoparticles. *NanoImpact, 9*, 42–50.

Savolainen, K., Alenius, H., Norppa, H., Pylkkänen, L., Tuomi, T., & Kasper, G. (2010). Risk assessment of engineered nanomaterials and nanotechnologies—A review. *Toxicology, 269*, 92–104.

Schmid, G. (2008). General features of metal nanoparticles physics and chemistry. In B. Corain, G. Schmid, & N. Toshima (Eds.), *Metal nanoclusters in catalysis and materials science, the issue o size control*. The Netherlands: Elsevier.

Scifo, L., Chaurand, P., Bossa, N., Avellan, A., Auffan, M., Masion, A., Angeletti, B., Kieffer, I., Labille, J., Bottero, J. Y., & Rose, J. (2018). Non-linear release dynamics for a CeO₂ nanomaterial embedded in a protective wood stain, due to matrix photo-degradation. *Environmental Pollution, 241*, 182–193.

Servin, A. D., Morales, M. I., Castillo-Michel, H., Hernandez-Viezcas, J. A., Munoz, B., Zhao, L., Nunez, J. E., Peralta-Videa, J. R., & Gardea-Torresdey, J. L. (2013). Synchrotron verification of TiO₂ accumulation in cucumber fruit: A possible pathway of TiO₂ nanoparticle transfer from soil into the food chain. *Environmental Science and Technology, 47*(20), 11592–11598.

Shang, L., Stockmar, F., Azadfar, N., & Nienhaus, G. U. (2013). Intracellular thermometry by using fluorescent gold nanoclusters. *Angew Chemie—Int. Ed., 52*, 11154–11157.

Sharma, B., Frontiera, R. R., Henry, A.-I., Ringe, E., & Van Duyne, R. P. (2012). SERS: Materials, applications, and the future. *Materials Today, 15*(1–2), 16–25.

Shvedova, A. A., Kisin, E. R., Mercer, R., Murray, A. R., Victor, J., Potapovich, A. I., Tyurina, Y. Y., Gorelik, O., Arepalli, S., Schwegler-berry, D., Hubbs, A. F., Antonini, J., Evans, D. E., Ramsey, D., Maynard, A., Kagan, V. E., Castranova, V., Baron, P., Johnson, V. J., Ku, B., Anna, A., Ashley, R., Ann, F., & Cas, V. (2010). Unusual inflammatory and fibrogenic pulmonary responses to single-walled carbon nanotubes in mice. *Molecular Physiology, 289*, 698–708.

Srivastava, V., Gusain, D., & Sharma, Y. C. (2015). Critical review on the toxicity of some widely used engineered nanoparticles. *Industrial and Engineering Chemistry Research, 54*(24), 6209–6233.

Stadler, T., Lopez-Garcia, G. P., Gitto, J. G., & Buteler, M. (2017). Nanostructured alumina: Biocidal properties and mechanism of action of a novel insecticide powder. *Bulletin of Insectology, 70*, 17–25.

Stewart, P. L. (2017). Cryo-electron microscopy and cryo-electron tomography of nanoparticles. *Wiley Interdisciplinary Reviews: Nanomedicine and Nanobiotechnology, 9*(2), 1–16.

Sui, J., Fu, Y., Zhang, Y., Ma, S., & Liang, G. (2018). Molecular mechanism for miR-350 in regulating of titanium dioxide nanoparticles in macrophage RAW264.7 cells. *Chemico-Biological Interactions, 280*, 77–85.

Sulheim, E., Baghirov, H., von Haartman, E., Bøe, A., Åslund, A. K. O., Mørch, Y., & de Lange Davies, C. (2016). Cellular uptake and intracellular degradation of poly(alkyl cyanoacrylate) nanoparticles. *Journal of Nanobiotechnology, 14*(1), 1–14.

Tafolla-Arellano, J. C., Báez-Sañudo, R., & Tiznado-Hernández, M. E. (2018). The cuticle as a key factor in the quality of horticultural crops. *Scientia Horticulturae, 232*, 145–152.

Tang, Z. X., & Lv, B. F. (2014). MgO nanoparticles as antibacterial agent: Preparation and activity. *Brazilian Journal of Chemical Engineering, 31*, 591–601.

Théoret, T., & Wilkinson, K. J. (2017). Evaluation of enhanced Darkfield microscopy and hyperspectral analysis to analyse the fate of silver nanoparticles in wastewaters. *Analytical Methods, 9*(26), 3920–3928.

Trickler, W. J., Lantz, S. M., Schrand, A. M., Robinson, B. L., Newport, G. D., Schlager, J. J., Paule, M. G., Slikker, W., Biris, A. S., Hussain, S. M., & Ali, S. F. (2012). Effects of copper nanoparticles on rat cerebral microvessel endothelial cells. *Nanomedicine, 7*(6), 835–846.

Tripathi, D. K., Singh, S., Singh, S., Pandey, R., Singh, V. P., Sharma, N. C., Prasad, S. M., Dubei, N. K., & Chauhan, D. K. (2017). An overview on manufactured nanoparticles in plants: Uptake, translocation, accumulation and phytotoxicity. *Plant Physiology and Biochemistry, 110*, 2–12.

Vítková, M., Komárek, M., Tejnecky, V., & Sillerová, H. (2015). Interactions of nano-oxides with low-molecular-weight organic acids in a contaminated soil. *Journal of Hazardous Materials, 293*, 7–14.

Wang, H. W., & Wang, J. W. (2017). How cryo-electron microscopy and X-ray crystallography complement each other. *Protein Science, 26*(1), 32–39.

Wang, J., Chen, C., Liu, Y., Jiao, F., Li, W., Lao, F., Li, Y., Li, B., Ge, C., Zhou, G., Gao, Y., Zhao, Y., & Chai, Z. (2008). Potential neurological lesion after nasal instillation of TiO$_2$ nanoparticles in the anatase and rutile crystal phases. *Toxicology Letters, 183*(1–3), 72–80.

Wang, T., Feng, Z., Wang, C., & He, N. (2018). Real-time investigation of interactions between nanoparticles and cell membrane model. *Colloids and Surfaces B: Biointerfaces, 164*, 70–77.

Wang, Z., Xie, X., Zhao, J., Liu, X., Feng, W., White, J. C., & Xing, B. (2012). Xylem- and phloem-based transport of CuO nanoparticles in maize (*Zea mays* L.). *Environmental Science and Technology, 46*, 4434–4441.

Xin, X., He, Z., Hill, M. R., Niedz, R. P., Jiang, X., & Sumerlin, B. S. (2018). Efficiency of biodegradable and pH-responsive Polysuccinimide nanoparticles (PSI-NPs) as smart nanodelivery systems in grapefruit: In vitro cellular investigation. *Macromolecular Bioscience, 18*(7), 1800159.

Yanik, F., & Vardar, F. (2018). Mechanism and interaction of nanoparticle-induced programmed cell death in plants. In M. Faisal, Q. Saquib, A. A. Alatar, & A. A. Al-Khedhairy (Eds.), *Phytotoxicity of nanoparticles* (pp. 175–196). Cham: Springer.

Zarschler, K., Rocks, L., Licciardello, N., Boselli, L., Polo, E., Pombo-Garcia, K., De Cola, L., Stephan, H., & Dawson, K. A. (2016). Ultrasmall inorganic nanoparticles: State-of-the-art and perspectives for biomedical applications. *Nanomedicine: Nanotechnology, Biology, and Medicine, 12*(6), 1663–1701.

Zhang, X., Li, W., & Yang, Z. (2015). Toxicology of nanosized titanium dioxide: An update. *Archives of Toxicology, 89*(12), 2207–2217.

Zhao, F., Zhao, Y., Liu, Y., Chang, X., Chen, C., & Zhao, Y. (2011). Cellular uptake, intracellular trafficking, and cytotoxicity of nanomaterials. *Small, 7*(10), 1322–1337.

Chapter 17
Nanomaterial Interaction and Cellular Damage: Involvement of Various Signalling Pathways

**Amit Kumar Singh, Prabhash Kumar Pandey, Astha Dwivedi,
Amit Kumar Sharma, Akhilesh Pandey, and Abhay Kumar Pandey**

Contents

1 Introduction

Nanomaterials are defined by the U.S. National Nanotechnology Initiative as materials that are 1–100 nm in size in at least one dimension. In recent years, nanotechnology has been involved in the creation and manipulation of materials at nanoscale levels to generate products that exhibit novel properties (Hussain et al.

Authors Amit Kumar Singh and Prabhash Kumar Pandey contributed equally.

A. K. Singh · P. K. Pandey · A. Dwivedi · A. K. Sharma · A. K. Pandey (✉)
Department of Biochemistry, University of Allahabad, Prayagraj, India
e-mail: akpandey@allduniv.ac.in; akpandey23@rediffmail.com

A. Pandey
Department of Neurology, Texas Tech University Health Sciences Centre, Garrison Institute on Aging, Lubbock, Texas, USA

© Springer Nature Switzerland AG 2021
N. Sharma, S. Sahi (eds.), *Nanomaterial Biointeractions at the Cellular, Organismal and System Levels*, Nanotechnology in the Life Sciences, https://doi.org/10.1007/978-3-030-65792-5_17

2005; Sanjay et al. 2014). The novel physical characteristics of nanomaterials can result in their having drastically different chemical and biological properties compared to the same material in bulk form. The unique chemical and biological properties of nanomaterials make them useful in many products for humans, including some in industry, agriculture, business, medicine, clothing, cosmetics, and food (Gonzalez et al. 2008; Brar et al. 2010; Ganguly et al. 2019). Nanotoxicology is an emerging field of research as a response to an exponential growth in the development and production of engineered nanoparticles (NPs) worldwide. The interactions of NPs with biological systems as well as their environmental and human health effects have not been fully explored (Gwinn and Vallyathan 2006). There is concern that research on the possible health risks of NPs is not keeping pace with the rapid growth in the number of NP products entering the industry and market. Due to their special physicochemical features, NPs biological behavior may differ from that of larger particles (Maurya et al. 2012). Nanotechnology research and development by industry and governments worldwide have been increasing dramatically. It has been expected that NPs will contribute approximately trillion dollars to the global economy (Nel et al. 2006; Xia et al. 2008). These particles may exist in aggregated or discrete form and can be tube like, spherical, hexagonal, or irregularly shaped. Since the use and manufacture of NPs are increasing day by day, humans and animals are more likely to be exposed through consumer products and the environment contamination. The rapid growth and commercialization of nanotechnology along with the health and safety recommendations for engineered nanomaterials are currently the area of greatest concern. With increase in production and use of nanomaterials increase, the possibilities of exposure of workers and the public to these NPs will surely increase (Stebounova et al. 2012). It has been seen that manufactured NPs can cause potential harmful effects on heath. Due to their unlimited utilization potential, metal and metal oxide NPs are now a major health concern. Metal and metal oxide NPs are widely used in the manufacturing of various commercial products and their applications are expected to massively expand during the next few years.

Oxidative stress (OS) has been implicated in NPs toxicity. OS elicits a wide variety of cellular events, such as apoptosis, cell cycle arrest, and the induction of antioxidant enzymes, and hence it is thought to be involved in nanoparticle toxicity (Fig. 17.1). Previous studies on NP toxicity with various cell types and NP types have reported that OS is one of the most important toxicity mechanisms related to NP exposure (Shvedova et al. 2003; Green and Howman 2005; Lin et al. 2006; Monteiller et al. 2007). Cellular redox homeostasis is carefully maintained by an elaborate antioxidant defense system, which includes antioxidant enzymes, proteins, and low molecular-weight scavengers. Excessive ROS production or a weakening of antioxidant defense could lead to OS. It is a state of redox disequilibrium that is defined as a decrease in the cellular glutathione (GSH)/glutathione disulfide (GSSG) ratio but functionally should be seen as a cellular stress response that activates a number of the redox-sensitive signalling cascades (Li et al. 2003; Sharma et al. 2019). The GSH/GSSG redox pair not only serves as the principal homeostatic regulator of redox balance but also functions as a sensor that triggers the stress responses that depending on the rate and

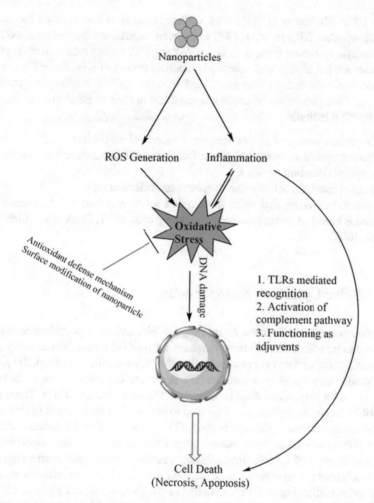

Fig. 17.1 Overview of signalling cascade-mediated nanotoxicity

level of change in this ratio could be protective or injurious in nature (Xiao et al. 2003; Kumar et al. 2015).

OS includes various deleterious processes resulting from an imbalance between protective antioxidants and damaging oxidants, reactive oxygen and nitrogen species (ROS and RNS, respectively). High levels of ROS or RNS are deleterious to all classes of cell components: lipids, proteins, nucleic acids, and other macromolecules. However, these species are also produced during normal physiological processes, reacting with cellular components, leading to the activation of intracellular signalling pathways and nuclear transcription factors, inducing gene expression and cell responses such as repair, adaptation, or transformation. Cumulatively these processes are described as redox signalling (Sen 1998; Allen and Tresini 2000; Forman and

Torres 2001; Sharma et al. 2017a, b). Currently, most of the reported methods to evaluate whether NP can induce OS have been ascertained by classical oxidative stress assays. However, these techniques designed for testing established drugs and chemicals are not always well adapted to evaluate impact of NPs. Indeed, some NPs can interfere with assay components and readout systems leading to conflicting reports and the generation of unreliable data due to their unusual physicochemical properties that include:

- High surface area, and as a consequence increased reactivity.
- Different optical properties that can interfere with fluorescence or visible light absorption detection systems.
- Increased catalytic activity due to enhanced surface energy.
- Magnetic properties that make them redox active and lead to interference with methods based on redox reactions (Pulskamp et al. 2007; Doak et al. 2009; Kroll et al. 2009).

2 Cellular Uptake of Nanoparticles

The advancement of NPs for a broad range of biomedical applications anticipates safer and more efficient resolutions to numerous medical issues. NPs are very small in size and some of them are also polar, but they cannot diffuse through the plasma membrane. The process by which the NPs enter into the cell is a crucial factor for deciding the toxicity, biomedical functions, and bio-distribution of NPs. The secured entry of NPs into the cells is a crucial step to accomplish high output of prognostic and therapeutic efficacy (Behzadi et al. 2017). Along with this, the intracellular destiny of NPs is crucial to their success. It is believed that NPs are important with respect to delivery of specific molecules such as drugs, genes, and contrast agents to the cytosol, nucleus, or other specific intracellular sites. However, efficient and controlled entry/trafficking of NPs into cells is a tough task. NP interactions with cell membranes and understanding of cellular uptake and trafficking mechanisms are very vital for the designing of harmless and effective nanomedicines. Addressing these issues in the light of NPs physicochemical properties, optimized cellular uptake, targeting, and trafficking can be appropriately achieved.

The exterior of NPs in biological fluids is changed by the process of biomolecule's adsorption on the surface (Walkey and Chan 2012). The adsorbed proteins on the surface of NPs are known as "protein corona." Protein corona coated NPs are available for the cellular uptake instead of native surfaces of NPs and thus the biological identity of NPs is recognized by the coated protein (type, quantity, and conformation) layers by the cells (Walczyk et al. 2010).

Some critical factors influencing biological recognition of the NPs are outlined below:

1. Physicochemical properties of NPs such as surface chemistry, size, shape, charge, surface hydrophilicity/hydrophobicity, and polydispersity collectively affect their biological attributes. These physicochemical features provide a sub-

stantial impact in regulating NPs uptake, endocytotic route, as well as the process of cytotoxicity.

2. The biological factors revealing the protein sources such as human or rat serum and the profusion of protein molecules are important factors.
3. Experimental factors such as change in the local temperature during the hyperthermic events, incubation period temperature, adsorption temperature, ionic strength, and osmolarity directly affect the properties of NPs.

Interaction of NPs with microenvironment present around the target cells is an important step for their biological recognition before their interaction with the outer target cell membranes. The microenvironment includes extracellular matrix (ECM), pH, fibrosis, etc. that can interfere with the interaction of the NPs with the cells, their entry, characteristics, and eventually alter their intracellular destiny. Furthermore, some other microenvironmental factors, e.g., vascular endothelial growth factor, matrix metalloproteinases, bradykinin, and prostaglandins, can also regulate the interaction of NPs and cell.

NPs interact with the membrane components or with the ECM when coming closer to the external surface of the cellular target. Endocytosis mainly facilitates the entry of the NPs inside the cell. After endocytosis of NPs, endocytic vesicles are formed and NPs are moved to specific intracellular sorting/trafficking compartments. Different endocytotic pathways can be used by the different cells to internalize the same NPs.

2.1 Intracellular Trafficking

The ultimate fate of cell and therapeutic or imaging effectiveness of NPs mainly depend on their intracellular trafficking. When NPs enter into the cells by the endocytosis process, intracellular trafficking process starts with the help of cellular endosomes network along with endoplasmic reticulum, Golgi apparatus, and lysosomes (Zhao et al. 2011). However, in other studies it has been reported that the endocytosed NPs leave the endocytic pathway at any stage and enter directly into the cell cytoplasm. In such cases NPs evade the fusion step with the lysosome, hence circumventing the lysosomal degradation process. As a result, NPs interact with the cytoplasm of the target cell or other intracellular compartment. In addition to this autophagy may also regulate the intracellular fate of NPs as it reduces the above-mentioned bypass process by recapturing NPs and enabling the process of lysosomal degradation.

There are numerous hidden factors such as cell vision and protein corona produce difficulties in cellular trafficking of NPs (Azhdarzadeh et al. 2015). Cell vision refers to the mechanisms/behavior that any specific cell can employ in response to NPs. Many proteins secreted by the cells may affect the biological recognition of NPs. Further, the type of applied protein on the NPs surface can also affect the intracellular trafficking (Schöttler et al. 2016). A recent finding regarding the biological identity of NPs revealed that different types of diseases could change the natural acceptance, intracellular fate, cellular uptake, and the toxicity of NPs (Hajipour et al. 2015). NP's entry and intracellular localization pathway affects the commencement of the cytotoxicity by them (Albanese et al. 2012).

3 Nanotoxicity and its Mechanism

Toxicity profiling of nanomaterial has been studied in various biological systems, cell lines, and different organisms including rodents, humans, and other aquatic species such as catfish and zebrafish (Wang et al. 2011), as well as in algae (Sohaebuddin et al. 2010). Nanomaterials of carbon and metallic origin are the most widely used variety of nanomaterials. Metallic nanoparticles of silver, gold, copper, nickel, cobalt, and aluminum have been widely studied in recent past years. Metallic nanoparticles are useful in several industrial purposes as additives in pharmaceutics, cosmetics, and as food colorants.

ZnO and TiO_2 NPs are used as additives in sunscreen lotion or fibrous component and the skin gets exposed to these metallic nanoparticles if applied topically as lotion or cream. In addition to the consumer products, the manufacture and use of NPs will definitely lead to increased occupational and environmental exposure. The toxicity of metal oxide (TiO_2, ZnO, CuO, CuZn, Fe_3O_4, and Fe_2O_3) NPs has been reported by several workers and the extent of toxicity depends on the type of nanomaterial (Yang and Watts 2005; Grassian et al. 2007; Jin et al. 2008; Karlsson et al. 2008; Ray et al. 2009). Zhu et al. (2013) studied the comparative toxicity of three nano-metal oxides of Cu, Cd, and Ti. Nano-CuO was found to cause the potent cytotoxicity and DNA damage via formation of 8-hydroxy-2-deoxyguanosine (8-OHdG), while nano-TiO_2 was the least toxic. At mechanistic level, it has been shown that nanotoxicity generates ROS which ultimately leads to OS in tissues (Gonzalez et al. 2008).

3.1 ROS Generation and Oxidative Stress

ROS, key signalling molecules during cell signalling and homeostasis, are reactive species of molecular oxygen. ROS constitute a pool of oxidative species including superoxide anion ($O_2^{\cdot}$), hydroxyl radical ($^{\cdot}OH$), hydrogen peroxide (H_2O_2), singlet oxygen ($^1O_2$), and hypochlorous acid (HOCl). ROS are generated intrinsically or extrinsically within the cell. Molecular oxygen generates $O_2^{\cdot-}$, the primary ROS via one-electron reduction catalyzed by nicotinamide adenine dinucleotide phosphate (NADPH) oxidase. Further reduction of oxygen may either lead to H_2O_2 or $^{\cdot}OH$ via dismutation and metal-catalyzed Fenton reaction, respectively (Risom et al. 2005). Some of the endogenous sources of ROS include mitochondrial respiration, inflammatory response, microsomes, and peroxisomes, while engineered nanomaterial and environmental pollutants act as exogenous ROS inducers. Physiologically, ROS are produced in trace amounts in response to various stimuli. Free radicals occur as essential by-products of mitochondrial respiration and transition metal ion-catalyzed Fenton-type reactions. Inflammatory phagocytes such as neutrophils and macrophages induce oxidative outburst as a defense mechanism towards environmental pollutants, tumor cells, and microbes.

The level of ROS generation by engineered nanomaterials is dependent on the chemical nature of the NPs (Gonzalez et al. 2008). Compared to their bulk-size counterparts, engineered nanomaterials possess a small size, high specific surface area, and high surface reactivity, leading to the production of higher levels of ROS, and resulting in cytotoxicity and genotoxicity (Oberdorster et al. 2005). A variety of nanomaterials has been found to induce toxicity mediated by ROS in many biological systems, such as human erythrocytes and skin fibroblasts (Li et al. 2012). A variety of NP including metal oxide particles induces ROS as one of the principal mechanisms of cytotoxicity (Risom et al. 2005). NPs have been reported to influence intracellular calcium concentrations, activate transcription factors, and modulate cytokine production via generation of free radicals (Huang et al. 2010c; Li et al. 2010).

Not all nano-metal oxide-induced toxicity is mediated by ROS. A good example comes from the study by Karlsson et al. (2008). They determined the cytotoxicity, DNA damage, and oxidative stress of different nano-metal oxides (CuO, TiO_2, ZnO, $CuZnFe_2O_4$, Fe_3O_4, and Fe_2O_3), carbon NPs, and multiwalled carbon nanotubes (CNTs) in human lung epithelial cell line A549. Among nano-metal oxides, nano-CuO was found to be the most powerful in inducing cytotoxicity, DNA damage, producing oxidative lesions, and considerably increasing intracellular ROS while nano-ZnO exhibited cytotoxicity and DNA damage response only. Nano-TiO_2, containing both rutile and anatase forms, merely produced DNA damage. Nano-$CuZnFe_2O_4$ was effective in inducing DNA lesions whereas Nano-Fe_3O_4 and nano-Fe_2O_3 displayed no or low cytotoxicity. CNTs caused cytotoxicity DNA damage. The study indicated that nano-CuO exhibited the highest cytotoxicity and genotoxicity, and is the only studied nanomaterial that induces ROS (Karlsson et al. 2008).

Huang et al. (2010a) proposed the hierarchical model of oxidative stress to explain a mechanism for NP-mediated OS. This model illustrated that cells and tissues responded to increasing levels of OS via antioxidant enzyme systems upon NP exposure. During mild OS conditions, transcriptional activation of phase II antioxidant enzymes takes place via nuclear factor (erythroid derived 2)-like 2 (Nrf2) induction. At an intermediate level, redox-sensitive mitogen-activated protein kinase (MAPK) and nuclear factor kappa-light-chain enhancer of activated B-cells (NF-Kβ) cascades mount a pro-inflammatory response. However, extremely toxic levels of OS result in mitochondrial membrane damage and electron chain dysfunction leading to cell death. Some of the crucial factors supporting the pro-oxidant effects of engineered nanomaterials include either the depletion of antioxidants or the increased production of ROS. Perturbation of the normal redox state contributes to peroxide and free radical production that has adverse effects on cell components including proteins, lipids, and DNA (Huang et al. 2010a). Depending on chemical reactivity, OS can produce DNA damage, lipid peroxidation, and activation of signalling networks associated with loss of cell growth, fibrosis, and carcinogenesis (Knaapen et al. 2004; Valko et al. 2006). Besides cellular damage, ROS can result from interactions of NP with several biological targets during cell respiration and metabolism, ischemia/reperfusion, inflammation, and metabolism of various NPs (Risom et al. 2005). Most significantly, the OS resulting from occupational nanomaterial exposures as well as experimental challenges with various NPs lead to airway inflammation and interstitial fibrosis (Donaldson et al. 2004; Nel 2005).

3.2 Nanoparticle-Induced Oxidative Stress

Generation of ROS and consequent oxidative insult in the form of OS are the chief culprit of NP-mediated toxicity. NPs have potential to enter inside cell membrane and exert various biological responses such as mitochondrial dysfunction, signalling cascade activation, apoptosis, and cell cycle arrest in the cytoplasm of cell (Asha Rani et al. 2008; Bhabra et al. 2009; Mahmoudi et al. 2011). NP-induced cell death (apoptosis and necrosis both) mainly depends on the NPs concentration, exposure duration, as well as on the type of cellular system investigated (Pan et al. 2009). Hydrophilic NPs such as TiO_2 have oncogenic character and could convert the progression of benign mouse fibro sarcoma cells into malignant cells via ROS generation (Onuma et al. 2009). Exposure of NPs to lysosome (acidic environment) induces ROS generation. Surplus ROS have ability to cleave the DNA strand and therefore affect the gene expression. Previous studies have revealed that NP-induced ROS generation either mediated or promoted the activation of the mitogen-activated protein kinase (MAPK) pathways. All types of MAPKs including growth factor-regulated and stress-regulated ERKs, c-Jun NH_2 terminal janus kinases (JNKs), and p38 MAPKs are involved in this signal transduction mechanism (Son et al. 2011). However the precise mechanism of ROS-induced MAPKs activation are not well defined. Gyoton and coworkers (1996) concluded that ROS activates the MAPKs via activation of growth factor receptors (Guyton et al. 1996). Ag-NPs possibly can induce the ROS generation and decrease the Akt and ERK activation; however Au-NPs reduce the EGF-dependent Akt and ERK activation, while superparamagnetic iron oxide (SPION) changed the EGF-mediated transcription and influenced cell proliferation, migration, and receptor expression (Comfort et al. 2011).

A dose-dependent toxicity profile of FeO_2 NPs has been shown to be mediated by increased oxidative stress (Naqvi et al. 2010). By using DCFDA assay after NP exposure, it was concluded that ROS is generated in concentration-dependent manner which eventually leads to cell death and injury. Lactate dehydrogenase assay further proved that NPs inflicted concentration- and time-dependent damage. NPs SPION causes cytotoxicity and oxidative stress by induction of redox-sensitive AP-1, and NF-κB signal transduction pathways (Murray et al. 2012).

CuO-NPs cause genotoxicity through the activation of p53 pathway in humans (Ahamed et al. 2010). In addition to this, Nrf2, NF-κB, and AP-1 were also found to be associated with CuO-NP exposure to HepG2 cells (Piret et al. 2012) and the toxicity pattern was in dose-dependent manner (Fahmy and Cormier 2009). TiO_2-NP exposure to U937 human monoblastoid cells causes apoptosis via mitochondrial dependent pathways. Kang et al. (2008) found that TiO_2-NPs cause ROS generation in lymphocytes, therefore activating the p53-mediated DNA damage checkpoint signals. TiO_2-NPs (600 μg/ml) exposed to fibroblast cell causes 80% viability within 24 hours (Jin et al. 2008). Sharma et al. (2009) showed that because of their smaller size zinc oxide NPs have ability to interact with DNA and reported DNA damage on exposure to human epidermal cell line (A431). If it is present as soluble NPs, it causes redox imbalance by the production of intracellular ROS. Experimental reports advocated that the cell endeavors to restore the redox imbalance via activation of transcription factor NF-κB which is accountable for the synthesis of enzymes involved in antioxidant defense pathways (Mercurio and Manning 1999; Christman et al. 2000).

3.2.1 DNA Damage

ROS generated inside the cell through the physiological process such as photosynthesis, respiration, and cell signalling are regulated by either the enzymes such as superoxide dismutase (SOD), catalase (CAT), and glutathione peroxidase or by the nonenzymatic antioxidants such as ascorbic acid, cysteine, glutathione, and bilirubin (Xu et al. 2016). Redox homeostasis can be altered by the free radical surge that can cause oxidative damage to the cells (Tapeinos and Pandit 2016). OS is a key element associated to nanotoxicity, coupled with altered cell motility, cell death, improper cell signalling, DNA damage, cytotoxicity, cancer initiation, and progression (Nel et al. 2006; Xia et al. 2008; Zhu et al. 2013).

NP-mediated DNA damages are the result of generation of free radicals such as hydroxyl radicals, which interact with the DNA to form a adduct 8-hydroxyl-20-deoxyguanosine (8-OHdG) that eventually leads to the DNA damage (Valavanidis et al. 2009). 8-OHdG is a significant biomarker of hydroxyl radical-induced DNA damage as its level is elevated during in vitro and in vivo exposure to NPs (Eblin et al. 2006; Inoue et al. 2006). An in vivo study by Song et al. (2012) reported that NPs (Ag, Ti, Fe, Cu) exposure causes nucleic acid damage-mediated genotoxicity. Moreover, ROS-induced lipid peroxidation associated mutations are also involved in NP-induced genotoxicity (Howden and Faux 1996; Turski and Thiele 2009; Shukla et al. 2011). Oxidative DNA damage is result of various mechanisms such as mutation, carcinogen attack, and aging-related diseases. ROS generation and OS triggered by NPs are crucial to DNA damage that improper cell signalling that and ultimately causes altered cell motility, apoptosis, and sometimes carcinogenesis (Fig. 17.1). Hence, the need of the hour to understand the mechanism by which NPs causes the adverse impact. DNA is a critical cellular target of ROS. Free radical-induced DNA damage involves sugar and base lesions, strand breaks (both single and double), DNA-protein crosslinking, as well as the formation of a basic site (Valko et al. 2006). Highly reactive radicals, such as hydroxyl radicals, can damage DNA quickly in the vicinity, whereas the less-reactive ROS may interact with DNA at a distance.

3.2.2 Cellular Damage

During OS, ROS generation surpasses the antioxidant defense system capability of the body to detoxify the reactive intermediates formed or to repair the cellular damage (Kumar et al. 2015; Sharma et al. 2017a, b). Free radicals generated in the system plays vital role in regulation of signal transduction, homeostasis, and defense mechanism (Kumar and Pandey 2013; Gupta et al. 2021). However under conditions of surplus production they inflict damage to cellular membranes as well as the macromolecules such as DNA, protein, and lipid causing harmful effects (Mishra et al. 2013; Kumar and Pandey 2015). Intrinsically ROS are generated through the cellular respiration, and during inflammatory response in peroxisome and microsomes. Environmental pollutant and NPs cause ROS generation through the intrinsic pathway (Fahmy and Cormier 2009; Sarkar et al. 2014). Phagocytic cells of the body exude free radicals as a defense mechanism in response to the microbes, environmental pollutants, and cancer cells (Risom et al. 2005). Metallic NPs are found to exhibit their cytotoxic property via ROS generation (Fig. 17.1). They alter the intracellular Ca^{2+} concentration, activate several transcription factors, as well as curb the cytokine production via ROS generation (Li et al. 2010; Huang et al. 2010a, b). To cope up with the NP exposure induced ROS surge, cellular systems activates their antioxidant defense mechanisms in the body (Sies 1991). Transcription factor Nrf2 is activated during the low level of OS as a defense mechanism. At moderate level of ROS generated, the MAP kinase and nuclear factor NF-κB induced pro-inflammatory response is activated. However, under severe OS mitochondrial dysfunction ensues via membrane damage and ultimately causes impaired electron transport system and cell death. Engineered NP exposure may lead to either increased ROS generation or riddance of antioxidant molecules (Valko et al. 2006; Buzea et al. 2007). In addition to the cellular damage, OS resulting from occupational NP exposures may exhibit adverse effect on organ system by causing airway inflammation and fibrosis (Donaldson et al. 2004; Nel 2005).

4 Nanoparticle and Inflammation

In several studies on single- as well as multiwalled CNTs and fullerene derivatives, it has been reported that the initiation of inflammation process occurs in various cell types (cultured monocyte-macrophage cells, alveolar and bronchial epithelial cells, and epidermal keratinocytes) (Khanna et al. 2015). The mechanistic study done by the computational model portrays that these CNTs and C60 fullerenes may be treated as pathogens by the Toll-like receptors which in response initiates the innate immune responses by the target body cells and as a result the inflammatory protein mediators such as chemokines and interleukins are secreted (Turabekova et al. 2014). Hypersensitivity and anaphylaxis reactions may be started by the activation of the complement system of the body on the exposure with the liposomes and other lipid-based NPs (Dobrovolskaia et al. 2008). The mechanism behind this event of nanotoxicity caused by the complement proteins has not yet been entirely revealed (Zolnik et al. 2010).

After initiation of the inflammation process, the by-products of inflammation like ROS, complement proteins, and receptor-induced apoptosis/necrosis cause toxicity which leads to the cell death (Fig. 17.1) (Wallach et al. 2014). However, these cascades are not fully explored in the context of cytotoxicity caused by the NPs (Khanna et al. 2015). A study on the lung injury as well as pulmonary fibrosis caused by the NPs clearly showed that ROS generated in the system activated the NF-κB and pro-inflammatory mediators such as IL-6, IL-8, IL-2, and TNF-α (Byrne and Baugh 2008).

Various metal oxide NPs such as cadmium, iron, silica, and zinc also cause toxicity by the NF-κB-mediated inflammatory cytokines (Pujalte et al. 2011). Roy et al. (2014) revealed in their study that the toxicity caused by the zinc oxide NPs is mainly due to the overexpression of Cox-2, iNOS, pro-inflammatory, and regulatory cytokine IL-10 which was induced by PI3-K signalling in the macrophages. Duan et al. (2014) also reported about the induction of inflammation and autophagy cascades via the PI3-K/Akt/mTOR pathways by the silica NPs.

Use of nanoscale molecular probes in diagnosis, treatment, and characterization of diseases opens a new platform for the technologies related to the imaging. These probes are mainly associated with inflammation. Early recognition of subclinical disease states by these probes can enhance the tailored therapies easily. Nanotechnology helps in the area of biomarkers for the diagnostic purposes and as a cure for progression of diseases related to the inflammation with the help of some techniques such as surface-enhanced Raman scattering, magnetic resonance imaging, and fluorescent quantum dots (Stevenson et al. 2011). These techniques along with sensitive probes can detect ultrasensitive inflammation.

5   Strategies to Overcome Nanoparticle-Mediated Toxicity

Overcoming toxicity of NPs is an important issue for perpetuating its beneficial effects. The toxicity is generated by the unavoidable interactions with the different cellular processes of the cell and biological compartments. Some methods given below are deployed to minimize the toxicity of nanoparticles.

5.1 *Masking of Nanomaterials to Reduce the Issue Related to the Toxicity*

Protection of NP's surface is an important step that can slow down the toxicity generated by the nanomaterials. The process of masking is used routinely for the above-related issue (Tran et al. 2009). Masking of NPs can be executed by the surface coating of NPs with the biocompatible hydrophilic polymers/surfactants or can be masked by formulation with biodegradable copolymers in association with hydrophilic polyethylene glycol, polyoxamer, polyethylene oxide, Tween 80, and poloxamine. This process of nanoparticle masking reduces the circulation time and toxicity of nanoparticles.

5.2   Incorporation of Targeting Moieties to NPs

Another approach to minimize the toxicity caused by the NPs is by incorporating the surface-bound targeting moieties with the NPs (Emerich and Thanos 2006). With these improved features, the targeted NPs show reduced toxicity at lower concentrations (Farokhzad et al. 2006). High efficacy of a drug with less toxicity can be achieved by the incorporation of peptide-targeted immunoliposomes or antibody (Immordino et al. 2006). The success of immune-targeted NPs or immunoliposomes depends on the selection of a perfect ligand for targeting. There are some features of ligand that should be considered while choosing the correct ligand for targeting. These include: (1) The number of steps during purification should be minimum and conjugate should be stable, and produced in vast quantities. (2) Incorporation of a ligand with the nanoparticles should be simple, and after their conjugation process either of the property of nanoparticles or ligand should not be affected.

Antibodies are a promising ligand among the tested ligands because they can be produced quickly by the hybridoma technology with the highly selective approach. This innovative targeting approach should be utilized for the targeted delivery into specific tumor cells or tumors with the help of identified targeting moiety. Several in vitro and in vivo studies revealed that the OS plays a central role in the nanotoxicity; therefore, an approach in the direction of minimizing the OS process will play a significant role for the reduction of nanotoxicity. The OS-mediated nanotoxicity can be minimized by the incorporation of ascorbic acid (vitamin C). Ascorbic acid is a very-well-known antioxidant, and it scavenges the free radicles, and thus reduces the oxidative process and ultimately the nanotoxicity (Niki 1991). Using the antioxidative properties of ascorbic acid with AgNP for the treatment of acute myeloid leukemia cells, it has been observed that the ROS production was almost negligible. This particular attribute of ascorbic acid also reduces the apoptosis cascades and damage in DNA and mitochondria caused by the AgNPs (Guo et al. 2013). Similar results were reported by the Ahamed et al. (2011) when they used ascorbic acid with the nickel ferrite NPs in the human lung epithelial (A549) cells; a depleted level of glutathione and moderation in ROS level was observed. These observations demonstrated that to overcome the toxicity caused by the nanomaterials, administration of vitamin C after NP exposure is a crucial step in reducing toxicity.

Quercetin, a natural flavonoid, is present in many plants and food items. Quercetin acts as an antioxidant as it can scavenge the free radicals. Supplementation of quercetin with Fe_2O_3-NPs lowers the oxidative injury and inflammation by enhancing the Bad phosphorylation and Nrf2 translocation through PI3-K/Akt dependent pathways

(Ahamed et al. 2011). Another study by Gonzalez-Esquivel et al. (2015) reported the role of quercetin in minimizing the TiO_2-NP-induced kidney and liver OS.

Pathways like MAPK, PI3-K, and NF-κB promote the nanomaterial-induced inflammation and, in result, different chemokines and pro-inflammatory cytokines are released which lead to cytotoxicity and finally cell death. To combat the inflammation an approach related to the modulation of Jun/AP-1 pathway components opens a new window for the therapeutic interventions. Therefore consideration of these signalling molecules might play a critical role in combating the NP-mediated inflammation (Schonthaler et al. 2011). These approaches provide upliftment of nanomaterial-based applications in the area of medicine.

Acknowledgments Amit Kumar Singh and Astha Dwivedi acknowledge CSIR-New Delhi for providing financial support in the form of Senior Research Fellowship. Prabhash K. Pandey gratefully acknowledges UGC-New Delhi for DSK Postdoctoral Fellowship. Authors also acknowledge the UGC-SAP and DST-FIST facilities of Department of Biochemistry, University of Allahabad, Prayagraj.

References

Ahamed, M., Akhtar, M. J., Siddiqui, M. A., Ahmad, J., Musarrat, J., Al-Khedhairy, A. A., AlSalhi, M. S., & Alrokayan, S. A. (2011). Oxidative stress mediated apoptosis induced by nickel ferrite nanoparticles in cultured a549 cells. *Toxicology, 283*, 101–108.

Ahamed, M., Siddiqui, M. A., Akhtar, M. J., Ahmad, I., Pant, A. B., & Alhadlaq, H. A. (2010). Genotoxic potential of copper oxide nanoparticles in human lung epithelial cells. *Biochemical and Biophysical Research Communications, 396*, 578.

Albanese, A., Tang, P. S., & Chan, W. C. (2012). The effect of nanoparticle size, shape, and surface chemistry on biological systems. *Annual Review of Biomedical Engineering, 14*, 1–16.

Allen, R. G., & Tresini, M. (2000). Oxidative stress and gene regulation. *Free Radical Biology & Medicine, 28*, 463–499.

Asha Rani, P. V., Low Kah Mun, G., Hande, M. P., & Valiyaveettil, S. (2008). Cytotoxicity and genotoxicity of silver nanoparticles in human cells. *ACS Nano, 3*, 279.

Azhdarzadeh, M., Saei, A. A., Sharifi, S., Hajipour, M. J., Alkilany, A. M., Sharifzadeh, M., Ramazani, F., Laurent, S., Mashaghi, A., & Mahmoudi, M. (2015). Nanotoxicology: Advances and pitfalls in research methodology. *Nanomedicine, 10*, 2931–2952.

Behzadi, S., Serpooshan, V., Tao, W., Hamaly, M. A., Alkawareek, M. Y., Dreaden, E. C., Brown, D., Alkilany, A. M., Farokhzad, O. C., & Mahmoudi, M. (2017). Cellular uptake of nanoparticles: Journey inside the cell. *Chemical Society Reviews, 46*, 4218–4244.

Bhabra, G., Sood, A., Fisher, B., Cartwright, L., Saunders, M., Evans, W. H., Surprenant, A., Lopez-Castejon, G., Mann, S., Davis, S. A., Hails, L. A., Ingham, E., Verkade, P., Lane, J., Heesom, K., Newson, R., & Case, C. P. (2009). Nanoparticles can cause DNA damage across a cellular barrier. *Nature Nanotechnology, 4*, 876.

Brar, S. K., Verma, M., Tyagi, R. D., & Surampalli, R. Y. (2010). Engineered nanoparticles in wastewater and wastewater sludge evidence and impacts. *Waste Management, 30*, 504–520.

Buzea, C., Pacheco, I. I., & Robbie, K. (2007). Nanomaterials and nanoparticles: Sources and toxicity. *Biointerphases, 2*(4), 17–71.

Byrne, J. D., & Baugh, J. A. (2008). The significance of nanoparticles in particle-induced pulmonary fibrosis. *McGill Journal of Medicine, 11*, 43–50.

Christman, J. W., Blackwell, T. S., & Juurlink, B. H. (2000). Redox regulation of nuclear factor kappa B: Therapeutic potential for attenuating inflammatory responses. *Brain Pathology, 10*, 153.

Comfort, K. K., Maurer, E. I., Braydich-Stolle, L. K., & Hussain, S. M. (2011). Interference of silver, gold, and iron oxide nanoparticles on epidermal growth factor signal transduction in epithelial cells. *ACS Nano, 5*, 10000.

Doak, S. H., Griffiths, S. M., Manshian, B., Singh, N., Williams, P. M., Brown, A. P., & Jenkins, G. J. (2009). Confounding experimental considerations in nanogenotoxicology. *Mutagenesis, 24*, 285–293.

Dobrovolskaia, M. A., Aggarwal, P., Hall, J. B., & McNeil, S. E. (2008). Preclinical studies to understand nanoparticle interaction with the immune system and its potential effects on nanoparticle biodistribution. *Molecular Pharmaceutics, 5*, 487–495.

Donaldson, K., Stone, V., Tran, C. L., Kreyling, W., & Borm, P. J. (2004). Nanotoxicology. *Occupational and Environmental Medicine, 61*(9), 727–728.

Duan, J., Yu, Y., Yu, Y., Li, Y., Wang, J., Geng, W., Jiang, L., Li, Q., Zhou, X., & Sun, Z. (2014). Silica nanoparticles induce autophagy and endothelial dysfunction via the PI3K/Akt/mTOR signaling pathway. *International Journal of Nanomedicine, 9*, 5131–5141.

Eblin, K., Bowen, M., Cromey, D., Bredfeldt, T., Mash, E. A., Lau, S., & Gandolfi, A. J. (2006). Arsenite and monomethylarsonous acid generate oxidative stress response in human bladder cell culture. *Toxicology and Applied Pharmacology, 217*, 7–14.

Emerich, D. F., & Thanos, C. G. (2006). The pinpoint promise of nanoparticle-based drug delivery and molecular diagnosis. *Biomolecular Engineering, 23*, 171–184.

Fahmy, B., & Cormier, S. A. (2009). Copper oxide nanoparticles induce oxidative stress and cytotoxicity in airway epithelial cells. *Toxicology In-Vitro, 23*, 1365–1371.

Farokhzad, O. C., Cheng, J., Teply, B. A., Sherifi, I., Jon, S., Kantoff, P. W., Richie, J. P., & Langer, R. (2006). Targeted nanoparticle-aptamer bioconjugates for cancer chemotherapy in vivo. *Proceedings of the National Academy of Sciences of the United States of America, 103*, 6315–6320.

Forman, J. H., & Torres, M. (2001). Redox signaling in macrophages. *Molecular Aspects of Medicine, 22*, 189–216.

Ganguly, R., Singh, A. K., Kumar, R., Gupta, A., Pandey, A. K., Pandey, A. K. (2019). Nanoparticles as modulators of oxidative stress. In: Nanotechnology in modern animal biotechnology, *Eds. P.K. Maurya and S. Singh, pp29–35*, Elsevier, St. Louis, Missouri.

Gonzalez, L., Lison, D., & Kirsch-Volders, M. (2008). Genotoxicity of engineered nanomaterials: A critical review. *Nanotoxicology, 2*, 252–273.

Gonzalez-Esquivel, A. E., Charles-Nino, C. L., Pacheco-Moises, F. P., Ortiz, G. G., Jaramillo-Juarez, F., & Rincon-Sanchez, A. R. (2015). Beneficial effects of quercetin on oxidative stress in liver and kidney induced by titanium dioxide (TiO2) nanoparticles in rats. *Toxicology Mechanisms and Methods, 25*, 1–10.

Grassian, V. H., O'Shaughnessy, P. T., Adamcakova-Dodd, A., Pettibone, J. M., & Thorne, P. S. (2007). Inhalation exposure study of titanium dioxide nanoparticles with a primary particle size of 2 to 5 nm. *Environmental Health Perspectives, 115*, 397–402.

Green, M., & Howman, E. (2005). Semiconductor quantum dots and free radical induced DNA nicking. *Chemical Communications, 121*, 121–123.

Guo, D., Zhu, L., Huang, Z., Zhou, H., Ge, Y., Ma, W., Wu, J., Zhang, X., Zhou, X., Zhang, Y., et al. (2013). Anti-leukemia activity of PVP-coated silver nanoparticles via generation of reactive oxygen species and release of silver ions. *Biomaterials, 34*, 7884–7894.

Gupta, A., Kumar, R., Ganguly, R., Singh, A. K., Rana, H. K., & Pandey, A. K. (2021). Antioxidant, anti-inflammatory and hepatoprotective activities of Terminalia bellirica and its bioactive component ellagic acid against diclofenac induced oxidative stress and hepatotoxicity. *Toxicology Reports, 8*, 44–52.

Guyton, K. Z., Liu, Y., Gorospe, M., Xu, Q., & Holbrook, N. J. (1996). Activation of mitogen-activated protein kinase by H2O2. Role in cell survival following oxidant injury. *The Journal of Biological Chemistry, 271*, 4138.

Gwinn, M. R., & Vallyathan, V. (2006). Nanoparticles: Health effects – Pros and cons. *Environmental Health Perspectives, 114*(12), 1818–1825.

Hajipour, M. J., Raheb, J., Akhavan, O., Arjmand, S., Mashinchian, O., Rahman, M., Abdolahad, M., Serpooshan, V., Laurent, S., & Mahmoudi, M. N. (2015). Personalized disease-specific protein corona influences the therapeutic impact of graphene oxide. *Nanoscale, 7*, 8978–8994.

Howden, P. J., & Faux, S. P. (1996). Fibre-induced lipid peroxidation leads to DNA adduct formation in salmonella typhimurium ta104 and rat lung fibroblasts. *Carcinogenesis, 17*, 413–419.

Huang, C., Aronstam, R. S., Chen, D., Yung, L. Y., & Bay, B. H. (2010a). Oxidative stress, calcium homeostasis, and altered gene expression in human lung epithelial cells exposed to ZnO nanoparticles. *Toxicology In Vitro, 24*(1), 45–55.

Huang, X., Zhuang, J., Teng, X., Li, L., Chen, D., Yan, X., & Tang, F. (2010b). The promotion of human malignant melanoma growth by mesoporous silica nanoparticles through decreased reactive oxygen species. *Biomaterials, 31*(24), 6142–6153.

Huang, Y., Wu, C., & Aronstam, R. (2010c). Toxicity of transition metal oxide nanoparticles: Recent insights from in vitro studies. *Materials, 3*(10), 4842–4859.

Hussain, S. M., Hess, K. L., Gearhart, J. M., Geiss, K. T., & Schlager, J. J. (2005). In vitro toxicity of nanoparticles in BRL 3A rat liver cells. *Toxicology, 19*, 975–983.

Immordino, M. L., Dosio, F., & Cattel, L. (2006). Stealth liposomes: Review of the basic science, rationale, and clinical applications, existing and potential. *International Journal of Nanomedicine, 1*, 297–315.

Inoue, K.-I., Takano, H., Yanagisawa, R., Hirano, S., Sakurai, M., Shimada, A., & Yoshikawa, T. (2006). Effects of airway exposure to nanoparticles on lung inflammation induced by bacterial endotoxin in mice. *Environmental Health Perspectives, 114*, 1325–1330.

Jin, C. Y., Zhu, B. S., Wang, X. F., & Lu, Q. H. (2008). Cytotoxicity of titanium dioxide nanoparticles in mouse fibroblast cells. *Chemical Research in Toxicology, 21*, 1871–1877.

Kang, S. J., Kim, B. M., Lee, Y. J., & Chung, H. W. (2008). Titanium dioxide nanoparticles trigger p53-mediated damage response in peripheral blood lymphocytes. *Environmental and Molecular Mutagenesis, 49*, 399.

Karlsson, H. L., Cronholm, P., Gustafsson, J., & Möller, L. (2008). Copper oxide nanoparticles are highly toxic: A comparison between metal oxide nanoparticles and carbon nanotubes. *Chemical Research in Toxicology, 21*, 1726–1732.

Khanna, P., Ong, C., Bay, B. H., & Baeg, G. H. (2015). Nanotoxicity: An interplay of oxidative stress, inflammation and cell death. *Nanomaterials, 5*, 1163–1180.

Knaapen, A. M., Borm, P. J. A., Albrecht, C., & Schins, R. P. F. (2004). Inhaled particles and lung cancer, part A: Mechanisms. *International Journal of Cancer, 109*(6), 799–809.

Kroll, A., Pillukat, M. H., Hahn, D., & Schnekenburger, J. (2009). Current in vitro methods in nanoparticle risk assessment: Limitations and challenges. *European Journal of Pharmaceutics and Biopharmaceutics, 72*, 370–377.

Kumar, S., Dwivedi, A., Kumar, R., & Pandey, A. K. (2015). Preliminary evaluation of biological activities and phytochemical analysis of *Syngonium podophyllum* leaf. *National Academy Science Letters, 38*(2), 143–146.

Kumar, S., & Pandey, A. K. (2013). Chemistry and biological activities of flavonoids: an overview. *The Scientific World Journal*, 162750.

Kumar, S., & Pandey, A. K. (2015). Free radicals: Health implications and their mitigation by herbals. *British Journal of Medicine Medical Research, 7*, 438–457.

Li, J. J., Muralikrishnan, S., Ng, C. T., Yung, L. Y., & Bay, B. H. (2010). Nanoparticle-induced pulmonary toxicity. *Experimental Biology and Medicine, 235*(9), 1025–1033.

Li, N., Hao, M., Phalen, R. F., Hinds, W. C., & Nel, A. E. (2003). Particulate air pollutants and asthma: A paradigm for the role of oxidative stress in PM induced adverse health effects. *Clinical Immunology, 109*, 250–265.

Li, Y., Yu, S., Wu, Q., Tang, M., Pu, Y., & Wang, D. (2012). Chronic Al2O3-nanoparticle exposure causes neurotoxic effects on locomotion behaviours by inducing severe ROS production and disruption of ROS defence mechanisms in nematode *Caenorhabditis elegans. Journal of Hazardous Materials*, 219–230.

Lin, W., Huang, Y. W., Zhou, X. D., & Ma, Y. (2006). In vitro toxicity of silica nanoparticles in human lung cancer cells. *Toxicology and Applied Pharmacology, 217*, 252–259.

Mahmoudi, M., Azadmanesh, K., Shokrgozar, M. A., Journeay, W. S., & Laurent, S. (2011). Effect of nanoparticles on the cell life cycle. *Chemical Reviews, 111*, 3407.

Maurya, A., Chauhan, P., Mishra, A., & Pandey, A. K. (2012). Surface functionalization of TiO_2 with plant extracts and their combined antimicrobial activities against *E. faecalis* and *E. coli*. *Journal of Research Updates in Polymer Science, 1*, 43–51.

Mercurio, F., & Manning, A. M. (1999). NF-kappaB as a primary regulator of the stress response. *Oncogene, 18*, 6163.

Mishra, A., Sharma, A. K., Kumar, S., Saxena, A. K., & Pandey, A. K. (2013). *Bauhinia variegata* leaf extracts exhibit considerable antibacterial, antioxidant, and anticancer activities. *BioMed Research International*. Article ID 915436.

Monteiller, C., Tran, L., MacNee, W., Faux, S., Jones, A., Miller, B., & Donaldson, K. (2007). The pro-inflammatory effects of low-toxicity low-solubility particles, nanoparticles and fine particles, on epithelial cells in vitro: The role of surface area. *Occupational and Environmental Medicine, 64*, 609–615.

Murray, A. R., Kisin, E., Inman, A., Young, S. H., Muhammed, M., Burks, T., Uheida, A., Tkach, A., Waltz, M., Castranova, V., Fadeel, B., Kagan, V. E., Riviere, J. E., Monteiro-Riviere, N., & Shvedova, A. A. (2012). *Cell Biochemistry and Biophysics*. https://doi.org/10.1007/s12013-012-9367-9.

Naqvi, S., Samim, M., Abdin, M., Ahmed, F. J., Maitra, A., Prashant, C., & Dinda, A. K. (2010). Concentration-dependent toxicity of iron oxide nanoparticles mediated by increased oxidative stress. *International Journal of Nanomedicine, 5*, 983.

Nel, A. (2005). Air pollution-related illness: Effects of particles. *Science, 308*(5723), 804–806.

Nel, A., Xia, T., Mädler, L., & Li, N. (2006). Toxic potential of materials at the nanolevel. *Science, 311*, 622–627.

Niki, E. (1991). Action of ascorbic acid as a scavenger of active and stable oxygen radicals. *American Journal of Clinical Nutrition, 54*, 1119s–1124s.

Oberdorster, G., Oberdorster, E., & Oberdorster, J. (2005). Nanotoxicology: An emerging discipline evolving from studies of ultrafine particles. *Environmental Health Perspectives, 113*, 823–839.

Onuma, K., Sato, Y., Ogawara, S., Shirasawa, N., Kobayashi, M., Yoshitake, J., Yoshimura, T., Iigo, M., Fujii, J., & Okada, F. (2009). Nano-scaled particles of titanium dioxide convert benign mouse fibrosarcoma cells into aggressive tumor cells. *The American Journal of Pathology, 175*, 2171.

Pan, Y., Leifert, A., Ruau, D., Neuss, S., Bornemann, J., Schmid, G., Brandau, W., Simon, U., & Jahnen-Dechent, W. (2009). Gold nanoparticles of diameter 1.4 nm trigger necrosis by oxidative stress and mitochondrial damage. *Small, 5*, 2067–2076.

Piret, J. P., Jacques, D., Audinot, J. N., Mejia, J., Boilan, E., Noël, F., Fransolet, M., Demazy, C., Lucas, S., Saout, C., & Toussaint, O. (2012). Copper(II) oxide nanoparticles penetrate into HepG2 cells, exert cytotoxicity via oxidative stress and induce pro-inflammatory response. *Nanoscale, 4*, 7168.

Pujalte, I., Passagne, I., Brouillaud, B., Tréguer, M., Durand, E., Ohayon-Courtès, C., & L'Azou, B. (2011). Cytotoxicity and oxidative stress induced by different metallic nanoparticles on human kidney cells. *Particle and Fibre Toxicology, 8*, 1–16.

Pulskamp, K., Diabate, S., & Krug, H. F. (2007). Carbon nanotubes show no sign of acute toxicity but induce intracellular reactive oxygen species in dependence on contaminants. *Toxicology Letters, 168*, 58–74.

Ray, P. C., Yu, H., & Fu, P. P. (2009). Toxicity and environmental risks of nanomaterials: Challenges and future needs. *Journal of Environmental Science and Health. Part C, Environmental Carcinogenesis & Ecotoxicology Reviews, 27*, 1–35.

Risom, L., Møller, P., & Loft, S. (2005). Oxidative stress-induced DNA damage by particulate air pollution. *Mutation Research, 592*(1–2), 119–137.

Roy, R., Parashar, V., Chauhan, L. K. S., Shanker, R., Das, M., Tripathi, A., & Dwivedi, P. D. (2014). Mechanism of uptake of ZnO nanoparticles and inflammatory responses in macrophages require PI3K mediated mapks signaling. *Toxicology In Vitro, 28*, 457–467.

Sanjay, S. S., Pandey, A. C., Kumar, S., & Pandey, A. K. (2014). Cell membrane protective efficacy of ZnO nanoparticles. *SOP Transactions on Nano-Technology, 1*, 21–29.

Sarkar, A., Ghosh, M., & Sil, P. C. (2014). Nanotoxicity: Oxidative stress mediated toxicity of metal and metal oxide nanoparticles. *Journal of Nanoscience and Nanotechnology, 14*, 730–743.

Schonthaler, H. B., Guinea-Viniegra, J., & Wagner, E. F. (2011). Targeting inflammation by modulating the Jun/AP-1 pathway. *Annals of the Rheumatic Diseases, 70*, 109–112.

Schöttler, S., Klein, K., Landfester, K., & Mailänder, V. (2016). Nanoscale, protein source and choice of anticoagulant decisively affect nanoparticle protein corona and cellular uptake. *Nanoscale, 8*, 5526–5536.

Sen, C. K. (1998). Redox signaling and the emerging therapeutic potential of thiol antioxidants. *Biochemical Pharmacology, 55*, 1747–1758.

Sharma, A. K., Sharma, U. K., & Pandey, A. K. (2017a). Protective effect of *Bauhinia variegata* leaf extracts against oxidative damage, cell proliferation and bacterial growth. *Proceedings of National Academy Sciences, India, Sect. B Biological Science, 87*, 45–51.

Sharma, U. K., Kumar, R., Gupta, A., Ganguly, R., Singh, A. K., Ojha, A., & Pandey, A. K. (2019). Ameliorative efficacy of eugenol against metanil yellow induced toxicity. *Food and Chemical Toxicology, 126*, 34–40.

Sharma, U. K., Sharma, A. K., Gupta, A., Kumar, R., Pandey, A., & Pandey, A. K. (2017b). Pharmacological activities of cinnamaldehyde and eugenol: Antioxidant, cytotoxic and anti-leishmanial studies. *Cellular and Molecular Biology, 63*(6), 73–78.

Sharma, V., Shukla, R. K., Saxena, N., Parmar, D., Das, M., & Dhawan, A. (2009). DNA damaging potential of zinc oxide nanoparticles in human epidermal cells. *Toxicology Letters, 185*, 211.

Shukla, R. K., Sharma, V., Pandey, A. K., Singh, S., Sultana, S., & Dhawan, A. (2011). Ros-mediated genotoxicity induced by titanium dioxide nanoparticles in human epidermal cells. *Toxicology In Vitro, 25*, 231–241.

Shvedova, A. A., Castranova, V., Kisin, E. R., Schwegler-Berry, D., Murray, A. R., Gandelsman, V. Z., Maynard, A., & Baron, P. J. (2003). Exposure to carbon nanotube material: Assessment of nanotube cytotoxicity using human keratinocyte cells. *Toxicology Environmental Health, 66*, 1909–1926.

Sies, H. (1991). Oxidative stress: Introduction. In H. Sies (Ed.), *Oxidative stress oxidants and antioxidants* (pp. 15–22). London: Academic Press.

Sohaebuddin, S. K., Thevenot, P. T., Baker, D., Eaton, J. W., & Tang, L. (2010). Nanomaterial cytotoxicity is composition, size, and cell type dependent. *Particle and Fibre Toxicology, 7*(22), 1–17.

Son, Y., Cheong, Y. K., Kim, N. H., Chung, H. T., Kang, D. G., & Pae, H. O. (2011). Mitogen-activated protein kinases and reactive oxygen species: How can ROS activate MAPK pathways? *Journal of Signal Transduction, 2011*, 792639.

Song, M.-F., Li, Y.-S., Kasai, H., & Kawai, K. (2012). Metal nanoparticle-induced micronuclei and oxidative DNA damage in mice. *Journal of Clinical Biochemistry and Nutrition, 50*, 211–216.

Stebounova, L. V., Morgan, H., Grassian, V. H., & Brenner, S. (2012). Health and safety implications of occupational exposure to engineered nanomaterials. *Reviews in Nanomedicine Nanobiotechnology, 4*, 310.

Stevenson, R., Hueber, A. J., Hutton, A., McInnes, I. B., & Graham, D. (2011). Nanoparticles and inflammation. *The Scientific World Journal, 11*, 1300–1312.

Tapeinos, C., & Pandit, A. (2016). Physical, chemical, and biological structures based on ros-sensitive moieties that are able to respond to oxidative microenvironments. *Advanced Materials, 28*.

Tran, M. A., Watts, R. J., & Robertson, G. P. (2009). Use of liposomes as drug delivery vehicles for treatment of melanoma. *Pigment Cell & Melanoma Research, 22*, 388–399.

Turabekova, M., Rasulev, B., Theodore, M., Jackman, J., Leszczynska, D., & Leszczynski, J. (2014). Immunotoxicity of nanoparticles: A computational study suggests that CNTs and C60 fullerenes might be recognized as pathogens by toll-like receptors. *Nanoscale, 6*, 3488–3495.

Turski, M. L., & Thiele, D. J. (2009). New roles for copper metabolism in cell proliferation, signalling, and disease. *The Journal of Biological Chemistry, 284,* 717–721.

Valavanidis, A., Vlachogianni, T., & Fiotakis, C. (2009). 8-hydroxy-2′-deoxyguanosine (8-OHDG): A critical biomarker of oxidative stress and carcinogenesis. *Journal of Environmental Science. Health Part C, 27,* 120–139.

Valko, M., Rhodes, C. J., Moncol, J., Izakovic, M., & Mazur, M. (2006). Free radicals, metals and antioxidants in oxidative stress-induced cancer. *Chemico-Biological Interactions, 160,* 1–40.

Walczyk, D., Bombelli, F. B., Monopoli, M. P., Lynch, I., & Dawson, K. A. (2010). What the cell "sees" in bionanoscience. *Journal of the American Chemical Society, 132,* 5761–5768.

Walkey, C. D., & Chan, W. C. W. (2012). Understanding and controlling the interaction of nanomaterials with proteins in a physiological environment. *Chemical Society Reviews, 41,* 2780–2799.

Wallach, D., Kang, T. B., & Kovalenko, A. (2014). Concepts of tissue injury and cell death in inflammation: A historical perspective. *Nature Reviews. Immunology, 14,* 51–59.

Wang, Y., Aker, W. G., Hwang, H. M., Yedjou, C. G., Yu, H., & Tchounwou, P. B. (2011). A study of the mechanism of in vitro cytotoxicity of metal oxide nanoparticles using catfish primary hepatocytes and human HepG2 cells. *Scientific Total Environment, 409,* 4753–4762.

Xia, T., Kovochich, M., Liong, M., Mädler, L., Gilbert, B., Shi, H., Yeh, J. I., Zink, J. I., & Nel, A. E. (2008). Comparison of the mechanism of toxicity of zinc oxide and cerium oxide nanoparticles based on dissolution and oxidative stress properties. *ACS Nano, 2,* 2121–2134.

Xiao, G. G., Wang, M., Li, N., Loo, J. A., & Nel, A. E. (2003). Use of proteomics to demonstrate a hierarchical oxidative stress response to diesel exhaust particle chemicals in a macrophage cell line. *The Journal of Biological Chemistry, 278,* 50781–50790.

Xu, Q., He, C., Xiao, C., & Chen, X. (2016). Reactive oxygen species (ROS) responsive polymers for biomedical applications. *Macromolecular Bioscience, 16*(5), 635–646.

Yang, L., & Watts, D. J. (2005). Particle surface characteristics may play an important role in phytotoxicity of alumina nanoparticles. *Toxicology Letters, 158,* 122–132.

Zhao, F., Zhao, Y., Liu, Y., Chang, X., Chen, C., & Zhao, Y. (2011). Cellular uptake, intracellular trafficking, and cytotoxicity of nanomaterials. *Small, 7,* 1322–1337.

Zhu, X., Hondroulis, E., Liu, W., & Li, C. Z. (2013). Biosensing approaches for rapid genotoxicity and cytotoxicity assays upon nanomaterial exposure. *Small, 9,* 1821–1830.

Zolnik, B. S., González-Fernández, A., Sadrieh, N., & Dobrovolskaia, M. A. (2010). Nanoparticles and the immune system. *Endocrinology, 151,* 458–465.

Index

© Springer Nature Switzerland AG 2021
N. Sharma, S. Sahi (eds.), *Nanomaterial Biointeractions at the Cellular,
Organismal and System Levels*, Nanotechnology in the Life Sciences,
https://doi.org/10.1007/978-3-030-65792-5